Probability and its Applications

A Series of the Applied Probability Trust

Editors: J. Gani, C.C. Heyde, T.G. Kurtz

Springer
New York
Berlin
Heidelberg
Barcelona
Budapest
Hong Kong
London
Milan
Paris
Santa Clara
Singapore
Tokyo

Probability and its Applications

Anderson: Continuous-Time Markov Chains.
Azencott/Dacunha-Castelle: Series of Irregular Observations.
Bass: Diffusions and Elliptic Operators.
Bass: Probabilistic Techniques in Analysis.
Choi: ARMA Model Identification.
Galambos/Simonelli: Bonferroni-type Inequalities with Applications.
Gani (Editor): The Craft of Probabilistic Modelling.
Grandell: Aspects of Risk Theory.
Gut: Stopped Random Walks.
Guyon: Random Fields on a Network.
Kallenberg: Foundations of Modern Probability.
Last/Brandt: Marked Point Processes on the Real Line.
Leadbetter/Lindgren/Rootzén: Extremes and Related Properties of Random Sequences and Processes.
Nualart: The Malliavin Calculus and Related Topics.
Rachev/Rüschendorf: Mass Transportation Problems. Volume I: Theory.
Rachev/Rüschendorf: Mass Transportation Problems. Volume II: Applications.
Resnick: Extreme Values, Regular Variation and Point Processes.
Shedler: Regeneration and Networks of Queues.
Todorovic: An Introduction to Stochastic Processes and Their Applications.

Svetlozar T. Rachev Ludger Rüschendorf

Mass Transportation Problems

Volume II: Applications

Springer

Svetlozar T. Rachev
Department of Statistics
University of California
Santa Barbara, CA 93106
USA

Ludger Rüschendorf
Institut für Mathematische Stochastik
University of Freiburg
79014 Freiburg
Germany

Library of Congress Cataloging-in-Publication Data
Rachev, S.T. (Svetlozar Todorov)
Mass transportation problems / Svetlozar T. Rachev, Ludger Rüschendorf.
p. cm. — (Probability and its applications)
Includes bibliographical references and indexes.
Contents: vol. 1. Theory — vol. 2. Applications.
ISBN 0-387-98350-3 (vol. 1 : hardcover). — ISBN 0-387-98352-X (vol. 2 : hardcover)
1. Transportation problems (Programming) I. Rüschendorf, Ludger, 1948- . II. Title. III. Series: Springer series in statistics. Probability and its applications.
QA402.6.R33 1998
519.7′2—dc21 97-34593

Printed on acid-free paper.

Production managed by Victoria Evarretta; manufacturing supervised by Thomas King.
Photocomposed pages prepared from the authors' LaTeX files.
Printed and bound by Maple-Vail Book Manufacturing Group, York, PA.
Printed in the United States of America.

9 8 7 6 5 4 3 2 1

ISBN 0-387-98352-X Springer-Verlag New York Berlin Heidelberg SPIN 10646816

To my wife Zoja
and
To my parents Nadezda
and Todor Rachevi.

Svetlozar (Zari) Rachev

To my wife Gabi.

Ludger Rüschendorf

Preface to Volume II

The second volume of the *Mass Transportation Problems* is devoted to applications in a variety of fields of applied probability, queueing theory, mathematical economics, risk theory, tomography, and others. In Volume I we encompassed the general mathematical theory of mass transportation, concentrating our attention on:

- the general duality theory of the transportation and transshipment problem;
- explicit optimality results;
- applications to minimal probability metrics, stochastic ordering, approximation and extension problems;
- applications to functional analysis and mathematical economics (the Debreu theorem, utility theory, dynamical systems, choice theory, and convex and nonconvex analysis were dicsussed in this context).

In Volume II we expand the scope of applications of mass transportation problems. Some of them arise from modifications of the admissible transportation plans. In fact, for applications to mathematical economics it is of interest to consider relaxations of the marginal constraints, such as upper or lower bounds on the supply and demand distributions, or additional constraints like capacity bounds for the transportation plans. In mathematical tomography the basic problem is to reconstruct the multivariate probability distribution based on some information about the marginal distributions in a certain finite number of directions. This information may be represented by additional constraints on the support functions or distributional moments, or it may be contained in only partial information on the marginals. Thus there is a close relationship between a class of problems in mathematical tomography and the classical theory on moment problems, which again can be viewed as a relaxation on the set of constraints in mass transportation problems. We discuss in detail applications to approximation problems for stochastic processes and to rounding problems based on moment-type characteristics. A particular example will be the approximation of queueing models. The minimal metrics allow us to compare various rounding rules and to determine optimal ones from an asymptotic point of view.

An important field of applications of mass transportation problems we shall consider in this second volume is to probabilistic limit theorems. This approach was introduced in the seventies by the Russian school of probability theory, headed by V.M. Zolotarev. By inherent regularity properties of probability metrics defined via certain mass transportation problems, there are streamlined proofs for central limit theorems on Banach spaces yielding sharp quantitative estimates of Berry–Esseen type for the convergence rate. The probability metric approach will be applied to general stable and operator stable limits theorems, martingale-type limit theorems, limit behavior of summability methods, and compound Poisson approximation. A particular application is to the classical problem in mathematical risk theory dealing with sharp approximation of the individual risk model by the collective risk model. The probability metric approach will also be applied to the quantitative asymptotics in rounding problems. A new field of application of probability metrics arising as solutions of mass transportation problems is the analysis of deterministic and stochastic algorithms. This research area is of increasing importance in computer science and various fields of stochastic modeling. Based on regularity properties of probability metrics, a general "contraction" method for the asymptotic analysis of algorithms has been developed. The contraction method has been applied successfully to a variety of search, sorting, and other tree algorithms. Furthermore, the recursive structure in iterated functions systems (image encoding), fractal measures, bootstrap statistics, and time series (ARCH) models has been analyzed by this method. It becomes clear that there are many interesting probabilistic applications of this method to be rigorously developed in the future.

In the final chapter we consider applications to stochastic differential equations (SDEs) and to convergence of empirical measures. SDEs will be interpreted as continuous recursive structures. From this point of view we provide a detailed discussion on the approximative solution of nonlinear stochastic differential equations of McKean–Vlasov type by interactive particle systems with application to the Kac theory of chaos propagation. The probability metrics approach allows us to establish approximation results for various modifications of the diffusion system, some of them of "nontraditional" type. In a general context we establish approximation results for empirical measures and give applications to the approximation of stochastic processes. As final applications we discuss a weak approximation of SDEs of Itô type by a combination of the time discretization methods of Euler and Milshtein with a chance discretization based on the strong invariance (embedding) principle. This approximation is given in terms of minimal L^p-metrics and thereby based on regularity properties of the solutions of the corresponding mass transportation problem.

Preface to Volume I

The subject of this book, mass transportation problems (MTPs), concerns the optimal transfer of masses from one location to another, where the optimality depends upon the context of the problem. Mass transportation problems appear in various forms and in various areas of mathematics and have been formulated at different levels of generality. Whereas the continuous case of the transportation problem may be cast in measure-theoretic terms, the discrete case deals with optimization over generalized transportation polyhedra. Accordingly, work on these problems has developed in several separate and independent directions.

The aim of this monograph is to investigate and to develop, in a systematic fashion, the *Monge–Kantorovich mass transportation problem (MKP)* and the *Kantorovich–Rubinstein transshipment problem (KRP)*. We consider several modifications of these problems known as the MTP with partial knowledge of the marginals and the MTP with additional constraints (MTPA). We also discuss extensively a variety of stochastic applications. In the first volume of *Mass Transportation Problems* we concentrate on the general mathematical theory of mass transportation. In Volume II we expand the scope of applications of mass transportation problems.

In 1781 Gaspard Monge proposed in simple prose a seemingly straightforward problem of optimization. It was destined to have wide ramifications. He began his paper on the theory of "clearings and fillings" as follows:

> When one must transport soil from one location to another, the custom is to give the name *clearing* to the volume of the soil that one

> must transport and the name *filling* ("remblai") to the space that it must occupy after transfer.
>
> Since the cost of transportation of one molecule is, all other things being equal, proportional to its weight and the interval that it must travel, and consequently the total cost of transportation being proportional to the sum of the products of the molecules each multiplied by the interval traversed; given the shape and position, the clearing and the filling, it is not the same for one molecule of the clearing to be moved to one or another spot of the filling. Rather, there is a certain distribution to be made of the molecules from the clearing to the filling, by which the sum of the products of molecules by intervals travelled will be the least possible, and the cost of the total transportation will be a *minimum*. (Monge, (1781, p. 666)).

In mathematical language Monge proposed the following nonlinear varational problem. Given two sets A, B of equal volume, find an optimal volume-preserving map between them; the optimality is evaluated by a cost function $c(x, y)$ representing the cost per unit mass for transporting material from $x \in A$ to $y \in B$. The optimal map is the one that minimizes the total cost of transferring the mass from A to B. Monge considered this problem with cost function equal to the Euclidean distance in $\mathbb{R}^d$: $c(x, y) = |x - y|$. Monge's problem turned out to be the prototype for a class of problems arising in various fields such as mathematical economics, functional analysis, probability and statistics, linear and stochastic programming, differential geometry, information theory, cybernetics, and matrix theory. The optimization function $\int_A c(x, t(x))\,dx$ is nonlinear in the transportation function t, and moreover, the set of admissible transportations is a nonconvex set. This explains why it took a long time until even existence results for optimal solutions could be established. The first general existence result was given in 1979 by Sudakov.

On the second page of his paper Monge himself had remarked that to obtain a minimum, the intervals traversed by two different molecules should not intersect. This simple observation applied to the discrete case—where there are only a finite number of molecules—leads to a "greedy" algorithm, the so-called northwest corner rule. The totality of mass transferences plans in the discrete case is a polytope that arises in the transportation problem of mathematical programming, where it is treated in specialized form as an assignment problem and in generalized form as a network-flow problem. The northwest corner rule solves transportation problems having a particular structure on the costs and is, moreover, at the heart of many seemingly different problems having an "easy" solution (cf. Hoffman (1961), Barnes and Hoffman (1985), Derigs, Goecke, and Schrader (1986), Hoffman and Veinott (1990), Olkin and Rachev (1991), and Rachev and Rüschendorf (1994); see also Burkard, Klinz, and Rudolf (1994) and the references therein).

The Academy of Paris offered a prize for the solution of Monge's problem, which was claimed by the differential geometer P. Appell (1884–1928), who established some geometric properties of optimal maps in the plane and in $\mathbb{R}^3$. But it took a long time until a real breakthrough in the transportation problem came, originating in the seminal 1942 paper of L.V. Kantorovich entitled "On the transfer of masses." Kantorovich stated the problem in a new, abstract, and in more easily accessible setting and without knowledge of Monge's work. Kantorovich learned of Monge's work only later (cf. his 1948 paper). In the Kantorovich formulation of the mass transportation problem (the so-called "continuous" MTP), the initial mass (the clearing) and the final mass (the filling) can be considered as probability measures on a metric space. The essential step in this formulation is the replacement of the class of transportation map by the wider class of generalized transportation plans, that are identifiable with the convex set of all probability measures on the product space with fixed marginals. The difficult nonlinear Monge problem was thereby replaced by a linear optimization problem over an abstract convex set. This made it possible to put this problem in the framework of linear optimization theory and encouraged the development of general duality theory for the solution of the Kantorovich formulation of the transportation problem as the basic tool. Accordingly, these problems and their generalizations will be referred to as *Monge–Kantorovich Mass Transportation Problems (MKPs)*.

Kantorovich's measure theoretic formulation made the problem accessible to various areas of the mathematical sciences and other scientific fields. Kantorovich himself received a Nobel Prize in Economics for related work in mathematical economics.(1) Here is a list of some references in the mathematical sciences:

- Functional analysis: Kantorovich and Akilov (1984)
- Probability theory: Fréchet (1951), Cambanis et al. (1976), Dudley (1976, 1989), Kellerer (1984), Rachev (1991c), Rüschendorf (1991)
- Statistics: Gini (1914, 1965), Hoeffding (1940, 1955), Kemperman (1987), Huber (1981), Bickel and Freedman (1981), Rüschendorf (1991)
- Linear and stochastic programming: Hoffman (1961), Barnes and Hoffman (1985), Anderson and Nash (1987), Burkard, Klinz and Rudolf (1994)
- Information theory and cybernetics: Wasserstein (1969), Gray et al. (1975), Gray and Ornstein (1979), Gray et al. (1980)
- Matrix theory: Lorentz (1953), Marcus (1960), Olkin and Pukelsheim (1982), Givens and Shortt (1984)

(1)L.V. Kantorovich together with T.C. Koopmans received the Nobel Memorial Prize in Economic Science in 1975 for "contributions to the theory of optimum allocation of resources"; see Dudley (1989, p. 342).

Many practical problems arising in various scientific fields have led mathematicians to solve MKPs: e.g., in

- Statistical physics: Tanaka (1978), Dobrushin (1979)
- Reliability theory: Barlow and Proschan (1975), Kalashnikov and Rachev (1990), Beneš (1985)
- Quality control: Jirina and Nedoma (1957)
- Transportation: Dantzig and Ferguson (1956)
- Econometrics: Shapley and Shubik (1972), Pyatt and Round (1985), Gretsky, Ostroy, and Zame (1992)
- Expert systems: Perez and Jirousek (1985)
- Project planning: Haneveld (1985)
- Optimal models for facility location: Ermoljev, Gaivoronski, and Nedeva (1983)
- Allocation policy: Rachev and Taksar (1992)
- Quality usage: Rachev, Dimitrov and Khalil (1992)
- Queueing theory: Rachev (1989), Anastassiou and Rachev (1992a, 1992b)

There are several surveys in the vast literature about MKP, among them Rachev (1984b), Rachev and Rüschendorf (1990), Burkard, Klinz, and Rudolf (1994), Cuesta-Albertos, Matrán, Rachev, and Rüschendorf (1996), and Gangbo and McCann (1996) related to dual solutions and applications of MKP; Shorack and Wellner (1985, Sect. 3.6) on optimal processes; Benes and Stepan (1987, 1991) on extremal mass transportation plans; Rüschendorf (1981, 1991, 1991a), Kellerer (1984), Rachev (1991c) on multivariate transportation problems; Dudley (1989) on distances in the space of measures; Talagrand (1992) and Yukich (1991) on matching problems.

In recent years, characterizations of the solutions of the Monge–Kantorovich problem have been given in terms of c-subgradients of generalized convex functions defined in terms of the cost functions $c(x, y)$ (cf. Knott and Smith (1984, 1992), Brenier (1987), Rüschendorf and Rachev (1990), Rüschendorf (1991, 1991a, 1995), Cuesta-Albertos, Matrán, Rachev, and Rüschendorf (1996), and Gangbo and McCann (1996)).

For the case of squared Euclidean costs $c(x, y) = |x - y|^2$, the generalized convexity property is equivalent to convexity, and c-subgradients are identical to the usual subgradients of convex analysis. From this characterization a series of explicit solutions of the transportation problem could be established. It also implies that the solutions of the MKP are under continuity assumptions given by mappings. Therefore, the solutions of the "easier" MKP imply as well the existence and characterizations of solutions of the original Monge problem, and so the MKP turns out to be the fundamental formulation of the transportation problem. For this reason, we concentrate in this book on the Kantorovich-type mass tranportation problems. For a discussion of interesting analytic aspects of the Monge problem, we refer to Gangbo and McCann (1996).

Another type of MTP appears in probability theory, even if it leaves the framework of probability measures as transportation plans. Its solutions are bounded measures on a product of two spaces with the difference of marginals equal to the difference of two given probability measures. It will be called the *Kantorovich–Rubinstein Problem (KRP)*, since the first results were obtained by Kantorovich and Rubinstein (1958). In its relation to the practical task of mass transportation it is sometimes referred to as the transshipment problem; see Kemperman (1983), and Rachev and Shortt (1990). The KRP has been developed to a great extent in the Russian school of probabilists and functional analysts, in particular by V.L. Levin, A.A. Milyutin, and A.M. Vershik and their students.

For metric cost functions the KRP coincides with the corresponding MKP; for general cost functions it can be reduced to the MKP for a corresponding reduced cost function. For the duality theory of the KRP a specific detailed theory with many results that are of value in themselves has been developed with wide-ranging applications to mathematical economics. For a different approach to the KRP as introduced in Dudley (1976) and as further extended in Rachev and Shortt (1990) we refer to the book of Rachev (1991c).

A problem related to both MKP and KRP is the *Mass Transportation Problem with Partial Knowledge of the Marginals (MTPP)*, which is expressed by stating finitely many moment conditions. Problems of this type were formulated and extensively studied by Rogosinski (1958), Kemperman (1983), and Kuznezova-Sholpo and Rachev (1989). Barnes and Hoffman (1985) considered mass tranportaion problems with capacity constraints on the admissible transportation plans as an example of *Mass Transportation Problems with Additional Constraints (MTPA)* (see Rachev (1991b) and Rachev and Rüschendorf (1994)).

In this book we give an extensive account of the duality theory of the MKP and the KRP, including the known results on explicit constructions and characterizations of optimal solutions.

In Chapters 2 and 3 we present important duality theorems for the Monge–Kantorovich problem based on work of H. Kellerer, L. Rüschendorf, S.T. Rachev, and D. Ramachandran.

In Chapters 4 and 5 we present basically work of V.L. Levin; we analyze measure-theoretic methods for infinite-dimensional linear programs developed in context with the KRP as well as applications to general utility theorems (the Debreu theorem), extension theorems, choice theory, and set-valued dynamical systems.(2)

In Chapters 6 and 8 we discuss new material on applications of the MKP and the KRP to the representation of ideal metrics and on various probabilistic approximation and limit theorems. This supplements the earlier results in this direction as described in the book of Rachev (1991) on probability metrics and stochastic models. In particular, we show that probability metrics allow us to find unified proofs for central limit theorems for martingales, (operator) stable limit theorems, and to more specific problems like compound Poisson approximation or rounding problems.

Chapter 7, the first chapter in the second volume, is concerned with modifications of the MKP by additional or relaxed constraints. We discuss various types of moment problems and applications to the tomography paradoxon and to the approximation of queueing systems. A wide range of applications of metrics based on the transportation problem has been established in recent years in connection with recursive stochastic equations. We discuss algorithms of informatics (sorting, searching, branching, search trees) as well as applications to the approximation of stochastic differential equations, to the propagation of the chaos property of particle systems with applications to the approximation of nonlinear PDEs, as well as to the rate of convergence of empirical measures, which is of interest for matching problems in Chapters 9 and 10.

From the technical point of view, MKPs can be subdivided into the discrete and continuous cases, according to the nature of their basic spaces and to the supports of the initial and the final masses. In the discrete case, the totality of the mass transference plans is the polytope that arises in the transportation problem of mathematical programming. There is, of course, a vast literature on the transportation problem, its specialization to the assignment problem, and its generalization to network flow problems. It turns out, as will be elaborated further in the book, that the northwest corner rule in the discrete case corresponds to a closed form for the solution in the continuous case. Indeed, the discrete analogue of a result known in the continuous case provides a new result in the discrete case; and its simple proof in the discrete case provides a new proof for the continuous case, see Rachev and Rüschendorf (1994c) and the references therein. Another approach in the discrete linear case prefers to exploit the special structure of supplies and demands (or clearings and fillings) and permits a particularly simple combinatorial algorithm for finding an optimal solution as developed

---

(2)These two chapters were written following closely the notes kindly provided to us by V.L. Levin.

by Balinski (1983), Balinski and Russakoff (1984), Balinski (1985, 1986), Goldfarb (1985), Kleinschmidt, Lee, and Schannath (1987), and Burkard, Klinz, and Rudolf (1994).

MTPs may be viewed as an analogue and a unifying framework of a problem considered by probabilists at the beginning of the twentieth century: *How does one measure the difference between two random quantities?* Many specific contributions to the analysis of this problem have been made, including Gini's (1914) notion of concordance, Kendall's τ, Spearman's ϱ, the analysis of greatest possible differences by Hoeffding (1940) and others, by Fréchet (1951, 1957), Robbins (1975), and Lai and Robbins (1976), and the generalizations of these results by Cambanis, Simons, and Stout (1976), Rüschendorf (1980), Tchen (1980), and Cambanis and Simons (1982). These (and others) offer piecemeal answers to basic questions that arise from different stochastic models; they give no guidance as to the question of what concept should be used where: There is no general theory underlying the diverse approaches. We refer to Kruskal (1958), Gini (1965), and Rachev (1984b, 1991c).

In this book we investigate, develop, and exploit the connections between the discrete and continuous versions of the mass transportation problems (MTP) as well as study systematically the relationships between the methods and results from different versions of the MTP. The MTPs are the basis of many problems related to the question of stability of stochastic models, to the question of whether a proposed model yields a satisfactory approximation to the phenomenon under consideration, and to the problem of approximation of stochastic and deterministic algorithms. It is our belief that MTPs hold great promise in stochastic analysis as well as in mathematical analysis. The MTP is full of connections with geometry, (partial) differential equations, (generalized) convex analysis, moment problems, infinite-dimensional linear programming, measurable choice theory, and extension problems, and it has many open problems. It has a great potential for a series of applications in several scientific fields.

This book grew out of joint work and lectures delivered by the authors at the Steklov Mathematical Institute, Universität Münster, Universität Freiburg, the Ecole Polytechnique, SUNY at Stony Brook, and the University of California, Santa Barbara, over many years. Many colleagues provided helpful suggestions after reading parts of the manuscript. All chapters were rewritten several times, and preliminary versions were circulated among friends, who eliminated many inaccuracies and obscurities. We would like to thank H.G. Kellerer, V.L. Levin, M. Balinski, D. Ramachandran, G.A. Anastassiou, M. Maejima, M. Cramer, I. Olkin, M. Gelbrich, W. Römisch, V. Beneš, L. Uckelmann, and many other friends and colleagues who encouraged us to complete the work. We are indebted to Mrs. M. Hattenbach and Ms. A. Blessing for their superb typing; the appearance of this monograph owes much to them. We are grateful to the publisher and especially to J. Kimmel for support and patience. We are particularly thankful to J. Gani for his invaluable suggestions concerning improvements of this work, his help with the organization of the material, and his encouragement to continue the project.

Finally, we thank the Alexander von Humboldt Foundation for its generous financial support of S.T. Rachev in 1995 and 1996, which made this joint work possible. (3)

(3)The work of S.T. Rachev was also partially supported by NSF Grants. The joint work of the authors was supported by NATO-Grant CRG900798.

Contents to Volume II

Contents to Volume I

7

Modifications of the Monge–Kantorovich Problems: Transportation Problems with Relaxed or Additional Constraints

In this chapter we study modifications of the usual transportation problem by allowing additional constraints on the admissible supply—resp. demand—distributions. In particular, we consider the case that the marginal distribution function of the supply is bounded below by a d.f. F_1, while the marginal d.f. of the demand is bounded above by a d.f. F_2. We also examine transportation plans with constraints of a local type concerning the densities of the marginals, and finally, we study transportation problems with additional moment-type constraints. For the solution of these problems we make use of some methods arising in the theory of marginal and moment problems, duality theory, and stochastic ordering results.

The next part is concerned with a solution of the tomography paradox. With respect to some weak metrics, two distributions are getting close if they coincide on an increasing number of directions. In the final sections we review results on the closeness of distributions under given moment-type characteristics and discuss applications to the rounding problem. Most of the results in these sections are contained in Rachev and Rüschendorf (1993, 1994c), Levin and Rachev (1989), Klebanov and Rachev (1995), and Anastassiou and Rachev (1992). A survey on related discrete transportation problems is given in Burkard, Klinz, and Rudolf (1994).

7.1 Mass Transportation Problem with Relaxed Marginal Constraints

For distribution functions F_1, F_2 let $\mathcal{F}(F_1, F_2)$ denote the set of all d.f.s F on $\mathbb{R}^2$ with marginals F_1, F_2 (i.e., $F(x,\infty) = F_1(x),\ F(\infty, y) = F_2(y)$). Then the transportation problem with nonnegative cost function $c(x,y)$, $x, y \in \mathbb{R}$, has the form

$$\text{minimize} \int_{\mathbb{R}^2} c(x,y)\,\mathrm{d}F(x,y) \quad \text{over all } F \in \mathcal{F}(F_1, F_2). \tag{7.1.1}$$

Usually, in the linear programming setting, F_1 is viewed as the supply distribution and F_2 as the demand distribution. Clearly, (7.1.1) is an infinite-dimensional analogue of the discrete transportation problem: Given $a_i \geq 0, b_j \geq 0,\ \sum_{i=1}^m a_i = \sum_{j=1}^n b_j$,

$$\text{minimize} \sum_{i=1}^m \sum_{j=1}^n c_{ij} x_{ij}, \text{ subject to the constraints} \tag{7.1.2}$$

$$\sum_{j=1}^n x_{ij} = a_i,\ 1 \leq i \leq m,\ \sum_{i=1}^m x_{ij} = b_j,\ j = 1, \ldots, n,\ x_{ij} \geq 0,\ \forall i, j.$$

Suppose $c(x,y)$ (resp. (c_{ij})) satisfies the "Monge" conditions, i.e., c is right continuous, and

$$c(x', y') - c(x, y') - c(x', y) + c(x, y) \leq 0 \quad \text{for all } x' \geq x, y' \geq y; \tag{7.1.3}$$

in the discrete case these conditions are of the form

$$c_{ij} + c_{i+1,j+1} - c_{i,j+1} - c_{i+1,j} \leq 0,\ \forall 1 \leq i < m, 1 \leq j < n. \tag{7.1.4}$$

Then the solution of (7.1.1), (7.1.4) is well known and based on the "northwest corner rule," which leads to a greedy algorithm; see Hoffman (1961). For (7.1.1) the solution is given by the d.f. F^*,

$$F^*(x,y) \quad = \quad \min\{F_1(x), F_2(y)\}. \tag{7.1.5}$$

F^* is the upper Fréchet bound, see (3.6.2). Recall that the Fréchet bounds provide the following characterization of $\mathcal{F}(F_1, F_2)$:

$$\begin{aligned} & F \in \mathcal{F}(F_1, F_2) \quad \text{if and only if} \\ F_*(x,y) \quad := \quad & (F_1(x) + F_2(y) - 1)_+ \ \leq\ F(x,y) \ \leq\ F^*(x,y) \end{aligned} \tag{7.1.6}$$

(here $(\cdot)_+ = \max(0, \cdot)$); the lower Fréchet bound yields the solution of the maximization problem corresponding to (7.1.1).

In terms of random variables an equivalent formulation of the transportation problem is the following:

$$\text{minimize } Ec(X,Y), \quad \text{subject to } F_X = F_1, F_Y = F_2, \tag{7.1.7}$$

where X, Y are random variables on a rich enough (e.g., atomless) probability space $(\Omega, \mathcal{A}, P)$. The solutions (7.1.5), resp. (7.1.6), then can be represented as distributions of r.v.s X^*, Y^*:

$$X^* = F_1^{-1}(U), \quad Y^* = F_2^{-1}(U) \quad (\text{for } (7.1.1), (7.1.5)), \tag{7.1.8}$$

resp.

$$X^* = F_1^{-1}(U), \quad Y^* = F_2^{-1}(1-U) \quad (\text{for } F_*), \tag{7.1.9}$$

where U is uniformly distributed on $(0,1)$, and $F_1^{-1}(u) = \inf\{y;\ F_1(y) \geq u\}$ is the generalized inverse of F_1.

We next consider the mass transportation problem (7.1.1), but with relaxed marginal constraints. For d.f.s F_1, F_2 the set

$$\begin{aligned} \mathcal{H}(F_1, F_2) \quad = \quad & \{F;\ F \text{ is a d.f. on } \mathbb{R}^2 \\ & \text{with marginal d.f.s } \widetilde{F}_1 \leq F_1,\ \widetilde{F}_2 \geq F_2\} \end{aligned} \tag{7.1.10}$$

of all d.f.s F with $\widetilde{F}_1(x) = F(x,\infty) \leq F_1(x), \forall x \in \mathbb{R}^1$, and $\widetilde{F}_2(y) = F(\infty, y) \geq F_2(y), \forall y \in \mathbb{R}^1$. We consider the transportation problem:

$$\text{minimize } \int_{R^2} c(x,y)\, \mathrm{d}F(x,y), \quad \text{subject to } F \in \mathcal{H}(F_1, F_2), \tag{7.1.11}$$

or, equivalently,

$$\text{minimize } Ec(X,Y), \quad \text{subject to } F_X \leq F_1, F_Y \geq F_2. \tag{7.1.12}$$

In the discrete case the problem is to

$$\text{minimize } \sum c_{ij} x_{ij}, \tag{7.1.13}$$

where for some "supplies" $s_1, \ldots, s_n,\ a_1 \leq s_1,\ a_1 + a_2 \leq s_1 + s_2, \ldots,$ and for some demands $d_1, \ldots, d_n,\ b_1 \geq d_1,\ b_1 + b_2 \geq d_1 + d_2, \ldots$ (a_i, b_i as in (7.1.2)). This describes a production process and a consumption process subject to some priorities (e.g., queueing priorities) with capacities $s_1, \ldots, s_n$ having the following property: Every remaining free capacity at stage i of the production (resp. consumption) process can be transferred to some of the next stages $i+1, \ldots, n$.

Theorem 7.1.1 *Let the cost function $c(x,y)$ be symmetric in x, y, let $c(x,y)$ satisfy the Monge condition* (7.1.3), *and let $c(x,x) = 0$ for all $x \in \mathbb{R}$. Set*

$$H^*(x,y) = \min\{F_1(x), \max\{F_1(y), F_2(y)\}\}, \quad x, y \in \mathbb{R}. \tag{7.1.14}$$

Then

(a) $H^* \in \mathcal{H}(F_1, F_2)$,

(b) H^* *solves the relaxed transportation problem* (7.1.11), (7.1.15)

(c) $$\int_{\mathbb{R}^2} c(x,y)\, \mathrm{d}H^*(x,y) = \int_0^1 c\left(F_1^{-1}(u), \min\left(F_1^{-1}(u), F_2^{-1}(u)\right)\right) \mathrm{d}u.$$

Remark 7.1.2 *Setting $G_1(y) = \max\{F_1(y), F_2(y)\}$, we see from Theorem 7.1.1 that the relaxed transportation problem* (7.1.11) *is equivalent to the transportation problem* (7.1.1) *with marginals F_1, G_1. In terms of random variables the solution can be expressed by the joint distribution of*

$$X^* = F_1^{-1}(U) \quad \textit{and} \qquad\qquad (7.1.16)$$
$$Y^* = G_1^{-1}(U) = \min\left(F_1^{-1}(U), F_2^{-1}(U)\right)$$

(cf. (7.1.8)*).*

Proof: The Monge condition implies that we can view the function $-c(x,y)$ as a "distribution function" corresponding to a nonnegative measure μ_c on $\mathbb{R}^2$. Let X, Y be any real r.v.s, and for $x, y \in \mathbb{R}^1$ set $x \vee y = \max\{x,y\}$, $x \wedge y = \min\{x,y\}$. Theorem 7.1.1 is a consequence of the following two claims.

Claim 7.1.3 (Cambanis, Simons, and Stout (1976); see also Dall'Aglio (1956) for the special case $c(x,y) = |x-y|^p$)

$$\begin{aligned} 2Ec(X,Y) = \int_{\mathbb{R}^2} (&P(X < x \wedge y, Y \geq x \vee y) \\ &+ P(X \geq x \vee y, Y < x \wedge y))\mu_c(\mathrm{d}x, \mathrm{d}y). \end{aligned} \tag{7.1.17}$$

For the proof of Claim 7.1.3 define the function $f(x,y,w) : \mathbb{R}^2 \times \Omega \to \mathbb{R}$ by

$$f(x,y,w) = \begin{cases} 1 & \text{if } X(w) < x, y \leq Y(w) \text{ or } Y(w) < x, y \leq X(w), \\ 0 & \text{otherwise.} \end{cases}$$

Using Fubini's theorem,

$$E_w \int_{\mathbb{R}^2} f(x,y,w)\mu_c(\,dx,\,dy) \;=\; \int_{\mathbb{R}^2} (E_w f(x,y,w))\mu_c(\,dx,\,dy). \quad (7.1.18)$$

Next, the symmetry of $c(x,y)$ and $c(x,x)=0$ yield

$$\begin{aligned}
&\int_{\mathbb{R}^2} f(x,y,w)\,d\mu_c && (7.1.19)\\
&= -\left[c\left(Y(w),Y(w)\right)+c\left(X(w),X(w)\right)-c\left(X(w),Y(w)\right)\right.\\
&\qquad\left.-c\left(Y(w),X(w)\right)\right]\\
&= 2c\left(X(w),Y(w)\right).
\end{aligned}$$

Clearly,

$$\begin{aligned}
&E_w f(x,y,w) && (7.1.20)\\
&\quad = P(X < x\wedge y, Y \ge x\vee y) + P(X\ge x\vee y, Y < x\wedge y).
\end{aligned}$$

Combining (7.1.18), (7.1.19), and (7.1.20) we obtain (7.1.17).

Claim 7.1.4 *Define* $X^* = F_1^{-1}(U), Y^* = \min\left(F_1^{-1}(U), F_2^{-1}(U)\right)$. *Then*

$$Ec(X^*,Y^*) \;=\; \min\left(Ec(X,Y); F_X \le F_1, F_Y \ge F_2\right), \qquad (7.1.21)$$

and the value of the expectation in (7.1.21) *is given by*

$$\begin{aligned}
Ec(X^*,Y^*) &= \frac{1}{2}\int_{\mathbb{R}^2} \max\left(0, F_2\left((x\wedge y)-\right) - F_1\left((x\vee y)-\right)\right)\mu_c(\,dx,\,dy)\\
&= \int_0^1 c\left(F_1^{-1}(t), \min\left(F_1^{-1}(t), F_2^{-1}(t)\right)\right)\,dt. \qquad (7.1.22)
\end{aligned}$$

For the proof of Claim 7.1.4 let X,Y be any r.v.s with d.f.s $F_X \le F_1$, $F_Y \ge F_2$. Using Claim 7.1.3 we obtain

$$\begin{aligned}
2Ec(X,Y) &\ge \int_{\mathbb{R}^2} P(X\ge x\vee y, Y < x\wedge y)\mu_c(\,dx,\,dy) \qquad (7.1.23)\\
&= \int_{\mathbb{R}^2} \left(P(Y < x\wedge y) - P(X < x\vee y, Y < x\wedge y)\right)\mu_c(\,dx,\,dy)\\
&\ge \int_{\mathbb{R}^2} (P(Y < x\wedge y) - \min\{P(X < x\vee y),
\end{aligned}$$

$$P(Y < x \wedge y)\}) \, \mu_c(\,dx,\, dy)$$

$$= \int_{\mathbb{R}^2} (P(Y < x \wedge y) - P(X < x \vee y))_+ \, \mu_c(\,dx,\, dy)$$

$$\geq \int_{\mathbb{R}^2} (F_2\,((x \wedge y)-) - F_1\,((x \vee y)-))_+ \, \mu_c(\,dx,\, dy).$$

Next, we check that the lower bound in (7.1.23) is attained for $X^* = F_1^{-1}(U), Y^* = \min\left(F_1^{-1}(U), F_2^{-1}(U)\right)$. In fact, by Claim 7.1.3 using $X^* \geq Y^*$ and $\{U < F_2(z)\} = \left\{F_2^{-1}(U) < z\right\}$ a.s. we get

$$2Ec(X^*, Y^*) \tag{7.1.24}$$

$$= \int_{\mathbb{R}^2} (P(X^* \geq x \vee y, Y^* < x \wedge y)$$

$$+P(X^* < x \wedge y, Y^* \geq x \vee y)) \, \mu_c(\,dx,\, dy)$$

$$= \int_{\mathbb{R}^2} P(X^* \geq x \vee y, Y^* < x \wedge y) \mu_c(\,dx,\, dy)$$

$$= \int_{\mathbb{R}^2} P\left(F_1^{-1}(U) \geq x \vee y, \min\left(F_1^{-1}(U), F_2^{-1}(U)\right) < x \wedge y\right) \mu_c(\,dx,\, dy)$$

$$= \int_{\mathbb{R}^2} P\left(F_1^{-1}(U) \geq x \vee y, F_2^{-1}(U) < x \wedge y\right) \mu_c(\,dx,\, dy)$$

$$= \int_{\mathbb{R}^2} P(U \geq F_1(x \vee y), U < F_2(x \wedge y))_+ \mu_c(\,dx,\, dy)$$

$$= \int_{\mathbb{R}^2} (F_2((x \wedge y)-) - F_1((x \vee y)-))_+ \mu_c(\,dx,\, dy).$$

Obviously, $F_{(X^*,Y^*)} = H^* \in \mathcal{H}(F_1, F_2)$, and the proof of Theorem 7.1.1 is complete. □

Remark 7.1.5 *The optimal coupling* (7.1.16) *leads to the following "greedy" algorithm for solving the finite discrete transportation problem with relaxed side conditions:*

$$\textit{minimize} \quad \sum_{i=1}^{n} \sum_{j=1}^{n} c_{ij} x_{ij} \tag{7.1.25}$$

$$\textit{subject to:} \quad x_{ij} \geq 0,$$

$$\sum_{s=1}^{j}\sum_{r=1}^{n} x_{rs} \geq \sum_{s=1}^{j} b_s =: G_j, \quad 1 \leq j \leq n,$$
$$\sum_{r=1}^{i}\sum_{s=1}^{n} x_{rs} \leq \sum_{r=1}^{i} a_r =: F_i, \quad 1 \leq i \leq n,$$

where the sum of the "demands" $\sum_{s=1}^{n} b_s$ equals the sum of the "supplies" $\sum_{r=1}^{n} a_r$, assuming that (c_{ij}) are symmetric, $c_{ii} = 0$, and c satisfies the Monge condition (7.1.4). *Set*

$$\begin{aligned} H_i &= \max(F_i, G_i), \quad 1 \leq i \leq n, \\ \delta_1 &= H_1, \ \delta_{i+1} = H_{i+1} - H_i, \quad 1 \leq i \leq n-1. \end{aligned} \tag{7.1.26}$$

Then (7.1.25) *is equivalent to the standard transportation problem* (7.1.2) *with side conditions* (a_i), (δ_i). *7In the following example we compare the solution of problem* (7.1.25) *with inequality constraints with the "greedy" solution of the standard transportation problem with equality constraints* (7.1.2). *For the problem with inequality constraints we first calculate the new artificial demands δ_j as in* (7.1.26) *and then apply the northwest corner rule.*

Example 7.1.6

							supply a_1	$F_i = \sum_{r=1}^{i} a_r$
y_{ij} x_{ij}	**20** *10*	*10*					*20*	*20*
							0	*20*
		20 *20*	**20** *10*	*10*			*40*	*60*
				20 *20*			*20*	*80*
				10 *10*			*10*	*90*
						10 *10*	*10*	*100*
demand b_1	*10*	*30*	*10*	*40*	*0*	*10*		
$G_j = \sum_{s=1}^{j} b_s$	*10*	*40*	*50*	*90*	*90*	*100*		
$H_j = F_j \vee G_j$	*20*	*40*	*60*	*90*	*90*	*100*		
$\delta_1 = H_1$, $\delta_{j+1} = H_{j+1} - H_j$	*20*	*20*	*20*	*30*	*0*	*10*	*"artificial" demands*	

x_{ij} = *solution of the standard transportation problem* (7.1.2), *using the classical northwest corner*

y_{ij} = *solution of the transportation problem with relaxed side conditions*

We next extend the solution to the nonsymmetric case. We relax the symmetry condition, assuming that for any x, y the functions $c(x,\cdot), c(\cdot, y)$ are unimodal:

$$\begin{aligned} c(x,y_1) \le c(x,y_2) \quad &\text{if} \quad x \le y_1 \le y_2 \quad \text{or} \quad y_2 \le y_1 \le x, \\ c(x_1,y) \le c(x_2,y) \quad &\text{if} \quad x_2 \le x_1 \le y \quad \text{or} \quad y \le x_1 \le x_2. \end{aligned} \tag{7.1.27}$$

Theorem 7.1.7 *If $c(x,x) = 0$ for all $x \in \mathbb{R}$ and c satisfies the Monge condition and the unimodality condtion (7.1.27), then the relaxed transportation problem*

$$\textit{minimize } Ec(X,Y) \quad \textit{subject to } F_X \ge F_1, \;\; F_Y \le F_2 \tag{7.1.28}$$

has a solution, given by the coupling

$$X^* = F_1^{-1}(U), \quad Y^* = \max\left(F_1^{-1}(U), F_2^{-1}(U)\right) \tag{7.1.29}$$

with joint distribution

$$F_{X^*,Y^*}(x,y) \;=\; \min\Big(F_1(x), \min\left(F_1(y), F_2(y)\right)\Big),$$

and the optimal value is given by

$$Ec(X^*,Y^*) \;=\; \int_0^1 c\left(F_1^{-1}(u), \max\left(F_1^{-1}(u), F_2^{-1}(u)\right)\right) du.$$

Proof: Let X, Y be r.v.s with $F_X \ge F_1, \;\; F_Y \le F_2$. Then by (7.1.8),

$$Ec(X,Y) \;\ge\; Ec\left(F_X^{-1}(U), F_Y^{-1}(U)\right). \tag{7.1.30}$$

Let $G(y) = \min(F_X(y), F_Y(y))$. Then $F_X^{-1} \le F_1^{-1}, F_Y^{-1} \ge F_2^{-1}$, and $G^{-1} = \max\left(F_X^{-1}, F_Y^{-1}\right)$. We now need the following

Claim 7.1.8

$$\int_0^1 c\left(F_X^{-1}(u), F_Y^{-1}(u)\right) du \;\ge\; \int_0^1 c\left(F_X^{-1}(u), G^{-1}(u)\right) du. \tag{7.1.31}$$

To show Claim 7.1.8 set (for a fixed $u \in (0,1)$), $x = F_X^{-1}(u)$, $y_1 = F_X^{-1}(u) \vee F_Y^{-1}(u) = G^{-1}(u)$, and $y_2 = F_Y^{-1}(u)$.

<u>Case 1:</u> $x < y_2$. In this case, $x \le y_1 \le y_2$, and therefore, the unimodality condition (7.1.27) implies $c(x,y_2) \ge c(x,y_1)$.

<u>Case 2:</u> $y_2 \le x$. In this case, $y_1 = x$, and therefore, $y_2 \le y_1 = x$. Again by the unimodality condition, $c(x,y_2) \ge c(x,y_1)$. So Claim 7.1.8 holds.

Claim 7.1.9 *The following bound holds for every coupling* (X,Y) :

$$\int_0^1 c\left(F_X^{-1}(u), F_Y^{-1} \vee F_X^{-1}(u)\right) du \tag{7.1.32}$$

$$\geq \int_0^1 c\left(F_1^{-1}(u), F_2^{-1}(u) \vee F_1^{-1}(u)\right) du.$$

For the proof define $\widetilde{x}_1 = F_X^{-1}(u)$, $\widetilde{x}_2 = F_Y^{-1}(u)$, $x_1 = F_1^{-1}(u)$, $x_2 = F_2^{-1}(u)$ for a fixed u. Then $\widetilde{x}_1 \leq x_1, x_2 \leq \widetilde{x}_2$.

$$\begin{aligned} &\text{If} \quad \widetilde{x}_1 < \widetilde{x}_2, \quad \text{then} \quad \widetilde{x}_1 \leq \widetilde{x}_2 \vee x_2 \leq \widetilde{x}_2. \\ &\text{if} \quad \widetilde{x}_1 \geq \widetilde{x}_2, \quad \text{then} \quad \widetilde{x}_1 = \widetilde{x}_1 \vee x_2 \geq \widetilde{x}_2. \end{aligned} \tag{7.1.33}$$

From (7.1.33) we obtain

Claim 7.1.10

$$c(\widetilde{x}_1, \widetilde{x}_1 \vee x_2) \;\geq\; c(x_1, x_1 \vee x_2). \tag{7.1.34}$$

For the proof of Claim 7.1.10 we use the relation $x_1 \geq \widetilde{x}_1$.

Case 1: $x_2 > x_1 > \widetilde{x}_1$. Then $c(\widetilde{x}_1, x_2) = c(\widetilde{x}_1, \widetilde{x}_1 \vee x_2) \geq c(x_1, x_2) = c(x_1, x_1 \vee x_2)$ by the unimodality condition.

Case 2:

(a) $x_1 \geq x_2 \geq \widetilde{x}_1$. Then, trivially, $c(\widetilde{x}_1, x_2) = c(\widetilde{x}_1, x_2 \vee \widetilde{x}_1) \geq c(x_1, x_1 \vee x_2) = c(x_1, x_1) = 0$.

(b) $x_1 \geq \widetilde{x}_1 \geq x_2$. Then again, $c(\widetilde{x}_1, \widetilde{x}_1) = c(\widetilde{x}_1, \widetilde{x}_1 \vee x_2) \geq c(x_1, x_1 \vee x_2) = c(x_1, x_1) = 0$.

Claims 7.1.8, 7.1.9, and 7.1.10 imply (7.1.28). □

Remark 7.1.11

(a) The unimodality assumption (7.1.27) *is quite natural from an application point of view. Note that the transportation problem in Theorem* 7.1.7 *is the same as in Theorem* 7.1.1 *(where only the indices 1,2 have been changed). We used this to demonstrate that the solution* F^* *is not unique. Without the symmetry, resp. the unimodality condition, the solution may differ substantially. Given a right continuous function* $f = f(y) \geq 0$, *consider the cost function* $c(x,y) = f(y)$. *Then*

c satisfies the Monge condition, and so (7.1.28) is equivalent to the following problem:

$$\text{minimize} \int f(y)\,\mathrm{d}F_Y(y) \quad \text{subject to } F_Y \le F_2. \tag{7.1.35}$$

Equivalently, we are seeking a d.f. $\widetilde{F}_2 \le F_2$ such that the distribution of f with respect to $\widetilde{F}_2$ has a minimal first moment. Obviously, the solution (7.1.31) of Theorem 7.1.7 is not a solution of (7.1.35).

(b) In the proof of Theorem 7.1.7, the assumption $c(x,x) = 0$ can be replaced with a weaker one,

$$c(x,x) \;\le\; c(x,y) \wedge c(y,x), \quad \text{for all } x,y \in \mathbb{R}. \tag{7.1.36}$$

7.2 Mass Transportation Problem with Fixed Sum (Difference) of the Marginals and with Stochastically Ordered Marginals

Consider a flow in a network with n nodes, $i = 1,\dots,n$, and let x_{ij} be the flow from node i to node j. Assume that for all nodes k the value of $\sum_i x_{ik} + \sum_j x_{kj}$ is fixed and equal to h_k. As motivation, suppose $a_i = \sum_{k=1}^n x_{ik}$, $b_i = \sum_{k=1}^n x_{ki}$ to be the amount of workload corresponding to the outflow, resp. to the inflow, in node i. Assume that the total work capacity at node i is given by h_i (in a certain time period). Then every admissible flow (x_{ij}) should satisfy the condition

$$h_i \;=\; a_i + b_i, \quad 1 \le i \le n. \tag{7.2.1}$$

Set $A(k) = \sum_{i=1}^k a_i$, $B(k) = \sum_{i=1}^k b_i$, and $H(k) = \sum_{i=1}^k h_i$. Then $h_k = A(k) + B(k) - (A(k-1) + B(k-1))$, and (7.2.1) is equivalent to

$$H(k) \;=\; A(k) + B(k), \quad 1 \le k \le n. \tag{7.2.2}$$

Let c_{ij} be the transportation cost of a unit from node i to node j. Then the problem is to minimize the total cost $\sum c_{ij}x_{ij}$ subject to the admissibility condition (7.2.1) and $x_{ij} \ge 0$.

The general formulation of this problem is the following. For two d.f.s A, B define $G(x) = \frac{1}{2}(A(x) + B(x))$. For a given cost function $c(x,y)$,

$$\text{minimize} \int_{\mathbb{R}^2} c(x,y)\,\mathrm{d}F(x,y), \quad \text{subject to } F \in \mathcal{F}_{A+B}. \tag{7.2.3}$$

Here $\mathcal{F}_{A+B}$ is the set of all d.f.s $F(x,y)$ with marginal d.f.s F_1, F_2 satisfying $F_1(x)+F_2(x) = A(x)+B(x)$.

Consider next the special case $c(x,y) = |x-y|$. Let X, Y be real r.v.s with joint d.f. F. Then by the triangle inequality,

$$E|X-Y| \;\leq\; \inf_{a\in\mathbb{R}^1}(E|X-a|+E|Y-a|). \tag{7.2.4}$$

Since $E|X-a|+E|Y-a| = \int |x-a|\,\mathrm{d}(F_X+F_Y)(x)$ depends only on the sum of the marginals, (7.2.3) is the best possible improvement of (7.2.4), provided that the sum of the marginal F_X+F_Y is known. Rachev (1984d) showed that

$$\sup\{E|X-Y|^p; F_X+F_Y = A+B\} = \int_0^1 \left|G^{-1}(t)-G^{-1}(1-t)\right|^p \mathrm{d}t, \quad p\geq 1. \tag{7.2.5}$$

The following result gives an explicit solution of the general problem in (7.2.3).

Proposition 7.2.1 *Suppose $c \geq 0$ is symmetric and satisfies the Monge condition:*

$$c(x',y')-c(x,y')-c(x',y)+c(x,y) \;\leq\; 0 \quad \forall x'\geq x,\;\; y'\geq y. \tag{7.2.6}$$

Then

$$\inf\left\{\int c(x,y)\,\mathrm{d}F(x,y); F\in\mathcal{F}_{A+B}\right\} = \int_0^1 c(G^{-1}(u),G^{-1}(u))\,du, \tag{7.2.7}$$

and

$$\sup\left\{\int c(x,y)\,\mathrm{d}F(x,y); F\in\mathcal{F}_{A+B}\right\} = \int_0^1 c(G^{-1}(u),G^{-1}(1-u))\,du. \tag{7.2.8}$$

The corresponding optimal pairs of r.v.s (couplings) are given by $(G^{-1}(U),G^{-1}(U))$, resp. $(G^{-1}(U),G^{-1}(1-U))$.

Proof: Since c is symmetric, we obtain for any $F\in\mathcal{F}_{A+B}$,

$$\begin{aligned}\int c(x,y)\,\mathrm{d}F(x,y) &= \int \frac{1}{2}(c(x,y)+c(y,x))\,\mathrm{d}F(x,y)\\ &= \int c(x,y)\,\mathrm{d}\frac{F(x,y)+F(y,x)}{2}.\end{aligned}$$

On the other hand, $F_s(x,y) = \frac{F(x,y)+F(y,x)}{2} \in \mathcal{F}(G,G)$. Consequently, we obtain (7.2.7), (7.2.8) by making use of (7.1.8) and (7.1.9) with $F_1 = F_2 = G$. □

We have the following analogue of the above proposition for nonsymmetric cost functions.

Proposition 7.2.2 *If $c(x,y)$ satisfies the Monge condition* (7.2.6) *and furthermore, $x_1 \le y \le x_2$ implies that $c(x_1,x_2) \ge c(y,y)$, then*

$$\inf\left\{\int c(x,y)\,\mathrm{d}F(x,y); F \in \mathcal{F}_{A+B}\right\} = \int_0^1 c(G^{-1}(u), G^{-1}(u))\,\mathrm{d}u. \qquad (7.2.9)$$

Proof: Applying the Monge conditions for every X, Y with $F_{X,Y} \in \mathcal{F}_{A+B}$, $Ec(X,Y) \ge Ec(F_X^{-1}(U), F_Y^{-1}(U))$. Since $F_X(x)+F_Y(x) = 2G(x)$, it follows that $F_X \wedge F_Y \le G \le F_X \vee F_Y$, and therefore, $F_X^{-1} \wedge F_Y^{-1} \le G^{-1} \le F_X^{-1} \vee F_Y^{-1}$. Consequently, we have $c(F_X^{-1}(U), F_Y^{-1}(U)) \ge c(G^{-1}(U), G^{-1}(U))$, which proves (7.2.9). □

Remark 7.2.3 *The marginals of the class $\mathcal{F}_{A+B}$ have largest and smallest elements, defined by*

$$F_1^*(x) = \begin{cases} 2\,G(x), & x < x_0, \\ 1, & x \ge x_0, \end{cases}$$

and

$$F_2^*(x) = \begin{cases} 2\,G(x) - 1, & x < x_0, \\ 1, & x \ge x_0, \end{cases}$$

with $x_0 := \inf\{y; 2G(y) \ge 1\}$. Note that there is no smallest d.f. in $\mathcal{F}_{A+B}$. To show this let $F_1(x), F_2(x)$ be the marginal d.f.s of the smallest elements $F \in \mathcal{F}_{A+B}$ and let G_1, G_2 be d.f.s such that $G_1(x)+G_2(x) = 2G(x)$. If the lower Fréchet bounds satisfy $(F_1(x)+F_2(y)-1)_+ \le (G_1(x)+G_2(y)-1)_+$, then $F_1 \le G_1$ and $F_2 \le G_2$, which implies that $F_1 = G_1, F_2 = G_2$. In particular, this implies that $(G^{-1}(U), G^{-1}(1-U))$ is in the general nonsymmetric case no longer a solution to the problem of maximizing $\int c(x,y)\,\mathrm{d}F(x,y)$ over the class $\mathcal{F}_{A+B}$. For example, let G be the d.f. of $\frac{1}{4}\sum_{i=1}^{4} \varepsilon_{(i)}$. Then $P_1 = P^{(G^{-1}(U), G^{-1}(1-U))} = \frac{1}{4}(\varepsilon_{(1,4)} + \varepsilon_{(2,3)} + \varepsilon_{(3,2)} + \varepsilon_{(4,1)})$, while $P_2 = P^{((F_1^*)^{-1}(U), (F_2^*)^{-1}(1-U))} = \frac{1}{2}(\varepsilon_{(1,4)} + \varepsilon_{(2,3)})$. For $c_1(x,y) =$

$1_{(-\infty,(3,2)]}(x,y)$, *we have* $E_{P_1}c_1 = \frac{1}{4}$, $E_{P_2}c_1 = 0$, *while for* $c_2 = 1_{[(2,3),\infty)}$, *we have* $E_{P_1}c_1 = \frac{1}{4}, E_{P_2}c_2 = \frac{1}{2}$. *Note that both functions,* $-c_1, -c_2$, *are Monge functions (but are not unimodal).*

We next consider the case where in the network example we fix the total outflow minus the inflow of each node. This problem is known in the literature as the minimal network flow problem (cf. for example Barnes and Hoffman (1985, Section 9) or Anderson and Nash (1987)). Assume that the outflow minus the inflow of each node is fixed; i.e., the following Kirchhoff equations hold

$$\sum_k x_{ik} - \sum_k x_{ki} \;=\; a_i - b_i \;=\; h_i \quad \text{for all } i,$$

or equivalently,

$$H(k) \;=\; A(k) - B(k), \quad 1 \le k \le n, \tag{7.2.10}$$

with $A(k) = \sum_{j=1}^{k} a_j$, $B(k) = \sum_{j=1}^{k} b_j$, and $H(k) = \sum_{j=1}^{k} h_j$. Consider now the general case: Let A, B be distribution functions and let $\mathcal{F}_{A-B}$ be the set of all "generalized" d.f.s of finite measures on $\mathbb{R}^2$ with marginals F_1, F_2 satisfying $F_1 - F_2 = A - B$. We consider the following transportation problem:

$$\text{minimize} \int c(x,y)\,\mathrm{d}F(x,y) \quad \text{subject to } F \in \mathcal{F}_{A-B}, \tag{7.2.11}$$

with $c(x,y)$ satisfying the Monge condition (7.2.6). To solve (7.2.11) we make use of the following dual representation (cf. (6.1.23)):

$$\begin{aligned} &\inf\left\{\int c(x,y)\,\mathrm{d}F(x,y);\;\; F \in \mathcal{F}_{A-B}\right\} \qquad\qquad (7.2.12)\\ &= \sup\left\{\int f\,\mathrm{d}(A-B)(x);\;\; f(x) - f(y) \le c(x,y), \quad \forall x,y\right\}. \end{aligned}$$

We first consider a particular type of cost function.

Proposition 7.2.4 *Let* $c(x,y) = |x-y|\max(1, h(|x-a|), h(|y-a|))$, *where* h *is a monotonically nondecreasing function on* $\mathbb{R}_+$. *Then*

$$\begin{aligned} &\inf\left\{\int c(x,y)\,\mathrm{d}F(x,y);\;\; F \in \mathcal{F}_{A-B}\right\} \qquad\qquad (7.2.13)\\ &= \int \max(1, h(|x-a|))|A-B|(x)\,\mathrm{d}x, \end{aligned}$$

provided that $h(|x-a|)$ *is locally integrable.*

Proof: We first note that the duality constraints condition $f(x) - f(y) \le c(x,y)$, for all x, y, holds if and only if f is absolutely continuous and moreover, $|f'(x)| \le \max(1, h(|x-y|))$ a.s. Consequently, by the dual representation (7.2.12), we obtain

$$\begin{aligned}
&\inf\left\{\int c(x,y)\, dF(x,y);\ F \in \mathcal{F}_{A-B}\right\} \\
&= \sup\left\{\int f\, d(A-B)(x);\ |f'| \le \max(1, h(|x-a|)),\quad \forall x\right\} \\
&= \sup\left\{\int f'(x)\, d(A-B)(x)\, dx;\ |f'| \le \max(1, h(|x-a|)),\quad \forall x\right\} \\
&= \int \max(1, h(|x-a|))|A-B|(x)\, dx.
\end{aligned}$$

□

To handle the general case set

$$c(x,y) = |x-y|\zeta(x,y) \quad \left(\text{i.e., } \zeta(x,y) = \frac{c(x,y)}{|x-y|}\right). \tag{7.2.14}$$

Theorem 7.2.5 *Assume that for any $x < t < y$, $\zeta(t,t) \le \zeta(x,y)$, $\zeta(x,y) = \zeta(y,x)$. Moreover, let $\zeta(x,y)$ be right continuous in y, and also assume that $t \to \zeta(t,t)$ is locally bounded. Then the optimal value in the minimization problem* (7.2.11) *is equal to*

$$\inf\left\{\int c(x,y)\, dF(x,y);\ F \in \mathcal{F}_{A-B}\right\} = \int \zeta(t,t)|A-B|(t)\, dt. \tag{7.2.15}$$

Proof: Let $\mathcal{F} = \{f;\ f(x) - f(y) \le c(x,y),\ \forall x,y\}$, and let $\mathcal{F}^* = \{f$ absolutely continuous and $|f'(t)| \le \zeta(t,t),\quad \forall t\}$. Then $\mathcal{F} \subset \mathcal{F}^*$, and for $f \in \mathcal{F}$ we have $\frac{f(x)-f(y)}{|x-y|} \le \zeta(x,y)$, and therefore, $\overline{\lim\limits_{y\to x}} \frac{f(x)-f(y)}{|x-y|} \le \zeta(x,x)$. Also,

$$\underline{\lim\limits_{y\to x}} \frac{f(x)-f(y)}{|x-y|} = -\overline{\lim} \frac{f(y)-f(x)}{|x-y|} \ge -\overline{\lim}\,\zeta(y,x) = -\zeta(x,x).$$

Since f is locally Lipschitz, it is absolutely continuous, so the inequalities above imply that $|f'(t)| \le \zeta(t,t)$ a.s. If, conversely, $f \in \mathcal{F}^*$, then $f(x) - f(y) = \int_x^y f'(t)\, dt$, and therefore, $|f(x) - f(y)| \le \int_x^y |f'(t)|\, dt \le \int_x^y \zeta(t,t)\, dt \le |x-y|\zeta(x,y) = c(x,y)$. The dual representation (7.2.12) again implies (7.2.13) (by the same arguments as in the proof of Proposition 7.2.4).

□

Next, we consider the following transportation problem with stochastically ordered marginals posed by Rogers (1992). Let F, G be real distribution functions, $F \leq_{st} G$; here as usual $\leq_{st}$ stands for the stochastic order. Let $C := \{(x,y) \in \mathbb{R}^2;\ x \leq y\}$, and let

$$M_C(F,G) := M(F,G) \cap \{\mu \in M^1(\mathbb{R}^2, \mathbb{B}^2);\ \mu(C) = 1\} \tag{7.2.16}$$

be the set of all measures with marginals F, G that are concentrated on the order cone C. The problem is to determine, for a given strictly convex function φ, the bound

$$\sup\left\{\int \varphi(x-y)\mu(\,dx,\,dy);\ \ \mu \in M_C(F,G)\right\}. \tag{7.2.17}$$

The motivation for problem (7.2.17) is to get a good monotone coupling of random walks $(S_n), (S_n')$ with $S_0' = x \geq X_0 = 0,\ \ S_n' \geq S_n$ for all n, and $S_n' = S_n$ for all large enough n.

Without the order restriction, a solution of (7.2.17) is given by the random variables $X = F^{-1}(U),\ Y = G^{-1}(1-U)$ for a uniform $(0,1)$ distributed r.v. U. It is intuitively clear that a solution of (7.2.17) should concentrate as much mass on the diagonal as possible. This is indeed true.

Theorem 7.2.6 (Rogers (1992)) *Each solution* (X,Y) *of* (7.2.17) *has the property that*

$$P(X=Y) \quad = \quad |F \wedge G| \quad = \int f \wedge g \,dm, \tag{7.2.18}$$

when $F = fm, G = gm$. *There exists a solution of* (7.2.17).

We next characterize the optimal solutions by an order-type relation.

Theorem 7.2.7 *Let* X, Y *be r.v.s with d.f.s* F, G *and* $X \leq Y$ *a.s. Then* (X,Y) *defines a solution of* (7.2.17) *iff*

$$X(\omega) \ <\ X(\omega') \ \leq\ Y(\omega) \ \leq\ Y(\omega') \quad \textit{implies } Y(\omega') \ =\ Y(\omega) \tag{7.2.19}$$

a.s. (for (ω, ω') *and with respect to the product measure).*

Proof: If (X,Y) is an optimal admissible coupling and if on a set of pairs (ω,ω') with positive measure $X(\omega) < X(\omega') \leq Y(\omega) < Y(\omega')$ holds, then let us define $Y'(\omega') := Y(\omega), Y'(\omega) := Y(\omega')$ and set $Y' = Y$ otherwise. Then Y' has d.f. G and $E\varphi(X - Y') > E\varphi(X - Y)$ because φ is strictly convex. Since there is essentially up to simultaneous rearrangements only one pair of r.v.s X, Y with d.f.s F, G satisfying the order relation (7.2.18), the opposite direction follows from the first part of the proof. □

In terms of measures $\mu \in M_C(F,G)$, the characterization of optimality of μ in (7.2.19) can be formulated as

$$\mu \otimes \mu(\{(x_1, y_1, x_2, y_2);\ x_1 < x_2 \le y_1 < y_2\}) = 0. \tag{7.2.20}$$

We remark that the characterization of optimal pairs in (7.2.19), resp. (7.2.20), implies the "maximal concentration on the diagonal" property in (7.2.18).

For finite discrete distributions one can explicitly construct optimal pairs with the ordering property given in (7.2.19). We consider at first the case of equiprobable atoms in each distribution. So let $\mu_1 = \frac{1}{n}\sum_{i=1}^n \varepsilon_{a_i}$, $\mu_2 = \frac{1}{n}\sum_{i=1}^n \varepsilon_{b_i}$ be the measures corresponding to F, G, where $a_1 \le \cdots \le a_n$, $b_1 \le \cdots \le b_n$, and $a_i \le b_i$ for all i. Problem (7.2.17) is equivalent to the following problem: Find a permutation $\pi \in \Upsilon_n$ such that

$$\sum_{i=1}^n \varphi(b_i - a_{\pi(i)}) \text{ is maximal.} \tag{7.2.21}$$

Here, the maximum is considered over all permuations $\pi \in \Upsilon_n$ such that $a_{\pi(i)} \le b_i, 1 \le i \le n$. Permutations with this property are called admissible permutations. An optimal admissible permutation is essentially unique (up to indices with equal values of a_i), and it is given in the following theorem.

Theorem 7.2.8 *Define $\pi^* \in \Upsilon_n$ inductively:*

$$\begin{aligned} \pi^*(1) &:= \max\{k \le n; a_k \le b_1\} \qquad (7.2.22)\\ \pi^*(k) &:= \max\{\ell \le n; \ell \notin \{\pi^*(1), \dots, \pi^*(k-1)\}, a_\ell \le b_k\},\ 2 \le k \le n. \end{aligned}$$

Then $\pi^* \in \Upsilon$ is the optimal admissible permutation.

Proof: Define on $\Omega = \{1, \dots, n\}$ (supplied with the uniform distribution P) random variables $X(i) := a_i$ and $Y(i) := b_{\pi^*(i)}, 1 \le i \le n$. Then $X \le Y$, since π^* is admissible and X, Y satisfy the order relation (7.2.19). Therefore, they are optimal couplings. Equivalently, π^* is the optimal admissible permutation. □

It is clear from the construction in Theorem (7.2.8) that up to a simultaneous permutation of the probability space, an optimal pair of r.v.s is essentially unique.

Remark 7.2.9 *Theorem 7.2.8 can be extended to the case that $\mu_1 = \sum_{i=1}^n p_i \varepsilon_{a_i}$, $\mu_2 = \sum_{i=1}^n q_i \varepsilon_{b_i}$ with rational p_i, q_i, by representing p_i, q_i in the formal equiprobable case. By an approximation argument—as given in Rogers (1992)—one can approximate the optimal couplings for F, G with* *couplings having compact support. The general case then can be approximated via the ordering criterion* (7.2.19) *using a truncation technique. Thus, applying Theorem* 7.2.8, *we are able to construct explicit approximate solutions in the general case.*

7.3 Mass Transportation Problems with Capacity Constraints

In this section we obtain explicit solutions of Monge–Kantorovich mass transportation problems with capacity constraints. The Hoeffding–Fréchet inequality is extended for bivariate distribution functions having fixed marginals and satisfying additional constraints. In the discrete case, our results lead to "greedy" algorithms similar to the classical northwest corner rule.

Let us start with recalling the abstract version of the MKP: Given two Borel measures μ and ν on a separable metric space S with equal total mass $\lambda = \mu(S) = \nu(S) < \infty$ and a measurable cost function c on $S \times S$, find

$$L_c(\mu,\nu) = \inf \int c(x,y) P(\mathrm{d}x, \mathrm{d}y), \tag{7.3.1}$$

$$U_c(\mu,\nu) = \sup \int c(x,y) P(\mathrm{d}x, \mathrm{d}y), \tag{7.3.2}$$

where the infimum and supremum are taken over all Borel measures P on $S \times S$ having projections (marginals)

$$P(\cdot \times S) = \mu(\cdot), \quad P(S \times \cdot) = \nu(\cdot). \tag{7.3.3}$$

As shown in Section 3.1, the explicit solutions of MKP are based on the Hoeffding–Fréchet inequality (referred to as upper and lower Fréchet bounds):

$$\max(0, F^\mu(x) + F^\nu(y) - \lambda) \le F^P(x,y) \le \min(F^\mu(x), F^\nu(y)), \tag{7.3.4}$$

for any P on $\mathbb{R}^2$ that satisfies (7.3.3) with $S = \mathbb{R}$. (In (7.3.4) and in the sequel, F^P stands for the distribution function of P.) If c is a lattice superadditive (equivalently, $-c$ is a Monge function):

$$c(x',y') + c(x,y) \ge c(x',y) + c(x,y') \quad \text{for all } x' \ge x, \ y' \ge y, \tag{7.3.5}$$

then under mild moment conditions on μ and ν the explicit values of L_c and U_c were given in Section 3.1.

In this section we consider two marginal problems with additional constraints on the joint distribution functions. Suppose μ and ν are two nonnegative Borel measures on $\mathbb{R}$, $\mu(\mathbb{R}) = \nu(\mathbb{R}) = \lambda < \infty$. Suppose $c : \mathbb{R}^2 \to$

$\mathbb{R}$ is a right-continuous Monge function generating a nonnegative measure on $\mathbb{R}^2$. Let σ be a nonnegative bounded Borel measure on $\mathbb{R}^2$. (Note that the total mass of σ may be different from λ.)

Problem I. *Find*

$$\textit{maximum} \int_{\mathbb{R}^2} c(x,y)P(\,\mathrm{d}x,\,\mathrm{d}y) \tag{7.3.6}$$

subject to the constraints

$$\begin{aligned} &P \textit{ is a nonnegative Borel measures on } \mathbb{R}^2 && (7.3.7)\\ &\qquad\qquad \textit{with marginals } \mu \textit{ and } \nu, \textit{ and} \\ &P((-\infty,x]\times(-\infty,y]) \;\le\; \sigma((-\infty,x]\times(-\infty,y]) && (7.3.8)\\ &\qquad\qquad \textit{for all } x,y\in\mathbb{R}. \end{aligned}$$

Problem II. *Find*

$$\textit{minimum} \int_{\mathbb{R}^2} c(x,y)P(\,\mathrm{d}x,\,\mathrm{d}y) \tag{7.3.9}$$

subject to (7.3.7) *and*

$$P((-\infty,x]\times[y,\infty)) \;\le\; \sigma((-\infty,x]\times[y,\infty)) \quad \textit{for all } x,y\in\mathbb{R}. \tag{7.3.10}$$

Problem I with discrete μ and ν was studied by Barnes and Hoffman (1985). Olkin and Rachev (1990) extended their results by completing the characterization of the "optimal feasible" $\widetilde{P}$; i.e., $\widetilde{P}$ satisfies (7.3.7), (7.3.8) and attains the maximum in (7.3.6). This method is extended to solve Problem II as well.

We start with a refinement of the Fréchet bounds (7.3.4). We shall do this by determining the exact bounds for a d.f. $F^P(x,y)$ with marginals F^μ and F^ν assuming that P satisfies the constraint (7.3.8) or (7.3.10). Then we shall apply the extended Fréchet bounds to solve Problems I and II. Whereas in the discrete case the solution of Problem I leads to the Barnes–Hoffman greedy algorithm, the solution of Problem II implies a new greedy algorithm for a transportation problem with capacity constraints (7.3.10).

We begin with some notation. For two nonnegative Borel measures μ and ν on $\mathbb{R}$ with equal total mass λ denote by $M(\mu,\nu)$ the set of all nonnegative Borel measures on $\mathbb{R}^2$ with projections μ and ν. Without loss of generality set $\lambda = 1$. Given a nonatomic probability space, the set $\mathcal{F}(A,B)$ of joint d.f.s $F(x,y) = F_{X,Y}(x,y) = P(X\le x, Y\le y)$ with fixed marginals $F_X = A$ and $F_Y = B$ is the set of d.f.s of the probability laws in $M(\mu, \nu)$. Thus, the Fréchet bounds (7.3.4) can be rewritten as

$$\max_{F \in \mathcal{F}(A,B)} F(x,y) = F^*(x,y) := \min(A(x), B(y)), \tag{7.3.11}$$

$$\max_{F \in \mathcal{F}(A,B)} G(x,y) = G^*(x,y) := \min(A(x), \overline{B}(y)), \tag{7.3.12}$$

where $\overline{B}(y) := \nu([y, \infty))$ and $G(x,y) := G_{X,Y}(x,y) := P(X \leq x, Y \geq y)$. Clearly, the laws corresponding to F^* and G^* are in $M(\mu, \nu)$. Furthermore, given a nonnegative bounded Borel measure σ on $\mathbb{R}^2$, set

$$\begin{aligned} F^\sigma(x,y) &:= \sigma((-\infty, x] \times (-\infty, y]), \\ G^\sigma(x,y) &:= \sigma((-\infty, x] \times [y, \infty)) \\ \mathcal{F}(A, B, F^\sigma) &:= \{F \in \mathcal{F}(A,B);\; F \leq F^\sigma\}, \\ \mathcal{G}(A, B, G^\sigma) &:= \{G_{X,Y};\; F_{X,Y} \in \mathcal{F}(A,B), G_{X,Y} \leq G^\sigma\}. \end{aligned} \tag{7.3.13}$$

Our objective in the next two theorems is to extend the Fréchet bounds; we shall characterize the bounds

$$\max_{F \in \mathcal{F}(A,B,F^\sigma)} F(x,y) =: \widetilde{F}(x,y), \quad x, y \in \mathbb{R}, \tag{7.3.14}$$

$$\max_{G \in \mathcal{G}(A,B,G^\sigma)} G(x,y) =: \widetilde{G}(x,y), \quad x, y \in \mathbb{R}, \tag{7.3.15}$$

and shall examine the conditions implying

$$\widetilde{F} \in \mathcal{F}(A, B, F^\sigma) \quad \text{and} \quad \widetilde{G} \in \mathcal{G}(A, B, G^\sigma). \tag{7.3.16}$$

Theorem 7.3.1 *If*

$$F^\sigma(x,y) \geq \max(0, A(x) + B(y) - 1), \tag{7.3.17}$$

then the maximum in (7.3.14) *is attained:*

$$\begin{aligned} \widetilde{F}(x,y) &= \inf_{\substack{t \leq x \\ s \leq y}} \{F^\sigma(t,s) + \mu((t,x]) + \nu((s,y])\} \\ &= \inf_{\substack{t \leq x \\ s \leq y}} \{F^\sigma(t,s) + \mu((t,x]) + \nu((s,y])\} \wedge (A(x) \wedge B(y)), \end{aligned} \tag{7.3.18}$$

and $\widetilde{F} \in \mathcal{F}(A, B, F^\sigma)$, *where* $\wedge := \min$.

Remark 7.3.2 *Condition* (7.3.17) *is necessary and sufficient for* $\mathcal{F}(A, B, F^\sigma) \neq \emptyset$, *cf. Fréchet (1951), Kellerer (1964).*

Remark 7.3.3 *The second equality in* (7.3.18) *follows from the fact that* $F^\sigma(t,s) = 0$ *for* $t = -\infty$ *or* $s = -\infty$.

Remark 7.3.4 *From* (7.3.18) $\widetilde{F}$ *is not greater than the Hoeffding–Fréchet upper bound* F^* (7.3.11).

Remark 7.3.5 *By* (7.3.4) *the maximum in* (7.3.11) *is attained for the pair* $X^* = A^-(U)$, $Y^* = B^-(U)$, *where* A^- *is the generalized inverse of* A, *and* U *is uniformly distributed on* $[0,1]$. *In contrast, for* $\widetilde{F}$ *given in* (7.3.18), *the explicit form of the optimal pair* $(\widetilde{X}, \widetilde{Y})$ *with joint d.f. given by* $\widetilde{F}$ *is not known. However, in the discrete case one can use the Barnes–Hoffman greedy algorithm to compute* $\widetilde{F}$. *Suppose* μ, ν, *and* σ *are discrete measures,*

$$\begin{aligned} a_i &:= \mu(\{x_i\}), \quad i \in M = \{1,2,\dots,m\}, \qquad (7.3.19) \\ b_j &:= \nu(\{y_j\}), \quad j \in N = \{1,2,\dots,\}, \\ \sum_{i\in M} a_i &= \sum_{j\in N} b_j = 1; \\ \sigma_{ij} &:= F^\sigma(x_i, y_j), \quad i \in M,\ j \in N. \qquad (7.3.20) \end{aligned}$$

Then

$$\widetilde{F}(x_i, y_j) = \sum_{r=1}^{i}\sum_{s=1}^{j} p_{rs}, \qquad (7.3.21)$$

where the probabilities p_{rs} *are determined by the following variant of the northwest corner rule (see Hoffman (1961), Barnes and Hoffman (1985)); in fact, we set*

$$\begin{aligned} p_{11} &:= \min(a_1, b_1, \sigma_{11}); \qquad (7.3.22) \\ p_{ij} &:= \min\left\{ a_i - \sum_{s=1}^{j-1} p_{is}, b_j - \sum_{r=1}^{i-1} p_{rj}, \sigma_{ij} - \sum_{\substack{r\le i \ \ s \le j \\ (r,s)\neq(i,j)}} p_{rs} \right\}, \end{aligned}$$

if p_{rs} *is determined for* $r \le i < m$ *and* $s \le j < n$, *and we let*

$$p_{ij} := \min\left\{ a_i - \sum_{s=1}^{j-1} p_{is}, b_j - \sum_{r=1}^{i-1} p_{rj} \right\}, \quad \text{if } i = m \text{ or } j = n.$$

In other words, taking discrete versions of μ, ν, *and* σ *in* (7.3.19) *one can apply the greedy algorithm* (7.3.22) *to approximate* $\widetilde{F}$ *in* (7.3.18) *by means of* (7.3.21).

Proof of Theorem 7.3.1: The proof is based on three assertions.

Claim 7.3.6 (Fréchet (1951)) *The condition* $F^\sigma(x,y) \ge H_-(x,y) = \max(0, A(x) + B(y) - 1)$ *is necessary and sufficient for* $\mathcal{F}(A, B, F^\sigma) \neq \emptyset$.

Suppose $\mathcal{F}(A, B, F^\sigma) \neq \emptyset$. *Then, by* (7.3.4)

$$H_-(x,y) \leq F(x,y) < F^\sigma(x,y), \quad F \in \mathcal{F}(A,B,F^\sigma). \tag{7.3.23}$$

On the other hand, if $H_- \leq F^\sigma$, *then* $H_- \in \mathcal{F}(A,B,F^\sigma)$.

Claim 7.3.7 $\widetilde{F}$ *defined by* (7.3.18) *has marginal d.f.s* A *and* B *and for all* $x, y \in \mathbb{R}$,

$$\sup_{F \in \mathcal{F}(A,B,F^\sigma)} F(x,y) \leq \widetilde{F}(x,y). \tag{7.3.24}$$

For any $F \in \mathcal{F}(A,B,F^\sigma)$ and any $t \leq x, s \leq y$, we have $F(x,y) \leq F^\sigma(t,s) + \mu((t,x]) + \nu((s,y])$, which clearly implies (7.3.24).

Invoking Remark (7.3.3), $\widetilde{F}(x,y) \leq H_+(x,y)$ where H_+ is the upper Hoeffding–Fréchet bound, $H_+(x,y) := \min(A(x), B(y))$. Since $\widetilde{F} \geq H_-$ (cf. (7.3.23), (7.3.24)), $\widetilde{F} \in [H_-, H_+]$ has marginals A and B.

Theorem 7.3.1 is now a consequence of the following assertion.

Claim 7.3.8 $\widetilde{F}$ *is a d.f.*

To this end, we choose $-\infty = x_0 < x_1 < \cdots < x_{m-1} < x_m = \infty$, $-\infty = y_0 < y_1 < \cdots < y_{n-1} < y_n = \infty$ *such that* $\mu((x_{i-1}, x_i)) < \varepsilon$, $\nu((y_{n-1}, y_1)) < \varepsilon$, *and* $\sigma((x_{i-1}, x_i) \times (y_{j-1}, y_j)) < \varepsilon$ *for all* $i \in M = \{1, \ldots, m\}$ *and* $j \in N = \{1, \ldots, n\}$. *Set* $a_i := \mu((x_{i-1}, x_i])$, $b_j := \nu((y_{j-1}, y_j])$, *and* $\sigma_{ij} := F^\sigma(x_i, y_j)$. *Consider the convex polygon*

$$\Bigg\{ \widetilde{p} = (p_{ij})_{\substack{i \in M \\ j \in N}};\ p_{ij} \geq 0,\ p_{i\cdot} := \sum_{j \in N} p_{ij} = a_i, \tag{7.3.25}$$

$$p_{\cdot j} := \sum_{i \in M} p_{ij} = b_j, \sum_{r=1}^{i} \sum_{s=1}^{j} p_{rs} \leq \sigma_{ij}, \quad \text{for all } i \in M, j \in N \Bigg\}$$

$$= \Bigg\{ \widetilde{p};\ p_{ij} \geq 0, p_{i\cdot} = a_i, p_{\cdot j} = b_j, \sum_{r=1}^{i} \sum_{s=1}^{j} p_{rs} \leq \sigma_{ij}, i = 1, \ldots, m-1,$$

$$j = 1, \ldots, n-1, \sum_{s=1}^{j} p_{\cdot s} \leq \sigma_{mj}, j \in N, \sum_{r=1}^{i} p_{r\cdot} \leq \sigma_{in}, i \in M \Bigg\}.$$

By the Fréchet condition (7.3.17) *(cf. Claim* 7.3.6*) the marginals of* F^σ *majorize* A *and* B, *respectively, and thus* $\sigma_{mj} \geq \sum_{s=1}^{j} p_{\cdot s}$ *and* $\sigma_{in} \geq \sum_{r=1}^{i} p_{r\cdot}$ *for all* $j \in N, i \in M$. *The polygon* (7.3.25) *becomes*

$$\Bigg\{ \widetilde{p};\ p_{ij} \geq 0,\ p_{i\cdot} = a_i,\ p_{\cdot j} = b_j \quad \text{for all } i \in M,\ j \in N, \tag{7.3.26}$$

$$\left.\sum_{r=1}^{i}\sum_{s=1}^{j} p_{rs} \leq \sigma_{ij} \quad \textit{for all } i=1,\ldots,m-1,\ j=1,\ldots,n-1\right\}.$$

Consider now the discrete analogue of $\widetilde{F}$ *in* (7.3.18):

$$\begin{aligned} d_{ij} &:= \min_{\substack{0\leq r\leq i\\ 0\leq s\leq j}} \{\sigma_{rs}+a_{r+1}+\cdots+a_i+b_{s+1}+\cdots+b_j\}, \\ d_{ij} &:= 0 \quad \textit{if } i=0 \textit{ or } j=0, \end{aligned} \tag{7.3.27}$$

where $\sigma_{rs}=0$ *if* $r=0$ *or* $s=0$. *Our aim now is to show that* $\widetilde{d}=(d_{ij})_{\substack{i\in M\\ j\in N}}$ *determines a bivariate d.f. with support on* $X\times Y, X=(x_i)_{i\in M}$, $Y=(y_j)_{j\in N}$.

Claim 7.3.9 *The greedy algorithm* (7.3.22) *is determined uniquely by* (7.3.27); *i.e.,*

$$d_{ij} := \sum_{r=1}^{i}\sum_{s=1}^{j} p_{rs}, \quad i\in N,\ j\in M. \tag{7.3.28}$$

Proof: Consider the discrete version of $\widetilde{F}$ (cf. (7.3.21), (7.3.25)). Let $\sigma_{r,s} := 0$ if $r=0$ or $s=0$, and define

$$\begin{aligned} d_{ij} &:= \min_{\substack{0\leq r\leq i\\ 0\leq s\leq j}} \{\sigma_{rs}+a_{r+1}+\cdots+a_i+b_{s+1}+\cdots+b_j\}, \\ d_{ij} &= 0 \quad \text{if } i=0 \text{ or } j=0. \end{aligned} \tag{7.3.29}$$

We need to check the equality

$$d_{ij} = \sum_{r=1}^{i}\sum_{s=1}^{j} p_{rs}, \quad i\in M,\ j\in N, \tag{7.3.30}$$

where the p_{ij}'s are determined by the greedy algorithm (7.3.22). If $i=j=1$, then $p_{11}=\min(a_1,b_1,\sigma_{11})$ (cf. (7.3.22)), and by (7.3.29)

$$d_{11} = \min\{\sigma_{11}+a_1+b_1,\ \sigma_{11}+a_1, \sigma_{10}+b_1, \sigma_{11}\} = p_{11}.$$

Suppose we have proved that

$$d_{1,j-1} = p_{11}+\cdots+p_{1,j-1}. \tag{7.3.31}$$

Then

$p_{11}+\cdots+p_{1j}$

$$\begin{aligned}
&= \sum_{s=1}^{j-1} p_{1s} + \min\left\{a_1 - \sum_{s=1}^{j-1} p_{1s},\ b_j,\ \sigma_{1j} - \sum_{s=1}^{j-1} p_{1s}\right\} \\
&= \min\{a_1,\ b_j + d_{1,j-1},\ \sigma_{1,j}\} \\
&= \min\{a_1,\ b_1 + \cdots + b_j,\ \sigma_{11} + b_2 + \cdots + b_j,\ \ldots, \sigma_{1,j-1} + b_j,\ \sigma_{i,j}\} \\
&= d_{1,j}.
\end{aligned}$$

These equalities hold due to (7.3.22), (7.3.31), and (7.3.29), respectively. By symmetry, $d_{i,1} = p_{11} + \cdots + p_{i1}$. Suppose next that

$$d_{rs} = \sum_{k=1}^{r} \sum_{l=1}^{s} p_{kl} \quad \text{for all } r \le i, s \le j,\ (r,s) \ne (i,j). \tag{7.3.32}$$

Then for $1 \le i \le m-1,\ 1 \le j \le n-1$,

$$\sum_{r=1}^{i} \sum_{s=1}^{j} p_{rs} = \min\left\{a_i + \sum_{r=1}^{i-1} \sum_{s=1}^{j} p_{rs},\ b_j + \sum_{r=1}^{i} \sum_{s=1}^{j-1} p_{rs},\ \sigma_{ij}\right\} = d_{ij},$$

where the equalities follow from (7.3.22) and (7.3.32). Thus

$$d_{ij} = \sum_{r=1}^{i} \sum_{s=1}^{j} p_{rs} \quad \text{for all } 1 \le i \le m-1,\ 1 \le j \le n-1. \tag{7.3.33}$$

Consider now the case $i = m$. Then,

$$\begin{aligned}
\sum_{r=1}^{m} p_{r1} &= \sum_{r=1}^{m-1} p_{r1} + \min\left\{a_m,\ b_1 - \sum_{r=1}^{m-1} p_{r1}\right\} \\
&= \min\{a_m + d_{m-1,1}, b_1\} = d_{m,1},
\end{aligned}$$

which follows from (7.3.22) and (7.3.33). Suppose that

$$d_{m,j-1} = \sum_{r=1}^{m} \sum_{s=1}^{j-1} p_{rs}. \tag{7.3.34}$$

Then using (7.3.22), (7.3.33), and (7.3.34), for $1 \le j \le n$,

$$\begin{aligned}
\sum_{r=1}^{m} \sum_{s=1}^{n} p_{rs} &= \min\left\{a_m + \sum_{r=1}^{m-1} \sum_{s=1}^{j} p_{rs}, b_j + \sum_{r=1}^{m} \sum_{s=1}^{j-1} p_{rs}\right\} \\
&= \min\{\sigma_{rs} + a_{r+1} + \cdots + a_m b_{s+1} + \cdots + b_j\},
\end{aligned}$$

for $0 \le r \le m,\ 0 \le s \le j,\ (r,s) \ne (m,j)$;

$$\sum_{r=1}^{m} \sum_{s=1}^{m} p_{rs} = d_{m,j}, \quad \text{for } r = m, s = j,\ \sigma_{m,j} = F^{\sigma}(\infty, y_j) \ge b_j.$$

Similarly, $d_{i,n} = \sum_{r=1}^{i} \sum_{s=1}^{n} p_{rs}$, for all $i \in M$, which proves Claim 7.3.9. □

The greedy algorithm (7.3.22) defines nonnegative p_{ij}'s (cf. Barnes and Hoffman (1985, Lemma 3.2)). Define the probability $P^{(\varepsilon)}$ on $X \times X$ by

$$P^{(\varepsilon)}((-\infty, x_i], (-\infty, y_j]) \quad := \quad d_{ij}, \quad i \in M, \ j \in N. \tag{7.3.35}$$

Similarly, $(a_i)_{i \in M}$ and $(b_i)_{i \in N}$ determine probabilites $\mu^{(\varepsilon)}$ and $\nu^{(\varepsilon)}$ with supports X and Y, respectively. If ϱ is the *Kolmogorov (uniform) distance*

$$\varrho(\mu, \nu) \quad := \quad \sup_{x \in \mathbb{R}} |F^{\mu}(x) - F^{\nu}(x)|, \tag{7.3.36}$$

then the sequences $\left(\mu^{(\varepsilon)}\right)_{\varepsilon > 0}$ and $\left(\nu^{(\varepsilon)}\right)_{\varepsilon > 0}$ are ϱ-relatively compact, and thus there exists $\varepsilon_n \downarrow 0$ such that

$$\varrho\left(\mu^{(\varepsilon_n)}, \mu\right) \to 0 \quad \text{and} \quad \varrho\left(\nu^{(\varepsilon_n)}, \nu\right) \to 0. \tag{7.3.37}$$

(For more facts on ϱ-relative compactness cf. Rachev (1984a) and Kakosjan, Klebanov, and Rachev (1988, Sec. 2.5).)

Similarly, by definition of $\sigma_{ij} := F^{\sigma}(x_i, y_j)$, $\sigma((x_{i-1}, x_i) \times (y_{j-1}, y_j)) < \varepsilon$ we have that $(\sigma_{ij})_{\substack{i \in M \\ i \in N}}$ determines a measure $\sigma^{(\varepsilon)}$ on $X \times Y$. Again, the family $\left(\sigma^{(\varepsilon)}\right)_{\varepsilon > 0}$ is ϱ-relatively compact. Thus, without loss of generality, we may assume that as $\varepsilon_n \to 0$,

$$\varrho\left(\sigma^{(\varepsilon_n)}, \sigma\right) \quad = \quad \sup_{x, y \in R} \left| F^{\sigma^{(\varepsilon_n)}}(x, y) - F^{\sigma}(x, y) \right| \to 0. \tag{7.3.38}$$

As in Claim 7.3.7, we conclude that $P^{(\varepsilon)}$ has marginals $\mu^{(\varepsilon)}$ and $\nu^{(\varepsilon)}$, and thus $\left(P^{(\varepsilon)}\right)_{\varepsilon > 0}$ is tight. By (7.3.37), (7.3.38), (7.3.26), and (7.3.18), there exists a subsequence $\{\varepsilon_n'\} \subset \{\varepsilon_n\}$ such that $P^{(\varepsilon_n')}$ weakly converges to a measure $\widetilde{P}$ with d.f. $\widetilde{F}$. The proof of Theorem 7.3.1 is now complete. □

The next theorem provides an explicit expression for the Fréchet type bound (7.3.15). We recall the notations (7.3.11)–(7.3.13).

Theorem 7.3.10 *Suppose $G^{\sigma}(x, y) := \sigma((-\infty, x] \times [y, \infty))$ (cf. (7.3.13)) satisfies the condition*

$$G^{\sigma}(x, y) \ \geq \ \max\left(0, A(x) - \mathring{B}(y)\right) \quad \left(\mathring{B}(y) := \nu((-\infty, y))\right). \tag{7.3.39}$$

Then the maximum in (7.3.15) *is attained, and*

$$\widetilde{G}(x,y) = \inf_{\substack{t\le x\\ s\ge y}} \{G^\sigma(t,s) + \mu((t,x] + v([y,s))\}. \tag{7.3.40}$$

Conclusions similar to those in Remarks 7.3.2–7.3.5 can be made. Here we shall only point out the greedy algorithm that can be used to approximate the optimal distribution $\widetilde{G}$. We use the notations (7.3.19) again and let

$$\lambda_{ij} := G^\sigma(x_i,y_j), \quad i\in M, \ j\in N. \tag{7.3.41}$$

Then, in this discrete case, $\widetilde{G}$ has the form

$$\widetilde{G}(x_i,y_j) = \sum_{r=1}^{i}\sum_{s=j}^{n} p_{rs}, \tag{7.3.42}$$

where the probabilites p_{ij} are determined by the following *southwest corner rule:*

$$p_{in} := \min\{a_1, b_n, \lambda_{1n}\}; \tag{7.3.43}$$

$$p_{ij} := \min\left\{a_i - \sum_{s=j+1}^{n} p_{is},\ b_j - \sum_{r=1}^{i-1} p_{rj}, \lambda_{ij} - \sum_{\substack{r\le i\\ (r,s)\ne(i,j)}}\sum_{s\ge j} p_{rs}\right\}, \tag{7.3.44}$$

if $i = m$ or $j = 1$.

if p_{rs} is determined for $r\le i\le m-1$ and $s\ge j>1$; moreover,

$$p_{ij} := \min\left\{a_i - \sum_{s=j+1}^{n} p_{is},\ b_j - \sum_{r=1}^{i-1} p_{rj}\right\}, \quad \text{if } i=m \text{ or } j=1. \tag{7.3.45}$$

We now give explicit solutions of the marginal problems I and II.

Theorem 7.3.11 *Suppose*

(i) $c:\mathbb{R}^2\to\mathbb{R}$ *is a right-continuous lattice superadditive function* ($-c$ *is a Monge function);*

(ii) μ *and* ν *are two Borel nonnegative measures on* $\mathbb{R}$ *with* $\mu(\mathbb{R}) = \nu(\mathbb{R}) = \lambda < \infty$ *and d.f.s* F^μ *and* F^ν, *and such that*

$$\int_{\mathbb{R}} c(x,y_0)\mu(\mathrm{d}x) + \int_{\mathbb{R}} c(x_0,y)\nu(\mathrm{d}y) < \infty \quad \textit{for some } x_0,y_0\in\mathbb{R}; \tag{7.3.46}$$

(iii) *σ is a nonnegative bounded Borel measure on $\mathbb{R}^2$ and*

$$F^{\sigma}(x,y) \ \geq \ \max(0, F^{\mu}(x) + F^{\nu}(y) - \lambda) \quad \textit{for all } x,y \in \mathbb{R}. \quad (7.3.47)$$

Then the maximum in (7.3.6) *is attained at the "optimal" measure $\widetilde{P}$. $\widetilde{P}$ satisfies the feasibility conditions* (7.3.7), (7.3.8) *and is determined by*

$$F^{\widetilde{P}}(x,y) \ := \ \inf_{\substack{t \leq x \\ s \leq y}} \{F^{\sigma}(t,s) + \mu((t,x]) + \nu((s,y])\}, \quad x,y \in \mathbb{R}. \quad (7.3.48)$$

Proof: We need Theorem 3.1.2 (cf. Cambanis, Simons, and Stout (1976, Theorem 1); see also Rachev (1991c, Section 7.3)). If (7.3.5) holds, then for measures P_1 and P_2 on $\mathbb{R}^2$ with marginals μ and ν,

$$F^{P_1} \ \leq \ F^{P_2} \quad \Rightarrow \quad \int_{\mathbb{R}^2} c \, \mathrm{d}P_1 \ \leq \ \int_{\mathbb{R}^2} c \, \mathrm{d}P_2, \quad (7.3.49)$$

which with an appeal to Theorem 7.3.1 yields the result. □

Remark 7.3.12 *The assumption* (7.3.46) *can be replaced by one of the following assumptions:*

(a) *$c(x,y)$ is symmetric, and $\int c(x,x)(\mu+\nu)(\,\mathrm{d}x) < \infty$.*

(b) *$c(x,y)$ is uniformly integrable for all P with marginals μ and ν.*

That (a) *implies* (7.3.49) *follows from Cambanis, Simons, and Stout (1976); that* (b) *implies* (7.3.49) *follows from Tchen (1980, Corollary 2.1); see also Rachev (1991c, Theorem 7.3.2).*

Remark 7.3.13 *Condition* (7.3.47) *guarantees that the set of feasible solutions P determined by* (7.3.7), (7.3.8) *is not empty.*

Remark 7.3.14 *If*

$$F^{\sigma}(x,y) \ \geq \ \min(F^{\mu}(x), F^{\nu}(y)) \ =: H_{+}(x,y), \quad (7.3.50)$$

then $F^{\widetilde{P}}$ in (7.3.48) *equals H_+ (cf. Remark* 7.3.3*). Thus, Theorem* 7.3.11 *(see also the next Theorem* 7.3.17*) can be considered as a generalization of Theorem 2 of Cambanis, Simons, and Stout (1976) and Corollary 2.2 of Tchen (1980). In this case, Hoffman's (1962) northwest corner rule gives a greedy algorithm to determine an "optimal" measure $\widetilde{P}$, provided that μ and ν have finite discrete support.*

Remark 7.3.15 *Consider the discrete version of Problem I (see (7.3.6)). Suppose $c(i,j)$, $i \in M, j \in N$, is a lattice superadditive sequence*

$$c(i,j) + c(i+1,j+1) \geq c(i,j+1) + c(i+1,j), \quad i = 1,\ldots,m-1, \; j = 1,\ldots,n-1. \tag{7.3.51}$$

Hoffman (1961) and Barnes and Hoffman (1985) treat $c(i,j)$ as the (negative) cost of shipping a unit commodity from origin i to destination j. Suppose the discrete measures μ and ν with supports M and N are given. Then $a_i = \mu\{i\}$ and $b_j = \nu\{j\}$ are interpreted as the amount of a product available at i and the amount required at destination j. Suppose the $(m-1) \times (n-1)$ matrix (σ_{ij}) satisfies

$$\sigma_{ij} \geq \max\left\{0, \sum_{r=1}^{i} a_r - \sum_{s=j+1}^{n} b_s\right\}, \tag{7.3.52}$$
$$\sigma_{ij} \leq \sigma_{is}, \; \sigma_{ij} \leq \sigma_{rj}, \; \sigma_{ij} + \sigma_{rs} \geq \sigma_{is} + \sigma_{rj}, \; r \geq i, s \geq j.$$

(These conditions are related to what is called a uniformly tapered matrix; see Marshall and Olkin (1979).) Barnes and Hoffman (1985) consider the following transportation problem:

$$\textit{maximize} \sum_{i\in M}\sum_{j\in N} c(i,j)p_{ij} \tag{7.3.53}$$

subject to

$$p_{ij} \geq 0, \; p_{i\cdot} = a_i, \; p_{\cdot j} = b_j \quad \textit{for all } i \in M, j \in N, \tag{7.3.54}$$
$$\sum_{r=1}^{i}\sum_{s=1}^{j} p_{ij} \leq \sigma_{ij}, \quad i = 1,\ldots,m-1, \; j = 1,\ldots,n-1.$$

Clearly, (7.3.54) *is a special case of Problem I. Following Barnes and Hoffman,* (7.3.54) *can be viewed as the capacity restrictions on the amount that can be shipped from the first i origins to the first j destinations. Theorem* 7.3.11 *is completed by showing that the greedy algorithm of Barnes and Hoffman (1985) for determining the solution $(p_{ij})_{\substack{i\in M\\ j\in N}}$ of* (7.3.53) *is also characterized by*

$$F^P(i,j) := \sum_{r=1}^{i}\sum_{s=1}^{j} p_{rs} \tag{7.3.55}$$
$$= \min_{\substack{0\leq r\leq i\\ 0\leq s\leq j}} \{\sigma_{rs} + a_{r+1} + \cdots + a_i + b_{s+1} + \cdots + b_j\},$$

$$\sigma_{rs} = \begin{cases} 0 & \textit{if } r = 0 \quad \textit{or } s = 0, \\ +\infty & \textit{if } r = m \quad \textit{or } s = n. \end{cases}$$

Remark 7.3.16 *One can determine the extremal value in* (7.3.6):

$$\max_{F\in\mathcal{F}(F^\mu,F^\nu;F^\sigma)} \int_{R^2} c\,\mathrm{d}F \;=\; \int_{\mathbb{R}^2} c\,\mathrm{d}F^{\widetilde{P}}, \tag{7.3.56}$$

where $F^{\widetilde{P}}$ *is given by* (7.3.48). *By* (7.3.46) *and Cambanis, Simons, and Stout (1976, p. 288, (9)),*

$$\int_{\mathbb{R}^2} c\,\mathrm{d}F \;= \tag{7.3.57}$$

$$\int_{\mathbb{R}} c(x,y_0)F^\mu(\,\mathrm{d}x) + \int_{\mathbb{R}} c(x_0,y)F^\nu(\,\mathrm{d}y) - c(x_0,y_0) + \int_{\mathbb{R}^2} B(x,y)\mu_c(\,\mathrm{d}x,\,\mathrm{d}y)$$

for any bivariate d.f. F with marginals F^μ *and* F^ν. *(Since* $F^{\widetilde{P}} \in \mathcal{F}(F^\mu, F^{\nu\cdot}, F^\sigma)$ *by Theorem* 7.3.1, (7.3.57) *can be used to compute the value of* $\int c\,\mathrm{d}F^{\widetilde{P}}$. *In* (7.3.57) *the points* x_0 *and* y_0 *are the same as in condition* (7.3.23), *the measure* μ_c *is generated by c (see condition (i) in Theorem* 7.3.11*), and we also assume that c is a nondecreasing function in both arguments.) Finally,*

$$B \;:=\; B_1 - B_2 \tag{7.3.58}$$

$$B_1(x,y) \;:=\; \begin{cases} 1 + F(x,y) - F^\mu(x) - F^\nu(y) & \text{if } x_0 < x, y_0 < y, \\ F(x,y) & \text{if } x \le x_0, y \le y_0, \\ 0 & \text{otherwise;} \end{cases}$$

$$B_2(x,y) \;:=\; \begin{cases} F^\mu(x) - F(x,y) & \text{if } x \le x_0, y_0 \le y, \\ F^\nu(y) - F(x,y) & \text{if } x_0 < x, y \le y_0, \\ 0 & \text{otherwise.} \end{cases}$$

Theorem 7.3.17 *Suppose conditions (i) and (ii) of Theorem* 7.3.11 *hold, and in addition*

(iii*) σ *is a nonnegative bounded Borel measure on* $\mathbb{R}^2$ *satisfying*

$$G^\sigma \;:=\; \sigma\left((-\infty,x],[y,\infty)\right) \;\ge\; \max\left(0, F^\mu(x) - \nu((-\infty,y))\right) \tag{7.3.59}$$

for all $x,y\in\mathbb{R}$. *Then, the minimum in* (7.3.9) *is attained at an optimal measure Q satisfying the feasibility conditions* (7.3.7) *and* (7.3.10); *Q is determined by*

$$G^Q(x,y) = Q((-\infty,x]\times[y,\infty)) = \inf_{\substack{t\le x\\ s\ge y}}\{G^\sigma(x,y)+\mu((t,x])+\nu([y,s))\}. \tag{7.3.60}$$

All the Remarks 7.3.12–7.3.16 can be easily reformulated regarding Theorem 7.3.17. In particular, consider the transportation problem

$$\text{minimize} \sum_{i\in M}\sum_{j\in N} c(i,j)p_{ij} \tag{7.3.61}$$

subject to

$$p_{ij} \ge 0,\ p_{i\cdot}=a_i,\ p_{\cdot j}=b_j \quad \text{for all } i\in M, j\in N, \tag{7.3.62}$$

$$\sum_{r=1}^{i}\sum_{s=j}^{n} p_{rs} \le \lambda_{ij},\ i=1,\dots,m-1,\ j=1,\dots,n-1.$$

Suppose $\sum_{i\in M}a_i=\sum_{j\in N}b_j$ and $c(\cdot,\cdot)$ is a lattice superadditive sequence (cf. (7.3.51)). Suppose also that $\lambda_{ij}\ge\max\left(0,\sum_{r=1}^{i}a_r-\sum_{s=1}^{j}b_s\right)$ and for any $r<i$ and $s>j$ the inequalities

$$\lambda_{rj}\le\lambda_{ij}\ge\lambda_{is},\quad \lambda_{ij}+\lambda_{rs}\ge\lambda_{is}+\lambda_{rj}\ge 0,\quad r<i, s>j, \tag{7.3.63}$$

hold. Then the greedy algorithm (7.3.43)–(7.3.45) realizes the minimum in (7.3.61). Moreover, the optimal p_{ij}'s are determined by $p_{ij}=f_{ij}-f_{i,j+1}-f_{i-1,j}+f_{i-1,j+1}$, where

$$f_{ij} := \min_{\substack{1\le r\le i\\ j\le s\le n}}\{\lambda_{rs}+(a_{r+1}+\cdots+a_i)+(b_j+\cdots+b_{s-1})\} \wedge\sum_{r=1}^{i}a_r\wedge\sum_{s=j}^{n}b_s. \tag{7.3.64}$$

The rest of this section is devoted to a generalization of the MKP with additional constraints stated in Problems I and II; see (7.3.6)–(7.3.10). The results are motivated by Hoffman and Veinott (1990), where the discrete version of the problem has been considered. We shall only state the results. The proofs are similar to those of Theorems 7.3.1 and 7.3.10 and will therefore be omitted.

The abstract form of the problem is the following.

Suppose that

(i) μ and ν are two nonnegative Borel measures on $\mathbb{R}$, $\mu(\mathbb{R})=\nu(\mathbb{R})=\lambda<\infty$;

(ii) L is a union of disjoint sublattices $L_i \subset \mathbb{R}^2$, $i \in S$, and the projections of L on each axis equal $\mathbb{R}$;

(iii) $(\sigma_i)_{i \in S}$ are nonnegative σ-finite Borel measures on L_i.

Then the problem is to find

$$\min \int_L c \, dP, \tag{7.3.65}$$

where the minimum is subjet to the following constraints:

(i) P's are nonnegative Borel measures on L with marginals μ and ν; (7.3.66)

(ii) $P(A) \le \sigma_i(A)$ for any $A = L_i \cap (-\infty, x] \times (-\infty, y]$, $(x, y) \in L_i$, $i \in S$. (7.3.67)

As before, see (7.3.1)–(7.3.3), the measures μ and ν are viewed as initial and final mass distributions, and P in (7.3.66), (7.3.67) are the (feasible) transportation plans. Here the generalization of problems I and II is that L describes the path of the transportation flow and σ_i's are capacity constraints on the cumulative supply–demand flow. Finally, $c : L \to R$ is a cost function, and therefore, the integral in (7.3.65) represents the total cost of mass transportation applying the plan P.

Suppose c is *subadditive* on the lattice L; that is; for all $x, y \in L$,

$$f(x) + f(y) \ \ge f(x \wedge y) + f(x \vee y).$$

Then we shall call a feasible plan of transportation achieving the minimum in (7.3.65) an *optimal measure* P^*.

As in problems I and II we start with extensions of the classical Hoeffding–Fréchet bounds (7.3.4), assuming that P meets the constraints (7.3.66) and (7.3.67), or their alternatives:

$$P \text{ is a nonnegative Borel measure on } \overline{L} = \sum_{i \in S} \overline{L}_i \tag{7.3.68}$$
$$\left(\overline{L}_i := \{(x, y);\ (x, -y) \in L\} \text{ with marginals } \mu \text{ and } \nu\right);$$

$$P(B) \ \le \ \sigma_i(B) \text{ for any } B = \overline{L}_i \cap ((-\infty, x] \times [y, \infty)). \tag{7.3.69}$$

The restriction on the support of P given in (7.3.66) has the form

$$L \ = \ \sum_{i \in S} L_i, \quad \text{where } S = \{0, 1, \ldots, s\},$$

or $S = \mathbb{N}$, and each sublattice L_i is a rectangle $(x_i^-, x_i^+] \times (y_i^-, y_i^+]$, where $x_0^- = y_0^- = -\infty$, $x_i^- < x_i^+$, $y_i^- < y_i^+$, $x_{i-1}^- \le x_i^- \le x_{i-1}^+ \le x_i^+$, $y_{i-1}^- \le y_i^- \le y_{i-1}^+ \le y_i^+$, $x_s^+ = y_s^+ = \infty$. Write $\mathcal{P}_L$ (resp. $\mathcal{P}_{\overline{L}}$) to denote the class of all P's on L with (7.3.66) and (7.3.67) (resp. (7.3.68), (7.3.69)). Recall that for any measure P on $\mathbb{R}^2$, F^P stands for the d.f. of P, and $G^P(x,y) = P((-\infty,x]\times[y,\infty))$. In the next two theorems we shall compute the bounds

$$F^*(x,y) = \max_{P\in\mathcal{P}_L} F^P(x,y) \tag{7.3.70}$$

and

$$G^*(x,y) = \max_{P\in\mathcal{P}_{\overline{L}}} G^P(x,y). \tag{7.3.71}$$

For $L = \mathbb{R}^2$ and $\sigma_i = +\infty$, F^* is indeed the *upper Hoeffding–Fréchet bound*

$$H_+(x,y) = \min\{F^\mu(x), F^\nu(y)\} \quad (F^\mu(x) := \mu((-\infty,x]) . \tag{7.3.72}$$

On the other hand, $G^*(x,y) = \min\{F^\mu(x), G^\nu(y)\}$ $(G^\nu(x) := \nu([y,\infty)))$ determines a measure with d.f.

$$H_-(x,y) = \max(0, F^\mu(x) + F^\nu(y) - \lambda), \tag{7.3.73}$$

which is the *lower Hoeffding–Fréchet bound.*

Theorem 7.3.18 *Suppose that $F^* : L \to \mathbb{R}$ is defined iteratively as follows:*

$$F^*(x,y) = \min_{\substack{-\infty<u\le x\le x_0^+\\ -\infty<v\le y\le y_0^+}} [\mu((u,x]) + \nu((v,y]) + F^{\sigma_0}(u,v)] \tag{7.3.74}$$

for $(x,y) \in L_0$, and

$$F^*(x,y) = \min_{\substack{x_i^-<u\le x\le x_i^+\\ y_i^-<v\le y\le y_i^+}} [\mu((u,x]) + \nu((v,y]) + F^{\sigma_i}(u,v) + F^*(x_{i-1}^+, v\wedge y_{i-1}^+) \vee F^*(x_{i-1}^+\wedge u, y_{i-1}^+)] \tag{7.3.75}$$

for $(x,y) \in L_i, i \ge 1$. Suppose also that F^* satisfies the inequalities

$$F^{\sigma_i}(x.y) + F^*(x_{i-1}^+, y) \vee F^*(x, y_{i-1}^+) \ge H_-(x,y) \tag{7.3.76}$$

for $(x,y) \in L_i$, $i \in S$. Then the equality (7.3.70) holds, and moreover, F^* is a d.f. of some $P^* \in \mathcal{P}_L$.

The proof is similar to that of Theorem 7.3.1. A slightly different approach (see Olkin and Rachev (1990)) can be used based on the following result of Topkis and Veinott (1973): The minimum of subadditive functions over a sublattice with respect to some variables is subadditive in the remaining variables; see also Hoffman and Veinott (1990) for the discrete version of Theorem 7.3.18.

In the next theorem we evaluate the bound G^* in (7.3.71). Recall that $\overline{L}_i := (x_i^-, x_i^+] \times [-y_i^+, -y_i^-)$, $\overline{L} := \sum_{i \in S} \overline{L}_i$.

Theorem 7.3.19 *Suppose that* $G^* : \overline{L} \to \mathbb{R}$ *is defined iteratively by*

$$G^*(x,y) = \min_{\substack{u \le v \\ v \ge y}} [\mu((u,x]) + \nu([y,v)) + G^{\sigma_0}(u,v)], \text{ for } (x,y) \in \overline{L}_0 \quad (7.3.77)$$

and

$$\begin{aligned} G^*(x,y) &= \min_{\substack{x_i^- < u \le x \le x_i^+ \\ -y_i^- > v \ge y \ge -y_i^+}} \big[\mu((u,x]) + \nu([y,v)) + G^{\sigma_i}(u,v) \quad (7.3.78) \\ &\quad + G^*(x_{i-1}^-, v \vee (-y_{i-1}^+)) \vee G^*(u \wedge x_{i-1}^+, -y_{i-1}^+)\big] \\ &\quad \text{for } (x,y) \in \overline{L}_i. \end{aligned}$$

Suppose also that G^* *satisfies the inequalities*

$$\begin{aligned} &G^\sigma(x,y) + G^*(x_{i-1}^+, y) \vee G^*(x, -y_{i-1}^+) \quad (7.3.79) \\ &\ge F^\mu(x) + G^\nu(y) - \lambda \quad \text{for any } (x,y) \in L_i, \quad i \in S. \end{aligned}$$

Then the inequality (7.3.71) *holds, and* G^* *defines* $P^* \in \mathcal{P}_{\overline{L}}$ *by* $G^*(x,y) = P^*((-\infty, x] \times [y, \infty))$ *for* $(x,y) \in \overline{L}$. *Condition* (7.3.79) *is necessary for* $P^* \in P_{\overline{L}}$.

Next, we shall formulate a multivariate analogue of Theorem 7.3.18. (In general, Theorem 7.3.19 does not admit a multivariate extension by the well-known reason that the lower Hoeffding–Fréchet bound for d.f.s on $\mathbb{R}^r$ ($r > 2$) with given one-dimensional projections does not generate a measure.)

Let $\mu = (\mu^{(1)}, \dots, \mu^{(r)})$ be a vector of r Borel nonnegative measures on $\mathbb{R}$ with one and the same total mass $\lambda < \infty$. Suppose L is a complete Borel sublattice on $\mathbb{R}^r$ whose projection on every axis $x^{(i)}$ ($i \in R := \{1, \dots, r\}$) is the entire real line $\mathbb{R}$. Suppose also that L is a union of disjoint nonempty sublattices L_i, $i \in S$ ($S = \{0, \dots, s\}$ or $S = \mathbb{N}$) and each L_i is a rectangle in $\mathbb{R}^r$,

$$L_i = (x_i^-, x_i^+] = \otimes_{j=1}^r \left(x_i^{(j)-}, x_i^{(j)+}\right], \quad i \in S,$$

with $x_0^- = -\infty$, $x_s^+ = +\infty$, $x_i^- < x_i^+$, $x_{i-1}^- \le x_i^- \le x_{i-1}^+ \le x_i^+$. (For representations of sublattices on a product of r lattices we refer to Veinott (1989, Section 4).)

Given σ_i, a nonnegative Borel measure on L_i, and a measure P on L with vector of one-dimensional projections μ we write $P_{L_i} \prec \sigma_i$ to denote that the restriction of P on L_i is *less concordant* that σ_i; that is, for any $x \in L_i$,

$$P((x_i^-, x]) \;\le\; \sigma_i((x_i^-, x]), \quad i \in S.$$

Note that in contrast with the usual definition of concordance (Kruskal (1958), Tchen (1980), Stoyan (1983)) we allow σ_i to have total mass different from that of P_{L_i}. For example, assuming that σ_i vanishes on a subset of L_i, we in fact impose additional restrictions on the support of P. Write $\mathcal{P}_L(\mu, \sigma)$ $(\sigma := (\sigma_i)_{i \in S})$ to denote the class of all P's on L possessing the properties

(i) P is a nonnegative Borel measure on L with vector of one-dimensional marginals μ. (7.3.80)

(ii) $P_{L_i} \prec \sigma_i$ for all $i \in S$. (7.3.81)

Define the mapping $F^* : L \to \mathbb{R}$ iteratively as follows: For $x \in L_0$, let

$$F^*(x) \;=\; \min_{-\infty < u \le x \le x_0^+} \left\{ \sum_{j \in R} \mu^{(j)}\left(\left(u^{(j)}, x^{(j)}\right]\right) + F^{\sigma_0}(u) \right\},$$

and for $x \in L_i$, let

$$F^*(x) \;=\; \min_{x_i^- < u \le x \le x_i^+} \left\{ \sum_{j \in R} \mu^{(j)}\left(\left(u^{(j)}, x^{(j)}\right]\right) + F^{\sigma_i}(u) + f_{L_i}(u) \right\},$$

where $f_{L_i}(u) = \max\left[F^*\left(x_i^{(1)-}, v_i^{(2)}, \ldots, v_i^{(r)}\right), F^*\left(v_i^{(1)}, x_i^{(2)}, \ldots, v_i^{(r)}\right), F^*\left(v_i^{(1)}, v_i^{(2)}, \ldots, x_i^{(r)-}\right)\right]$ and $v_i^{(j)} := u^{(j)} \wedge x_{i-1}^{(j)+}$, $u \in L_i$. Denote by H_- the lower Hoeffding–Fréchet bounds for multivariate d.f. with prescribed one-dimensional marginals:

$$H_-(u) \;=\; \max\left(0, \sum_{j \in \mathbb{R}} \mu^{(j)}\left(\left(-\infty, u^{(j)}\right]\right) - (r-1)\lambda\right).$$

Theorem 7.3.20 *Suppose F^* defined above satisfies the inequality*

$$F^{\sigma_i}(u) + f_{L_i}(u) \;\; \geq \;\; H_-(u) \quad \textit{for every } u \in L_i, \; i \in S.$$

Then

$$\max_{P \in \mathcal{P}_L} F^P \;\; = \;\; F^*,$$

and F^* is a d.f. of some $P^* \in \mathcal{P}_L$.

The proof is similar to that of Theorem 7.3.1. It requires a multivariate analogue of the greedy algorithm similar to that in Barnes and Hoffman (1985) and Olkin and Rachev (1990). A multivariate version of Hoffman's (1963) northwest corner rule is given in Balinski and Rachev (1989), where the interplay between greedy algorithms and MKPs is emphasized.

We are now ready to state the solution of MKP (7.3.65) with constraints (7.3.66) and (7.3.67).

Theorem 7.3.21 *Suppose that the assumptions of Theorem* 7.3.18 *hold, and $c : L \to \mathbb{R}$ is subadditive and left-continuous on L with*

$$\int_{\mathbb{R}} c(x, y_0)\mu(\,\mathrm{d}x) + \int_{\mathbb{R}} c(x_0, y)\nu(\,\mathrm{d}y) \;\; > \;\; -\infty \tag{7.3.82}$$

for some $(x_0, y_0) \in L$. Then the minimum in (7.3.65) *is attained at $P^* \in \mathcal{P}_L$ defined in Theorem* 7.3.18.

The next theorem gives the solution of the following MKP:

$$\text{minimize} \int_{\overline{L}} c \,\mathrm{d}P \tag{7.3.83}$$

under the constraints (7.3.68), (7.3.69), and assuming that c is *superadditive on $\overline{L}$*; that is, $(-c)$ is subadditive on L.

Theorem 7.3.22 *Suppose that the assumptions of Theorem* 7.3.19 *hold, and $c : \overline{L} \to \mathbb{R}$ is superadditive and right-continuous with*

$$\int_{\mathbb{R}} c(x, y_0)\mu(\,\mathrm{d}x) + \int_{\mathbb{R}} c(x_0, y)\nu(\,\mathrm{d}y) \;\; < \;\; \infty. \tag{7.3.84}$$

Then the minimum in (7.3.84) *is attained at $P^* \in \mathcal{P}_{\overline{L}}$ defined in Theorem* 7.3.19.

The proof of the above two theorems is the same as that of Theorem 7.3.11.

Example 7.3.23 (The discrete case) *Suppose μ and ν are discrete measures with supports $I = \{1, \dots, m\}$ and $J = \{1, \dots, n\}$ and $L = L_0 + \cdots + L_s$ is a sublattice of $I \times J$ with projections I and J, respectively. Then Theorem* 7.3.18 *corresponds to the main theorem in Hoffman and Veinott (1990).*

Example 7.3.24 (MKP with capacity constraints) *Suppose that in Theorems* 7.3.21 *and* 7.3.22, *$L = \overline{L} = L_0 = \overline{L_0} = \mathbb{R}^2$. Then we obtain the solution of problems I and II; see* (7.3.6)–(7.3.10). *In fact, Theorems* 7.3.21 *and* 7.3.22 *reduce to Theorems* 7.3.11 *and* 7.3.17.

We shall complete this section with another possible extension of Problems I and II. Consider a finite measure μ on $(\mathbb{R}^2, \mathcal{B}^2)$ and define for two probability measures P_1, P_2 on $(\mathbb{R}^1, \mathcal{B}^1)$ and $A_i \times B_i \in \mathcal{B}^1 \otimes \mathcal{B}^1$, $i \in I$,

$$\begin{aligned} & M^{\mu}(P_1, P_2) \qquad (7.3.85) \\ = & \left\{P \in M^1(P_1, P_2);\ P(A_i \times B_i) \leq \mu(A_i \times B_i),\ i \in I\right\}, \end{aligned}$$

where $M^1(P_1, P_2)$ denotes the set of all probability measures P on $\mathbb{R}^2$ with marginals P_1, P_2. As in Theorem 7.3.1 (see (7.3.17)), we assume that

$$\mu(A_i \times B_i) \ \geq \ (P_1(A_i) + P_2(B_i) - 1)_+. \qquad (7.3.86)$$

Theorem 7.3.25 *Under the assumption* (7.3.86) *let us define*

$$\begin{aligned} P^*(A \times B) = & \inf_{\substack{A_i \subset A \\ B_i \subset B}} \{\mu(A_1 \times B_i) + (P_1(A) - P_1(A_i)) + (P_2(B) - P_2(B_i))\} \\ & \wedge \min(P_1(A), P_2(B)), \quad A, B \in \mathcal{B}^1. \qquad (7.3.87) \end{aligned}$$

Then

$$h_\mu(A \times B) \ := \ \sup\{P(A \times B);\ P \in M^\mu(P_1, P_2)\} \ \leq \ P^*(A \times B). \ (7.3.88)$$

If P^* determines a measure, then

$$h_\mu(A \times B) \ = \ P^*(A \times B), \quad \textit{and } P^* \textit{ is a solution of } (7.3.87). \qquad (7.3.89)$$

Remark 7.3.26 *The proof of Theorem* 7.3.25 *is similar to that of Theorem* 7.3.1. *In contrast to Theorem* 7.3.1 *it allows us to consider "local" bounds in the transportation problem. Observe that in the finite discrete case, bounds of the type*

$$x_{ij} \ \leq \ \mu_{ij} \quad \textit{for some } (i,j) \qquad (7.3.90)$$

are of this "local" type. As far as we know, in the literature there is no result concerning the solution of (7.3.90) *with local bounds. See the next section for a possible approach to the problem.*

7.4 Local Bounds for the Transportation Plans

While in the preceding three sections the additional constraints were formulated mainly in terms of the d.f.s, we now consider "local" constraints formulated in terms of the probability densities. These local-type restrictions are stronger than those in the previous section, and generally they are much more difficult to handle; see Remark 7.3.26.

Our first result deals with a transportation problem with "indicator" cost function

$$c(x,y) = I(x \neq y) = \begin{cases} 1 & \text{if } x \neq y, \\ 0 & \text{if } x = y; \end{cases} \tag{7.4.1}$$

i.e., the cost of transportation is one for any unit mass that has to be moved, and zero otherwise. The cost function c does not satisfy a Monge-type condition. We formulate this transportation problem on a general measure space $(S,\mathcal{U})$ assuming only that

$$\{(x,y);\ x \neq y\} \in \mathcal{U} \otimes \mathcal{U}. \tag{7.4.2}$$

Let $M_f(S), M_f(S\times S)$ be the set of all finite measures on $(S,\mathcal{U})$, respectively $(S\times S,\mathcal{U}\otimes\mathcal{U})$, and for $\mu \in M_f(S\times S)$, let $\pi_i\mu, i = 1,2$, denote the marginals of μ. (This transportation problem leads to an extension of Dobrushin's result on optimal couplings.)

Theorem 7.4.1 (Optimal couplings with local restrictions) *Assume that* (7.4.2) *holds and let* $\mu_1, \mu_2 \in M_f(S)$ *with* $\mu_1(S) \leq \mu_2(S)$. *Then*

(a) $$\begin{aligned} \inf\{\mu(\{(x,y); x \neq y\});\ \mu \in M_f(S\times S),\ \pi_1\mu \geq \mu_1,\ \pi_2\mu \leq \mu_2\} \quad (7.4.3) \\ = \lambda^-(S) := \sup_{C\in\mathcal{U}}(\mu_1(C) - \mu_2(C)). \end{aligned}$$

(b) *Moreover, the infimum in* (7.4.3) *is attained at*

$$\mu^*(A\times B) = \gamma(A\cap B) + \frac{\lambda^-(A)\lambda^+(B)}{\lambda^+(S)}, \tag{7.4.4}$$

where $\lambda^+(A) = \sup_{C\subset A}(\mu_2 - \mu_1)(C), \lambda^-(A) = \sup_{C\subset A}(\mu_1 - \mu_2)(C)$ *and* $\gamma(A) = \mu_2(A) - \lambda^+(A) = \mu_1(A) - \lambda^-(A)$.

Proof: For any $\mu \in M_f(S\times S)$,

$$\begin{aligned} \mu(x\neq y) &\geq \sup_C \mu(C\times(S\setminus C)) = \sup_C\{\mu(C\times S) - \mu(C\times C)\} \\ &\geq \sup_C\{\mu(C\times S) - \mu(S\times C)\} \geq \sup_C\{\mu_1(C) - \mu_2(C)\} \\ &= \sup_C\{\lambda^-(C) - \lambda^+(C)\} = \lambda^-(\operatorname{supp}\lambda^-) = \lambda^-(S). \end{aligned}$$

On the other hand,

$$\begin{aligned}\mu^*(A \times S) &= \gamma(A) + \lambda^-(A)\lambda^+(S)/\lambda^+(S) = \mu_1(A) \quad \text{and} \\ \mu^*(S \times B) &= \gamma(B) + \lambda^-(S)\lambda^+(B)/\lambda^+(S) \le \gamma(B) + \lambda^+(B) = \mu_2(B).\end{aligned}$$

Finally, we have

$$\begin{aligned}\mu^*(x \neq y) &= \int I(x \neq y)(\gamma(\mathrm{d}x, \mathrm{d}y) + \lambda^-(\mathrm{d}x)\lambda^+(\mathrm{d}y)/\lambda^+(S)) \\ &= \int I(x \neq y)\ \lambda^-(\mathrm{d}x)\lambda^+(\mathrm{d}y)/\lambda^+(S) \\ &= \lambda^-(S)\lambda^+(S)/\lambda^+(S) = \lambda^-(S).\end{aligned}$$

□

Consider next some finite measures μ_1, μ_2 on $\mathbb{R}$ with densities h_1, h_2 with respect to a dominating measure μ on $\mathbb{R}^1$. Define

$$\mathcal{P}_{\mu_1}^{\mu_2} := \{P \in M^1(\mathbb{R}^2, \mathcal{B}^2);\ \pi_1 P \geq \mu_1, \pi_2 P \leq \mu_2\}. \tag{7.4.5}$$

Any $P \in \mathcal{P}_{\mu_1}^{\mu_2}$ has marginals $P_1 = \pi_1 P, P_2 = \pi_2 P$ with densities $f_1 \geq h_1$ and $f_2 \leq h_2$ with respect to μ. We assume that $1 = \mu_1(\mathbb{R}^1) \leq \mu_2(\mathbb{R}^1)$; i.e., μ_1 is a probability measure, and so $f_1 = h_1$.

Theorem 7.4.2 *Let* $z_0 = \inf\left\{y;\ \int_{(y,\infty)} h_2\,\mathrm{d}\mu \leq 1\right\}$,

$$f_2^*(y) = \begin{cases} h_2(y) & \text{if } y > z_0, \\ \dfrac{1 - \int_{(z_0,\infty)} h_2(u)\,\mathrm{d}u}{\mu(z_0)} & \text{if } y = z_0 \text{ and } \mu\{z_0\} > 0, \\ 0 & \text{otherwise,} \end{cases} \tag{7.4.6}$$

and let P^* *be the corresponding probability measure with* μ*-density* f_2^**. Then the following characterizations of the optimal coupling hold:*

(a) $\sup\{\overline{F}_P(x,y); P \in P_{\mu_1}^{\mu_2}\} = 1 - \max(F_{\mu_1}(x), F_{P^*}(y))$, *for all* x, y *where* $\overline{F}_P(x,y) = P([x,\infty) \times [y,\infty))$ *is the survival function.*

(b) *The sup in (a) is attained for the distribution* $F^* = F_{X^*,Y^*}$, *where* $X^* = F_{\mu_1}^{-1}(U), Y^* = F_{P^*}^{-1}(U)$.

(c) *If c is a cost function that is componentwise antitone and satisfies the Monge condition (cf.* (7.1.3), (7.1.4)*), then*

$$\inf\left\{\int c(x,y)\,\mathrm{d}F_P(x,y);\ \ P \in \mathcal{P}_{\mu_1}^{\mu_2}\right\} = \int c(x,y)\,\mathrm{d}F^*(x,y). \qquad (7.4.7)$$

Proof: (a), (b) For $P \in \mathcal{P}_{\mu_1}^{\mu_2}$ with marginals F_{μ_1}, G_2, we know that $\overline{F}_P(x,y) \le P\left(F_{\mu_1}^{-1}(U) \ge x, G_2^{-1}(U) \ge y\right) = P(U \ge \max(F_{\mu_1}(x), G_2(y))) = 1 - \max(F_{\mu_1}(x), G_2(y))$. By the definition of P^*, $F_{P^*}(y) \le G_2(y)$ for all y, and therefore, $\overline{F}_P(x,y) \le 1 - \max(F_{\mu_1}(x), F_{P^*}(y))$.

(c) Applying (a), (b), and Theorem 3.1.2 we obtain (7.4.7). The conditions on the cost function c were studied by Rüschendorf (1980). In that terminology $(-c)$ is a Δ-monotone function. Applying the results in Rüschendorf (1980), it is easy to check that (c) follows from (a), (b). □

The "antitone" assumption in (c) of Theorem 7.4.2 does not have a transparent interpretation in terms of cost functions. Moreover, under some additional assumptions on the bounding measures we can construct solutions for more "natural" cost functions. Again, let μ_1 have densities h_i with respect to μ, $1 = \mu_1(\mathbb{R}^1) \le \mu_2(\mathbb{R}^1)$.

Theorem 7.4.3 *Assume that for some* $y_0 \in \mathbb{R}_1$,

$$h_1(u) \le h_2(u) \text{ for } u < y_0 \quad \text{and} \quad h_1(u) \ge h_2(u) \text{ for } u \ge y_0. \qquad (7.4.8)$$

Define

$$x_0 = \inf\left\{y;\ \int_{(y,\infty)} h_1(u)\,\mathrm{d}\mu(u) \ge \int_{(y,\infty)} h_2(u)\,\mathrm{d}\mu(u)\right\},$$

and let

$$f_2(u) := \begin{cases} h_2(u) & \text{if } u > x_0, \\[2ex] \dfrac{\displaystyle\int_{[x_0,\infty)} h_1(u)\,\mathrm{d}\mu(u) - \int_{(x_0,\infty)} h_2(u)\,\mathrm{d}\mu(u)}{\mu(x_0)} & \text{if } u = x_0 \text{ and } \mu\{x_0\} > 0, \\[2ex] h_1(u) & \text{if } u < x_0. \end{cases} \qquad (7.4.9)$$

Then for any cost function c satisfying the Monge condition (7.1.3) and the unimodality condition (7.1.27) we have

$$\inf\left\{\int c(x,y)\,\mathrm{d}f_P(x,y);\;\; P\in\mathcal{P}_{\mu_1}^{\mu_2}\right\}=\int_0^1 c\left(F_{\mu_1}^{-1}(u),F_2^{-1}(u)\right)\,\mathrm{d}u, \qquad (7.4.10)$$

where F_2 is the d.f. of the measure with density f_2 with respect to μ. The optimal distribution is determined by the r.v.s $X^*=F_{\mu_1}^{-1}(U), Y^*=F_2^{-1}(U)$.

Proof: Invoking the Monge condition, for any $P\in\mathcal{P}_{\mu_1}^{\mu_2}$ with marginals F_{μ_1}, G_2, we have $\int c(x,y)\,\mathrm{d}F_P(x,y)\geq\int_0^1 c\left(F_{\mu_1}^{-1}(u),G_2^{-1}(u)\right)\,\mathrm{d}u$.

By the definition of F_2,

$$\begin{array}{rclll} G_2(y) & \geq & F_2(y) \;\geq\; F_{\mu_1}(y) & \text{for all } y\geq x_0, & \text{and} \\ F_2(y) & = & F_{\mu_1}(y) \quad \text{for all } y\leq x_0; & & \end{array} \qquad (7.4.11)$$

in fact, (7.4.11) implies that $F_{\mu_1}^{-1}(u)\geq F_2^{-1}(u)\geq G_2^{-1}(u)$ for $u>F_2(x_0)$ and $F_2^{-1}(u)=F_{\mu_1}^{-1}(u)$ for $u\leq F_2(x_0)$. Our assumptions on c imply that $c\left(F_{\mu_1}^{-1},G_2^{-1}(u)\right)\geq c\left(F_{\mu_1}^{-1}(u),F_2^{-1}(u)\right)$ for all u. □

Remark 7.4.4 *It is not difficult to extend the solution of Theorem 7.4.3 to the case $\mu_1(\mathbb{R}^1)<1$ and to the case $f_1\geq h_1, f_2\leq h_2$ for the densities (here, f_1 and f_2 are the marginal densities of an admissible plan P), if we still keep the assumption (7.4.8). To see this, choose x_0 as in (7.4.9), and define*

$$f_2(x)=\begin{cases} h_2(x) & \text{if } x>z_0,\\[4pt] \dfrac{1-\displaystyle\int_{(z_0,\infty)}h_2(x)\,\mathrm{d}\mu(x)}{\mu(z_0)} & \text{if } x=z_0 \text{ and } \mu(z_0)>0,\\[4pt] 0 & \text{otherwise,}\end{cases} \qquad (7.4.12)$$

where $z_0=\inf\left\{x;\;\displaystyle\int_{(x,\infty)}h_2(x)\,\mathrm{d}\mu(x)\leq 1\right\}$. *Define next*

$$y_0=\inf\left\{y;\;\int_{(y,\infty)}h_2(x)\,\mathrm{d}\mu(x)\leq\int_{(y,\infty)}h_1(x)\,\mathrm{d}\mu(x)\right\}$$

and

$$f_1(x) = \begin{cases} h_1(x) & \text{if } x > y_0, \\ f_2(x) & \text{if } x < y_0, \\ \dfrac{\displaystyle\int_{[y_0,\infty)} (h_2(x) - h_1(x))\, \mathrm{d}\mu(x)}{\mu(y_0)} & \text{if } \mu(y_0) > 0. \end{cases} \tag{7.4.13}$$

Then for a cost function c defined as in Theorem 7.4.3, *we have*

$$\begin{aligned} &\inf\left\{\int c(x,y)\,\mathrm{d}F_P(x,y);\ \ \pi_1 P \geq \mu_1, \pi_2 P \leq \mu_2\right\} \qquad (7.4.14) \\ &= \int_0^1 c\left(F_1^{-1}(u), F_2^{-1}(u)\right)\, \mathrm{d}u, \end{aligned}$$

where F_i have densities f_i with respect to $\mu, i = 1, 2$.

Let us return to the comment we made in Remark 7.3.26. Consider transportation problems with local upper bounds on the transportation plans $x_{ij} \leq \mu_{ij}$ in the discrete case, while $P \leq \mu$ for some finite measure μ in the general case. The following framework allows us to handle quite general transportation problems.

On a measurable space $(X, \mathcal{B})$, let $\mathcal{B}_i \subset \mathcal{B}$, be sub-$\sigma$-algebras, $2 \leq i \leq n, P_i \in M^1(X, \mathcal{B}_i)$. Further, let μ be a finite measure on $(X, \mathcal{B})$, and define

$$M_\mu := \left\{P \in M^1(X, \mathcal{B});\ \ P/\mathcal{B}_i = P_i, 1 \leq i \leq n, P \leq \mu\right\}. \tag{7.4.15}$$

Assume that $M_\mu \neq \emptyset$ and define the set of generalized transportation plans with local upper bound μ as follows:

$$U_\mu(\varphi) := \inf\left\{U(\varphi_0) + \int h\,\mathrm{d}\mu;\ \ h \geq 0,\ \ \varphi_0 + h \geq \varphi\right\}, \tag{7.4.16}$$

where

$$U(\varphi_0) := \inf\left\{\sum_{i=1}^n \int f_i\,\mathrm{d}P_i;\ \ f_i \in L^1(\mathcal{B}_i, P_i),\ \ \varphi_0 \leq \sum_{i=1}^n f_i\right\}.$$

We view U as the dual operator for the "pure" transportation problem, and typically,

$$\sup\left\{\int \varphi_0\,\mathrm{d}P;\ \ P \in M(P_1, \ldots, P_n)\right\} = U(\varphi_0) \tag{7.4.17}$$

will hold (cf. Chapter 2). The duality principle allows us to infer the corresponding "minimization" problem. Similarly, U_μ is the dual operator for the local majorized transportation problem. A linear operator S is majorized by U_μ,

$$S \le U_\mu \text{ if and only if } S \ge 0,\ \ S/\mathcal{B}_i = P_i,\ \ 1 \le i \le n, \text{ and } S \le \mu. \tag{7.4.18}$$

Therefore, the approach developed in Chapter 2 yields the duality theorem

$$U_\mu(\varphi) = \sup\left\{\int \varphi \, \mathrm{d}P;\ P \in M_\mu\right\} =: M_\mu(\varphi) \tag{7.4.19}$$

for any upper semicontinuous or uniformly approximable integrable functions φ in the case of a compactly approximable measure space $(X, \mathcal{B}_i, P_i)$ with countable topological basis. In some sense (7.4.19) gives the duality result for the general case of order restrictions as considered, for example, in Sections 3.5 and 5.5. In particular, we obtain upper bounds $M_\mu(\varphi) \le \sum_{i=1}^n f_i \, \mathrm{d}P_i$ for any admissible system of functions (f_i) with $\varphi \le \sum_{i=1}^n f_i$.

We next consider the question of more explicit evaluations of the dual operator U_μ for the case $\varphi = 1_B$, $B \in \mathcal{B}$; or equivalently, we wish to establish sharp upper Fréchet bounds in the class M_μ. Define $M_\mu(B) := M_\mu(1_B)$, and assume the duality (7.4.19) for $\phi = 1_B$.

Theorem 7.4.5

$$M_\mu(B) = \sup_{P \in M(P_1,\dots,P_n)} P \wedge \mu(B), \tag{7.4.20}$$

where $P \wedge \mu$ is the infimum in the lattice of measures.

Proof: From (7.4.19),

$$\begin{aligned} M_\mu(B) &= \inf\{\mu(h) + U(\varphi);\ h \ge 0, \varphi + h \ge 1_B\} \\ &= \inf\{\mu(h) + U(1_B - h);\ 0 \le h \le 1_B\}. \end{aligned} \tag{7.4.21}$$

To see the second equality in (7.4.4), take $\varphi = (1_B - h)_+$. Thus, $0 \le \varphi$, and it is possible to assume that $h \le 1_B$. Next, we make use of the "integration" approach in Strassen (1965),

$$\begin{aligned} M_\mu(B) &= \inf_{0 \le h \le 1_B} \sup_{P \in M(P_1,\dots,P_n)} \{\mu(h) + P(1_B - h)\} \\ &= \inf_{0 \le h \le 1_B} \sup_P \left\{ \int_0^1 \mu(h > t) \, \mathrm{d}t + \int_0^1 P(1_B - h \ge 1 - t) \, \mathrm{d}t \right\}. \end{aligned} \tag{7.4.22}$$

With $C_t := \{h > t\} \subset B$ we see that

$$\{x;\ h(x) \le 1_B(x) - 1 + t\} \;=\; \{x \in B;\ h(x) \le t\} \;=\; B \setminus C_t.$$

Therefore,

$$\begin{aligned} M_\mu(B) &= \inf_{0 \le h \le 1_B} \sup_P \int_0^1 (\mu(C_t) + P(B \setminus C_t))\,\mathrm{d}t \\ &\ge \sup_P \inf_{C \subset B} \{\mu(C) + P(B \setminus C)\} \;=\; \sup_P \mu \wedge P(B). \end{aligned}$$

On the other hand,

$$\begin{aligned} M_\mu(B) &= \sup\{P(B);\ P \in M(P_1, \ldots, P_n), P \le \mu\} \\ &= \sup\{P \wedge \mu(B);\ P \in M(P_1, \ldots, P_n), P \le \mu\} \\ &\le \sup_{P \in M(P_1, \ldots, P_n)} P \wedge \mu(B). \end{aligned}$$

□

Theorem 7.4.5 allows us to reduce the problem of the majorized Fréchet boundsto a problem of "usual" Fréchet bounds, but for a more complicated functional. It remains an open problem to determine more explicit formulas for $M_\mu(B)$ in the general case.

7.5 Closeness of Measure with Joint Marginals on a Finite Number of Directions

In this section we follow the work of Kakosjan and Klebanov (1984), Khalfin and Klebanov (1990), Klebanov and Rachev (1995a, 1995b, 1995c), on the application of marginal problems to computer and diffraction tomography. Here, estimates of the closeness between probability measures defined on $\mathbb{R}^n$ that have the same marginals on a finite number of arbitrary directions will be provided. The estimates show that the probability laws get closer in a certain metric when the number of coinciding marginals increases. The results offer a solution to the computer tomography paradox stated in Gutman, Kemperman, Reeds, and Shepp (1991).

We start with some historical remarks and with the statement of the problem. Let Q_1 and Q_2 be a pair of probabilities on $\mathbb{R}$, i.e., probability measures defined on the Borel σ-field of $\mathbb{R}$. Lorentz (1949) studied conditions for the existence of a probability density function $g(\cdot)$ on $\mathbb{R}^2$ taking only two values, 0 or 1, and having Q_1 and Q_2 as marginals. In his 1961 paper Kellerer generalized this result and gave necessary and sufficient conditions for the existence of a density $f(\cdot)$ on $\mathbb{R}^2$ that satisfies the inequalities $0 \leq f(\cdot) \leq 1$ and has Q_1 and Q_2 as marginals (see also Strassen (1965) and Jacobs (1987)). Fishburn et al. (1990) were able to show that Kellerer's and Lorentz's conditions are equivalent; i.e., for any density $0 \leq f \leq 1$, on $\mathbb{R}^2$ there exists a density taking the values 0 and 1 only that has the same marginals. In general, similar results hold for probability densities on $\mathbb{R}^m, m \geq 2$, when the $(m-1)$-dimensional marginals are prescribed. Gutmann et al. (1991) show that for any probability density $0 \leq f \leq 1$ on $\mathbb{R}^m$ and for any finite number of directions, there exists a probability density taking the values $0, 1$ only that has the same marginals in the chosen directions. It follows that densities having the same marginals in a finite number of arbitrary directions may differ considerably in the uniform metric between densities, which is indeed a very *strong* metric; recall that convergence in the uniform metric implies convergence in total variation.

The goal in this section is to show that under moment-type conditions, measures having a "large" number of coinciding marginals are close to each other in the *weak*metrics.(1) The method is based on techniques used in the classical moment problem. On the other hand, most of our results will make use of relationships between different probability metrics, analyzed in the monograph by Kakosjan, Klebanov, and Rachev (1988), referred to below as KKR (1988). The key idea in showing that measures with a large number of common marginals are close to each other in the weak metrics is best understood by comparing three results. The first is the theorem of Gutman et al. (1991) mentioned above.

The second (see Karlin and Studden (1966, p. 265)) states that if a finite number of moments $\mu_1, \ldots, \mu_n$ of a function f, $0 \leq f \leq 1$, are given, then there exists a function g that takes the values 0 or 1 only and possesses the moments $\mu_1, \ldots, \mu_n$.

Finally, the third result (see KKR (1988, pp. 170–197)) gives estimates of the closeness in terms of a weak metric (the so-called λ-metric) on $\mathbb{R}$ for measures having a finite number of common moments. Of course, since the condition of common marginals seems to be more restrictive than the condition of equal moments, one should be able to construct a similar estimate expressed in terms of the common marginals only. Furthermore, the technique should be similar to that used here.

For simplicity, let us consider the 2-dimensional case. Let $\theta_1, \ldots, \theta_n$ be n unit vectors in the plane and P_1, P_2 be two probabilities on $\mathbb{R}^2$ having the same marginals in the directions $\theta_1, \ldots, \theta_n$. To estimate the distance

---

(1) Here *weak* metric stands for a metric metrizing the weak convergence in the space of probability measures on a Euclidean space.

between P_1 and P_2, various weak metrics can be used; however, it seems that the λ-metric is the most convenient for this purpose. This metric is defined as follows (see, for example, Zolotarev (1986)): Let

$$\varphi_i(t) = \int_{\mathbb{R}^2} e^{i(t,x)} P_i(dx), \quad i = 1,2,$$

be the characteristic function of P_i. Then define the λ-distance between P_1 and P_2 as

$$\lambda(P_1, P_2) = \min_{T>0} \max \left\{ \max_{\|t\|\le T} |\varphi_1(t) - \varphi_2(t)|, \frac{1}{T} \right\}; \tag{7.5.1}$$

here $(\cdot,\cdot)$ is the inner product and $\|\cdot\|$ is the Euclidean norm. Clearly, λ metrizes the weak convergence.

Our first result concerns the important case where one of the probability measures considered has compact support.

Lemma 7.5.1 *Let $\theta_1, \ldots, \theta_n$ be $n \ge 2$ unit vectors in $\mathbb{R}^2$, no two of which are collinear. Let the support of the probability P_1 be a subset of the unit disk, and let the probability P_2 have the same marginals as P_1 in the directions $\theta_1, \ldots, \theta_n$. Set*[(2)]

$$s = 2\left[\frac{n-1}{2}\right]. \tag{7.5.2}$$

Then

$$\lambda(P_1, P_2) \le \left(\frac{2}{s!}\right)^{\frac{1}{s+1}}. \tag{7.5.3}$$

Remark 7.5.2 *We can replace the right-hand side of* (7.5.3) *by C/s, where C is a constant; note that as $s \to \infty$, $\left(\frac{2}{2!}\right)^{\frac{1}{s+1}} \sim e/s$. The difference $\left(\frac{2}{s!}\right)^{\frac{1}{s+1}} - \frac{e}{s}$ is plotted in figures* 7.1 *and* 7.2.

Proof of Lemma 7.5.1: The λ-metric is invariant under rotations of the coordinate system, so without loss of generality we assume that

(a) the directions $\theta_j (j = 1, \ldots, n)$ are not parallel to the axis;

(b) there exists at least one pair of directions, say θ_{j1} and θ_{j2}, such that $\theta_{j1} = (a,b), \theta_{j2} = (a,-b)$, where $a \ne 0, b \ne 0$; i.e., the vectors θ_{j1} and θ_{j2} are symmetric about the horizontal axis.

---

(2) Here and in what follows $[r]$ denotes the integer part of the number r.

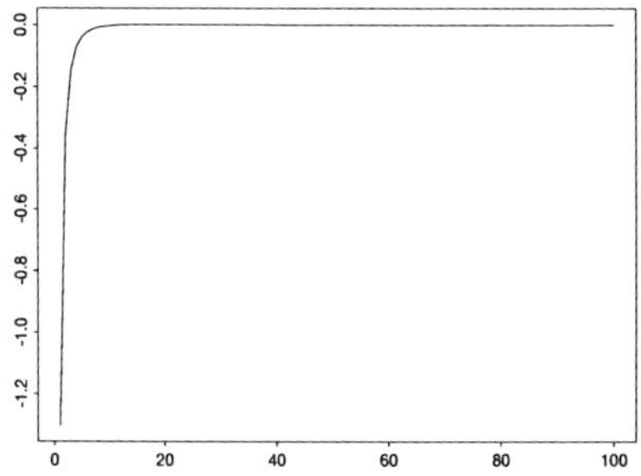

FIGURE 7.1. Plot of the difference $(2/s!)^{1/(s+1)} - e/s$ for $s = 1, \dots, 100$

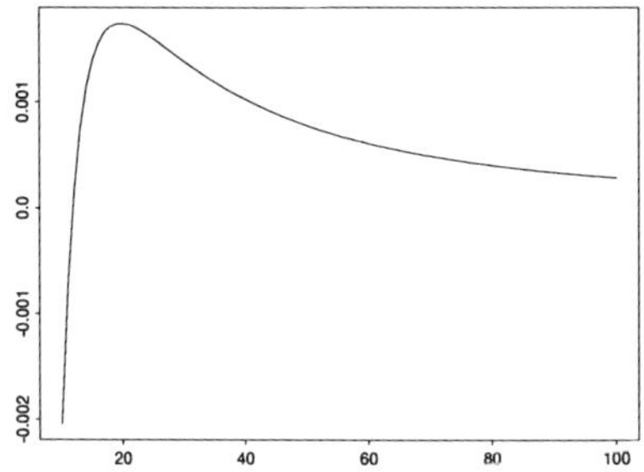

FIGURE 7.2. Plot of the difference $(2/s!)^{1/(s+1)} - e/s$ for $s = 10, \dots, 100$

The law P_1 has bounded support, and so, since the marginals on the directions $\theta_1, \dots, \theta_n$ of P_1 and P_2 coincide, then for all $j = 1, \dots, n$,

$$\int_{\mathbb{R}^2} (x, \theta_j)^k P_1(\,dx) \;=\; \int_{\mathbb{R}^2} (x, \theta_j)^k P_2(\,dx). \tag{7.5.4}$$

To see that P_2 has moments of any order, consider (7.5.4) with $j = j_1, j = j_2$, and $x = (x_1, x_2)$. Then

$$\int_{\mathbb{R}^2} (x_1 a \pm x_2 b)^k (P_1 - P_2)(\,dx) \;=\; 0,$$

$$\int_{\mathbb{R}^2} \left[(x_1 a + x_2 b)^k + (x_1 a - x_2 b)^k\right] (P_1 - P_2)(\,dx) \;=\; 0, \tag{7.5.5}$$

and all integrals are finite. If k is even, then

$$(ax_1 + bx_2)^k + (ax_1 - bx_2)^k \;\geq\; a^k x_1^k + b^k x_2^k,$$

and thus (7.5.5) implies the existence of all moments of P_2 of even order.

The next step is to show that all moments of P_1 and P_2 of order $\leq n-1$ agree. Set

$$\mu_{r,t}(P_\ell) = \int_{\mathbb{R}^2} x_1^r x_2^t P_\ell(\mathrm{d}x), \quad \ell = 1,2.$$

Then setting $\theta_j = (u_j, v_j)$ in (7.5.4) yields

$$\sum_{\ell=0}^{k} \binom{k}{\ell} u_j^\ell v_j^{k-\ell} \left[\mu_{\ell,k-\ell}(P_1) - \mu_{\ell,k-\ell}(P_2)\right] = 0,$$

$j = 1, \ldots, n$; $k \geq 0$. Now, setting $z_j = v_j/u_j$ in the last equation leads to

$$\sum_{\ell=0}^{k} \binom{k}{\ell} z_j^{k-\ell} \left[\mu_{\ell,k-\ell}(P_1) - \mu_{\ell,k-\ell}(P_2)\right] = 0, \tag{7.5.6}$$

$j = 1, \ldots, n$. Since no two of the directions $\theta_1, \ldots, \theta_n$ are collinear, the points $z_1, \ldots, z_2$ are distinct. Hence from (7.5.6) we find that the following polynomial of degree k of the variable z,

$$\sum_{l=0}^{k} \binom{k}{\ell} z^{k-\ell} \left[\mu_{\ell,k-\ell}(P_1) - \mu_{\ell,k-\ell}(P_2)\right], \tag{7.5.7}$$

has n distinct roots $z_1, \ldots, z_n$. If $n \geq k+1$, then this is possible only if all coefficients of (7.5.7) are equal to zero, that is, $\mu_{\ell,k-\ell}(P_1) = \mu_{\ell,k-\ell}(P_2)$, $\ell = 0, \ldots, k$; $k = 0, \ldots, n-1$. So, for any unit vector t, and $k = 0, 1, \ldots, n-1$,

$$\int_{\mathbb{R}^2} (t,x)^k P_1(\mathrm{d}x) = \int_{\mathbb{R}^2} (t,x)^k P_2(\mathrm{d}x). \tag{7.5.8}$$

Denote by $P_\ell^{(t)}$ the marginal of P_ℓ ($\ell = 1,2$) in the direction t, and by $\varphi_\ell(\tau; t)(\tau \in \mathbb{R})$ its characteristic function. By assumption, the support of $P_1^{(t)}$ is in the segment $[-1, 1]$. Then (7.5.8) is equivalent to

$$\varphi_1^{(k)}(\tau; t)|_{\tau=0} = \varphi_2^{(k)}(\tau; t)|_{\tau=0}, \quad k = 0, \ldots, n-1, \tag{7.5.9}$$

where $\varphi_\ell^{(k)}(\tau; t)$ is the kth derivative of $\varphi_\ell(\tau; t)$ with respect to τ ($\ell = 1, 2$). A Taylor expansion now gives

$$\varphi_1(\tau; t) - \varphi_2(\tau; t) \tag{7.5.10}$$

$$= \sum_{k=0}^{s-1} \frac{\varphi_1^{(k)}(0; t) - \varphi_2^{(s)}(0; t)}{k!} \tau^k + \frac{\varphi_1^{(k)}(\tilde{\tau}; t) - \varphi_2^{(s)}(\tilde{\tau}; t)}{s!} \tau^s$$

for some $\widetilde{\tau} \in (0, \tau)$. From (7.5.9), the first sum on the right-hand side of (7.5.10) is equal to zero. Since s is an even number,

$$|\varphi_\ell^{(s)}(\widetilde{\tau}; t)| \le \int_{\mathbb{R}} z^s P_\ell^{(t)}(\mathrm{d}x) = \int_{-1}^{1} z^s P_1^{(t)}(\mathrm{d}z) \le 1, \quad \ell = 1, 2.$$

Thus for all $\tau \in \mathbb{R}$,

$$|\varphi_1(\tau; t) - \varphi_2(\tau; t)| \le 2\frac{\tau^s}{s!}.$$

Choose $T = (\frac{s!}{2})^{\frac{1}{s+1}}$;(3) then

$$\sup_{|\tau| \le T} |\varphi_1(\tau; t) - \varphi_2(\tau; t)| \le \left(\frac{2}{s!}\right)^{\frac{1}{s+1}}.$$

□

Corollary 7.5.3 *Let $\theta_1, \ldots, \theta_n, n \ge 2$, be directions in $\mathbb{R}^2$ no two of which are collinear. Suppose that the marginals of the probabilities P_1 and P_2 with respect to the directions $\theta_1, \ldots, \theta_n$ have moments up to the even order $k \le n - 1$. Then the marginals of P_1 and P_2 with respect to any direction t have the same moments up to order k.*

Corollary 7.5.4 *Lemma* 7.5.1 *still holds if we replace the assumption that P_1 and P_2 have coinciding marginals with respect to the directions $\theta_j (j = 1, \ldots, n)$ with the assumption that these marginals have the same moments up to order $n - 1$.*

To prove our main result we must relax the condition that the support of P_1 is compact, assuming only the existence of all moments together with Carleman's conditions for the definiteness of the moments problem.

Set

$$\mu_k = \sup_{\theta \in S^1} \int_{\mathbb{R}^2} (x, \theta)^k P_1(\mathrm{d}x), \quad k = 0, 1, \ldots,$$

where S^1 is the unit circle, and let

$$\beta_s = \sum_{j=1}^{(s-1)/2} \mu_{2j}^{-\frac{1}{2j}},$$

where the number s is determined in Lemma 7.5.1; see (7.5.2).

(3) This choice of T is optimal, since $2\frac{T^s}{s!} = \frac{1}{T}$; see the definition (7.5.1) of λ-metric.

Theorem 7.5.5 *Let $\theta_1, \dots, \theta_n$ be $n \geq 2$ directions in $\mathbb{R}^2$, no two of which are collinear. Suppose that the measure P_1 has moments of any order. Suppose also that the marginals of the measures P_1 and P_2 in the directions $\theta_1, \dots, \theta_n$ have the same moments up to order $n-1$. Then there exists an absolute constant C such that*[4]

$$\lambda(P_1, P_2) \;\leq\; C\beta_s^{-\frac{1}{4}} \left(\mu_0 + \sqrt{\mu_2}\right)^{1/4}.$$

Proof: Let t be an arbitrary vector of the unit circle. From Corollary 7.5.3 we have that the marginals $P_1^{(t)}$ and $P_2^{(t)}$ have the same moments up to order s. From KKR (1988, p. 180) and Klebanov and Mkrtchian (1980), it follows that

$$\lambda\left(P_1^{(t)}, P_2^{(t)}\right) \;\leq\; C \left(\sum_{j=1}^{(s-1)/2} \mu_{2j}(t)^{-1/(2j)} \right)^{-1/4} \left(\mu_0(t) + \sqrt{\mu_2(t)}\right)^{1/4},$$

where $\mu_k(t) = \int_{-\infty}^{\infty} u^k P_i^{(t)}(du)$, $k = 0, \dots, s, i = 1, 2$. The theorem now follows from the obvious inequality

$$\mu_{2j}(t) \;\leq\; \mu_{2j} \quad (j = 0, 1, \dots, s/2). \qquad \square$$

Let us now consider the situation where the marginals of P_1 and P_2 in the directions $\theta_1, \dots, \theta_n$ are not the same but are close in the metric λ.

Theorem 7.5.6 *Let $\theta_1, \dots, \theta_n, n \geq 2$, be directions in $\mathbb{R}^2$, no two of which are collinear. Suppose that the supports of the measures P_1 and P_2 are in the unit disk, and that P_1 and P_2 have ε-coinciding marginals with respect to the directions $\theta_j (j = 1, \dots, n)$; i.e.,*

$$\lambda\left(P_1^{(\theta_j)}, P_2^{(\theta_j)}\right) \tag{7.5.11}$$

$$:= \min_{T>0} \max \left\{ \max_{|\tau| \leq T} |\varphi_1(\tau; \theta_j) - \varphi_2(\tau; \theta_j)|, 1/T \right\} \;\leq\; \varepsilon.$$

Then there exists a constant C depending on the directions $\theta_j (j = 1, \dots, n)$ such that for sufficiently small $\varepsilon > 0$, we have

$$\lambda(P_1, P_2) \;\leq\; C\left(1/\ln\left(\frac{1}{\varepsilon}\right) + 1/s\right), \tag{7.5.12}$$

where $s = 2\left[\frac{n-1}{2}\right]$.

---

[4] That is, C is independent of s, P_1, and P_2.

Proof: Set $\psi_j(\tau) := \varphi_1(\tau;\theta_j) - \varphi_2(\tau;\theta_j), j = 1,\ldots,n$. For $0 < \varepsilon \le 1$ we have $\sup_{|\tau|\le 1} |\psi_j(\tau)| \le \varepsilon$, cf. (7.5.11). Since the supports of the measures $P_1^{(\theta_j)}$ and $P_2^{(\theta_j)}$ are subsets of $[-1,1]$, for any even number $k \ge 2$ we have

$$\sup_{|\tau|\le 1} |\psi_j^{(k)}(\tau)| \;\le\; \frac{\left(|\varphi_1^{(k)}(0;\theta_j)| + |\varphi_2^{(k)}(0;\theta_j)|\right)}{k} \;\le\; \frac{2}{k!}. \qquad (7.5.13)$$

Now we apply Corollary 1.5.1 in KKR (1988), which states that there exist constants $C_{\ell k}$ such that

$$\begin{aligned} &\sup_{|\tau|\le 1} |\varphi_j^{(\ell)}(\tau)| &(7.5.14)\\ &\le\; C_{\ell k}\left\{\sup_{|\tau|\le 1} |\varphi_j(\tau)|\right\}^{\frac{k-\ell}{k}} \left\{\sup_{|\tau|\le 1} |\varphi_j^{(k)}(\tau)|\right\}^{\frac{1}{k}}, \quad \ell = 0,1,\ldots,k. \end{aligned}$$

Choosing $k \ge 2s, \ell \le s$, and applying (7.5.13), we obtain

$$\sup_{|\tau|\le 1} |\varphi_j^{(\ell)}(\tau)| \;\le\; C_s\varepsilon^{1/2}, \quad \ell = 0,1,\ldots,s; \;\; j = 1,\ldots,n,$$

where C_s is a new constant depending on s only. In particular,

$$|\varphi_1^{(\ell)}(0;\theta_j) - \varphi_2^{(\ell)}(0;\theta_j)| \;\le\; C_s\varepsilon^{1/2}, \quad \ell = 0,1,\ldots,s; \;\; j = 1,\ldots,n,$$

or equivalently,

$$\left|\int_{\mathbb{R}^2} (x,\theta_j)^k (P_1 - P_2)(\,\mathrm{d}x)\right| \;\le\; C_s\varepsilon^{1/2}, \qquad (7.5.15)$$

$k = 0,1,\ldots,s$; $j = 1,\ldots,n$. Following the notation in Lemma 7.5.1, we can rewrite (7.5.15) in the form for $k = 0,\ldots,s$ and $j = 1,\ldots,n$,

$$\left|\sum_{\ell=0}^{k}\binom{k}{\ell} u_j^\ell v_j^{k-\ell}\left[\mu_{\ell,k-\ell}(P_1) - \mu_{\ell,k-\ell}(P_2)\right]\right| \;\le\; C_s\varepsilon^{1/2}.$$

Thus, setting

$$\mathcal{R}_{kj} \;=\; \sum_{\ell=0}^{k}\binom{k}{\ell} z_j^{k-\ell}\left[\mu_{\ell,k-\ell}(P_1) - \mu_{\ell,k-\ell}(P_2)\right], \qquad (7.5.16)$$

$k = 2,\ldots,s$; $j = 1,\ldots,n$; $z_j = v_j/u_j$, we obtain

$$|\mathcal{R}_{kj}| \;\le\; \widetilde{C}\varepsilon^{1/2}, \qquad (7.5.17)$$

where $\widetilde{C}$ depends on the directions $\theta_1,\ldots,\theta_n$ only. For any fixed k ($k = 2,\ldots,s$) consider

(i) the matrix A_k with elements $a_{\ell j}^{(k)} = \binom{k}{\ell-1} z_j^{k-(\ell-1)}, \quad \ell, j = 1, \ldots, k+1$;

(ii) the vector B_k with elements

$$b_\ell^{(k)} = \mu_{\ell-1,k-\ell+1}(P_1) - \mu_{\ell-1,k-\ell+1}(P_1), \quad \ell = 1, \ldots, k+1;$$

(iii) the vector D_k with elements $d_j = \mathcal{R}_{kj}, \; j = 1, \ldots, k+1$.

Then (7.5.16) has the form $A_k B_k = D_k \; (k = 1, \ldots, s-1)$, while (7.5.17) yields $\|D_k\| \le \widetilde{C}\varepsilon^{1/2}$. The matrices A_k are invertible, and so

$$\|B_k\| \le \|A_k^{-1}\| \, \|D_k\| \le \overline{C}\varepsilon^{1/2}, \tag{7.5.18}$$

where the constant $\overline{C}$ depends on the directions $\theta_1, \ldots, \theta_n$ only. Inequality (7.5.18) shows that the first $s-1$ moments of the two-dimensional distributions are close when $\varepsilon > 0$ is sufficiently small. Such an evaluation of closeness holds for the first $s-1$ moments of the marginals corresponding to an arbitrary direction t; i.e.,

$$\left| \int_{\mathbb{R}^2} (x,t)^k (P_1 - P_2)(dx) \right| \le \overline{C}\varepsilon^{1/2},$$

$k = 0, \ldots, s-1$. Now we have

$$\begin{aligned} |\varphi_1(\tau;t) - \varphi_2(\tau;t) &\le \left| \sum_{j=0}^{s-1} \frac{\varphi_1^{(j)}(0;t) - \varphi_2^{(j)}(0;t)}{j!} \tau^j \right| + \frac{2}{s!}|\tau|^s \\ &\le \sum_{j=0}^{s-1} \frac{\overline{C}\varepsilon^{1/2}}{j!}|\tau|^j + \frac{2|\tau|^s}{s!} \le \overline{C}\varepsilon^{1/2} e^{|\tau|} + \frac{2|\tau|^s}{s!}. \end{aligned}$$

Choose $T = \min\left\{ \ln\left[1 + \left(\overline{C}\varepsilon^{1/2}\right)^{1/2}\right], \left(\frac{s!}{2}\right)^{\frac{1}{s-1}} \right\}$.

Since t is arbitrary on the unit circle, we obtain

$$\begin{aligned} \lambda(P_1, P_2) &\le \max\left\{ \overline{C}^{1/2}\varepsilon^{1/4} + \overline{C}\varepsilon^{1/2} + \left(\frac{2}{s!}\right)^{\frac{1}{s-1}}, \; 1/T \right\} \\ &\le C\left[1/\ln(1/\varepsilon) + 1/s\right], \end{aligned}$$

which proves the theorem. □

Remark 7.5.7 *The statement in Theorem* 7.5.6 *still holds if instead of the* ε*-coincidence of the marginals as in* (7.5.11)*, we require the* ε*-coincidence of the moments up to order* s *of these marginals.*

Theorems 7.5.5 and 7.5.6 can be generalized for probability measures defined on $\mathbb{R}^m$. However, we cannot choose the directions $\theta_1, \ldots, \theta_n$ in an arbitrary way. Furthermore, to obtain the same order of precision in $\mathbb{R}^m, m > 2$, corresponding to the n directions in $\mathbb{R}^2$, we need n^{m-1} directions. The results can be obtained by induction on the dimension m.

We define next the set of directions we are going to use. Choose $n \geq 2$ distinct real numbers $u_1, \ldots, u_n$, all different from zero, and first construct the set of n two-dimensional vectors $(1, u_1), (1, u_2), \ldots, (1, u_n)$. Then construct n^2 three-dimensional vectors $(1, u_{j_1}, u_{j_2}), j_1, j_2 = 1, \ldots, n$. Repeating this process, by the last step we have constructed a set of m-dimensional vectors

$$(1, u_{j_1}, u_{j_2}, \ldots, u_{j_{m-1}}), \quad j_\ell = 1, \ldots, n; \ \ell = 1, \ldots, m-1. \tag{7.5.19}$$

Denote these m-dimensional vectors by $\theta_1, \ldots \theta_N$, where $N = n^{m-1}$. These inductive arguments lead to the following extensions of Theorems 7.5.5 and 7.5.6.

Theorem 7.5.8 *The results in Theorems* 7.5.5 *and* 7.5.6 *still hold if we consider the measures* P_1 *and* P_2 *in* $\mathbb{R}^m$, *and we choose as directions the* $N = n^{m-1}$ *vectors in* (7.5.19). *Further,* $s = 2[(n-1)/2]$.

To prove this, it is sufficient to note that instead of the m-dimensional vectors, we can first consider a pair of one-dimensional probabilities; the first component is the distribution of the inner product of the projections of the vector x and the vector θ_j upon the $(m-1)$-dimensional subspace, while the second is the law of the last coordinate of the vector x. This allows us to decrease the dimensionality by one. To complete the proof it is sufficient to apply inductive arguments.

The bounds of the deviation between probability measures with coinciding marginals offers a solution to the computer tomography paradox as stated in Gutman et al. (1991): *"It implies that for any human object and corresponding projection data there exist many different reconstructions, in particular, a reconstruction consisting only of bone and air (density 1 or 0), but still having the same projection data as the original object. Related nonuniqueness results are familar in tomography and are usually ignored because CT machines seem to produce useful images. It is likely that the 'explanation' of this apparent paradox is that point reconstruction in tomography is impossible."* Lemma 7.5.1 shows that although the densities of the probability measures P_1 and P_2 (given that such densities exist) can be quite distant from each other for any "large" number of coinciding marginals, yet the measures P_1 and P_2 themselves are close in the weak metric λ.

Khalfin and Klebanov (1990) have analyzed this paradox and obtained some bounds for the closeness of probability measures with coinciding marginals for specially chosen directions for the case of uniform distance between the smoothed densities of these measures.

In tomography the observations are, in fact, integrals of body densities along some straight lines. Using quadratic formulas enables us to evaluate the moments of a set of marginals; these in turn make it possible to apply the results in this section (see Remark 7.5.7) to evaluate the precision of the reconstruction for densities. The classical theory of *moments* makes it possible to give numerical methods for reconstructing the probability measures using the moments (see, for example, Ahiezer (1961)).

7.6 Moment Problems with Applications to Characterization of Stochastic Processes, Queueing Theory, and Rounding Problems

The theory of moment has a long history, which originated in the pioneering works of Shohat and Tamarkin (1943), Hoeffding (1955), Rogosinsky (1958), Ahiezer and Krein (1962), Karlin and Studden (1966), Kemperman (1968). It was also in the 1950s and '60s that moment theory became a separate mathematical discipline. Currently, it is appropriate to talk about the moment problems as beeing a whole range of problems with applications to many mathematical theories. We refer to the monograph of Annastassiou (1993) for a recent survey on the developments in the theory of moments.

In this section we present some applications of moment theory to probabilistic-statistical models. The results presented here are due to Anastassiou and Rachev (1992). For the proofs of the theorems, which are only stated but not proved in this section, we refer to Anastassiou (1993).

First we shall state results on the following five moment problems:

Moment problem 1: *Find*

$$\sup_{\mu} \int_S |x-y|^p \mu(\mathrm{d}x, \mathrm{d}y), \quad S \subset \mathbb{R}^2, \ p \geq 1, \tag{7.6.1}$$

and

$$\inf_{\mu} \int_S |x-y|^p \mu(\mathrm{d}x, \mathrm{d}y), \quad S \subset \mathbb{R}^2, \tag{7.6.2}$$

where the supremum (resp. infimum) is taken over the set of all probability measures μ with support in S having fixed marginal moments

$$\int_S x^i \mu(\mathrm{d}x, \mathrm{d}y) = \alpha_i, \quad \int_S y^i \mu(\mathrm{d}x, \mathrm{d}y) = \beta_i, \quad i = 1, 2, \ldots, n. \tag{7.6.3}$$

Remark 7.6.1 *Problem* (7.6.2) *with fixed marginal distributions*

$$\mu(\cdot \times S) = \mu_1(\cdot), \quad \mu(S \times \cdot) = \mu_2(\cdot) \tag{7.6.4}$$

is indeed the L_p-Kantorovich problem on mass transportation (see Chapters 2 and 3).

Moment problem 2: *For given $x_0 \in \mathbb{R}$ and positive α find the Kantorovich radius*

$$\sup E|X - x_0|^\alpha, \tag{7.6.5}$$

where the supremum is over all random variables X with fixed moments $EX = p$ and $EX^2 = q$.

Remark 7.6.2 *Problem 2 will be used in approximation of complex queueing models by means of deterministic models.*

Moment problem 3: *Find*

$$\sup \int_A [t]_c \mu(\,dt), \quad A = [0,a] \ \ or \ \ [0,\infty), \tag{7.6.6}$$

and

$$\inf \int_A [t]_c \mu(\,dt), \quad A = [0,a] \ \ or \ \ [0,\infty), \tag{7.6.7}$$

over the set of all probability measures with support A having fixed rth moment

$$\int_A t^r \mu(\,dt) = d_r, \quad r > 0, d_r > 0, \tag{7.6.8}$$

where for a given nonnegative x the c-rounding ($0 \le c \le 1$) of x is defined by

$$[x]_c = \begin{cases} m & \text{if } m \le x \le m + c, \\ m+1 & \text{if } m + c < x \le m + 1. \end{cases}$$

Remark 7.6.3 *Moment problem 3 can be applied to the problems of rounding and apportionment; see Mosteller, Youtz, and Zahn (1967), Diaconis and Freedman (1979), Balinski and Young (1982), and Balinski and Rachev (1993). In the apportionment theory, $c = 0$ corresponds to the Adams method; $c = 1/2$ corresponds to the Webster method (or conventional rounding, or Mosteller–Youtz–Zahn "broken stick" rule of rounding); $c = 1$ corresponds to the Jefferson method; see Balinski and Young (1982).*

Moment problem 4: *Find* (7.6.6) *and* (7.6.7) *subject to* (7.6.8) *and*

$$\int_A t\mu(\mathrm{d}t) = d_1. \tag{7.6.9}$$

We next consider some infinite-dimensional analogues of the moment problems 1 and 2.

Let $C[0,1]$ be the space of continuous functions on $[0,1]$ with the usual sup-norm $\|x\|$, and let $\mathcal{X}(C[0,1])$ be the space of r.v.s on a nonatomic probability space (Ω, A, P) with values in $C[0,1]$. Let $\mathcal{M}$ be the class of all strictly increasing continuous functions $f : [0,\infty] \to [0,\infty]$, $f(0) = 0$, $f(\infty) = \infty$. Finally, let T be a set of finitely many points in $[0,1]$,

$$0 \le t_1 < t_2 < \cdots \le t_N \le 1. \tag{7.6.10}$$

Moment problem 5: *Given* $h, g_i \in \mathcal{M}(i = 1, \ldots, N)$ *find*

$$\inf Eh(\|X - Y\|), \tag{7.6.11}$$

where the infimum is over the set of all possible joint distributions of X *and* Y *subject to the moment constraints*

$$Eg_i(|X(t_i)|) = a_i, \quad Eg_i(|Y(t_i)|) = b_i. \tag{7.6.12}$$

Remark 7.6.4 *This problem can be interpreted as follows. Having observations of two random processes (more precisely, we suppose the moments* (7.6.12) *are known), the goal is to evaluate the minimal possible distance* $Eh(\|X - Y\|)$ *between the processes* X *and* Y. *We shall determine the minimum in* (7.6.11) *and show that essentially this minimum can be achieved.*

7.6.1 Moment Problems and Kantorovich Radius

In this section we state the solutions of moment problems 1 and 2; the proofs are given in Anastassiou and Rachev (1992) and the monograph Anastassiou (1993). Let $S = [a,b] \times [c,d] \subset \mathbb{R}^2$ and $\varphi(x_1, x_2) = |x_1 - x_2|^p$, $p \ge 1$. Suppose $(\alpha, \beta) \in S$, and denote by $U = U(\varphi, \alpha, \beta)$ the supremum in (7.6.1) subject to

$$\int_S x_1 \mu(\mathrm{d}x_1, \mathrm{d}x_2) = \alpha, \quad \int_S x_2 \mu(\mathrm{d}x_1, \mathrm{d}x_2) = \beta. \tag{7.6.13}$$

Theorem 7.6.5 *The supremum* U *in* (7.6.1) *is given by*

$$U = D\delta + T, \tag{7.6.14}$$

where

$$
\begin{aligned}
D &:= \left[|b-d|^p+|a-c|^p\right]-\left[|b-c|^p+|a-d|^p\right],\\
T &:= (1-B)|b-c|^p+(B+C-1)|a-c|^p+(1-C)|a-d|^p,\\
B &:= \frac{b-\alpha}{b-a}, \quad C := \frac{d-\beta}{d-c},\\
\delta &:= \max(0,1-B-C).
\end{aligned}
$$

Remark 7.6.6 *Since* φ *is convex on any* $S\subset \mathbb{R}^2$, *then given* (7.6.13), $\inf_\mu \int_S \varphi\, d\mu = |\alpha-\beta|^p$.

Next, consider nonbounded regions S. Namely, define for $b\ge 0$ the following stripes in $\mathbb{R}^2$:

$$
\begin{aligned}
S_1^b &:= \{(x,y);\ y=x+b',\ \text{where } 0\le b'\le b\},\\
S_2^b &:= \{(x,y);\ y=x-b',\ \text{where } 0\le b'\le b\},\\
S^b &:= S_1^b\cup S_2^b.
\end{aligned}
$$

We extend Theorem 7.6.5 to this type of unbounded region.

Theorem 7.6.7 *Assume that* $0<p\le 1$.

(i) *If* $S=S_1^b$ *or* S_2^b, $(\alpha,\beta)\in S$, *then the supremum* U *in* (7.6.1) *is equal to*

$$
U := U(\varphi,\alpha,\beta) = |\alpha-\beta|^p.
$$

(ii) *If* $S=S^b$, *then* $U=b^p$.

(iii) *Let* L *be the lower bound* (7.6.2) *subject to* (7.6.13). *Then if* $S=S_1^b$ *or* $S=S_2^b$ *or* $S=S^b$, $(\alpha,\beta)\in S$, *we have*

$$
L := L(\varphi;\alpha,\beta) = b^{p-1}|\alpha-\beta|.
$$

Next, consider another type of stripe in $\mathbb{R}^2$: For $b,\gamma>0$,

$$
\begin{aligned}
S_1^b &:= \{(x,y);\ y=x+b',\ \text{where } 0\le b'\le b\},\\
S_2^\gamma &:= \{(x,y);\ y=x-\gamma',\ \text{where } 0\le \gamma'\le \gamma\},\\
S^{b,\gamma} &:= S_1^b\cup S_2^\gamma.
\end{aligned}
$$

Theorem 7.6.8 *Let* $p\ge 1$.

(i) *If* $S=S_1^b$, $(\alpha,\beta)\in S$, *then* $U:=U(\varphi,\alpha,\beta)=b^{p-1}(\beta-\alpha)$.

(ii) *If* $S = S_2^{\gamma}$, $(\alpha,\beta) \in S$, *then* $U = U(\varphi,\alpha,\beta) = \gamma^{p-1}(\alpha-\beta)$.

(iii) *If* $S = S^{b,\gamma}$, $(\alpha,\beta) \in S$, *then* $U = \dfrac{(b^p - \gamma^p)(\beta - \alpha - b) + b^p(b+\gamma)}{b+\gamma}$.

Next, we shall state the explicit solutions of Moment problem 2. So we will be interested in the following problem. For given $x_0 \in \mathbb{R}$, $\alpha > 0$, $p \in \mathbb{R}$, $q > 0$ $(p^2 \le q)$, $-\infty \le a < b \le +\infty$, find the *Kantorovich radius*

$$\begin{aligned} K &:= K(x_0;\alpha,p,q,a,b) \\ &:= \sup\{E|X-x_0|^\alpha;\ X \in [a,b] \text{ a.s.},\ EX = p,\ EX^2 = q\}. \end{aligned} \tag{7.6.15}$$

Theorem 7.6.9 (Case (A): $\alpha \ge 2$, $-\infty < a < b < +\infty$) *Let* $x_0 = (a+b)/2$, $a \le p \le b$, *and* $0 \le q \le b^2 + (a+b)(p-b)$. *Then the Kantorovich radius* K *admits the following bound:*

$$K \le \left(\frac{b-a}{2}\right)^{\alpha-2}\left[q - p(a+b) + \frac{(a+b)^2}{4}\right].$$

Moreover, if there exist $\lambda_1, \lambda_2 \ge 0$, $\lambda_1 + \lambda_2 \le 1$, *such that*

$$\begin{aligned} p &= \frac{a+b}{2} + \frac{b-a}{2}(\lambda_1 - \lambda_2), \\ q &= \frac{(a+b)^2}{4} + \frac{b^2-a^2}{2}(\lambda_1-\lambda_2) + \frac{(b-a)^2}{4}(\lambda_1+\lambda_2), \end{aligned}$$

then

$$K = \left(\frac{b-a}{2}\right)^{\alpha-2}\left[q - p(a+b) + \frac{(a+b)^2}{4}\right].$$

The next theorem gives an analogue of Theorem 7.6.9 when the X's in (7.6.15) have unbounded support.

Theorem 7.6.10 (Case (B): $0 < \alpha \le 2$, $a = -\infty$, $b = +\infty$) *For any* $x_0 \in \mathbb{R}, p \in \mathbb{R}, q > 0, p^2 \le q$, *the Kantorovich radius* K *is given by*

$$K = K(x_0;\alpha;p,q) = (q - 2x_0 p + x_0^2)^{\alpha/2}.$$

The rest of the results in this section treat various versions of Theorems 7.6.9 and 7.6.10.

Theorem 7.6.11 (Case (C): $0 < \alpha \le 2$, $-\infty < a < b < +\infty$) *For any* $x_0 \in (a,b); p \in \mathbb{R}, p^2 \le q$,

$$K = \sup\{E|X-x_0|^\alpha;\ X \in [a,b] \ a.s., EX = p, EX^2 = q\}.$$

Set $P = p - x_0$, $Q = q - 2x_0p + x_0^2$, $A(x_0) = a - x_0$, $B(x_0) = b - x_0$, $C(x_0) = \min(-A(x_0), B(x_0))$.

(i) *If* $0 \le Q \le C^2(x_0)$, *then* $K = Q^{\alpha/2}$.

(ii) *If* $Q > C^2(x_0)$ *and* $(A(x_0) + B(x_0))P - Q - A(x_0)B(x_0) \ge 0$, *then* $K \le Q^{\alpha/2}$.

Theorem 7.6.12 (Case (D): $1 \le \alpha \le 2$, $-\infty < a < b \le +\infty$) *For any* $p \in \mathbb{R}$, $p^2 \le q$, $a \le p \le b$, *set* $P = p - a$, $Q = q - 2ap + a^2$, $B = b - a$. *Suppose* $Q \le BP$. *Then*

$$K \; := \; \sup\{E|X-a|^\alpha;\; X \in [a,b], EX = p, EX^2 = q\} \; = \; p^{2-\alpha}Q^{\alpha-1}.$$

Theorem 7.6.13 (Case (E): $1 \le \alpha \le 2$, $-\infty \le a < b < +\infty$) *For any* $p \in \mathbb{R}$, $p^2 \le q$, $a \le p \le b$, *set* $P = p - b$, $Q = q - 2bp + b^2$, $\theta = a - b$. *Suppose* $Q \le \theta P$. *Then*

$$K \; := \; \sup\{E|X-b|^\alpha;\; X \in [a,b], EX = p, EX^2 = q\} \; = \; p^{2-\alpha}Q^{\alpha-1}.$$

7.6.2 Moment Problems Related to Rounding Proportions

Here we state results on explicit solutions of Moment problems 3 and 4; see (7.6.6) and (7.6.9). To this end recall the definition of *c-rounding* ($0 \le c \le 1$): For any $x \ge 0$,

$$[x]_c = \begin{cases} m, & \text{if} \quad m \le x < m + c, \\ m+1, & \text{if} \quad m + c \le x < m+1, \end{cases} \qquad m \in \mathbb{N} \cup \{0\},$$

$$= \begin{cases} 0, & \text{if} \quad 0 \le x < c, \\ m, & \text{if} \quad m - 1 + c < x \le m + c, \end{cases} \qquad m \in \mathbb{N}.$$

The next four theorems deal with problem 3.[(5)]

Theorem 7.6.14 *Let* $c \in (0,1)$, $r > 0$, $0 < a < \infty$, $d > 0$, *and*

$$U \; := \; U_{[\cdot]_c}(a,r,d) \; := \; \sup\{E[X]_c;\; 0 \le X \le a \text{ a.s.}, (EX^r)^{1/r} = d\}.$$

Set $n = [a]$.

(I) *If* $n + c < a$, $n + c \le d \le a$, *then* $U = n + 1$.

(II) *If* $n + c \ge a$, $n - 1 + c \le d \le a$, *then* $U = n$.

---

(5) In the sequel, the underlying probability space is assumed to be nonatomic, and thus the space of laws of nonnegative r.v.s coincides with the space of all Borel probability measures on $\mathbb{R}_+$.

(III) *If* $0 < a \le c$, *then* $U = 0$.

(IV) *If* $0 < r \le \dfrac{\ln 2}{\ln(1+1/c)} (< 1)$, $n + c < a$, *and* $0 \le d \le n + c$, *then* $U = (n+1)d^r(n+c)^{-r}$.

(V) *If* $0 < r \le \dfrac{\ln 2}{\ln(1+1/c)}$, $n + c \ge a$, *and* $0 \le d \le n - 1 + c$, *then* $U = nd^r(n-1+c)^{-r}$.

(VI) *If* $r \ge 1$ *and* $0 \le d \le c$, *then* $U = d^r c^{-r}$.

(VII) *Suppose* $r \ge 1$. *If either*

(a) $n + c < a$ *and* $k \in \{1, \ldots, n\}$ *is determined by* $k - 1 + c \le d < k + c$, *or*

(b) $n + c \ge a$ *and* $k \in \{1, \ldots, n\}$ *is determined by* $k - 1 + c \le d < k + c$,

$$\text{then } U = k + \frac{d^r - (k-1+c)^r}{(k-c)^r - (k-1+c)^r} \le 1 - c + d.$$

The next theorem extends Theorem 7.6.14 to the case $a = +\infty$.

Theorem 7.6.15 *Let* $0 < c < 1$, $r > 0$, $d > 0$, *and* $U := U_{[\cdot]_c}(r, d) := \sup\{E[X]_c;\ X \ge 0 \text{ a.s.}, (EX^r)^{1/r} = d\}$.

(I) *If* $0 < r < 1$, *then* $U = +\infty$.

(II) *If* $r \ge 1$ *and* $0 \le d \le c$, *then* $U = d^r c^{-r}$.

(III) *Suppose* $r \ge 1$. *Define* $k \in N$ *by* $k - 1 + c \le d < k + c$. *Then*

$$U = k + \frac{d^r - (k-1+c)^r}{(k+c)^r - (k -+ c)^r} \le 1 - c + d.$$

The next two theorems are versions of the previous two; here we consider the lower bounds in c-rounding.

Theorem 7.6.16 *Let* $c \in (0,1), r > 0, 0 < a < \infty, d > 0$, *and* $L := L_{[\cdot]_c}(a, r, d) := \inf\{E[X]_c;\ 0 \le X \le a \text{ a.s.}, (EX^r)^{1/r} = d\}$.

Set $n = [a]$.

(I) *If* $0 < d \le c$, *then* $L = 0$.

(II) *If* $c < a \le 1 + c$ *and* $c \le d \le a$, *then* $L = (d^r - c^r)/(a^r - c^r)$.

(III) *Let* $0 < r \leq 1$, $n + c < a$, *and determine* $k \in \{0, 1, \ldots, n-1\}$ *by* $k + c \leq d < k + 1 + c$. *Then*

$$L = k + \frac{d^r - (k+c)^r}{(k+1+c)^r - (k+c)^r}.$$

(IV) *If* $0 < r \leq 1$, $n + c < a$, *and* $n + c \leq d \leq a$, *then*

$$L = n + \frac{d^r - (n+c)^r}{a^r - (n+c)^r}.$$

The case $a = +\infty$ is extended as follows.

Theorem 7.6.17 *Let* $c \in (0,1)$, $r > 0$, $d > 0$, *and* $L := L_{[\cdot]_c}(r,d) := \inf\{E[X]_c : X \geq 0 \text{ a.s.}, (EX^r)^{1/r} = d\}$.

(I) *If* $r > 1$, *then* $L = 0$.

(II) *If* $0 \leq r \leq 1$, $0 < d \leq c$, *then* $L = 0$.

(III) *If* $r = 1$, $c \leq d < \infty$, *then* $L = d - c$.

(IV) *If* $0 < r \leq 1$, *define* $k \in \mathbb{N} \cup \{0\}$ *by* $k + c \leq d < k + 1 + c$. *Then*

$$L := k + \frac{d^r - (k+c)^r}{(k+1+c)^r - (k+c)^r}.$$

Next we pass to Moment problem 4, see (7.6.6)–(7.6.9). For simplicity we shall consider only special cases of c-rounding. For the general case we refer to Anastassiou and Rachev (1992) and Anastassiou (1993).

First consider the *conventional (Webster) rounding*, or *MYZ-rounding*, $[x] := [x]_1$.

Theorem 7.6.18 *Let* $a > 0$, $0 < r \neq 1$, $d_1 > 0$, $d_r > 0$, *and* $U = U_{[\cdot]}(a, r, d_1, d_2) := \sup\{E[X];\ 0 \leq X \leq a \text{ a.s. and } EX = d_1, EX^r = d_r\}$.

Let $\theta = [a]$.

(I) *Set*

$$\Delta_{r,a} := a^r + \frac{a^r - \theta^r}{a - \theta}(d_1 - \theta).$$

Suppose that $a \neq \theta, \theta \leq d_1 \leq a$, *and either* $r > 1, d_1^r \leq d_r \leq \Delta_{r,a}$, *or* $0 < r < 1$ *and* $\Delta_{r,a} \leq d_r \leq d_1^r$. *Then* $U = \theta$.

(II) *Suppose* $0 < \theta \neq a$ *and there are* $\lambda_1, \lambda_2 \geq 0$ *with* $\lambda_1 + \lambda_2 \leq 1$ *and such that* $d_1 = \lambda_1\theta + \lambda_2 a$ *and* $d_r = \lambda_1\theta^r + \lambda_2 a^r$. *Then*

$$U = \frac{(\theta^r - a^r)d_1 + (a-\theta)d_r}{a(\theta^{r-1} - a^{r-1})}.$$

(III) *Let* $\theta \geq 1$ *and suppose there exists* $k \in \{0, 1, \ldots, \theta - 1\}$ *such that* $k \leq d_1 < k+1$ *and either* $r > 1$ *and* $d_{r,k} := k^r + [(k+1)^r - k^r](d_1 - k) \leq d_r \leq \theta^{r-1}d_1$ *or* $0 < r < 1$ *and* $\theta^{r-1}d_1 \leq d_r \leq d_{r,k}$. *Then* $U = d_1$.

For the case $a = +\infty$ we have the following version of the above theorem.

Theorem 7.6.19 *Let* $0 < r \neq 1$, $d_1 > 0$, $d_r > 0$, *and* $U := U_{[\cdot]}(r, d_1, d_r) := \sup\{E[X];\ X \geq 0$ *a.s. and* $EX = d_1, EX^r = d_r\}$.

Suppose there exists a nonnegative integer k *such that* $k \leq d_1 \leq k+1$ *and either* $r > 1$ *and* $d_r \geq d_{r,k} := k^r + [(k+1)^r - k^r](d_1 - k)$ *or* $0 < r < 1$ *and* $0 < d_r \leq d_{r,k}$. *Then* $U = d_1$.

If we change in Theorems 7.6.18 and 7.6.19 the upper bound U to the corresponding lower bound, we obtain the following two theorems.

Theorem 7.6.20 *Let* $a > 0$, $0 < r \neq 1$, $d_1 > 0$, $d_r > 0$, *and* $L := L_{[\cdot]}(a, r, d_1, d_r) := \inf\{E[X];\ 0 \leq X \leq a$ *a.s.*, $EX = d_1, EX^r = d_r\}$.

(I) *Suppose there exist* t_1, t_2, λ *such that* $0 \leq t_1 \leq t_2 \leq 1$, *and* $d_1 = (1-\lambda)t_1 + \lambda t_2$ *and* $d_r = (1-\lambda)t_1^r + \lambda t_2^r$. *Then* $L = 0$.

(II) *If* $0 < a \leq 1$, *then* $L = 0$.

(III) *If* $1 < a < 2$ *and there exist* $\lambda_1, \lambda_2 > 0$ *with* $\lambda_1 + \lambda_2 \leq 1$ *and such that* $d_1 = \lambda_1 + \lambda_2 a$, $d_r = \lambda_1 + \lambda_2 a^r$, *then* $L = \dfrac{(d_r - d_1)}{(a^r - a)}$.

From now on assume that $a \geq 2$, *and let* $\theta = [a]$.

(IV) *Suppose* $\Delta_\theta := \dfrac{(\theta - 1)(a^r - a)}{(\theta^r - \theta)} \leq \theta$.

(i) *If* $d_1 = \lambda_1 + \lambda_2\theta$, $d_r = \lambda_1 + \lambda_2\theta^r$ *for some* $\lambda_1, \lambda_2 \geq 0$, $\lambda_1 + \lambda_2 \leq 1$, *then* $L = \Delta_\theta$.

(ii) *If* $d_1 = \lambda_1\theta + \lambda_2 a$ *and* $d_r = \lambda_1\theta^r + \lambda_2\theta^r$ *for some* $\lambda_1, \lambda_2 \geq 0$, $\lambda_1 + \lambda_2 \leq 1$, *then*

$$L = \frac{(\theta - a(\theta-1))d_r - (\theta^{r+1} - a^r(\theta-1))d_1}{\theta a^r - a\theta^r}.$$

(V) *Suppose* $\Delta_\theta > \theta$.

(i) *If* $d_1 = \lambda_1+\lambda_2\theta+\lambda_3 a$ *and* $d_r = \lambda_1+\lambda_2\theta^r+\lambda_3 a^r$ *for some* $\lambda_1, \lambda_2, \lambda_3 \geq 0,\ \lambda_1 + \lambda_2 + \lambda_3 = 1$, *then*

$$L = \frac{(\theta-1)(\theta-a+1)(d_r-1) - ((\theta^r-1)\theta - (a^r-1)(\theta-1))(d_1-1)}{(\theta-1)(a^r-1) - (a-1)(\theta^r-1)}.$$

(ii) *If* $d_1 = \lambda_1 + \lambda_2 a, d_r = \lambda_1 + \lambda_2 a^r$ *for some* $\lambda_1, \lambda_2 \geq 0,\ \lambda_1 + \lambda_2 \leq 1$, *then* $L = \dfrac{\theta(d_r - d_1)}{a^r - a}$.

(VI) *Suppose* $\theta > 1$ *and one of the following holds.*

(i) $r > 1$ *and there exists* $k \in \{1, \ldots, \theta - 1\}$ *such that* $k \leq d_1 k + 1$ *and*

$$\begin{aligned} d_{r,k} \;&:=\; k^r + ((k+1)^r - k^r)(d_1 - k) \;\leq\; d_r \\ &\leq\; \Delta_{r,\theta} := 1 + \frac{(\theta^r - 1)(d_1 - 1)}{\theta - 1}; \end{aligned}$$

(ii) $0 < r < 1$ *and there exists* $k \in \{1, \ldots, \theta-1\}$ *such that* $k \leq d_1 \leq k+1$ *and* $\Delta_{r,k} \leq d_r \leq d_{r,k}$.

Then $L = d_1 - 1$.

The special case $a = +\infty$ is treated as follows.

Theorem 7.6.21 *Let* $0 < r \neq 1,\ d_1 > 0,\ d_r > 0$, *and* $L := L_{[\cdot]}(r, d_1, d_r) := \inf\{E[X] : X \geq 0 \text{ a.s.}, EX = d_1, EX^r = d_r\}$.

(I) *Suppose there exist* $0 \leq t_1 \leq t_2 \leq 1,\ 0 \leq \lambda \leq 1$ *such that* $d_1 = (1-\lambda)t_1 + \lambda t_2$ *and* $d_r = (1-\lambda)t_1^r + \lambda t_2^r$. *Then* $L = 0$.

(II) *Suppose* $0 < r < 1$ *and either* $d_1 = \lambda_1 + \lambda_2,\ d_r = \lambda_1,\ \lambda_1 \geq 0,\ \lambda_1 + \lambda_2 \leq 1$ *or* $d_1 \geq 1,\ 0 < d_1 \leq 1$. *Then* $L = d_1 - d_r$.

(III) *Suppose* $0 < r < 1$ *and for some integer* $k,\ k \leq d_1 < k+1$, *and* $1 \leq d_r \leq (k^r + ((k+1)^r - k^r))(d_1 - k)$. *Then* $L = d_1 - 1$.

(IV) *Suppose* $r > 1$ *and either* $d_1 = \lambda_1,\ d_r = \lambda_1 + \lambda_2,\ \lambda_i \geq 0,\ i = 1, 2,\ \lambda_1 + \lambda_2 \leq 1$ *or* $0 < d_1 \leq 1$ *and* $1 \leq d_r$. *Then* $L = 0$.

(V) *Suppose* $r > 1$ *and for some integer* $k,\ k \leq d_1 < k+1$, *and* $(k^r + ((k+1)^r - k^r))(d_1 - k) \leq d_r$. *Then* $L = d_1 - 1$.

Similar results are valid for Adams ($c = 0$) and Jefferson ($c = 1$) rules of roundings; see Anastassiou and Rachev (1992) and Anastassiou (1993).

7.6.3 Closeness of Random Processes with Fixed Moment Characteristics

The moment problems we are going to consider in this section may be viewed as extensions of Moment problem 1 (on page 52) for measures μ generated by the joint distribution of random processes. Namely, let the class $\mathcal{M}$, the space $\mathcal{X}(C[0,1])$, and the set $T = \{t_1, \dots, t_N\}$ be defined by (7.6.10). The subject of this section is the following general version of problem (7.6.10)–(7.6.12).

Moment problem 6: Given $h, g_{i,j} \in \mathcal{M}$ $(i = 1, \dots, N,\ j = 1, \dots, n)$ find the set of valued $Eh(||X - Y||)$ subject to the moment constraints

$$Eg_{i,j}(|X(t_i)|) = a_{i,j}, \quad Eg_{i,j}(|Y(t_i)|) = b_{i,j}, \quad i = 1, \dots, N. \tag{7.6.16}$$

In particular, determine the bounds

$$\inf Eh(||X - Y||), \ \sup Eh(||X - Y||), \tag{7.6.17}$$

given the constraints (7.6.16).

One interpretation of the problem is as follows: Suppose we observe two continuous processes X and Y, only through the "window" T. Suppose for each point of the "window" we know some moment characteristics of X and Y. The problem is to determine the possible deviations between the processes outside the window. In particular, "What is the minimal distance between X and Y with given moment information (7.6.16)?" is just a special case of Moment problem 7.6.3.

We start with the case $n = 1$ in (7.6.16); i.e., given $h, g_i \in \mathcal{M}(i = 1, \dots, N)$ and assuming that

$$Eg_i(|X(t_i)|) = a_i, \quad Eg_i(|Y(t_i)|) = b_i, \quad i = 1, \dots, N, \tag{7.6.18}$$

we are interested in the range of $Eh(||X - Y||)$. The solution of this problem will be given under some assumptions of the following type:

Assumption $A(h, g)$: $h \circ g^{-1}(t)$ $(t \geq 0)$ is a convex function (here and in the sequel, f^{-1} stands for the inverse of $f \in \mathcal{M}$).

Assumption $B(g)$: $g^{-1}(Eg(|\xi + \eta|)) \leq g^{-1}\ (Eg(|\xi|)) + g^{-1}(E(|\eta|))$ for any $\xi, \eta \in \mathcal{X}(R)$ (here, $\mathcal{X}(R)$ is the set of all real-valued r.v.s.).

Assumption $C(g)$: $Eg(|\xi + \eta|) \leq Eg(|\xi|) + Eg(|\eta|)$ for any $\xi, \eta \in \mathcal{X}(R)$.

Assumption $D(h, g)$: $\lim_{t \to \infty} h(t)/g(t) = 0$.

Remark 7.6.22 *Take the most interesting case:* $h(t) = t^p$, $g(t) = t^q$ $(p > 0,\ q > 0)$. *Then*

$$\begin{aligned} A(h,g) &\Leftrightarrow p \geq q, \\ B(g) &\Leftrightarrow q \geq 1, \\ C(g) &\Leftrightarrow q \leq 1, \\ D(h,g) &\Leftrightarrow q > p. \end{aligned}$$

Now let $T = \{0 \leq t_1 \leq \cdots \leq t_N \leq 1\}$ and take $\overline{a} = (a_1, \ldots, a_N) \in \mathbb{R}_+^N, \overline{b} = (b_1, \ldots, b_N) \in \mathbb{R}_+^N$ and $\overline{g} = (g_1, \ldots, g_N) \in \mathcal{M}$ to be fixed vectors. Denote by $\mathcal{X}(T, \overline{g}, \overline{a})$, the space of all $X \in \mathcal{X}(C[0,1])$ satisfying the marginal moment conditions $Eg_i(|X(t_i)|) = a_i$ $(i = 1, \ldots, N)$, and let

$$\begin{aligned} I\{h, \overline{g}, T, \overline{a}, \overline{b}\} &:= \inf\{Eh(\|X - Y\|); \\ &\qquad X \in \mathcal{X}(T, \overline{g}, \overline{a}),\ Y \in \mathcal{X}(T, \overline{g}, \overline{b})\}. \end{aligned} \tag{7.6.19}$$

In the next four theorems we describe the exact range of values of $Eh(\|X - Y\|)$ under different conditions of type A–D.

Theorem 7.6.23 *Let* $A(h, g_i)$ *and* $B(g_i)$ *hold for any* $i = 1, 2, \ldots, N$. *Then*

(i) $$I\{h, \overline{g}, T, \overline{a}, \overline{b}\} = \sup_{1 \leq i \leq N} h\left(|g_i^{-1}(a_i) - g_i^{-1}(b_i)|\right); \tag{7.6.20}$$

(ii) *for any* $\nu \geq I\{h, g, T, \overline{a}, \overline{b}\}$ *there exist random processes* $X_\nu \in \mathcal{X}(T, \overline{g}, \overline{a})$ *and* $Y_\nu \in \mathcal{X}(T, \overline{g}, \overline{b})$ *such that*

$$Eh(\|X_\nu - Y_\nu\|) = \nu. \tag{7.6.21}$$

Proof: We shall split the proof into three claims.

Claim 1: $I\{h, \overline{g}, T, \overline{a}, \overline{b}\} \geq \sup_{1 \leq i \leq N} \phi_i(a_i, b_i)$, where $\phi_i(a_i, b_i) := h(|g_i^{-1}(a_i) - g_i^{-1}(b_i)|)$.

Proof of Claim 1: Let $X, Y \in \mathcal{X}(C[0,1])$ and $\xi := g(|X(t_i) - Y(t_i)|)$. Then, by the Jensen's inequality and $A(h, g_i)$,

$$\begin{aligned} h^{-1}(Eh(\|X - Y\|)) &\geq h^{-1}(Eh(|X(t_i) - Y(t_i)|)) \\ &= h^{-1}(Eh \circ g_i^{-1}(\xi)) \geq h^{-1} \circ h \circ g_i^{-1} E(\xi) \\ &= g_i^{-1} E(\xi). \end{aligned}$$

Further, by $B(g_i)$,

$$\begin{aligned} h \circ g_i^{-1}(E(\xi)) &\geq h\left[(\|g_i^{-1}(Eg_i(|X(t_i)|)) - (\|g_i^{-1}(Eg_i(|Y(t_i)|))\right] \\ &= h(|g_i^{-1}(a_i) - g_i^{-1}(b_i)|), \end{aligned}$$

which proves the claim.

Claim 2: The infimum in the left-hand side of (7.6.19) is attained, and (7.6.20) holds.

Proof of Claim 2: Define $\widetilde{X}, \widetilde{Y} \in \mathcal{X}(C[0,1])$ to be random polygonal lines with vertices at points $0, t_1, \dots, t_n, 1$ given by

$$\begin{cases} \widetilde{X}(t_i,\omega) = g^{-1}(a_i), & \widetilde{Y}(t_i,\omega) = g^{-1}(b_i), \quad i = 1,\dots,N, \\ \widetilde{X}(0,\omega) = \widetilde{Y}(0,\omega) = 0, & \text{if } t_1 > 0, \\ \widetilde{X}(1,\omega) = \widetilde{Y}(1,\omega) = 0 & \text{if } t_N < 1, \ \omega \in \Omega. \end{cases} \tag{7.6.22}$$

For any $\omega \in \Omega$,

$$\left\| \widetilde{X}(\cdot,\omega) - \widetilde{Y}(\cdot,\omega) \right\| = \sup_{1 \le i \le N} |g_i^{-1}(a_i) - g_i^{-1}(b_i)|,$$

and hence

$$Eh(||\widetilde{X} - \widetilde{Y}||) = \sup_{1 \le i \le N} \phi_i(a_i, b_i). \tag{7.6.23}$$

Further, by (7.6.22),

$$\widetilde{X} \in \mathcal{X}(T, \overline{g}, \overline{a}), \quad \widetilde{Y} \in \mathcal{X}(T, \overline{g}, \overline{b}). \tag{7.6.24}$$

Invoking (7.6.23), (7.6.24), and Claim 1, we complete the proof of the claim.

Claim 3: (ii) is satisfied.

Proof of Claim 3: Let $\tau \in (0,1), t \notin T$. Define the r.v.s X_ν and Y_ν in $\mathcal{X}(C[0,1])$ as follows:

$$X_\nu(t) = \widetilde{X}(t), \quad Y_\nu(t) = \widetilde{Y}(t) \quad \text{for } t = 0, t_1, \dots, t_N, 1, \tag{7.6.25}$$

where $\widetilde{X}, \widetilde{Y}$ are determined by (7.6.22), and

$$X_\nu(\tau) = h^{-1}(\nu), \quad Y_\nu(\tau) = 0. \tag{7.6.26}$$

Next, let $X_\nu(t), Y_\nu(t)$ be a random polygonal lines with vertices at $0, t_1, \dots, t_N, 1$ and τ.

Making use of Claim 2, we have

$$||X_\nu - Y_\nu|| = h^{-1}(\nu) \ge \sup_{1 \le i \le N} |g_i^{-1}(a_i) - g_i^{-1}(b_i)|,$$

and thus (7.6.21) holds. □

Theorem 7.6.24 *Let $A(h, g_i)$ and $C(g_i)$ hold for any $i = 1, \dots, N$. Then*

(i) $$I\{h, \overline{g}, T, \overline{a}, \overline{b}\} = \sup_{1 \le i \le N} h \circ g_i^{-1}(|a_i - b_i|),$$

(ii) *for any $\nu \ge I\{h, \overline{g}, T, \overline{a}, \overline{b}\}$ there exist $X_{n\nu} \in \mathcal{X}(T, \overline{g}, \overline{a})$ and $Y_{n\nu} \in \mathcal{X}(T, \overline{g}, \overline{b})$ such that $Eh(||X_{n\nu} - Y_{n\nu}||) \to Y$ as $n \to \infty$.*

Proof:
Claim 1: $I\{h, \overline{g}, T, \overline{a}, \overline{b}\} \ge \sup_{1 \le i \le N} \varphi_i(a_i, b_i)$, where $\varphi_i(a_i, b_i) := h \circ g_i^{-1}(|a_i - b_i|)$.

Proof of Claim 1: Let $X, Y \in \mathcal{X}(C[0,1])$. Then, as in Claim 1 of Theorem 7.6.23, by $A(h, g_i), C(g_i)$, and Jensen's inequality, we have

$$\begin{aligned} h^{-1}(Eh(||X - Y||)) &\ge g_i^{-1} E[g_i(|X(t_i) - Y(t_i)|)] \\ &\ge g_i^{-1}(|Eg_i(|X(t_i)|) - Eg_i(|Y(t_i)|)|) \\ &= h^{-1} \circ \varphi_i(a_i, b_i), \end{aligned}$$

which proves the claim.

Claim 2: For any $\varepsilon > 0$ there exists a pair $(X_\varepsilon, Y_\varepsilon) \in \mathcal{X}(T, \overline{g}, \overline{a}) \times \mathcal{X}(T, \overline{g}, \overline{b})$ such that

$$Eh(||X_\varepsilon - Y_\varepsilon||) = \sup_{1 \le i \le N} h \circ g_i^{-1}(|a_i - b_i| + \varepsilon) \sup_{1 \le i \le N} \frac{|a_i - b_i|}{|a_i - b_i| + \varepsilon}. \tag{7.6.27}$$

Proof of Claim 2: Without loss of generality we can assume that $a_i > b_i$, $i = 1, \dots, N$. Let $p_i := (a_i - b_i)/(a_i - b_i + \varepsilon)$ and $q_i := 1 - p_i$, $i = 1, \dots, N$. We rearrange the indices i so that $p_1 \le p_2 \le \dots \le p_N$. Choose sets $A_i \in \mathcal{A}$ such that $A_1 \subset \dots \subset A_N$ and $P(A_i) = p_i$ by using the assumption of (Ω, A, P) being a nonatomic space. More precisely, since (Ω, A, P) is nonatomic, then for any $B \in \mathcal{A}$ and any $\lambda \in [0, P(B)]$ there exists $C = C(B, \lambda) \in \mathcal{A}, C \subseteq B$ such that $P(C) = \lambda$ (see Loeve (1977, p. 101)). Then the required sets $A_k (k = 1, \dots, N)$ are given by

$$A_k = C(A_{k+1}, P_k), \quad k = 1, \dots, N \quad (A_{N+1} := \Omega).$$

Further, for any $\omega \in \Omega$, define

$$X_\varepsilon(t_i, \omega) := \begin{cases} c_i := g_i^{-1}(a_i - b_i + \varepsilon), & \text{if } \omega \in A_i, \\ d_i := g_i^{-1}(b_i / q_i), & \text{if } \omega \notin A_i, \end{cases}$$

and

$$Y_\varepsilon(t_i, \omega) := \begin{cases} 0, & \text{if } \omega \in A_i, \\ d_i, & \text{if } \omega \notin A_i. \end{cases}$$

We define $(t, X_\varepsilon(t))_{t\in[0,1]}$ to be a random polygonal line with vertices $(t_i, X_\varepsilon(t_i))$ and let $X_\varepsilon(0,\omega) = 0$ if $t_i > 0$ and $X_\varepsilon(1,\omega) = 0$ if $t_N < 1$ for any $\omega \in \Omega$. Analogously define the process $Y_\varepsilon(t)$. Then $Eg_i(|X_\varepsilon(t_i)|) = g_i(c_i)p_i + g_i(d_i)q_i = a_i$, and $Eg_i(|Y_\varepsilon(t_i)|) = g_i(d_i)q_i = b_i$ for any $i = 1, \ldots, N$; i.e., $X_\varepsilon \in \mathcal{X}(T, \overline{g}, \overline{a})$ and $Y_\varepsilon \in \mathcal{X}(T, \overline{g}, \overline{b})$. Further,

$$Eh(\|X_\varepsilon - Y_\varepsilon\|) = Eh[\sup_{t\in T} |X_\varepsilon(t) - Y_\varepsilon(t)|], \tag{7.6.28}$$

where

$$|X_\varepsilon(t_i,\omega) - Y_\varepsilon(t_i,\omega)| = \begin{cases} c_i, & \text{if } \omega \in A_i, \\ 0, & \text{if } \omega \notin A_i. \end{cases} \tag{7.6.29}$$

Since $g \in \mathcal{M}$, then for any $i = 1, \ldots, N-1$, $p_i \le p_{i+1} \Leftrightarrow a_i - b_i \le a_{i+1} - b_{i+1} \Leftrightarrow c_i \le c_{i+1}$; i.e., $c_1 \le c_2 \le \cdots \le c_N$. Hence, by (7.6.29) and $A_1 \subset A_2 \subset \cdots \subset A_N$,

$$\sup_{t\in T} |X_\varepsilon(t,\omega) - Y_\varepsilon(t,\omega)| = \begin{cases} c_N, & \text{if } \omega \in A_N, \\ 0, & \text{if } \omega \notin A_N. \end{cases} \tag{7.6.30}$$

Now, (7.6.28) and (7.6.30) imply that $Eh(\|X_\varepsilon - Y_\varepsilon\|) = h(c_N)p_N$, which is in fact the right-hand side of equality (7.6.27), and thus the claim is proved.

Claims 1 and 2 prove the desired equality (i).

Claim 3: (ii) is satisfied.

Proof of Claim 3: Let $\tau \in (0,1), \tau \notin T$. Using the same notations as in Claims 1 and 2 we define

$$\begin{aligned} X_\nu(t_i,\omega) &:= \mathcal{X}_\varepsilon(t_i,\omega), \quad Y_\nu(t_i,\omega) := Y_\varepsilon(t_i,\omega), \quad \omega\in\Omega, \\ X_\nu(\tau,\omega) &= \begin{cases} h^{-1}(\nu), & \text{if } \omega \in A_N, \\ 0, & \text{if } \omega \notin A_N, \end{cases} \\ Y_\nu(\tau,\omega) &= 0 \quad \text{for any } \omega\in\Omega, \text{ and } \varepsilon > 0 \text{ is chosen so small that} \\ \nu > h(c_N) &= \sup_{1\le i\le N} h\circ g_i^{-1}(|a_i - b_i| + \varepsilon). \end{aligned}$$

We define the random broken lines X_ν and Y_ν with vertices $X_\nu(t_i), Y_\nu(t_i)$, $i = 1, \ldots, N$ and $X_\nu(0) = Y_\nu(0)$ if $t_i > 0; X_\nu(1) = Y_\nu(1) = 0$ if $t_N < 1$. Hence, as in Claim 2 we conclude that

$$\|X_\nu(\cdot,\omega) - Y_\nu(\cdot,\omega)\| = \begin{cases} \max(c_N, h^{-1}(\nu)), & \text{if } \omega\in A_N, \\ 0, & \text{if } \omega\notin A_N. \end{cases}$$

Hence, $Eh(\|X_\nu - Y_\nu\|) = \nu p_N = \nu \frac{a_N - b_N}{a_N - b_N + \varepsilon}$. This proves the claim. □

Theorem 7.6.25 *Let $D(h, g_i)$ hold for any $i = 1, \dots, N$. Then*

$$I\{h, \overline{g}, T, \overline{a}, \overline{b}\} = 0, \tag{7.6.31}$$

and for any $\nu > 0$ there exist random processes $X_{n\nu} \in \mathcal{X}(T, \overline{g}, \overline{a})$ and $Y_{n\nu} \in \mathcal{X}(T, \overline{g}, \overline{b})$ such that $Eh(||X_{n\nu} - Y_{n\nu}||) \to \nu$.

Proof:
Claim 1: For any $n = 1, 2, \dots$ there exist $X_n \in \mathcal{X}(T, \overline{g}, \overline{a})$ and $Y_n \in \mathcal{X}(T, \overline{g}, \overline{b})$ such that

$$Eh(||X_n - Y_n||) \leq \sum_{i=1}^{n} \left(h(nb_i) \frac{b_i}{g_i(nb_i)} + h(na_i) \frac{a_i}{g_i(na_i)} \right). \tag{7.6.32}$$

Since $g_i \in \mathcal{M}$ for n large enough, say $n \geq n_0$, we can define disjoint sets A_{in}, B_{in}, C_{in}, and such that $A_{in} + B_{in} + C_{in} = \Omega$ and

$$P(A_{in}) = c_{in} := \frac{a_i}{g_i(na_i)}, \quad P(B_{in}) = d_{in} := \frac{b_i}{g_i(nb_i)}.$$

Now, for any $i = 1, \dots, N$, $n \geq n_0$, define

$$\begin{cases} X_n(t_i, \omega) = na_i, \quad Y_n(t_i, \omega) = 0, & \text{if } \omega \in A_{in}, \\ X_n(t_i, \omega) = 0, \quad Y_n(t_i, \omega) = nb_i, & \text{if } \omega \in B_{in}, \\ X_n(t_i, \omega) = Y_n(t_i, \omega) = 0, & \text{if } \omega \in C_{in}. \end{cases} \tag{7.6.33}$$

Then $Eg_i[|X_n(t_i)] = g_i(na_i)c_{in} = a_i$ and $Eg_i[|Y_n(t_i)|] = g_i(nb_i)d_{in} = b_i$; i.e., $X \in \mathcal{X}(T, \overline{g}, \overline{a})$ and $Y \in \mathcal{X}(T, \overline{g}, \overline{b})$. Further, we define the random broken lines $X_n(t), Y_n(t)$ ($t \in [0, 1]$) in the way we already did in Theorems 7.6.23 and 7.6.24. Without loss of generality we can assume that

$$a_1 \leq a_2 \leq \cdots \leq a_N \leq b_1 \leq b_2 \leq \cdots \leq b_N.$$

Then

$$||X_n - Y_n|| = \sup_{t \in T} |X_n(t) - Y_n(t)|, \tag{7.6.34}$$

$$= \begin{cases} nb_N & \text{if } \omega \in B_{N,n}, \\ nb_{N-1} & \text{if } \omega \in B_{N-1,n} \setminus B_{N,n}, \\ \vdots & \\ nb_1 & \text{if } \omega \in B_{1,n} \setminus \bigcup_{j=2}^{N} B_{j,n}, \\ na_N & \text{if } \omega \in A_{N,n} \setminus \bigcup_{j=1}^{N} B_{j,n}, \\ \vdots & \\ na_1 & \text{if } \omega \in A_{1,n} \setminus \left[\bigcup_{j=2}^{N} A_{j,n} \cup \bigcup_{j=1}^{N} B_{j,n} \right], \\ 0 & \text{if } \omega \notin \bigcup_{j=1}^{N} A_{j,n} \cup \bigcup_{j=1}^{N} B_{j,n}. \end{cases}$$

Hence,

$$Eh(||X_n - Y_n||) \leq \sum_{j=1}^{N} h(nb_{j,n})d_{j,n} + \sum_{j=1}^{N} h(na_{j,n})c_{j,n},$$

which proves (7.6.32) and the claim.

By $D(h, g_i)$ $(i = 1, \ldots, N)$ it follows that the right-hand side of (7.6.32) goes to 0 as $n \to \infty$. Hence, (7.6.31) holds true.

Claim 2: (ii) is valid.

Let $\tau \in (0,1)$, $t \notin T$, and define $\nu_n = \dfrac{\nu}{1 - P\left(\bigcap_{j=1}^{N} C_{j,n}\right)}$, $X_{n,\nu}(\tau) = h^{-1}(\nu_n)$, $Y_{n,\nu}(\tau) = 0$. We define the random broken line $X_{n,\nu}(t)$ with vertices $X_{n,\nu}(t_j) = X_n(t_j)$ (see (7.6.33)) and $X_{n,\nu}(\tau)$ (cf. Claim 2 of Theorem 7.6.23). Following the same notations as in Claim 1, we have

$$h(nb_i)P\left(B_{i,n} \setminus \bigcup_{j=i+1}^{N} B_{j,n}\right) \leq h(nb_i)d_{in} \to 0 \quad \text{as } n \to \infty \tag{7.6.35}$$

and

$$h(na_i)P\left[A_{i,n} \setminus \left[\bigcup_{j=i+1}^{N} A_{j,n} \cup \bigcup_{j=1}^{N} B_{j,n}\right]\right] \leq h(na_i)c_{in} \to 0 \tag{7.6.36}$$

as $n \to \infty$. Hence, by (7.6.34)–(7.6.36), for n large enough,

$$
\begin{aligned}
&Eh(\|X_{N,\nu} - Y_{N,\nu}\|) \\
&= \sum_{i=1}^{N} \max(\nu_n, h(nb_i)) P\left(B_{i,n} \setminus \bigcup_{j=i+1}^{N} B_{j,n}\right) \\
&\quad + \sum_{i=1}^{N} \max(\nu_n, h(na_i)) P\left(A_{i,n} \setminus \left[\bigcup_{j=i+1}^{N} A_{j,n} \cup \bigcup_{j=1}^{N} B_{j,n}\right]\right) \\
&= \sum_{j=1}^{N} \nu_n P\left(B_{i,n} \setminus \bigcup_{j=i+1}^{N} B_{j,n}\right) \\
&\quad + \sum_{i=1}^{N} \nu_n P\left(A_{i,n} \setminus \left[\bigcup_{j=i+1}^{N} A_{j,n} \cup \bigcup_{j=1}^{N} B_{j,n}\right]\right) \\
&= \nu_n P\left(\bigcup_{j=1}^{N} (A_{j,n} \cup B_{j,n})\right).
\end{aligned}
$$

□

Theorem 7.6.26 *For any $g_i \in \mathcal{M}$ $(i = 1, \ldots, N)$,*

(i) $$I\{0, \overline{g}, \overline{a}, \overline{b}\} := \inf\{P(X \neq Y);\ X \in \mathcal{X}(T, \overline{g}, \overline{a}), y \in \mathcal{X}(T, \overline{g}, \overline{b})\} = 0; \tag{7.6.37}$$

(ii) *for any $\nu \in (0,1)$ there exists a sequence $(X_{n\nu}, Y_{n\nu}) \in \mathcal{X}(T, \overline{g}, \overline{a}) \times \mathcal{X}(T, \overline{g}, \overline{b})$ such that*

$$P(X_{n\nu} \neq Y_{n\nu}) \to \nu \quad \text{as } n \to \infty.$$

Proof: (i) Let $c_n \in \mathcal{A}$ and $P(c_n) = \frac{1}{n}$. For any $i = 1, \ldots, N$ define

$$\begin{cases} X_n(t_i, \omega) := g_i^{-1}(na_i), \quad Y_n(t_i, \omega) := g_i^{-1}(nb_i), & \text{if } \omega \in C_n, \\ X_n(t_i, \omega) := Y_n(t_i, \omega) = 0, & \text{if } \omega \notin C_n. \end{cases} \tag{7.6.38}$$

Then (7.6.38) determines the random polygonal lines $X_n \in \mathcal{X}(T, \overline{g}, \overline{a})$ and $Y_n \in \mathcal{X}(T, \overline{g}, \overline{b})$. Since $X_n(t_i, \omega) = Y_n(t_i, \omega) = 0$ whenever $\omega \notin C_n$ and $i = 1, \ldots, N$, then $X_n(t, \omega) = Y_n(t, \omega) = 0$ if $\omega \notin C_n$ and $t \in [0, 1]$. Hence,

$$P(X_n = Y_n) \geq P(\Omega \setminus C_n) = \frac{n-1}{n} \to 1$$

as desired.

(ii) Let $0 < \nu < 1$ and $\tau \in (0,1) \setminus T$. Choose $A \in \mathcal{A}$ with $P(A) = \nu$ and let

$$\begin{cases} X_{n\nu}(\tau,\omega) = 1, \quad Y_{n\nu}(\tau,\omega) = 0, & \text{for any } \omega \in A, \\ X_{n\nu}(\tau,\omega) = Y_{n\nu}(\tau,\omega) = 0, & \text{for any } \omega \notin A, \\ X_{n\nu}(t_i) = X_n(t_i), \quad Y_{n\nu}(t_i) = Y_n(t_i), & i = 1,\dots,N, \end{cases} \tag{7.6.39}$$

where $X_n(t_i)$ and $Y_n(t_i)$ are given by (7.6.38). We construct the random broken lines $X_{n\nu}$ and $X_{n\nu}$ by using (7.6.39) (cf. Claim 3 of Theorem 7.6.23). From the implications

$$X_{n\nu}(\cdot,\omega) = Y_{n\nu}(\cdot,\omega) \Leftrightarrow \begin{cases} X_{n\nu}(t_i,\omega) = Y_{n\nu}(t_i,\omega), & i = 1,\dots,N, \\ X_{n\nu}(\tau,\omega) = Y_{n\nu}(\tau,\omega) \end{cases}$$

$$\Leftrightarrow \quad \omega \notin A \cap C_n,$$

it follows that

$$P(X_{n\nu} = Y_{n\nu}) = P(\Omega \setminus (A \cup C_n)) \to 1 - \nu,$$

which proves (ii), and the theorem as well. □

As a consequence of Theorems 7.6.23–7.6.26 we obtain the following solution of Moment problem 7.6 for $q_i(t) = t^{q_i}$ $(y_i > 0)$ and $h(t) = t^p$ $(P \geq 0$, with the convention $t^0 := I\{t \neq 0\})$.

Corollary 7.6.27 *Let* $\overline{q} = (q_1, \dots, q_N)$ $(q_i > 0), p \geq 0$, *and*

$$I\{p,\overline{q},\overline{a},\overline{b}\} := \inf\{E\|X - Y\|^p;\ X, Y \in \mathcal{X}(C[0,1]),\ E|X(t_i)|^{q_i} = a_i,\ E|Y(t_i)|^{q_i} = b_i,\ i = 1,\dots,N\}.$$

Then

$$I\{p,\overline{q},\overline{a},\overline{b}\} = \begin{cases} \sup\limits_{1\leq i\leq N} \left|a_i^{1/q_i} - b_i^{1/q_i}\right|^p, & \text{if } p \geq q_i \geq 1,\ i = 1,\dots,N, \\ \sup\limits_{1\leq i\leq N} |a_i - b_i|^{p/q_i}, & \text{if } p \geq q_i,\ 0 < q_i < 1,\ i = 1,\dots,N. \\ 0, & \text{if } 0 \leq p \leq q_i,\ i = 1,\dots,N. \end{cases} \tag{7.6.40}$$

Moreover, for any $\nu > I\{p, \overline{q}, \overline{a}, \overline{b}\}$ *there exists a sequence* $(X_{n\nu}, Y_{n\nu}) \in \mathcal{X}([0,1])$ *such that*

$$E||X_{n\nu} - Y_{n\nu}||^p \quad \rightarrow \quad \nu \quad \text{as } n \to \infty$$

and

$$E|X(t_i)|^{q_i} \; = \; a_i, \quad E|Y(t_i)|^{q_i} = b_i, \quad i = 1, \ldots, N.$$

Remark 7.6.28 *Corollary* 7.6.27 *gives an explicit expression for* $I\{p, \overline{q}, \overline{a}, \overline{b}\}$ *if* p *and* q_i *are subject to certain inequalities (cf.* (7.6.40)*) or if* $q_i = q$ *for all* $i = 1, \ldots, N$*. The problem of an explicit description of* $I\{p, \overline{q}, \overline{a}, \overline{b}\}$ *for any* $p \geq 0$ *and* $q_i > 0$ *is still open.*

7.6.4 Approximation of Queueing Systems with Prescribed Moments

In this section we discuss applications of Moment problem 1 (on page 52) to the problem of best approximation of a queueing system with known moment characteristics. As an example, suppose our "real" queueing system is of type $G|G|1|\infty$ (for some acquaintance with the usual notations in queueing theory we refer to Borovkov (1984), Kalashnikov and Rachev (1990)). For this system, the sequences of nonnegative r.v.s (possibly dependent and nonidentically distributed) $\overline{e} = \{e_n\}_{n \in \mathbb{N}}, \overline{s} = \{s_n\}_{n \in \mathbb{N}}$ ($\mathbb{N} = (1, 2, \ldots)$) are viewed as sequences of interarrival and service times. Looking at $\overline{e}$ and $\overline{s}$ as "input" of laws, we define (as the "output" flow) the sequence of waiting times

$$w_1 \;=\; 0, \quad w_{n+1} \;=\; (w_n + s_n - e_n)_+, \quad n \in \mathbb{N}, \tag{7.6.41}$$

where $(\cdot)_+ = \max(0, \cdot)$. Since the distribution of $\overline{w} = \{w_n\}_{n \in \mathbb{N}}$ is not known, the aim is to approximate, model, or simulate the "real" system determined by the triplet $(\overline{e}, \overline{s}, \overline{w})$ with a "simpler" queueing model $(\overline{e}^*, \overline{s}^*, \overline{w}^*)$. Assuming that the marginal distributions (the laws of e_i, s_i) are known, Borovkov (1984, Chapter 4) and Kalashnikov and Rachev (1990) examine different approximating models $(\overline{e}^*, s^*, \overline{w}^*)$ and estimate the possible discrepancy between the "real" system $(\overline{e}, \overline{s}, \overline{w})$ and the "ideal" model $(\overline{e}^*, \overline{s}^*, \overline{w}^*)$. Further, we shall relax the constraints "the laws of e_i's and s_i's are known" by "certain moment characteristics of e_i's and s_i's are fixed." In this setup the solutions of Moment problem 1 are used in cases when the "ideal" model is not deterministic, say $G|G|1|\infty$ but with simpler structure. We invoke Moment problem 2 (on page 53) when the approximation model has some deterministic components, like $D|G|1|\infty$ (i.e., e_j^*'s are constants), or $D|D|1|\infty$ (i.e., e_j^*'s and s_j^*'s are constants).

Summarizing, we shall consider here the following two problems:

(a) *Bounds for the deviation of output characteristics of two dependent queueing models.*

(b) *Approximation of queueing systems by deterministic-type queueing models.*

Consider the following problem, which occurs in investigations stability of queueing models (see Kalashnikov and Rachev (1990, Chapter 5)). Suppose two queueing models of type $G|G|1|\infty$, $(\overline{e}, \overline{s}, \overline{w})$ and $(\overline{e}^*, \overline{s}^*, \overline{s}^*)$, with dependent characteristics are given. Here $\overline{e} = \{e_n\}_{n\in\mathbb{N}}$, $\overline{s} = \{s_n\}_{n\in\mathbb{N}}$, $\overline{w} = \{w_n\}_{n\in\mathbb{N}}$ are, respectively, the sequences of interarrival, service, and waiting times. Assume that the components $d_j, s_j, j \in \mathbb{N}$ of the "input flows" $\overline{e}$ and $\overline{s}$ are *dependent and nonidentically distributed.*

The "output" flow is given by the sequence of waiting times (7.6.41). Suppose that the distribution of e_j (resp. s_j) is concentrated on a compact interval $[a_j, b_j]$ (resp. $[c_j, d_j]$). While this assumption is quite natural from the practical point of view, it is not used frequently in the literature, simply because it is easier to analyze queueing models with input distributions having unbounded support. We make similar assumptions for the model $(\overline{e}^*, \overline{s}^*, \overline{w}^*)$; in particular, it is assumed that $a_j^* \le e_j^* \le b_j^*$, $c_j^* \le s_j^* \le d_j^*$ a.s. for all $j \in \mathbb{N}$. The input pairs (e_j, e_j^*), (s_j, s_j^*) of the two models are *arbitrarily mutually dependent*, the distributions of e_j's, e_j^*'s, s_j's, s_j^*'s are unknown. We assume that only the moments

$$Ee_j = \alpha_j, \quad Ee_j^* = \alpha_j^*, \quad Es_j = \beta_j, \quad Es_j^* = \beta_j^* \tag{7.6.42}$$

are given. Our problem to find a sharp bound for the deviation between the waiting times in both models.

Let $\varphi_k(\overline{e}_{n-k,n-1}, \overline{s}_{n-k,n-1})$ $(\overline{e}_{n-k,n-1} := (e_{n-k}, \ldots, e_{n-1}), \overline{s}_{n-k,n-1} := (s_{n-k}, \ldots, s_{n-1}))$ be the waiting time for the nth arrival, assuming that the system is "free" at the moment $n - k$. In other words,

$$\begin{aligned} &\varphi(\overline{e}_{n-k,n-1}, \overline{s}_{n-k,n-1}) \\ &:= \max\left[0, e_{n-1} - s_{n-1}, (e_{n-1} - s_{n-1}) + (e_{n-2} - s_{n-2}), \ldots, \right. \\ &\qquad \left. (e_{n-1} - s_{n-1}) + (e_{n-2} - s_{n-2}) + \cdots + (e_{n-k} - s_{n-k})\right]. \end{aligned} \tag{7.6.43}$$

As a measure of deviation between the waiting times of the both systems we shall use

$$\delta_p(T) = \sup_{n\in\mathbb{N}} \max_{1\le k\le T} L_p[\varphi_k(\overline{e}_{n,n+k-1}, \overline{s}_{n,n+k-1}), \varphi_k(\overline{e}^*_{n,n+k-1}, \overline{s}^*_{n,n+k-1})],$$

where $p \ge 1$ and $T \ge 2$ are fixed, and

$$L_p(X, Y) := \{E|X - Y|^p\}^{1/p}, \ p \ge 1, \ X, Y \in \mathcal{X}(R). \tag{7.6.44}$$

For random vectors we extend (7.6.44) as follows:

$$L_p(\overline{X}, \overline{Y}) = \{E||\overline{X} - \overline{Y}||^p\}^{1/p}, \ \overline{X}, \overline{Y} \in \mathcal{X}(R^T), \tag{7.6.45}$$

where $||(x_1,\ldots,x_T)|| = \sum_{i=1}^T |x_j|$. Since φ_k is a Lipschitz function with respect to the Minkowski norm $||\cdot||$, we have that for any $k=1,\ldots,T$,

$$\begin{aligned} & L_p[\varphi_k(\overline{e}_{n,n+k-1},\overline{s}_{n,n+k-1}),\varphi_k(\overline{e}^*_{n,n+k-1},\overline{s}^*_{n,n+k-1})] \\ & \quad \le \; L_p[(\overline{e}_{n,n+T-1},\overline{s}_{n,n+T-1}),(\overline{e}^*_{n,n+T-1},\overline{s}_{n,n+T-1})] \\ & \qquad + L_p[(\overline{e}^*_{n,n+T-1},\overline{s}_{n,n+T-1}),(\overline{e}^*_{n,n+T-1},\overline{s}^*_{n,n+T-1})] \\ & \quad \le \; \sum_{j=n}^{n+T-1} \left[L_p(e_j,e_j^*)+L_p(s_j,s_j^*)\right]. \end{aligned} \tag{7.6.46}$$

Now we invoke Theorem 7.6.5 to obtain sharp estimates of $L_p(e_j,e_j^*)$ and $L_p(s_j,s_j^*)$. Namely,

$$L_p(e_j,e_j^*) \;\le\; (D_j\delta_j+s_j)^{1/p}, \tag{7.6.47}$$

where

$$\begin{aligned} D_j \;&:=\; D_j(a_j,b_j,a_j^*,b_j*) \\ &:=\; |b_j-b_j^*|^p+|a_j-a_j^*|^p-|b_j-a_j^*|^p-|a_j-b_j^*|^p; \end{aligned} \tag{7.6.48}$$

$$\begin{aligned} T_j \;&:=\; T_j(a_j,a_j^*,b_j,b_j^*,\alpha_j,\alpha_j^*) \\ &:=\; (1-B_j)|b_j-a_j^*|^p+(B_j+C_j-1)|a_j-a_j^*|^p+(1-C_j)|a_j-b_j^*|^p; \end{aligned} \tag{7.6.49}$$

$$B_j \;:=\; \frac{b_j-\alpha_j}{b_j-a_j}, \qquad C_j \;:=\; \frac{b_j^*-\alpha_j^*}{b_j^*-a_j^*}; \tag{7.6.50}$$

$$\delta_j \;:=\; \delta_j(a_j,a_j^*,b_j,b_j^*,\alpha_j,\alpha_j^*) \;:=\; \max(0,1-B_j-C_j). \tag{7.6.51}$$

Remark 7.6.29 *If e_j and e_j^* are unknown, i.e., $a_j=a_j^*=0$, $b_j=b_j^*=+\infty$, then*

$$\sup\{L_p(d_j,e_j^*);\; Ee_j=\alpha_j, Ee_j^*=\alpha_j^*\} \;=\infty$$

(cf. Kuznezova-Sholpo and Rachev (1989)).

In a similar way,

$$L_p(s_j,s_j^*) \;\le\; (\widetilde{D}_j\widetilde{\delta}_j+\widetilde{T}_j)^{1/p}, \tag{7.6.52}$$

where $\widetilde{D}_j,\widetilde{\delta}_j,\widetilde{T}_j$ are defined by (7.6.48)–(7.6.51), exchanging b_j with d_j, b_j^* with d_j^*, a_j with c_j, and a_j^* with c_j^*. In this way we have proved the following theorem.

Theorem 7.6.30 *For any $p\ge 1$ and $T=2,3,\ldots,$*

$$\delta_p(T) \;\le\; T\sup_{j\ge 1}\left[(D_j\delta_j+T_j)^{1/p}+(\widetilde{D}_j\widetilde{\delta}_j+\widetilde{T}_j)^{1/p}\right]. \tag{7.6.53}$$

The estimate is sharp or nearly sharp, since the inequalities (7.6.53) and (7.6.52) are the best possible bounds under the moment assumptions (7.6.42) (cf. Theorem 7.6.5), and also, the inequality (7.6.46) cannot be improved in the set of all possible input flows $\overline{e}, \overline{e}^*, \overline{s}, \overline{s}^*$.

Next, we shall consider a much more general case than the single-channel models discussed above. Suppose the dynamics of a queueing system are determined by the transformation $\mathcal{F}$ from the set $\mathcal{U}$ of input flows U to the set $\mathcal{V}$ of output flows V. Let V_0 represent the output at moment zero; V_0 is assumed to be an ℓ-dimensional vector; i.e., $V_0 \in \mathfrak{X}(R^\ell)$. It is quite general to assume that the input and the output flows have the form $U = (V_0, U_0, U_1, \ldots)$ and $V = (V_0, V_1, \ldots)$, where $U_j \in \mathfrak{X}(R^k)$. We endow $\mathcal{U}$ and $\mathcal{V}$ with the norms

$$||U||_{\mathcal{U}} := \sum_{j=0}^{\infty} 2^{-j} ||U_j||_{k,p} + ||V_0||_{\ell,p} \tag{7.6.54}$$

and

$$||V||_{\mathcal{V}} := \sum_{j=0}^{\infty} 2^{-j} ||V_j||_{\ell,p}, \tag{7.6.55}$$

where $p \geq 1$,

$$\begin{aligned} ||U_j||_{k,p} &:= (E||U_j||_k^p)^{1/p}, \\ ||U_j||_k &= ||(U_j^{(1)}, \ldots, U_j^{(k)})|| = |U_j^{(1)}| + \cdots + |U_j^{(k)}|, \end{aligned}$$

and $||V_j||_{\ell,p}$ is defined in a similar way. Suppose the transformation $\mathcal{F} : \mathcal{U} \to \mathcal{V}$ is determined by the set of mappings

$$\mathcal{F}_j : R^\ell \times R^{kj} \to R^\ell, \quad j \in \mathbb{N}, \tag{7.6.56}$$

such that the output at "time" j is defined recursively:

$$V_j = \mathcal{F}_j(V_0, U_0, \ldots, U_{j-1}). \tag{7.6.57}$$

A smoothness assumption on $\mathcal{F}_j$ is given by the Lipschitz condition

$$||F_j(\alpha_0, \beta_0, \ldots, \beta_{j-1}||_\ell \leq c_j \left[||\alpha_0||_\ell + \sum_{j=0}^{j-1} ||\beta_j||_k \right]. \tag{7.6.58}$$

A reasonably large number of queueing models meet conditions (7.6.56)–(7.6.58). Among them are the single-channel models $G|G|1|\infty$, the multichannel models $G|G|J|\infty$, and the multichannel–multiphased model $(G|G|J_1) \to (G|J_2) \to \cdots \to (G|J_n)$ (cf. Kalashnikov and Rachev (1990,

Chapter 5)). By (7.6.55), (7.6.57), and (7.6.58), $||V_j||_{\ell,p} \leq c_j \left[||V_0||_{\ell,p} + \sum_{i=0}^{j-1} ||U_i||_{k,p}\right]$, and thus

$$\begin{aligned} ||V||_{\mathcal{V}} &\leq 2c_j ||U||_{\mathcal{U}} \qquad (7.6.59) \\ &\leq 2c_j \left[\sum_{i=1}^{\ell} ||V_0^{(i)}||_{\ell,p} + \sum_{i=1}^{k} \sum_{j=0}^{\infty} 2^{-1} ||V_j^{(i)}||_{\ell,p} \right]. \end{aligned}$$

Combining (7.6.59) with Theorem 7.6.5 gives us a sharp bound on the deviation of two queueing models $V = \mathcal{F}U, V^* = \mathcal{F}U^*$, whose dynamics are determined by (7.6.54)–(7.6.58).

Theorem 7.6.31 *Suppose*

$$V = \mathcal{F}U, \quad U \in \mathcal{U}, \quad V \in \mathcal{U} \qquad (7.6.60)$$

is a queueing model satisfying (7.6.54)–(7.6.58) *such that*

$$\begin{aligned} a_0^{(i)} &\leq V_0^{(i)} \leq b_0^{(i)} \text{ a.s.,} \quad EV_0^{(i)} = L_0^{(i)}, \quad i = 1, \ldots, \ell, \\ c_j^{(i)} &\leq V_j^{(i)} \leq d_j^{(i)} \text{ a.s.,} \quad EV_j^{(i)} = \beta_j^{(i)}, \quad j = 0, 1, \ldots, \quad i = 1, \ldots, k. \end{aligned}$$

In addition to model (7.6.60) *consider the same type model indexed by* $*$ *and satisfying the above two sets of inequalities with constants indexed by* $*$*. Then*

$$\begin{aligned} &||V - V^*||_{\mathcal{V}} \\ &\leq 2c_j \left[\sum_{i=1}^{\ell} \left(D_0^{(i)} \delta_0^{(i)} + T_0^{(i)} \right)^{1/p} + \sum_{i=1}^{k} \sum_{j=0}^{\infty} 2^{-j} \left(\widetilde{D}_j^{(i)} \widetilde{\delta}_j^{(i)} + \widetilde{T}_j^{(i)} \right)^{1/p} \right], \end{aligned}$$

where the D's, δ's, and T's are determined by the same formula as in (7.6.58)–(7.6.60)*, and*

$$\begin{aligned} D_0^{(i)} &= D_i \left(a_0^{(i)}, b_0^{(i)}, a_0^{(i)*}, b_0^{(i)*} \right), \\ T_0^{(i)} &= T_i \left(a_0^{(i)}, b_0^{(i)}, a_0^{(i)*}, b_0^{(i)*}, \alpha_0^{(i)}, \alpha_0^{(i)*} \right), \\ \delta_0^{(i)} &= \delta_i \left(a_0^{(i)}, b_0^{(i)}, a_0^{(i)*}, b_0^{(i)*}, \alpha_0^{(i)}, \alpha_0^{(i)*} \right), \\ \widetilde{D}_j^{(i)} &= D_j \left(c_j^{(i)}, d_j^{(i)}, c_j^{(i)*}, d_j^{(i)*} \right), \\ \widetilde{T}_j^{(i)} &= T_j \left(c_j^{(i)}, d_j^{(i)}, c_j^{(i)*}, d_j^{(i)*}, \beta_j^{(i)}, \beta_j^{(i)*} \right), \\ \widetilde{\delta}_j^{(i)} &= \delta_j \left(c_j^{(i)}, d_j^{(i)}, c_j^{(i)*}, d_j^{(i)*}, \beta_j^{(i)}, \beta_j^{(i)*} \right). \end{aligned}$$

The rest of this section deals with Problem (b) (on page 72). Suppose again that the "real" queueing system is determined by the triplet $(\overline{e}, \overline{s}, \overline{w})$, where $\overline{w}$ is given by the recursive equation (7.6.41). Often in practice one models the random input characteristics by replacing their random values with constants, usually equal to the corresponding means. In doing so, it is natural to investigate the deviation between the "real" output $\overline{w}$ and the modeled ("ideal") output $\overline{w}^*$. (In the sequel, all quantities related to the approximating model will have the same notations as in the "real" system but superscribed with $*$.)

The deviation between $\overline{w}$ and $\overline{w}^*$ will be expressed by the *Kantorovich metric* ℓ_p, defined here as follows: For $X, Y \in \mathcal{X}(\mathbb{R}^\infty)$,

$$\begin{aligned} \ell_p(X,Y) &:= \ell_p(P^X, P^Y) && (7.6.61)\\ &:= \min\{L_p(\widetilde{X},\widetilde{Y});\ \widetilde{X},\widetilde{Y} \in \mathcal{X}(\mathbb{R}^\infty), \widetilde{X} \stackrel{d}{=} X, \widetilde{Y} \stackrel{d}{=} Y\}, \quad p > 0, \end{aligned}$$

where $L_p(\widetilde{X},\widetilde{Y}) := \left[E\, d^p(\widetilde{X},\widetilde{Y})\right]^q$, $q = \min(1, 1/p)$ is the L_p-metric. In the above definition, the space $\mathcal{X}(\mathbb{R}^\infty)$ consists of all random sequences taking values in the metric space $(\mathbb{R}^\infty, d)$, where $d(\overline{x},\overline{y}) := \sum_{j=1}^\infty 2^{-j}\|x_j - y_j\|$. Since we have assumed that the underlying probability space is not atomic, the minimum in the right-hand side of (7.6.61) is equal to

$$\min\left\{ \int_{\mathbb{R}^\infty\times\mathbb{R}^\infty} \mathrm{d}(x,y)P(\,\mathrm{d}x,\,\mathrm{d}y); P\text{s are probabilities on } \mathbb{R}^\infty \times \mathbb{R}^\infty \right.$$
$$\left. \text{with fixed projections } P^X \text{ and } P^Y \right\}.$$

For $\overline{X}^{(n)} = \left(X_1^{(n)}, X_2^{(n)}, \ldots\right) \in \mathcal{X}_p(\mathbb{R}^\infty)$, $\overline{X} = (X_1, X_2, \ldots) \in \mathcal{X}_p(\mathbb{R}^\infty)$, we have $\ell_p\left(\overline{X}^{(n)}, \overline{X}\right) \geq 2^{-j}\ell_p\left(X_j^{(n)}, X_j\right)$, and thus $\ell_p(\overline{X}^{(n)}, \overline{X}) \to 0$ implies the weak convergence of any j-component $X_j \stackrel{d}{=} X$ and $E|X_j^{(n)}|^p \to E|X_j|^p$.

Further, we consider two types of approximating queues $D|G|1|\infty$ (i.e., e_j^* are constants) and $D|D|1|\infty$ (i.e., e_j^* and s_j^* are constants). Similar results can be obtained if one examines the model $G|D|1|\infty$ (i.e., s_j^* are constants) as an approximation of the "real" queue $G|G|1|\infty$.

In both queues $D|G|1|\infty$ and $G|G|1|\infty$, the sequences of service times $\overline{s}^*$ and $\overline{s}$ consist of dependent nonidentically distributed random variables. The next lemma shows that the outputs for the ideal and real models meet a lower bound of deviation if $\overline{s}^*$ is chosen to have independent components.

Let $\varepsilon > 0$ and $X = (X_1, X_2, \ldots) \in \mathcal{X}_p(\mathbb{R}^\infty)$. The components of X are said to be (ℓ_p, ε)-*independent* if

$$\mathrm{IND}(X) := \ell_p(X, \underline{X}) \leq \varepsilon, \tag{7.6.62}$$

where the $\underline{X}_i$'s (the components if $\underline{X}$) are independent and $\underline{X}_i \stackrel{d}{=} X_i$ ($i \in \mathbb{N}$).

Lemma 7.6.32 *Let the approximating model be of type $D|G|1|\infty$. Assume that the sequences $\overline{e}$ and $\overline{s}$ of the queueing model $G|G|1|\infty$ are independent. Then*

$$\begin{aligned} & \frac{1}{2}\ell_p(\overline{w},\overline{w}^*) \\ \leq\ & \mathrm{IND}(\overline{s}) + \mathrm{IND}(\overline{s}^*) + \sum_{j=1}^{\infty} 2^{-j}(\ell_p(e_j,e_j^*) + \ell_p(s_j,s_j^*)). \end{aligned} \tag{7.6.63}$$

Proof: Using the recursive equations (7.6.41) for $\overline{w}$ and $\overline{w}^*$, we obtain $\frac{1}{2}d(\overline{w},\overline{w}^*) \leq d(\overline{e},\overline{e}^*) + d(\overline{s},\overline{s}^*)$. Hence, $\frac{1}{2}L_p(\overline{w},\overline{w}^*) \leq L_p(\overline{e},\overline{e}^*) + L_p(\overline{s},\overline{s}^*)$. Since $\overline{e}$ and $\overline{s}$ (resp. $\overline{e}^*$ and $\overline{s}^*$) are independent, we have, passing to the minimal metrics, that

$$\frac{1}{2}\ell_p(\overline{w},\overline{w}^*) \ \leq\ \ell_p(\overline{e},\overline{e}^*) + \ell_p(\overline{s},\overline{s}^*). \tag{7.6.64}$$

By (7.6.61) and since e_j ($j \in \mathbb{N}$) are constants, we obtain the bound

$$\ell_p(\overline{e},\overline{e}^*) \ =\ \sum_{j=1}^{\infty} 2^{-j} L_p(e_j,e_j^*) \ =\ \sum_{j=1}^{\infty} 2^{-j}\ell_p(e_j,e_j^*). \tag{7.6.65}$$

To estimate $\ell_p(\overline{s},\overline{s}^*)$ in (7.6.64) we use the (ℓ_p,ε)-independence characteristic defined in (7.6.62):

$$\ell_p(\overline{s},\overline{s}^*) \ \leq\ \mathrm{IND}(\overline{s}) + \mathrm{IND}(\overline{s}^*) + \ell_p(\underline{\overline{s}},\underline{\overline{s}}^*), \tag{7.6.66}$$

where $\underline{\overline{s}}$ (resp. $\underline{\overline{s}}^*$) has independent components and $\underline{s}_j \stackrel{d}{=} s_j$ (resp. $\underline{s}_j^* \stackrel{d}{=} s_j^*$). We now invoke the "regularity" property of the Kantorovich metric:

$$\ell_p\left(\sum_{n=1}^{\infty} X^{(n)}, \sum_{n=1}^{\infty} Y^{(n)}\right) \ \leq\ \sum_{n=1}^{\infty} \ell_p\left(X^{(n)}, Y^{(n)}\right) \tag{7.6.67}$$

for sequences $\{X^{(n)}\}_{n\geq 1} \subset \mathcal{X}_p(\mathbb{R}^\infty)$, $\{Y^{(n)}\}_{n\geq 1} \subset \mathcal{X}_p(\mathbb{R}^\infty)$ of independent components. Let E^j be a sequence with components all equal to zero except for the jth component, which equals 1. Then by (7.6.67),

$$\begin{aligned} \ell_p(\underline{\overline{s}},\underline{\overline{s}}^*) \ &=\ \ell_p\left(\sum_{j=1}^{\infty} \underline{\overline{s}}E^j, \sum_{j=1}^{\infty} \underline{\overline{s}}^* E^j\right) \\ &\leq\ \sum_{j=1}^{\infty} \ell_p\left(\underline{\overline{s}}\overline{E}^j, \underline{\overline{s}}^*\overline{E}^j\right) \ =\ \sum_{j=1}^{\infty} 2^{-j}\ell_p(s_j,s_j^*). \end{aligned} \tag{7.6.68}$$

Combining (7.6.64), (7.6.65), (7.6.66), and (7.6.68) proves the lemma. □

The estimate (7.6.63) suggests that the approximating model should be chosen with $\overline{s}^*$ having independent components. If this is the case, then $\mathrm{IND}(\overline{s}^*) = 0$, and the first problem is to estimate $\mathrm{IND}(\overline{s})$.

Lemma 7.6.33 (a) *Suppose that the only information known about the "real" service times are the moments*

$$ES_j^{q_1} = \beta_j, \quad Es_j^{q_2} = \gamma_j, \quad j \in \mathbb{N}, \tag{7.6.69}$$

and that the support of F_{s_j} is $[0,\infty)$. Then

$$\mathrm{IND}(\overline{s}) \le \sum_{j=1}^{\infty} 2^{-j}\Delta_j, \tag{7.6.70}$$

where

$$\Delta_j := \begin{cases} \left[2\beta_j^{1/q_1}\right]^{pq}, & \text{if } 0 < p \le q_1, \ 1 \le q_1 < q_2, \\ +\infty, & \text{if } 0 < q_1 < q_2 < p \text{ and } \beta_j^{1/q_1} = \gamma_j^{1/q_2}, \end{cases} \tag{7.6.71}$$

and $q = \min(1, 1/p)$.

(b) *Suppose the support F_{s_j} is the compact interval $[c_j, d_j]$, and $\beta_j = Es_j$. Then* (7.6.70) *holds with*

$$\Delta_j = \left(\widetilde{D}_j\widetilde{\delta}_j + \widetilde{T}_j\right)^{1/p}, \quad p \ge 1, \ \text{where } \widetilde{D}_j = -2(d_j - c_j)^p, \tag{7.6.72}$$

$$\widetilde{T}_j = 2\left(1 - \frac{d_j - \beta_j}{d_j - c_j}\right), \text{ and } \widetilde{\delta}_j = \max\left(0, 1 - 2\frac{d_j - \beta_j}{d_j - c_j}\right).$$

Proof: Assertion (a) follows from Corollary 2 of Kuznezova-Sholpo and Rachev (1989) and (b) from Theorem 7.6.5. □

Lemma 7.6.34 *Suppose that for every $j \in \mathbb{N}$ the first two moments of e_j are known:*

$$m_j := Ee_j, \quad m_j^{(2)} := Ee_j^2, \quad \sigma_j^2 := \mathrm{Var}\, e_j, \tag{7.6.73}$$

and let $a_j \le e_j \le b_j$ a.s.

(i) *If* $p \geq 2$ *and* $-\infty < a_j < b_j < \infty$ *and if* e_j^* *is chosen to be the midpoint of* $[a_j, b_j]$, *then*

$$\ell_p(e_j, e_j^*) \;\leq\; \left[\left[\frac{b_j - a_j}{2}\right]^{p-2} \left[m_j^{(2)} - m_j(a_j + b_j) + e_j^{*2}\right]\right]^{1/p}. \quad (7.6.74)$$

(ii) *Suppose* $0 < p \leq 2$ *and either* $-\infty = a_j, +\infty = b_j$, *or* $-\infty < a_j < b_j < \infty$ *and*

$$\sigma_j \;\leq\; \min[m_j - a_j, b_j - m_j]. \quad (7.6.75)$$

Then the "optimal" d_j^* *for the approximating model is given by* $e_j^* = m_j$, *and in this case*

$$\ell_p(e_j, e_j^*) \;\leq\; \sigma_j^{pq}, \quad q \;=\; \min(1, 1/p). \quad (7.6.76)$$

Proof: This follows from Theorems 7.6.10, 7.6.11, and 7.6.12 after some obvious arguments. The estimates (7.6.74) and (7.6.76) are sharp. □

Lemma 7.6.35 (a) *If* $0 < p \leq q_1, 1 \leq q_1 < q_2$, *then*

$$\sup\left\{\ell_p(s_j, s_j^*);\; n_j = Es_j^{q_1},\; n_j^{(2)} = Es_j^{q_2},\; n_j^* = Es_j^{*q_1},\; n_j^{*(2)} = Es_j^{*q_2}\right\}$$
$$= \left(n_j^{1/q_1} + n_j^{*1/q_1}\right)^{pq}, \quad q \;=\; \min(1, 1/p). \quad (7.6.77)$$

(b) *Suppose* $p \geq 1, c_j \leq s_j \leq d_j, c_j^* \leq s_j^* \leq d_j^*$ a.s., *and* $n_j = Es_j, n_j^* = Es_j^*$. *Then*

$$\ell_p(s_j, s_j^*) \;\leq\; \left(\widetilde{D}_j \widetilde{\delta}_j + \widetilde{T}_j\right)^{1/p}, \quad (7.6.78)$$

where $\widetilde{D}_j = D_j(c_j, d_j, c_j^*, d_j^*)$, $\widetilde{\delta}_j = \delta_j(c_j, d_j, c_j^*, d_j^*, n_j, n_j^*)$, $\widetilde{T}_j = T_j(c_j, d_j, c_j^*, d_j^*, n_j, n_j^*)$ *are given by* (7.6.48)–(7.6.51).

Proof: Assertion (a) follows from Corollary 2 of Kuznezova-Sholpo and Rachev (1989) and (b) from Theorem 7.6.5. □

Lemmas 7.6.32–7.6.35 lead us to the main result.

Theorem 7.6.36 *Let the approximating queueing model be of type* $D|G|1|\infty$. *Assume that the sequences* $\overline{e}$ *and* $\overline{s}$ *of the "real" queueing model are independent. Then the Kantorovich metric between the sequences of waiting* *times of the "approximating" and "real" models is bounded as follows:*

$$
\begin{aligned}
&\ell_p(\overline{w},\overline{w}^*) \qquad\qquad (7.6.79)\\
&\leq\; 2\,\mathrm{IND}(\overline{s}) + 2\,\mathrm{IND}(\overline{s}^*) + \sum_{j=1}^{\infty} 2^{-j+1}(\ell_p(e_j,e_j^*) + \ell_p(s_j,s_j^*)).
\end{aligned}
$$

Each term in the right-hand side of (7.6.79) *can be estimated as follows:*

(a) *An appropriate choice for the approximating sequence of service times will be* $\mathrm{IND}(s^*) = 0$.

(b) *If* (7.6.69) *holds, a bound for* $\mathrm{IND}(\overline{s})$ *is given by* (7.6.70).

(c) *If the means and variances of the* e_j*'s are known, then* $\ell_p(e_j,e_j^*)$ *can be estimated from above by* (7.6.74).

(d) *The last term in* (7.6.78), $\ell_p(s_j,s_j^*)$, *can be estimated by* (7.6.77) *(resp.* (7.6.78)*), provided that the corresponding moment conditions hold.*

In the next theorem we shall omit the restriction that $\overline{e}$ and $\overline{s}$ are independent, but we shall assume that the approximating model is of completely deterministic type $D|D|1|\infty$.

Theorem 7.6.37 *If the approximation queueing model is of type* $D|D|1|\infty$, *then*

$$
\ell_p(\overline{w},\overline{w}^*) \;\leq\; \sum_{j=1}^{\infty} 2^{-j+1}(\ell_p(e_j,e_j^*) + \ell_p(s_j,s_j^*)). \qquad (7.6.80)
$$

If the first moments of e_j *and* s_j *are fixed, then* $\ell_p(e_j,e_j^*)$ *and* $\ell_p(s_j,s_j^*)$ *can be estimated as in Lemma* 7.6.34.

The proof is similar to that of Theorem 7.6.36.

7.6.5 Rounding Random Numbers with Fixed Moments

In this part we shall discuss the interplay between Moment problems 3 and 4 (on pages 53, 54) and the problem of rounding of random proportions. Given a vector $X = (X_1,\ldots,X_n)$ of r.v.s consider the sum $X_1+\cdots+X_n$. If the X_i's are uniformly distributed on the simplex $\{(s_i)\geq 0;\;\; s_1+\cdots+s_n = 1\}$, then they can be treated as proportions, and clearly $S_n := X_1+\cdots+X_n = 1$. If S_n^* is the sum of conventional roundings $[X_1]_{1/2}+\cdots+[X_n]_{1/2}$, then Mosteller, Youtz, and Zahn (1967), Diaconis and Freedman (1979), and Balinski and Rachev (1993) have estimated the probability that $S_n =$

S_n^*. Here we shall examine the closeness between S_n and S_n^* in the case of i.i.d. observations X_i where *only* one or two moments are known.

Suppose $\{X_i\}_{i\in\mathbb{N}}$ are nonnegative i.i.d. r.v.s with known moments

$$EX_1 = d_1, \quad EX_1^r = d_r. \tag{7.6.81}$$

The c-rounding $[\cdot]_c (c \in [0,1])$ (see Section 7.6.2) gives us the sequence of i.i.d. roundings $\{[X_i]_c\}_{i\in\mathbb{N}}$. Let $V_i := [X_i]_c - X_i$ be the *rounding error*, and $S_{n,c} = \sum_{i=1}^n V_i$ is the *total rounding error*. Then the *normalized rounding error* $n^{-1}S_{n,c}$ converges by the LLN to $E[X_1]_c - d_1$. Our objective here is to find sharp bounds for the distribution function of $n^{-1}S_{n,c}$ subject to (7.6.81).

In other words, for a suitably chosen metric μ in the distribution functions space, the problem is to determine the "radius" of the set of probabilistic laws, i.e.,

$$\begin{aligned}\mathcal{D}_n = D_n(\mu) := \sup\{\mu\left(n^{-1}S_{n,c}, E[X]_c - d_1\right); \\ EX = d_1,\ EX^r = d_r\}.\end{aligned} \tag{7.6.82}$$

In (7.6.82), X has the same distribution as the X_i's, and thus $E[X]_c - d_1 = EV$, where $V \overset{d}{=} V_1$.

Clearly, there is a great variety of metrics μ from which one can choose in (7.6.82). We shall consider two metrics, one especially designed for the problem, one the *ideal metric* $\theta_s (s > 1)$ and the other the Lévy metric L.

Note that Theorems 7.6.18–7.6.21 provide us with sharp bounds for $E[X]_c - d_1$, in the case of the conventional rounding $c = \frac{1}{2}$.[(6)] In fact, with $[X] = [X]_{1/2}$,

$$\sup\{E[X] - d_1;\ EX = d_1, EX^r = d_r\} = U - d_1, \tag{7.6.83}$$

$$\inf\{E[X] - d_1;\ EX = d_1, EX^r = d_r\} = L - d_1, \tag{7.6.84}$$

where the exact values of U and L are given in Theorems 7.6.18–7.6.21. Next we can chose μ in the definition of $\mathcal{D}_n = \mathcal{D}_n(\mu)$ to be the Lévy metric

$$\begin{aligned}L(X,Y) = \inf\{\varepsilon > 0;\ F_X(x-\varepsilon) - \varepsilon \le F_Y(x) \le F_X(x+\varepsilon) + \varepsilon \\ \text{for all } x \in \mathbb{R}\},\end{aligned}$$

and thus for the distribution function F_n of $n^{-1}S_{n,c}$ we obtain the following bounds:

$$0 \le F_n(x) \le \mathcal{D}_n \quad \text{for } 0 \le x \le L - \mathcal{D}_n, \tag{7.6.85}$$

$$0 \le F_n(x) \le 1 \quad \text{for } L - \mathcal{D}_n \le x \le U - \mathcal{D}_n, \tag{7.6.86}$$

$$1 - \mathcal{D}_n \le F_n(x) \le 1 \quad \text{for } U + \mathcal{D}_n \le x. \tag{7.6.87}$$

---

(6) The general case of $c \le 1$ was treated in Anastassiou and Rachev (1992).

From Theorems 7.6.18–7.6.21 it follows that the above bounds are sharp.

Our next step is to find a good estimate for $\mathcal{D}_n = \mathcal{D}_n(L)$. To this end we first estimate $\mathcal{D}_n(\theta_s)$ for

$$\theta_s(X,Y) = \sup |Ef(X) - Ef(Y)|. \tag{7.6.88}$$

Here, the supremum is taken over all bounded functions f on $\mathbb{R}$ with q-integrable second derivative $\int |f''|^q \leq 1,\ 1 < s < 2,\ q = \frac{1}{2-s}$. The next lemma shows that the θ_s-radius $\mathcal{D}_n(\theta_s) = O(n^{1-s})$ for all $1 < s < 2$. We use the notation $\vee := \max$.

Lemma 7.6.38 *For* $1 < s < 2$,

$$\mathcal{D}_n(\theta_s) \leq \widetilde{c} n^{1-s}, \quad \widetilde{c} := \frac{2}{s}(c \vee (1-c))^s.$$

Proof: For any X and Y with equal means,

$$\begin{aligned}\theta_s(X,Y) &\leq \frac{1}{s}\kappa_s(X,Y) \\ &:= \inf\{E|\widetilde{X}|\widetilde{X}|^{s-1} - \widetilde{Y}|\widetilde{Y}|^{s-1};\ \widetilde{X} \stackrel{d}{=} X,\ \widetilde{Y} \stackrel{d}{=} Y\}.\end{aligned}$$

Therefore, from the ideality of θ_s [7]

$$\begin{aligned}\mathcal{D}_n(\theta_s) &\leq n^{1-s}\theta_s(V, EV) \leq n^{1-s}\frac{1}{s}\kappa_s(V, EV) \\ &= n^{1-s}\frac{1}{s}E\big|V|V|^{s-1} - (EV)|EV|^{s-1}\big| \\ &\leq n^{1-s}\frac{1}{s}.\end{aligned}$$

The latter follows since $|V| = |X - [x]_c| \in (0, c \vee (1-c))$. □

In the next theorem we bound $\mathcal{D}_n = \mathcal{D}_n(L)$ in (7.6.82) as $O\left[n^{-\frac{s-1}{1+s}}\right]$.

Theorem 7.6.39 *For any* $1 < s < 2,\ 0 < c < 1$,

$$\mathcal{D}_n(L) \leq (4\widetilde{c})^{\frac{1}{1+s}} n^{\frac{1-s}{1+s}},$$

where the constant $\widetilde{c}$ *is defined as in Lemma* 7.6.38.

[7] θ_s is an ideal metric of order $s > 0$; that is, $\theta_s(\sum_i c_i X_i, \sum_i c_i Y_i) \leq \sum_i |c_i|^s \theta_s(X_i, Y_i)$, for all independent X_i, Y_i and constants $c_i \in \mathbb{R}$.

Proof: The following claim was proved by Grigorevski and Shiganov (1976) for the case $s = 2$; i.e., in (7.6.88) the functions f have a.e. f'' and $|f''| \leq 1$ a.e.; see also Maejima and Rachev (1987) and Rachev and Rüschendorf (1992).

Claim: For any $1 < s < 2$,

$$\theta(X,Y) \;\geq\; \frac{1}{4}L^{1+s}(X,Y).$$

Proof of the Claim: Let $L(X,Y) > \varepsilon$. Then there exists x_0 such that either $F_X(x_0) > F_Y(x_0+\varepsilon)+\varepsilon$ or $F_Y(x_0) > F_Y(x_0+\varepsilon)+\varepsilon$. Say the first inequality takes place. Define

$$f_0(x) \;:=\; \begin{cases} 1 & \text{for } x \leq x_0; \\ 1 - \left(\dfrac{2(x-x_0)}{\varepsilon}\right)^2 & \text{for } x_0 < x \leq x_0 + \dfrac{\varepsilon}{2}; \\ -1 + \left(\dfrac{2(x_0+\varepsilon+x)}{\varepsilon}\right)^2 & \text{for } x_0 + \dfrac{\varepsilon}{2} \leq x < x_0 + \varepsilon; \\ 1 & \text{for } x \geq x_0 + \varepsilon. \end{cases}$$

Observe that $|f_0(x)| \leq 1, f''(x)$ exists a.e., and

$$||f_0''||_q \;=\; \left[\int_{x_0}^{x_0+\varepsilon} |f_0''(x)|^q \,\mathrm{d}x\right]^{1/q} \;=\; 8\varepsilon^{-s} \;=:\; c(\varepsilon)$$

for $q = \dfrac{1}{2-s}$. Recalling the definition of θ_s, we have

$$\begin{aligned}
\theta_s(X,Y) \;&\geq\; \left| E\left[\frac{f_0(X)}{c(\varepsilon)}\right] - E\left[\frac{f_0(Y)}{c(\varepsilon)}\right]\right| \\
&=\; \frac{1}{c(\varepsilon)}\left|\int (f_0(x)+1)\,\mathrm{d}\left[F_X(x) - F_Y(x)\right]\right| \\
&=\; \frac{1}{c(\varepsilon)}\left|\int_{-\infty}^{x_0} (f_0(x)+1)\,\mathrm{d}F_X(x) + \int_{x_0}^{\infty} (f_0(x)+1)\,\mathrm{d}F_X(x)\right. \\
&\qquad \left. - \int_{-\infty}^{x_0+\varepsilon} (f_0(x)+1)\,\mathrm{d}F_Y(x) - \int_{x_0+\varepsilon}^{\infty} (f_0(x)+1)\,\mathrm{d}F_Y(x)\right| \\
&\geq\; \frac{1}{c(\varepsilon)}\left[\int_{-\infty}^{x_0} (f_0(x)+1)\,\mathrm{d}F_X(x) - \int_{x_0+\varepsilon}^{\infty} (f_0(x)+1)\,\mathrm{d}F_Y(x)\right] \\
&\geq\; \frac{2}{c(\varepsilon)}\left[F_X(x_0) - F_Y(x_0+\varepsilon)\right]
\end{aligned}$$

$$
\begin{aligned}
&\geq \frac{2\varepsilon}{c(\varepsilon)} \\
&= \frac{1}{4}\varepsilon^{1+s}.
\end{aligned}
$$

Letting $\varepsilon \to L(X, Y)$ completes the proof of the claim.

Now the desired estimate follows from Lemma 7.6.38 and the claim. □

8

Application of Kantorovich-Type Metrics to Various Probabilistic-Type Limit Theorems

We have discussed already in detail the Kantorovich metric as the solution of mass transportation and mass transshipment problems with a metric cost function; cf. Section 2.5 and Chapter 4. In Chapter 7 we studied generalized transshipment problems, leading to extensions of the Kantorovich metric to encompass a variety of ideal probability metrics. This chapter is devoted to applications of these metrics to the rate of convergence problem in the central limit theorem (CLT) and different summability methods for random vectors. We also discuss applications to the asymptotics of various rounding rules.

8.1 Rate of Convergence in the CLT with Respect to the Kantorovich Metric

In this section, we investigate bounds for the rate of convergence in the CLT with respect to the Kantorovich metric for random variables with values in separable Banach spaces. In the first part, the rate in stable limit theorems for sums of i.i.d. random variables is considered. The method of proof is an extension of the Bergström convolution method. All assumptions regarding the domain of attraction are given in a metric form. In the second part an extension is given to the martingale case. The proof is based on smoothing properties of suitable conditonal versions of the Kantorovich metric. Smoothing inequalities for the Kantorovich metric will be established, and the Bergström convolution method (cf. Zolotarev (1977, 1979, 1983, 1986), Senatov (1980), Sazonov (1981), Rachev and Yukich (1989, 1991), Rachev (1991c)) will be extended to the case of stable limit theorems and at the same time to the Kantorovich metric. All assumptions concerning the domain of attraction and the order of convergence are described in terms of finiteness conditions for certain convolution-type metrics. As a consequence of the results for the Kantorovich metric, one obtains rate of convergence results in stable limit theorems for martingales with respect to the Prohorov metric.(1)

We start with the rate of convergence in the i.i.d. case. Consider a separable Banach space $(U, \|\cdot\|)$ and the space $\mathcal{X}(U)$ of U-valued r.v.s defined on a rich enough probability space. The r.v. $\vartheta \in \mathcal{X}(U)$ is said to be α-stable $(0 < \alpha \le 2)$ if

$$n^{-1/\alpha} \sum_{i=1}^{n} \vartheta_i \stackrel{d}{=} \vartheta, \tag{8.1.1}$$

where the ϑ_i's are i.i.d. copies of ϑ. We are interested in the rate of convergence of the normalized sum

$$Z_n = n^{-1/\alpha} \sum_{i=1}^{n} X_i \tag{8.1.2}$$

of i.i.d. r.v.s to ϑ with respect to the Kantorovich metric:

$$\ell_1(X,Y) := \sup\{|E(f(X) - f(Y))|;\ f : U \to \mathbb{R} \text{ bounded}, \quad |f(x) - f(y)| \le \|x - y\|\}. \tag{8.1.3}$$

Recall from Chapters 2 and 4 that ℓ_1-convergence is equivalent to convergence in distribution and convergence of the moments $E\|\cdot\|$ (existence assumed); moreover, for $U = \mathbb{R}$, $\ell_1(X,Y) = \int |F_X(x) - F_Y(x)|\, dx$. The Prohorov metric

$$\pi(X,Y) := \inf\{\varepsilon > 0;\ P(X \in A) \le P(Y \in A^\varepsilon) + \varepsilon, \quad \text{for all Borel sets } A \text{ in } U\} (A^\varepsilon := \{x;\ |x - A|\} < \varepsilon) \tag{8.1.4}$$

and the Kantorovich metric satisfy the well-known inequality

$$\pi^2 \le \ell_1, \tag{8.1.5}$$

(1) Some results in the literature are formulated more generally but use bounds involving moments of order ≥ 2 and, therefore, are restricted to the Gaussian case. For some recent literature we refer to Bolthausen (1982), Häussler (1988), Bentkus et al. (1990), and Rackauskas (1990). Our method involves various extensions of an idea in Gudynas (1985) on suitably conditioned versions of probability metrics. The results in this section are based on Rachev and Rüschendorf (1994a).

which is in fact an immediate consequence of the Strassen and Kantorovich theorems. In particular, ℓ_1-convergence rates imply convergence rates for π.

Theorem 8.1.1 *For any* $0 < \alpha < 1$,

$$\ell_1(Z_n, \vartheta) \;\leq\; n^{1-1/\alpha}\ell_1(X_1, \vartheta). \tag{8.1.6}$$

Proof: The result follows from (8.1.1)–(8.1.3) and the contraction properties of ℓ_1; in fact, if ϑ_i are i.i.d. copies of ϑ, then

$$E\left|Z_n - n^{-1/\alpha}\sum_{i=1}^{n}\vartheta_i\right| \;\leq\; n^{1-1/\alpha}E|X_1 - \vartheta_1|. \tag{8.1.7}$$

Next, we take in both sides of (8.1.7) the infimum over all joint distributions P^{X_1,ϑ_1} with fixed marginals P^{X_1} and P^{ϑ}. The result is (8.1.6) as desired. □

Note that (8.1.6) is a general property for every ideal metric of order 1; see for example Zolotarev (1979). Recall that a probability metric μ is said to be ideal of order r if

$$\mu(X+Z, Y+Z) \;\leq\; \mu(X,Y) \tag{8.1.8}$$

for all r.v.s X, Y, Z such that Z is independent of (X,Y) and

$$\mu(cX, cY) \;=\; |c|^r\mu(X,Y) \quad \text{for all } c \in \mathbb{R}, \tag{8.1.9}$$

see Sections 6.3 and 6.4.

Consider next the rate of convergence in

$$\overline{\ell}_1(Z_n, \vartheta) \;\to\; 0 \tag{8.1.10}$$

for $1 < \alpha \leq 2$. Define the following ideal (smoothing) Kantorovich metric of order $r > 1$:

$$\overline{\ell}_r(X,Y) \;=\; \sup_{h>0} h^{r-1}\ell_1(X+h\vartheta, Y+h\vartheta), \quad r > 1, \tag{8.1.11}$$

and

$$\overline{\sigma}_r(X,Y) \;=\; \sup_{h>0} h^{r}\sigma(X+h\vartheta, Y+h\vartheta), \quad r > 0. \tag{8.1.12}$$

Here ϑ in (8.1.11) and (8.1.12) is assumed to be independent of X and Y, and σ is the total variation metric:

$$\begin{aligned}\sigma(X,Y) &= \sup\{|E(f(X) - f(Y))|;\ f : U \to [0,1] \text{ continuous}\} \qquad (8.1.13)\\ &= 2\sup_{A\in\mathcal{B}(U)}|P(X\in A) - P(Y\in A)|.\end{aligned}$$

Note that $\overline{\ell}_r$ and $\overline{\sigma}_r$ are ideal metrics of order r. Throughout this section $\overline{\ell}_r$ stands for the smoothed ℓ_1-metric of order r. The notion ℓ_p has been used in previous sections for the minimal L_p-metric. So, we have increased the level of "ideality" for ℓ_1 and σ (recall that ℓ_1 is an ideal metric of order 1, while σ is ideal of order 0) by appropriate smoothing; see (8.1.11) and (8.1.12). The next theorem provides an estimate of the convergence rate in (8.1.10).

In what follows C stands for an absolute constant that can be different in different places. Set $\ell_1 = \ell_1(X_1, \vartheta), \overline{\ell}_r = \overline{\ell}_r(X_1, \vartheta), \sigma = \sigma(X_1, \vartheta), \overline{\sigma}_r = \overline{\sigma}_r(X_1, \vartheta)$. We always assume $r > 0$. The results in this section are due to Rachev and Rüschendorf (1994).

Theorem 8.1.2 *Suppose that*

(a) $E\|\vartheta\| < \infty$;

(b) $\ell_1 + \overline{\ell}_r + \sigma_1 + \overline{\sigma}_r < \infty$.

Then

$$\ell_1(Z_n, \vartheta) \ \leq \ C\left(n^{1-r/\alpha}\overline{\ell}_r + \tau_r n^{-1/\alpha}\right), \tag{8.1.14}$$

where

$$\tau_r \ = \ \max\left(\ell_1, \sigma_1, \overline{\sigma}_r^{1/(r-\alpha)}\right). \tag{8.1.15}$$

Remark 8.1.3 *Zolotarev (1986, §5.4) provides a similar bound for $\ell_1(Z_n, \vartheta)$ in the normal univariate case. Zolotarev's bound contains ζ_r-metrics in the right-hand side of* (8.1.14)*, which can be easily estimated from above in the normal case. In the stable case, however, we need more refined bounds. The problem of finiteness of $\overline{\sigma}_r$ was discussed in Rachev and Yukich (1989) (see also Section* 8.3*); for the finiteness of $\overline{\ell}_r$ see the next corollary. Further in this section the sum of any random variables $X+Y$ means $\widetilde{X}+\widetilde{Y}$, where $\widetilde{X}$ and $\widetilde{Y}$ are independent and $\widetilde{X} \stackrel{d}{=} X, \widetilde{Y} \stackrel{d}{=} Y$. ϑ, and ϑ_i are defined as in* (8.1.1) *and satisfy (a).*

Proof: The proof is similar to that of Theorem 8.1.16, further in this section, which we shall give in detail. Here we give only a short sketch of the proof. It uses the following two properties of the metrics $\ell_1, \overline{\ell}_r, \overline{\sigma}_r$; see Zolotarev (1986, §5.4).

Smoothing Property 1. For any $X, Y \in \mathcal{X}(U)$,

$$\ell_1(X, Y) \ \leq \ \ell_1(X + \varepsilon\vartheta, Y + \varepsilon\vartheta) + 2\varepsilon E\|\vartheta\|. \tag{8.1.16}$$

Smoothing Property 2. For any X, Y, Z, W independent,

$$\ell_1(X+Z, Y+Z) \leq \ell_1(Z, W)\sigma(X, Y) + \ell_1(X+W, Y+W). \quad (8.1.17)$$

Next, let $m = [\frac{n}{2}]$; then by (8.1.16),

$$\begin{aligned}
\ell_1(Z_n, \vartheta_1) &\leq \ell_1(Z_n + \varepsilon\vartheta, \vartheta_1 + \varepsilon\vartheta) + C\varepsilon \\
&\leq \ell_1\left(Z_n + \varepsilon\vartheta, \frac{\vartheta_1 + X_1 + \cdots + X_n}{n^{1/\alpha}} + \varepsilon\vartheta\right) \\
&\quad \cdot \sum_{j=1}^{m} \ell_1\left(\frac{\vartheta_1 + \cdots + \vartheta_j + X_{j+1} + \cdots + X_n}{n^{1/\alpha}} + \varepsilon\vartheta,\right. \\
&\qquad \left.\frac{\vartheta_1 + \cdots + \vartheta_{j+1} + X_{j+2} + \cdots + X_n}{n^{1/\alpha}} + \varepsilon\vartheta\right) \\
&\quad + \ell_1\left(\frac{\vartheta_1 + \cdots + \varepsilon_{m+1} + X_{m+2} + \cdots + X_n}{n^{1/\alpha}} + \varepsilon\vartheta, \vartheta_1 + \varepsilon\vartheta\right) \\
&= I_0 + \sum_{j=1}^{m} I_j + I_{m+1}.
\end{aligned}$$

By (8.1.17),

$$\begin{aligned}
I_0 \leq \ell_1 &\left(\frac{X_2 + \cdots + X_n}{n^{1/\alpha}}, \frac{\vartheta_2 + \cdots + \vartheta_n}{n^{1/\alpha}}\right) \sigma\left(n^{-1/\alpha}X_1 + \varepsilon\vartheta, n^{-1/\alpha}\vartheta_1 + \varepsilon\vartheta\right) \\
&+ \ell_1\left(\frac{X_1 + \vartheta_2 + \cdots + \vartheta_n}{n^{1/\alpha}} + \varepsilon\vartheta, \frac{\vartheta_1 + \cdots + \vartheta_n}{n^{1/\alpha}} + \varepsilon\vartheta\right).
\end{aligned}$$

Similar upper bounds are obtained for $I_j, 1 \leq j \leq m+1$. Some of the terms obtained in this way can be estimated using the ideality properties of the metrics. For example, a term of the following form, $\Delta = (m+1)\ell_1\left(\frac{X_1 + \vartheta_2 + \cdots + \vartheta_n}{n^{1/\alpha}}, \vartheta\right)$, can be estimated by

$$\begin{aligned}
\Delta &= (m+1)\ell_1\left(n^{-1/\alpha}X_1 + \left(\frac{n-1}{n}\right)^{1/\alpha}\vartheta, n^{-1/\alpha}\vartheta_1 + \left(\frac{n-1}{n}\right)^{1/\alpha}\vartheta\right) \\
&\leq (m+1)\left(\frac{n}{n-1}\right)^{\frac{r-1}{\alpha}} \overline{\ell}_r(n^{-1/\alpha}X_1, n^{-1/\alpha}\vartheta) \\
&\leq (m+1)\left(\frac{n}{n-1}\right)^{\frac{r-1}{\alpha}} n^{-r/\alpha}\overline{\ell}_r \leq Cn^{1-r/\alpha}\overline{\ell}_r,
\end{aligned}$$

where in the first inequality we use the obvious relation

$$\overline{\ell}_r(X, Y) \geq h^{r-1}\ell_1(X + h\vartheta, Y + h\vartheta).$$

For terms of the form $B_j := \ell_1\left(\frac{X_1+\cdots+X_j}{j^{1/\alpha}}, \frac{\vartheta_1+\cdots+\vartheta_j}{j^{1/\alpha}}\right)$ we use an induction argument to get the bound $B_j \le C(\overline{\ell}_r j^{1-r/\alpha} + \tau_r j^{-1/\alpha})$. For more details see the proof of Theorem 8.1.16. □

Corollary 8.1.4 *Suppose that $U = \mathbb{R}^K$ and that ϑ has a Fréchet differentiable density p_ϑ and let*

$$C(\vartheta) \quad = \quad \sup_{\|z\|\le 1} \int |p'_\vartheta(y)(z)|\,\mathrm{d}z \quad < \quad \infty. \tag{8.1.18}$$

Suppose that $E\|\vartheta\| < \infty$ and $\ell_1 + \overline{\ell}_r < \infty$. Then

$$\ell_1(Z_n, \vartheta) \quad \le \quad C(n^{1-r/\alpha}\overline{\ell}_r + \tau_r^* n^{-1/\alpha}), \tag{8.1.19}$$

where $\tau_1^* = \max(\ell_1, \overline{\ell}_r^{\,1/(r-\alpha)})$.

For an integer $r, \overline{\ell}_r$ can be estimated from above by the ζ_r-metric (see Zolotarev (1983, p. 294)): $\overline{\ell}_r \le C\zeta_r$ if $\sup_{\|z\|\le 1} \int |p_\vartheta^{(r)}(y)(z)|\,\mathrm{d}z$ is finite. We shall discuss the finiteness of $\overline{\ell}_r$ in Section 8.5 in more detail.

Proof: *Claim 1.* For any X and $Y \in \mathcal{X}(\mathbb{R}^k)$ and $\delta > 0$,

$$\sigma(X + \delta\vartheta, Y + \delta\vartheta) \quad \le \quad C(r)\delta^{-r}\overline{\ell}_r(X, Y), \tag{8.1.20}$$

with $C(r) = 2^{(2-3)/\alpha}C(\vartheta)$.

To prove the claim we first use the obvious bound

$$\sigma(X + \delta\vartheta, Y + \delta\vartheta) \quad \le \quad \delta^{-r}\overline{\sigma}_r(X, Y). \tag{8.1.21}$$

Next, we show that for any $\delta > 0$,

$$\sigma(X + \delta\vartheta, Y + \delta\vartheta) \quad \le \quad \delta^{-1}C(\vartheta)\ell_1(X, Y). \tag{8.1.22}$$

Indeed, by the ideality of σ and ℓ_1 it is enough to show (8.1.22) for $\delta = 1$. Then

$$\sigma(X + \vartheta, Y + \vartheta) \quad \le \quad \sup_{|f|\le 1} \left| \int \overline{f}(x)(P_X(\mathrm{d}x) - P_Y(\mathrm{d}x)) \right|,$$

where $\overline{f}(x) = \int f(x + y)p_\vartheta(y)\,\mathrm{d}y$. Since $|f| \le 1$,

$$\|\overline{f}'(x)\| \quad = \quad \sup_{\|z\|\le 1} |\overline{f}'(x)(z)| \quad \le \quad \sup_{\|z\|\le 1} \int |p'_\vartheta(y)(z)|\,\mathrm{d}y \quad =: \quad C(\vartheta),$$

and thus $|\overline{f}(x) - \overline{f}(y)| \leq C(\vartheta)\|x - y\|$, which obviously implies (8.1.22). To show (8.1.19) we use (8.1.21), (8.1.22), and the following bound:

$$\begin{aligned}\overline{\sigma}_r(X,Y) &= \sup_{h>0} h^r \sigma(X + h\vartheta, Y + h\vartheta) \\ &\leq \sup_{h>0} h^r \ell_1(X + 2^{-1/\alpha}h\vartheta, Y + 2^{-1/\alpha}h\vartheta)\frac{2^{1/\alpha}}{h}C(\vartheta) \\ &= C(\vartheta)2^{r/\alpha}\overline{\ell}_r(X,Y).\end{aligned}$$

This completes the proof of the claim as well as that of (8.1.19). □

Remark 8.1.5 (Rate of convergence in the CLT for random elements with LePage representation) *Consider a symmetric α-stable U-valued random variable ϑ with LePage representation*

$$\vartheta \stackrel{d}{=} \sum_{j=1}^{\infty} \Gamma_j^{-1/\alpha} \eta_j Y_j, \tag{8.1.23}$$

where

(i) *Y_j' are i.i.d. with $E\|Y_1\|^r < \infty$;*

(ii) *η_j are i.i.d. symmetric real-valued random variables with $\|\eta_1\|_\alpha = (E|\eta_1|^\alpha)^{1/\alpha} < \infty$;*

(iii) *(Γ_j) is a sequence of successive times of jump of a standard Poisson process;*

(iv) *we assume that the three sequences are independent; see Ledoux and Talagrand (1991, Sect. 5.1) and Samorodnitsky and Taqqu (1994).*

Suppose X has a similar representation

$$X \stackrel{d}{=} \sum_{j=1}^{\infty} \Gamma_j^{-1/\alpha} \eta_j^* Y_j^*, \tag{8.1.24}$$

where (Y_j^*) and (η_j^*) are chosen as in (i) and (ii) with the only difference that they are not identically distributed. Write Z_n, the normalized sum of i.i.d. copies X_i as in (8.1.2). *Then Theorem* 8.1.2 *yields the following rate of convergence of Z_n to ϑ in the ℓ_1-metric.*

Corollary 8.1.6 *Let $1 \vee \alpha < r < 2$, and $E||Y_1||^r + \sup_{j\geq 1} E||Y_j^*||^r + E||\eta_1||_r + \sup_{j\geq 1} E||\eta_j^*||_r < \infty$. Then*

$$\ell_1(Z_n, \vartheta) \leq C(n^{1-r/\alpha}\ell_r^* + \tau_r^* n^{-1/\alpha}), \tag{8.1.25}$$

where $\ell_r^* := \sup_{j\geq 1}(\overline{\ell}_r(Y_j^*, Y_1) + \overline{\ell}_r(\eta_j^* Y_1, \eta_1 Y_1))$ *and* $\tau_r^* = \max\left\{\ell_1^*, \sigma_1^*, \sigma_r^{*1/(r-\alpha)}\right\}$ *with* $\sigma_r^* := \sup_{j\geq 1}(\overline{\sigma}_r(Y_j^*, Y_1) + \overline{\sigma}_r(\eta_j^* Y_1, \eta_1 Y_1))$.

Proof: In view of (8.1.14), (8.1.15) we need only show the finiteness of $\overline{\sigma}_r$ and $\overline{\ell}_r$. For $\overline{\sigma}_r = \overline{\sigma}_r(X, \vartheta)$ we use the ideality of order r and the asymptotics $E\Gamma_j^{-r/\alpha} \sim j^{-r/\alpha} (j \to \infty)$ to obtain

$$\begin{aligned} \overline{\sigma}_r(X, \vartheta) &= \sum_{j\geq 1} E\Gamma_j^{-r/\alpha} \overline{\ell}_r(\eta_j Y_j, \eta_j^* Y_j^*) \\ &\leq \left(\sum_{j\geq 1} E\Gamma_j^{-r/\alpha}\right) \sup_{j\geq 1}\{E|\eta_j^*|^r\, \overline{\sigma}_r(Y_j^*, Y_j) + \overline{\sigma}_r(\eta_j^* Y_j, \eta_j Y_j)\} \\ &\leq C \sup_{j\geq 1}(\overline{\sigma}_r(Y_j^*, Y_1) + \overline{\sigma}_r(\eta_j^* Y_1, \eta_1 Y_1)). \end{aligned}$$

The same type estimate is valid for $\overline{\ell}_r$. □

Since in the LePage representations Y_j, Y_j^* can have any high enough moment, examples with finite ℓ_r^* and τ_r^* can be readily constructed. Take, for example, U to be a Hilbert space with basis $(h_m)_{m\geq 1}$, and set $Y_j^* \stackrel{d}{=} \sum_{m\geq 1} \zeta_{j,m}^* h_m$, $Y_1 \stackrel{d}{=} \sum_{m\geq 1} \zeta_m h_m$, where $(\zeta_m)_{m\geq 1}$, $(\zeta_{j,m}^*)_{m\geq 1}$ are sequences of independent random variables. Then, by the ideality of $\overline{\sigma}_r$,

$$\overline{\sigma}_r(Y_j^*, Y_1) \leq \sum_{m\geq 1} \overline{\sigma}_r(\zeta_{j,m}^*, \zeta_m) \leq C \sum_{m\geq 1} \kappa_r(\zeta_{j,m}^*, \zeta_m), \tag{8.1.26}$$

where κ_r is the rth pseudomoment,

$$\kappa_r(\zeta^*, \zeta) = r \int |x|^{r-1} |F_{\zeta^*}(x) - F_\zeta(x)| dx,$$

see Zolotarev (1983). Similarly,

$$\overline{\ell}_r(Y_j^*, Y_1) \leq C \sum_{m\geq 1} \kappa_r(\zeta_{j,m}^*, \zeta_m). \tag{8.1.27}$$

The same example is valid if we relax the independence assumption to "independence in finite blocks," requiring only that $(\zeta_{1+\ell}, \ldots, \zeta_{L+\ell})$, $\ell = 0, L, 2L, \ldots$, are independent.

Remark 8.1.7 (Finite-dimensional approximation) *An alternative use of the explicit upper bounds of the smoothing metrics in the finite-dimensional case is to combine Theorem* 8.1.2 *with an approximation step by the finite-dimensional case. To be concrete, let* X, Y *be* $C(S)$ *valued processes,* (S, d)

a totally bounded metric space. For $\varepsilon > 0$ let V_ε denote a finite covering ε-net and let $P_\varepsilon X = (X_t)_{t\in V_\varepsilon}$ be the corresponding finite-dimensional approximation of $X = (X_t)_{t\in S}$. If

$$E \sup_{d(s,t)\le\varepsilon} |X_s - X_t| \ \le\ a(\varepsilon) \tag{8.1.28}$$

and

$$E \sup_{d(s,t)\le\varepsilon} |Y_s - Y_t| \ \le\ b(\varepsilon),$$

then

$$\ell_1(X,Y) \ \le\ \ell_1(P_\varepsilon X, P_\varepsilon Y) + a(\varepsilon) + b(\varepsilon). \tag{8.1.29}$$

So we can combine fluctuation inequalities (8.1.28) *with the finite-dimensional bounds derived in* (8.1.14) *for the normalized sum Z_n in order to choose an optimal rate of approximation $\varepsilon = \varepsilon(n) \to 0$. A general and simple useful tool to derive fluctuation inequalities as in* (8.1.28) *is Pollard's lemma, which applied to* (8.1.28) *yields*

$$E \sup_{d(s,t)\le\varepsilon} |X_s - X_t| \ \le\ \sqrt{N_\varepsilon}\sqrt{\max_{1\le i\le N_\varepsilon} E \sup_{\substack{d(s,t_i)<\varepsilon\\ d(t,t_i)<\varepsilon}} |X_s - X_t|}\,, \tag{8.1.30}$$

where $N_\varepsilon = \mathrm{card}\,(V_\varepsilon)$ and $V_\varepsilon = \{t_i,\ 1\le i\le N_\varepsilon\}$.

The case $\alpha = 1$ requires special consideration. We shall state a variant of Theorem 8.1.2 that will cover the case $\alpha = 1$ but requires additional smoothing conditions on the law of the X_i's. The next theorem is based on the following lemma (see Rachev and Yukich (1989) or Rachev (1991c, Ch. 14)).

Lemma 8.1.8 *Let $0 < \alpha \le 2$, $r > \alpha$, $a_r = \frac{1}{2^{1+r/\alpha}(2^{r/\alpha}+3^{r/\alpha})}$, and $A_r = A_r(a) = 2^{-r/\alpha}a_r$. Suppose*

$$\delta_0 \ :=\ \delta_0(X_1,\vartheta) \ :=\ \max(\sigma,\vartheta_r) \ \le\ a_r. \tag{8.1.31}$$

Then for any $n \ge 1$

$$\sigma(Z_n,\vartheta) \ \le\ A_r\delta_0 n^{1-r/\alpha} \ \le\ 2^{-r/\alpha}n^{1-r/\alpha}. \tag{8.1.32}$$

Theorem 8.1.9 *Suppose condition* (8.1.23) *holds and*

$$\overline{\tau}_r \ :=\ \max(\ell_1,\overline{\ell}_r) \ <\ \infty. \tag{8.1.33}$$

Then for $\frac{1}{2} < \alpha \le 2$ and $r > \alpha$,

$$\ell_1(Z_n,\vartheta) \ \le\ B_{r,\alpha}\overline{\tau}_r n^{1-r/\alpha}, \tag{8.1.34}$$

where $B_{r,\alpha} \ge 8^{(r-1)/\alpha} + 2(2^{r/\alpha}+3^{r/\alpha})$.

The proof uses the following analogue of (8.1.17): For any independent X, Y, Z, W,

$$\ell_1(X+Z, Y+Z) \;\leq\; \ell_1(X,Y)\sigma(Z,W) + \ell_1(X+W, Y+W). \quad (8.1.35)$$

The proof is similar to that of the smoothing inequality in Zolotarev (1986, §5.4) (see also Rachev (1991c, Theorem 15.2.2)) and thus is omitted.

The theorem is of interest for $1 \leq \alpha \leq 2$, as for $0 < \alpha < r < 1$ we get from (8.1.6), $\ell_1(Z_n, \vartheta) \leq n^{1-1/\alpha}\overline{\tau}_r$.

Our next objective is the extension of Theorem 8.1.2 to the martingale case. Let $(\Omega, \mathcal{A}, P)$ be a rich enough probability space, $(\mathcal{F}_i)$ an increasing sequence of sub σ-algebras of $\mathcal{A}$, and let $(X_i, \mathcal{F}_i)$ be an adapted martingale difference sequence with values in a separable Banach space $(U, \|\cdot\|)$; that is, $E(X_i|\mathcal{F}_{i-1}) = 0$ a.s., $i \in \mathbb{N}$. For a given probability metric μ and a sub σ-algebra $\mathcal{G} \subset \mathcal{A}$ define the *$\mathcal{G}$-dependence metric* $\mu(\cdot\|\mathcal{G})$ by

$$\mu(X,Y\|\mathcal{G}) \;=\; \sup_{V\in\mathcal{G}} \mu(X+V, Y+V), \quad (8.1.36)$$

where $V \in \mathcal{G}$ denotes that V is a $\mathcal{G}$-measurable random variable. This notion generalizes an idea due to Gudynas (1985).

Lemma 8.1.10 *If μ is homogeneous of order r, that is,*

$$\mu(cX, cY) \;\leq\; |c|^r \mu(X,Y), \quad (8.1.37)$$

then $\mu(\cdot\|\mathcal{G})$ also is homogeneous of order r.

We shall use the following metrics: $\overline{\ell}_r(\cdot\|\mathcal{G})$, $\overline{\sigma}_r(\cdot\|\mathcal{G})$, where $\overline{\ell}_r$, $\overline{\sigma}_r$ are respectively the smoothed Kantorovich metric and the total variation metric (cf. (8.1.11), (8.1.12)).

Lemma 8.1.11 *Let the regular conditional distributions $P_{X|\mathcal{G}}, P_{Y|\mathcal{G}}$ exist. Then*

$$\overline{\ell}_r(X,Y\|\mathcal{G}) \;\leq\; E\overline{\ell}_r(P_{X|\mathcal{G}}, P_{Y|\mathcal{G}}) \quad (8.1.38)$$

and

$$\overline{\sigma}_r(X,Y\|\mathcal{G}) \;\leq\; E\overline{\sigma}_r(P_{X|\mathcal{G}}, P_{Y|\mathcal{G}}). \quad (8.1.39)$$

Proof: Let ϑ be independent of X, Y, and $\mathcal{G}$. Then

$$\begin{aligned}\overline{\ell}_r(X,Y\|\mathcal{G}) &= \sup_{V\in\mathcal{G}} \overline{\ell}_r(X+V, Y+V) \\ &= \sup_{\|f\|_L\leq 1} \sup_{h>0} \sup_{V\in\mathcal{G}} h^{r-1}\big|E(f(X+V+h\vartheta)\end{aligned}$$

$$
\begin{aligned}
& \qquad\qquad\qquad\qquad -f(Y+V+h\vartheta))\big| \\
\leq\ & E \sup_{\|f\|_L \leq 1} \sup_{h>0} \sup_{V\in\mathcal{G}} h^{r-1}\big|E(f(X+V+h\vartheta)|\mathcal{G}) \\
& \qquad\qquad\qquad\qquad -E(f(Y+V+h\vartheta)|\mathcal{G})\big| \\
=\ & E \sup_{\|f\|_L \leq 1} \sup_{h>0} \sup_{V\in\mathcal{G}} h^{r-1}\big|E(f_V(X+h\vartheta)|\mathcal{G}) \\
& \qquad\qquad\qquad\qquad -E(f_V(Y+h\vartheta)|\mathcal{G})\big|,
\end{aligned}
$$

where $f_V(\cdot) = f(\cdot + V)$ is the translation by V, which is again a Lipschitz function, and where $\|f\|_L = \sup_{x\neq y} \frac{|f(x)-f(y)|}{\|x-y\|}$ is the Lipschitz norm. We arrive at

$$
\begin{aligned}
\overline{\ell}_r(X,Y\|\mathcal{G}) \ \leq\ & E \sup_{\|f\|_L\leq 1} \sup_{h>0} h^{r-1}\big|E(f(X+h\vartheta)|\mathcal{G}) - E(f(Y+h\vartheta)|\mathcal{G})\big| \\
=\ & E\overline{\ell}_r(P_{X|\mathcal{G}}P_{Y|\mathcal{G}}).
\end{aligned}
$$

The proof for the metric $\overline{\sigma}_r$ is similar. □

As a consequence we obtain the following regularity property of $\overline{\ell}_r$ and $\overline{\sigma}_r$.

Lemma 8.1.12 *Let $(X_i, \mathcal{F}_i)$ be a stochastic sequence and $(\mathcal{G}_i)$ a decreasing sequence of sub σ-algebras such that Y_j are $\mathcal{G}_i$-measurable for $j \geq i$. Suppose that the following condition holds:*

(c) *X_i and $\mathcal{G}_{i+1}$ are conditionally independent given $\mathcal{F}_{i-1}$, and Y_i and $\mathcal{G}_{i+1}$ are conditionally independent given $\mathcal{F}_{i-1}$.*

Then, for $c_i \in \mathbb{R}$,

$$
\overline{\ell}_r\left(\sum_{i=1}^n c_i X_i, \sum_{i=1}^n c_i Y_i\right) \ \leq\ \sum_{i=1}^n |c_i|^r\, E\overline{\ell}_r(P_{X_i|\mathcal{F}_{i-1}}, P_{Y_i|\mathcal{F}_{i-1}}) \tag{8.1.40}
$$

and

$$
\overline{\sigma}_r\left(\sum_{i=1}^n c_i X_i, \sum_{i=1}^n c_i Y_i\right) \ \leq\ \sum_{i=1}^n |c_i|^r\, E\overline{\sigma}_r(P_{X_i|\mathcal{F}_{i-1}}, P_{Y_i|\mathcal{F}_{i-1}}), \tag{8.1.41}
$$

assuming that the conditional distributions exist.

Proof: By Lemma 8.1.10,

$$\begin{aligned}
\overline{\ell}_r & \left(\sum_{i=1}^n c_i X_i, \sum_{i=1}^n c_i Y_i\right) \\
& \le \sum_{i=1}^n \overline{\ell}_r(c_1X_1 + \cdots + c_iX_i + c_{i+1}Y_{i+1} + \cdots + c_nY_n, \\
& \qquad\qquad c_1X_1 + \cdots + c_{i-1}X_{i-1} + c_iY_i + \cdots + c_nY_n) \\
& \le \sum_{i=1}^n \sup_{h>0} h^{r-1} \sup_{V \in \mathcal{F}_{i-1} \vee \mathcal{G}_{i+1}} \ell_1(c_iX_i + V + h\vartheta, c_iY_i + V + h\vartheta) \\
& = \sum_{i=1}^n \overline{\ell}_r(c_iX_i, c_iY_i \| \mathcal{F}_{i-1} \vee \mathcal{G}_{i+1}) \\
& \le \sum_{i=1}^n |c_i|^r \, \overline{\ell}_r(X_1, Y_1 \| \mathcal{F}_{i-1} \vee \mathcal{G}_{i+1}),
\end{aligned}$$

where $\mathcal{F}_{i-1} \vee \mathcal{G}_{i+1}$ is the σ-algebra generated by $\mathcal{F}_{i-1}$ and $\mathcal{G}_{i+1}$. From Lemma 8.1.11 and the conditional independence assumption,

$$\begin{aligned}
\overline{\ell}_r(X_i, Y_i \| \mathcal{F}_{i-1} \vee \mathcal{G}_{i+1}) & \le E\overline{\ell}_r(P_{X_i|\mathcal{F}_{i-1}\vee\mathcal{G}_{i+1}}, P_{Y_i|\mathcal{F}_{i-1}\vee\mathcal{G}_{i+1}}) \\
& = E\overline{\ell}_r(P_{X_i|\mathcal{F}_{i-1}}, P_{Y_i|\mathcal{F}_{i-1}}).
\end{aligned}$$

As for the metric $\overline{\sigma}_r$, the proof is similar. □

Remark 8.1.13 *If Y_i are independent of $\mathcal{F}_{i-1}$, $EY_i = 0$, then*

$$\begin{aligned}
\overline{\ell}_r(P_{X_i|\mathcal{F}_{i-1}}, P_{Y_i}) & \le C_r\zeta_r(P_{X_i|\mathcal{F}_{i-1}}, P_{Y_i}) \\
& \le C_r\kappa_r(P_{X_i|\mathcal{F}_{i-1}}, P_{Y_i}),
\end{aligned} \tag{8.1.42}$$

where ζ_r is the Zolotarev metric and κ_r is the pseudo-difference moment (cf. (8.1.26) and Rachev (1991, p. 377)). In the α-stable case $1 < \alpha < 2$ and $r = 2$, the finiteness of κ_r implies that

$$E(X_i|\mathcal{F}_{i-1}) = EY_i = 0, \tag{8.1.43}$$

which is fulfilled in the martingale case. In the normal case $\alpha = 2$ and $r = 3$, the finiteness of ζ_r implies in the Euclidean case that the conditional covariance

$$\text{Cov}\,(X_i|\mathcal{F}_{i-1}) = \text{Cov}\,(Y_i) \tag{8.1.44}$$

is almost surely constant. This and related conditions have been assumed in several papers on the martingale convergence theorem (cf. Basu (1976), Dvoretzky (1970), Bolthausen (1982), Butzer et al. (1983), Häussler (1988), Rackauskas (1990)).

Lemma 8.1.14

$$\overline{\ell}_r(X,Y) \ \geq \ (\ell_1(X,Y))^r \left(\frac{r-1}{r}\right)^{r-1} \frac{1}{r} 2E\|\vartheta\|. \tag{8.1.45}$$

Proof: By the triangle inequality and from the definition of $\overline{\ell}_r$,

$$\begin{aligned} \overline{\ell}_r(X,Y) &\geq \ell_1(X+\varepsilon\vartheta, Y+\varepsilon\vartheta)\varepsilon^{r-1} \\ &\geq (\ell_1(X,Y) - 2\varepsilon E\|\vartheta\|)\varepsilon^{r-1} \ := \ \varphi(\varepsilon). \end{aligned}$$

Maximizing $\varphi(\varepsilon)$ with respect to ε, we obtain (8.1.45). □

In the next step we extend the smoothing inequality (8.1.17).

Lemma 8.1.15 *Suppose that X, Z, Y, W are random variables with values in U such that (X, Z) is independent of (Y, W) and Y, W are independent. Then*

$$\begin{aligned} \ell_1(X+Z, Y+Z) &\leq \ell_1(Z,W)\sigma(X,Y) + \ell_1(X+W, Y+W) \\ &\quad + \ell_1(Z+X, \widetilde{Z}+X), \end{aligned} \tag{8.1.46}$$

where $\widetilde{Z} \stackrel{d}{=} Z$ and $\widetilde{Z}$ is independent of X, and

$$\begin{aligned} \sigma(X+Z, Y+Z) &\leq \sigma(Z,W)\sigma(X,Y) + \sigma(X+W, Y+W) \\ &\quad + \sigma(Z+X, \widetilde{Z}+X). \end{aligned} \tag{8.1.47}$$

Proof: By the triangle inequality,

$$\begin{aligned} \ell_1(X+Z, Y+Z) &= \sup_{\|f\|_L \leq 1} |E[(f(X+Z) - f(X+W)) \\ &\qquad - (f(Y+Z) - f(Y+W))]| \\ &\quad + \sup_{\|f\|_L \leq 1} |E(f(X+W) - f(Y+W))|. \end{aligned}$$

Furthermore,

$$\begin{aligned} &|E[f(X+Z) - f(X+W) - (f(Y+Z) - f(Y+W))]| \\ &= \left|\int (E(f(Z+x)|X=x) - Ef(W+x))\, \mathrm{d}P_X(x) \right. \\ &\qquad \left. - \int (Ef(Z+x) - Ef(W+x))\, \mathrm{d}P_Y(x)\right| \\ &\leq \left|\int (E(f(Z+x)|X=x) - Ef(Z+x))\, \mathrm{d}P_X(x)\right| \\ &\quad + \left|\int Ef(Z+x)(\mathrm{d}P_X(x) - \mathrm{d}P_Y(x))\right. \end{aligned}$$

$$
\begin{aligned}
& - \int Ef(W+x)(\,\mathrm{d}P_X(x) - \mathrm{d}P_Y(x))\Big| \\
\leq \ & \ell_1(Z+X, \widetilde{Z}+X) + \ell_1(Z,W)\sigma(X,Y).
\end{aligned}
$$

The proof of (8.1.47) is similar. □

The last term in (8.1.46) is a measure of dependence of Z, X, which disappears if Z, X are independent.

Making use of the smoothing properties, we next extend Theorem 8.1.2 to the martingale case. Let $(X_i, \mathcal{F}_i)$ be a martingale difference sequence, $Z_n = n^{1/\alpha}\sum_{j=1}^n X_j$, and as in (8.1.4) let ϑ, ϑ_i be independent, α-stable distributed r.v.s. For $r > \alpha$ we define

$$
\begin{aligned}
\overline{\ell}_r &= \sup_j \overline{\ell}_r(X_j, \vartheta_j), \ \ \tau_r = \sup_j E\overline{\ell}_r(P_{X_j|\mathcal{F}_{j-1}}, P_{\vartheta_j}), \ \ \widetilde{\ell}_r = \overline{\ell}_r \vee \tau_r, \qquad (8.1.48)\\
\widetilde{\tau}_r &= \sup_j E\overline{\ell}_r(P_{X_j|\mathcal{F}_{j-1}}, P_{X_j}), \ \ \widehat{\tau}_r = \sup_j E\overline{\ell}_r(P_{X_j|\widehat{\mathcal{G}}_{j+1}}, P_{X_j}),
\end{aligned}
$$

where $\widehat{\mathcal{G}}_{j+1} = \sigma(X_{j+1}, X_{j+2}, \ldots)$, and $\overline{\sigma}_r = \sup_j \overline{\sigma}_r(X_j, \vartheta_j)$, the conditional distributions, are assumed to exist.

Theorem 8.1.16 *Suppose that $E\|\vartheta\| < \infty$. Then*

$$
\ell_1(Z_n, \vartheta) \ \leq \ C(n^{1-r/\alpha}\widetilde{\ell}_r + n^{-1/\alpha}t_r), \qquad (8.1.49)
$$

where $t_r = \max\left(\ell_1, \sigma_1, \overline{\sigma}_r^{\frac{1}{r-\alpha}}, \widehat{\tau}_r^{\frac{1}{r-\alpha}}, \widetilde{\tau}_1\right)$.

Proof: Applying (8.1.16) we shall estimate $\ell_1(Z_n + \varepsilon\vartheta,\ \vartheta_1 + \varepsilon\vartheta)$. Set $m = [\frac{n}{2}]$. Then

$$
\begin{aligned}
& \ell_1\left(Z_n + \varepsilon\vartheta, n^{-1/\alpha}\sum_{i=1}^n \vartheta_i + \varepsilon\vartheta\right) \\
& \quad \leq \ \ell_1\left(Z_n + \varepsilon\vartheta, n^{-1/\alpha}(\vartheta_1 + X_2 + \cdots + X_n) + \varepsilon\vartheta\right) \\
& \qquad + \sum_{j=1}^m \ell_1\Big(n^{-1/\alpha}(\vartheta_1 + \cdots + \vartheta_j + X_{j+1} + \cdots + X_n) + \varepsilon\vartheta, \\
& \qquad\qquad n^{-1/\alpha}(\vartheta_1 + \cdots + \vartheta_{j+1} + X_{j+2} + \cdots + X_n) + \varepsilon\vartheta\Big) \\
& \qquad + \ell_1\Big(n^{-1/\alpha}(\vartheta_1 + \cdots + \vartheta_{m+1} + X_{m+2} + \cdots + X_n) + \varepsilon\vartheta,
\end{aligned}
$$

$$n^{-1/\alpha}\sum_{j=1}^{n}\vartheta_j+\varepsilon\vartheta\Bigg)$$

$$=: \quad I_0+\sum_{j=1}^{m}I_j+I_{m+1}.$$

From the generalized smoothing inequality (8.1.46),

$$\begin{aligned}
I_0 &= \ell_1\left(\frac{X_2+\cdots+X_n}{n^{1/\alpha}}+\left(\frac{X_1}{n^{1/\alpha}}+\varepsilon\vartheta\right),\frac{X_2+\cdots+X_n}{n^{1/\alpha}}+\left(\frac{\vartheta_1}{n^{1/\alpha}}+\varepsilon\vartheta\right)\right)\\
&\le \ell_1\left(\frac{X_2+\cdots+X_n}{n^{1/\alpha}},\frac{\vartheta_2+\cdots+\vartheta_n}{n^{1/\alpha}}\right)\sigma\left(n^{-1/\alpha}X_1+\varepsilon\vartheta,n^{-1/\alpha}\vartheta_1+\varepsilon\vartheta\right)\\
&\quad+\ell_1\left(\frac{X_1+\vartheta_2+\cdots+\vartheta_n}{n^{1/\alpha}}+\varepsilon\vartheta,\frac{\vartheta_1+\cdots+\vartheta_n}{n^{1/\alpha}}+\varepsilon\vartheta\right)\\
&\quad+\ell_1\left(\frac{X_1+\cdots+X_n}{n^{1/\alpha}}+\varepsilon\vartheta,\frac{X_1+\cdots+X_{n-1}+\widetilde{X}_n}{n^{1/\alpha}}+\varepsilon\vartheta\right)\\
&=: \Delta_1+\Delta_2+\Delta_3,
\end{aligned}$$

where $\widetilde{X}_n\overset{d}{=}X_n$ and $\widetilde{X}_n$ is independent of $X_1,\dots,X_{n-1},\vartheta$. Similarly,

$$\begin{aligned}
\sum_{j=1}^{m}I_j &\le \sum_{j=1}^{m}\ell_1\left(\frac{X_{j+2}+\cdots+X_n}{n^{1/\alpha}},\frac{\vartheta_{j+2}+\cdots+\vartheta_n}{n^{1/\alpha}}\right)\\
&\quad\cdot\sigma\left(\frac{\vartheta_1+\cdots+\vartheta_j+X_{j+1}}{n^{1/\alpha}}+\varepsilon\vartheta,\frac{\vartheta_1+\cdots+\vartheta_{j+1}}{n^{1/\alpha}}+\varepsilon\vartheta\right)\\
&\quad+\sum_{j=1}^{m}\ell_1\left(\frac{\vartheta_1+\cdots+\vartheta_j+X_{j+1}+\vartheta_{j+2}+\cdots+\vartheta_n}{n^{1/\alpha}}+\varepsilon\vartheta,\right.\\
&\qquad\qquad\left.\frac{\sum_{j=1}^{n}\vartheta_j}{n^{1/\alpha}}+\varepsilon\vartheta\right)\\
&\quad+\sum_{j=1}^{m}\ell_1\left(\frac{\vartheta_1+\cdots+\vartheta_j+X_{j+1}+\cdots+X_n}{n^{1/\alpha}}+\varepsilon\vartheta,\right.\\
&\qquad\qquad\left.\frac{\vartheta_1+\cdots+\vartheta_j+\widetilde{X}_{j+1}+X_{j+2}+\cdots+X_n}{n^{1/\alpha}}+\varepsilon\vartheta\right)\\
&=: \Delta_4+\Delta_5+\Delta_6.
\end{aligned}$$

We first estimate Δ_5. By the ideality of $\overline{\ell}_r$,

$$\Delta_5 = \sum_{j=1}^{m}\ell_1\left(\frac{X_{j+1}}{n^{1/\alpha}}+\left(\frac{n-1}{n}\right)^{1/\alpha}\vartheta_1+\varepsilon\vartheta,\right. \tag{8.1.50}$$

$$
\begin{aligned}
& \qquad \frac{\vartheta_{j+1}}{n^{1/\alpha}} + \left(\frac{n-1}{n}\right)^{1/\alpha} \vartheta + \varepsilon\vartheta\Bigg) \\
\leq\ & \sum_{j=1}^{m} \ell_1 \left(\frac{X_{j+1}}{n^{1/\alpha}} + \left(\frac{n-1}{n}\right)^{1/\alpha} \vartheta, \frac{\vartheta_{j+1}}{n^{1/\alpha}} + \left(\frac{n-1}{n}\right)^{1/\alpha} \vartheta \right) \\
\leq\ & C n^{1-r/\alpha} \overline{\ell}_r.
\end{aligned}
$$

Similarly, by Lemma 8.1.12,

$$
\begin{aligned}
\Delta_7 \ =:\ & I_{m+1} \\
\leq\ & \ell_1 \left(\left(\frac{m+1}{n}\right)^{1/\alpha} \vartheta + \frac{X_{m+1} + \cdots + X_n}{n^{1/\alpha}}, \right. \\
& \qquad \left. \left(\frac{m+1}{n}\right)^{1/\alpha} \vartheta + \frac{\vartheta_{m+1} + \cdots + \vartheta_n}{n^{1/\alpha}} \right) \\
\leq\ & \left(\frac{m+1}{n}\right)^{(1-r)/\alpha} n^{-r/\alpha} \sum_{j=m+1}^{n} E\overline{\ell}_r \left(P_{X_i|\mathcal{F}_{i-1}}, P_{\vartheta_i}\right) \\
\leq\ & C n^{1-r/\alpha} \tau_r,
\end{aligned} \tag{8.1.51}
$$

and in the same way as for Δ_5, we obtain

$$
\Delta_2 \ \leq\ C n^{1-r/\alpha} \overline{\ell}_r. \tag{8.1.52}
$$

The remaining terms are dealt with by induction. Assume next that for $j < n$,

$$
\ell_1 \left(\frac{X_1 + \cdots + X_j}{j^{1/\alpha}}, \frac{\vartheta_1 + \cdots + \vartheta_j}{j^{1/\alpha}} \right) \ \leq\ B \left(\widetilde{\ell}_r j^{1-r/\alpha} + t_r j^{-1/\alpha} \right) \tag{8.1.53}
$$

and let

$$
\varepsilon \ =\ A \max \left(\sigma_1, \overline{\sigma}_r^{1/(r-\alpha)}, \overline{\ell}_r^{1/(r-\alpha)}, \widehat{\tau}_r^{1/(r-\alpha)} \right) n^{-1/\alpha} \tag{8.1.54}
$$

with a constant $A \geq 0$ that we shall fix later in the proof. Then

$$
\begin{aligned}
\Delta_1 \ \leq\ & BC(n^{1-r/\alpha} \widetilde{\ell}_r + n^{-1/\alpha} t_r) \varepsilon^{-1} n^{-1/\alpha} \sigma_1(X_1, \vartheta) \\
\leq\ & \frac{1}{A} BC(n^{1-r/\alpha} \widetilde{\ell}_r + n^{-1/\alpha} t_r).
\end{aligned} \tag{8.1.55}
$$

In the same way,

$$
\begin{aligned}
\Delta_4 \ \leq\ & CB \left(\overline{\ell}_r (n-m-2)^{1-r/\alpha} + t_r (n-m-2)^{-1/\alpha} \right) \\
& \cdot \sum_{j=1}^{\infty} \left(\frac{j}{n} + \varepsilon^{\alpha} \right)^{-r/\alpha} \overline{\sigma}_r \left(\frac{X_{j+1}}{n^{1/\alpha}}, \frac{\vartheta_{j+1}}{n^{1/\alpha}} \right)
\end{aligned} \tag{8.1.56}
$$

$$
\begin{aligned}
&\leq CB\left(\widetilde{\ell}_r n^{1-r/\alpha} + t_r n^{-1/\alpha}\right) \\
&\quad \cdot \sum_{j=1}^{\infty} \frac{\overline{\sigma}_r(X_{j+1}, \vartheta_{j+1})}{\left(j + A^\alpha \overline{\sigma}_r(X_{j+1}, \vartheta_{j+1})^{\alpha/(r-\alpha)}\right)^{r/\alpha}} \\
&\leq CB(\widetilde{\ell}_r n^{1-r/\alpha} + t_r n^{-1/\alpha})/A^{r-\alpha},
\end{aligned}
$$

using that $\varepsilon^\alpha \geq A^\alpha \overline{\sigma}_r{}^{\alpha/(r-\alpha)} n^{-1}$. To estimate Δ_3 we apply the $\mathcal{G}$-dependence metric $\ell_1(\cdot \| \mathcal{F}_{n-1})$:

$$
\begin{aligned}
\Delta_3 &\leq \ell_1\left(\frac{X_n}{n^{1/\alpha}} + \varepsilon\vartheta, \frac{\widetilde{X}_n}{n^{1/\alpha}} + \varepsilon\vartheta \| \mathcal{F}_{n-1}\right) \qquad (8.1.57) \\
&\leq \ell_1\left(\frac{X_n}{n^{1/\alpha}}, \frac{\widetilde{X}_n}{n^{1/\alpha}} \| \mathcal{F}_{n-1}\right) \leq n^{-1/\alpha} E\ell_1(P_{X_n|\mathcal{F}_{n-1}}, P_{X_n}) \\
&\leq n^{-1/\alpha}\widetilde{\tau}_1.
\end{aligned}
$$

Finally, we estimate Δ_6 as follows:

$$
\begin{aligned}
\Delta_6 &= \sum_{j=1}^{m} \ell_1\left(\frac{X_{j+1}}{n^{1/\alpha}} + \frac{X_{j+2} + \cdots + X_n}{n^{1/\alpha}} + \left(\frac{j}{n} + \varepsilon^\alpha\right)^{1/\alpha} \vartheta,\right. \\
&\qquad \left. \frac{\widetilde{X}_{j+1}}{n^{1/\alpha}} + \frac{X_{j+2} + \cdots + X_n}{n^{1/\alpha}} + \left(\frac{j}{n} + \varepsilon^\alpha\right)^{1/\alpha} \vartheta\right) \\
&\leq \sum_{j=1}^{m} \left(\frac{j}{n} + \varepsilon^\alpha\right)^{(1-r)/\alpha} \\
&\quad \cdot \overline{\ell}_r\left(\frac{X_{j+1}}{n^{1/\alpha}} + \frac{X_{j+2} + \cdots + X_n}{n^{1/\alpha}}, \frac{\widetilde{X}_{j+1}}{n^{1/\alpha}} + \frac{X_{j+2} + \ldots + X_n}{n^{1/\alpha}}\right) \\
&\leq \sum_{j=1}^{m} \frac{(j + n\varepsilon^\alpha)^{(1-r)/\alpha}}{n^{(1-r)/\alpha}} n^{-r/\alpha} \overline{\ell}_r(X_{j+1}, \widetilde{X}_{j+1} \| \widehat{\mathcal{G}}_{j+2}) \\
&\leq \widetilde{\tau}_r n^{-1/\alpha} \sum_{j=1}^{m} (j + nA^\alpha \widehat{\tau}_r^{\alpha/(r-\alpha)} n^{-1})^{(1-r)/\alpha} \\
&\leq Cn^{-1/\alpha} \frac{1}{A^{r-1-\alpha}} \widehat{\tau}_r^{1/(r-\alpha)}.
\end{aligned}
$$

Gathering all the inequalities, we obtain

$$
\begin{aligned}
\ell_1(Z_n, \vartheta) &\leq C_1 \frac{B}{A}\left(n^{1-r/\alpha}\widetilde{\ell}_r + n^{-1/\alpha} t_r\right) + C_2 n^{1-r/\alpha}\overline{\ell}_r + C_3 n^{-1/\alpha}\widetilde{\tau}_1 \\
&\quad + C_4 \frac{B}{A^{r-\alpha}}\left(n^{1-r/\alpha}\widetilde{\ell}_r + n^{-1/\alpha} t_r\right) + C_5 n^{1-r/\alpha}\overline{\ell}_r
\end{aligned}
$$

$$+ C_6 \frac{1}{A^{r-1-\alpha}} \widehat{\tau}_r^{1/(r-\alpha)} n^{-1/\alpha} + C_7 n^{1-r/\alpha} \tau_r$$
$$+ 2E\|\vartheta\| n^{-1/\alpha} \max\left(\sigma_1, \overline{\sigma}_r^{\,1/(r-\alpha)}, \overline{\ell}_r^{\,1/(r-\alpha)}, \widehat{\tau}_r^{1/(r-\alpha)}\right).$$

Choose A large enough such that $\frac{C_1}{A} + \frac{C_4}{A^{r-\alpha}} \le \frac{1}{2}$ and then choose B large enough such that $C_2 + C_3 + C_5 + C_6 A^{1+\alpha-r} + C_7 + 2E\|\vartheta\| \le \frac{B}{2}$. Thus we obtain (8.1.49). □

8.2 Application to Stable Limit Theorems

Zolotarev (1976) introduced the ζ_r-metric as an extension of the Kantorovich metric. For any pair of random vectors X, Y on $\mathbb{R}^k$ and $r = m + a$ it is defined by

$$\zeta_r(X, Y) = \sup\{|E(f(X) - f(Y))|;\ f \in \mathcal{F}_r\}, \tag{8.2.1}$$

where $\mathcal{F}_r$ is the class of all functions $f : \mathbb{R}^k \to \mathbb{R}$ such that $\|f^{(m)}(x) - f^{(m)}(y)\| \le \|x - y\|_1^a$, $0 < a \le 1$; $f^{(m)}$ denotes the mth Fréchet derivative of f supplied with the usual supremum norm of multilinear functionals (cf. Zolotarev (1986, Section 6.3) and Rachev (1991, p. 264)), and $\|x - y\|_1$ denotes the L_1-norm in $\mathbb{R}^k$. Indeed, ζ_1 is merely the Kantorovich metric; see (8.1.3). ζ_r is ideal of order r and therefore is suitable for analyzing the rate of convergence in various central limit theorems. (The definition of an ideal metric was given in (8.1.8) and (8.1.9).)

A disadvantage of ζ_r is that only for integers r can ζ_r be estimated by difference pseudomoments from above, while for $r \notin \mathbb{N}$ the known upper estimates involve absolute moments $b_r = \max(E\|X\|^r,\ E\|Y\|^r)$ or absolute pseudomoments of order r and therefore are not suitable for approximation by stable distributions of order $\alpha < 2$. In $\mathbb{R}$ an alternative ideal metric of order r that does not have this drawback of ζ_r was found by Maejima and Rachev (1987) and applied to prove convergence to self-similar processes; see also Rachev (1991c, Section 17.1).

In this section we introduce a new ideal metric $\vartheta_{s,p}$ (with respect to summation of independent random vectors in $\mathbb{R}^k$), which generalizes the construction in Maejima and Rachev (1987). This ideal metric has the following properties. It is ideal of order $r = s - 1 + \frac{1}{p}$. It can be estimated from above by a Zolotarev-type metric and, what is more important, by a pseudo difference moment, which allows applications to stable distributions. Finally, it can be bound from below by the Lévy metric, and thus $\vartheta_{s,p}$ describes weak convergence of distributions. The degree of ideality of this metric does not depend on the dimension. This is an important property, which is not satisfied by some obvious generalizations of one-dimensional ideal metrics of order greater than 1, see Sections 6.1, 6.3. We shall establish relations between $\vartheta_{1,p}$ and $\vartheta_{s,p}$ and prove various smoothing inequalities. In the second part of this section we give an application to the rate of convergence in stable limit theorems. The upper bounds in the limit theorem are formulated in metric terms. We establish some new results ensuring the finiteness of these bounds and apply these results to show that random vectors in a neighborhood of the LePage decomposition of a stable law satisfy the central limit theorem with rate. Further applications are to the convergence of summability methods of i.i.d. random vectors and to the approximation by compound Poisson distributions. All these applications are based on the thorough analysis of the metric properties of ideal metrics having a structure close to that of the Kantorovich metric. The results in this section are due to Rachev and Rüschendorf (1992).

We start with the construction of the $\vartheta_{s,p}$-metric. Let $X, Y \in \mathcal{X}(\mathbb{R}^k)$, the class of k-dimensional random vectors, and define for $s \in \mathbb{N}$, $1 \le p \le \infty$,

$$\vartheta_{s,p}(X,Y) = \sup\{|E(f(X) - f(Y))|;\ f \in G_{s,p}\}. \tag{8.2.2}$$

Here $G_{s,p}$ is the class of functions $f : \mathbb{R}^k \to \mathbb{R}$, such that for $1 \le i_1 \le \cdots \le i_s \le k$ and for $1 \le j \le k$, $x = (x_1, \ldots, x_{j-1}, x_{j+1}, \ldots, x_k) \in \mathbb{R}^{k-1}$, and q with $\frac{1}{p} + \frac{1}{q} = 1$,

$$\begin{aligned} &\|D^s_{i_1,\ldots,i_s} f\|_{q,j}(x) \qquad\qquad (8.2.3)\\ &= \begin{cases} \left(\displaystyle\int_{\mathbb{R}} |D^s_{i_1,\ldots,i_s} f(x_1, \ldots, x_j, \ldots, x_k)|^q \, dx_j \right)^{1/q} & \text{if } q < \infty, \\ \operatorname*{ess\,sup}_{x_j \in \mathbb{R}} |D^s_{i_1,\ldots,i_s} f(x_1, \ldots, x_j, \ldots, x_k)| & \text{if } q = \infty \end{cases} \\ &\le 1 \quad \text{a.s. with respect to the Lebesgue measure.} \end{aligned}$$

Lemma 8.2.1 *For any $1 \le p \le \infty$ and $s \in \mathbb{N}$, the metric $\vartheta_{s,p}$ is an ideal metric of order $r = s - 1 + \frac{1}{p}$.*

Proof: If $f \in G_{s,p}$ and $z \in \mathbb{R}^k$, then $f(\cdot + z) \in G_{s,p}$ and hence $\vartheta_{s,p}(X + Z, Y + Z) \le \vartheta_{s,p}(X,Y)$ for any Z independent of X and Y. Further, when $q < \infty$, for any $c \in \mathbb{R}$, $x \in \mathbb{R}^{k-1}$, $1 \le j \le k$, $f_c(x) := f(cx)$,

$$\|D^s_{i_1,\ldots,i_s} f_c\|_{q,j}(x) = \left[\int_{\mathbb{R}} |D^s_{i_1,\ldots,i_s} f_c(x_1, \ldots, x_k)|^q \, dx_j \right]^{1/q} \tag{8.2.4}$$

$$= |c|^{s-1/q} \left[\int\limits_{\mathbb{R}} |D^s_{i_1,\ldots,i_s} f(y_1,\ldots,y_k)|^q \, dy_j \right]^{1/q}$$

$$= |c|^r \, \|D^s_{i_1,\ldots,i_s} f\|_{q,j},$$

which yields the ideality of $\vartheta_{s,p}$ of order r. The case $q = \infty$ can be handled similarly. □

Remark 8.2.2 *Note that the direct generalization of the Maejima–Rachev (1987) construction leads to*

$$\widetilde{\vartheta}_{s,p}(X,Y) = \sup\left\{ |E(f(X) - f(Y))|; \left(\int \|f^{(s)}(x)\|^q \, dx \right)^{1/q} \le 1 \right\}, \quad (8.2.5)$$

which is an ideal metric of order $s - \frac{k}{q} = s - k(1 - \frac{1}{p})$. *This unpleasant dependence on the dimensionality is avoided in the definition of* $\vartheta_{s,p}$ *by the restriction to one-dimensional integration in* (8.2.3).

We next show that $\vartheta_{s,p}$ is estimated from above by the following modification $\overline{\zeta}_r$ of the ζ_r-metric:

$$\overline{\zeta}_r(X,Y) = \sup\{|E(f(X) - f(Y))|; \ f \in \overline{\mathcal{F}}_r\}, \quad (8.2.6)$$

where $\overline{\mathcal{F}}_r$ is the class of functions $f : \mathbb{R}^k \to \mathbb{R}$ with

$$\|f^{(m)}(x) - f^{(m)}(y)\| \le d_\alpha(x,y) := \sum_{i=1}^{k} |x_i - y_i|^a, \quad (8.2.7)$$

$m = 0, 1, \ldots,$ $a \in (0,1]$, and $r = m + a$. In fact, (8.2.7) is equivalent to

$$|f(x) - f(y)| \le d_\alpha(x,y), \ \forall x,y \in \mathbb{R}^k, \quad \text{if } m = 0, \quad (8.2.8)$$

and to

$$\sup_{1 \le i_1 \le \cdots \le i_m \le k} |D^m_{i_1,\ldots,i_m} f(x) - D^m_{i_1,\ldots,i_m} f(y)| \le d_a(x,y), \ \text{if } m \ge 1.$$

Since $\|x - y\|_1^a \le d_a(x,y)$, we have

$$\zeta_r \le \overline{\zeta}_r. \quad (8.2.9)$$

Lemma 8.2.3 (a) *For any integer* r,

$$\vartheta_{r,1} = \overline{\zeta}_r = \zeta_r. \quad (8.2.10)$$

(b) *For any $r > 0$, if $\vartheta_{s,p}(X,Y) < \infty$, then*

$$\vartheta_{s,p}(X,Y) \;\le\; \overline{\zeta}_r \;\le\; \frac{\Gamma(1+a)}{\Gamma(1+r)}\overline{v}_r(X,Y), \tag{8.2.11}$$

where $r = s - 1 + \frac{1}{p} = m + a$, and

$$\overline{v}_r(X,Y) \;:=\; \int_{\mathbb{R}^k} d_a(x,0) \sum_{1\le i_1,\cdots,i_m\le k} |x_{i_1}\cdots x_{i_m}|\,|P_X - P_Y|(\,dx). \tag{8.2.12}$$

Proof: (a) It is enough to check that $G_{r,1} = \overline{\mathcal{F}}_r$. If $f \in G_{r,1}$, then

$$\begin{aligned}
&\sup_{1\le i_1,\ldots,i_{r-1}\le k} |D^{r-1}_{i_1,\ldots,i_{r-1}} f(x) - D^{r-1}_{i_1,\ldots,i_{r-1}} f(y)| \\
&\le \sup_{1\le i_1,\ldots,i_{r-1}\le k} \left\{ \left| \int_{x_1}^{y_1} D^r_{i_1,\ldots,i_{r-1},1} f(t,x_2,\ldots,x_k)\,dt \right| \right. \\
&\quad + \left| \int_{x_2}^{y_2} D^r_{i_1,\ldots,i_{r-1},2} f(y_1,t,x_3,\ldots,x_k)\,dt \right| \\
&\quad \left. + \cdots + \left| \int_{x_k}^{y_k} D^r_{i_1,\ldots,i_{r-1},k} f(y_1,\ldots,y_{k-1},t)\,dt \right| \right\} \\
&\le \|x-y\|_1 \;=:\; d_1(x,y); \text{ i.e., } f \in \overline{\mathcal{F}}_r.
\end{aligned}$$

Conversely, for $f \in \overline{\mathcal{F}}_r$ we have from (8.2.8),

$$\begin{aligned}
&\sup_{1\le i_1,\ldots,i_r\le k} |D^r_{i_1,\ldots,i_r} f(x)| \\
&= \sup_{1\le i_1,\ldots,i_r\le k} \lim_{y_r\to x_r} \frac{|D^{r-1}_{i_1,\ldots,i_{r-1}}(f(x_1,\ldots,x_k) - f(x_1,\ldots,y_r,\ldots,x_k))|}{|x_r - y_r|} \\
&\le 1 \text{ a.s.; i.e., } f \in G_{r,1}.
\end{aligned}$$

(b) If $f \in G_{s,p}$ and $1 < p \le \infty$, then similarly to the proof in (a),

$$\begin{aligned}
&\sup_{1\le i_1,\ldots,i_{s-1}\le k} |D^{s-1}_{i_1,\ldots,i_{s-1}} f(x) - D^{s-1}_{i_1,\ldots,i_{s-1}} f(y)| \\
&\le \sup_{1\le i_1,\ldots,i_{s-1}\le k} \sum_{i=1}^{k} \int_{x_i}^{y_i} |D^s_{i_1,\ldots,i_{s-1},i} f(y_1,\ldots,t,x_{i+1},\ldots,x_k)|\,dt
\end{aligned} \tag{8.2.13}$$

$$\leq \sum_{i=1}^{k} \|D^s_{i_1,\ldots,i_s} f\|_{q,i}(y_1,\ldots,y_{i-1},x_{i+1},\ldots,x_k)|x_i - y_i|^{1/p}$$
$$\leq d_a(x,y).$$

For the second part of (b) first note that $\vartheta_{s,p}(X,Y) < \infty$ implies that for any $1 \leq i_1 \leq \cdots \leq i_j \leq k$, $j \leq s-1$,

$$E(X_{i_1}\cdots X_{i_j} - Y_{i_1}\cdots Y_{i_j}) = 0. \tag{8.2.14}$$

This follows, by taking $f_c(x) = c x_{i_1}\cdots x_{i_j}$, $j \leq s-1$, and the obvious inequality $\vartheta_{s,p}(X,Y) \geq \sup_{c>0} |E f_c(X) - E f_c(Y)|$. Following the argument in the first part of this proof, (8.2.14) is also a consequence of the condition $\overline{\zeta}_r(X,Y) < \infty$. We obtain from the Taylor expansion and applying (8.2.14) with $m = s-1$ that

$$|E(f(X) - f(Y))|$$
$$= \Bigg|\sum_{1\leq i_1,\ldots,i_m\leq k} \int_0^1 \frac{(1-t)^{m-1}}{(m-1)!} \cdot E\left(D^m_{i_1,\ldots,i_m} f(tX)X_{i_1}\cdots X_{i_m} - D^m_{i_1\ldots i_m} f(tY)Y_{i_1}\cdots Y_{i_m}\right) dt\Bigg|$$
$$\leq \sum_{1\leq i_1,\ldots,i_m\leq k} \int_0^1 \frac{(1-t)^{m-1}}{(m-1)!} \left(\int_{\mathbb{R}^k} |D^m_{i_1,\ldots,i_m} f(tx)x_{i_1}\cdots x_{i_m} - D^m_{i_1,\ldots,i_m} f(t0)x_{i_1}\cdots x_{i_m}|\,|P_X - P_Y|(\,dx)\right) dt$$
$$\leq \sum_{1\leq i_1,\ldots,i_m\leq k} \int_0^1 \frac{(1-t)^{m-1}}{(m-1)!} \cdot \left(\int_{\mathbb{R}^k} d_a(tx,t0)|x_{i_1}\cdots x_{i_m}|\,|P_X - P_Y|(dx)\right) dt$$
$$\leq \frac{1}{(m-1)!}\left(\int_0^1 (1-t)^{m-1}t^a\,dt\right) \cdot \sum_{1\leq i_1,\ldots,i_m\leq k} \int_{\mathbb{R}^k} d_a(x,0)|x_{i_1}\cdots x_{i_m}|\,|P_X - P_Y|(\,dx)$$
$$= \frac{\Gamma(1+\alpha)}{\Gamma(1+r)}\,\overline{v}_r(X,Y).$$

□

Remark 8.2.4 (a) *Since* $\|x\|_1^r = \left(\sum_{i=1}^k |x_i|\right)^r = \left(\sum_{i=1}^k |x_i|\right)^m \left(\sum_{i=1}^k |x_i|\right)^a$ $\le \sum_{1\le i_1,\dots,i_m\le k} |x_{i_1}\dots x_{i_m}| \left(\sum_{i=1}^k |x_i|^a\right)$, *and on the other hand* $k\|x\|_1^r \ge$ $\sum_{1\le i_1,\dots,i_m\le k} |x_{i_1}\dots x_{i_m}| \sum_{i=1}^k |x_i|^a$, *we have*

$$v_r(X,Y) := \int_{\mathbb{R}^k} \|x\|_1^r |P_X - P_Y|(\mathrm{d}x) \ \le\ \overline{v}_r(X,Y) \ \le\ k\, v_r(X,Y). \quad (8.2.15)$$

(b) *By the same arguments as in* (a) *we have*

$$\zeta_r \ \le\ \overline{\zeta}_r \ \le\ k\zeta_r. \quad (8.2.16)$$

In particular, by (8.2.11), $\vartheta_{s,p}$ *is also estimated from above by the Zolotarev metric* ζ_r *(up to a constant).*

The following theorem gives an estimate of $\vartheta_{s,p}$ in terms of certain pseudomoments, which allows one to apply $\vartheta_{s,p}$ to stable distributions. For random vectors X, Y with densities u_X, u_Y, define

$$\alpha_{s,p}(X,Y)$$
$$= \sum_{i=1}^{k} \sum_{1\le i_1,\dots,i_s\le k} \int_{\mathbb{R}^{k-1}} \left[\int_{\mathbb{R}} |y_{i_1}\dots y_{i_s}| \int_0^1 \left| t^{-s-k} \frac{(1-t)^{s-1}}{(s-1)!} (u_X - u_Y) \right.\right.$$
$$\left.\left. \cdot \left(\frac{y_1}{t},\dots,\frac{y_k}{t}\right) \mathrm{d}t \right| |y_i|^p \,\mathrm{d}y_i \right]^{1/p} \mathrm{d}y_i \dots \mathrm{d}y_{i-1}\,\mathrm{d}y_{i+1}\dots \mathrm{d}y_k.$$

If $k = 1$, then after some transformations we obtain

$$\begin{aligned} \alpha_{s,p}(X,Y) &= \|F_{s,X} - F_{s,Y}\|_p \quad (8.2.17) \\ &= \left(\int \left| \int_{-\infty}^{x} \frac{(x-t)^{s-1}}{(s-1)!} \mathrm{d}(F_X - F_Y)(t) \right|^p \mathrm{d}x \right)^{1/p}, \\ F_{s,X}(x) &:= \frac{1}{(s-1)!} E(x-X)_+^{s-1} \end{aligned}$$

(see Maejima and Rachev (1987), Rachev and Rüschendorf (1990)). Indeed, $\alpha_{s,p}$ is an ideal metric of order r. Representation (8.2.17) shows that $\alpha_{s,p}$ depends only on the difference pseudomoments. A similar representation holds also for $k \ge 1$.

Theorem 8.2.5 $\alpha_{s,p}$ *is the upper bound for* $\vartheta_{s,p}$; *i.e.,*

$$\vartheta_{s,p} \ \le\ \alpha_{s,p}. \quad (8.2.18)$$

Proof: By (8.2.14) and the Taylor expansion,

$$
\begin{aligned}
&E(f(X) - f(Y)) \\
&= \int_{\mathbb{R}^k} \sum_{1 \le i_1 \le \cdots \le i_s \le k} \int_0^1 \frac{(1-t)^{s-1}}{(s-1)!} \\
&\qquad \cdot \left[D^s_{i_1,\ldots,i_s} f(tx_1,\ldots,tx_k) x_{i_1} \cdots x_{i_s} \right] \mathrm{d}t \, \mathrm{d}(F_X - F_Y)(x) \\
&= \sum_{1 \le i_1,\ldots,i_s \le k} \int_0^1 \frac{(1-t)^{s-1}}{(s-1)!} \left[\int_{\mathbb{R}^k} D^s_{i_1,\ldots,i_s} f(y_1,\ldots,y_k) y_{i_1} \ldots y_{i_s} \right. \\
&\qquad \left. \cdot (u_X - u_Y)\left(\frac{y}{t}\right) t^{-s-k} \, \mathrm{d}y \right] \mathrm{d}t.
\end{aligned}
$$

This implies, by making use of Hölder's inequality, that

$$
\begin{aligned}
&|E(f(X) - f(Y))| \\
&\le \sum_{i=1}^k \sum_{j=0}^{s-1} \sum_{1 \le i_1,\ldots,i_j \le k, i_j \ne i} \int_{\mathbb{R}^{k-1}} |y_{i_1} \cdots y_{i_j}| \int_{\mathbb{R}} \left[D_{i_1,\ldots,i_j,i,\ldots,i} f(y) \right. \\
&\qquad \cdot \left| \int_0^1 t^{-s-k} \frac{(1-t)^{s-1}}{(s-1)!} (u_X - u_Y)\left(\frac{y}{t}\right) \mathrm{d}t \right| \\
&\qquad \left. \cdot |y_i|^{s-j} \, \mathrm{d}y_i \right] \mathrm{d}y_1 \ldots \mathrm{d}y_{i-1} \, \mathrm{d}y_{i+1} \ldots \mathrm{d}y_k \\
&\le \sum_{i=1}^k \sum_{j=0}^{s-1} \sum_{1 \le i_1,\ldots,i_j \le k, i_j \ne i} \int_{\mathbb{R}^{k-1}} |y_{i_1} \cdots y_{i_j}| \| D_{i_1,\ldots,i_j,i,\ldots,i} f \|_{q,i}(y) \\
&\qquad \cdot \left\| \int_0^1 t^{-s-k} \frac{(1-t)^{s-1}}{(s-1)!} (u_X - u_Y)\left(\frac{y}{t}\right) |y_i|^{s-j} \, \mathrm{d}t \right\|_p \\
&\qquad \mathrm{d}y_1 \ldots \mathrm{d}y_{i-1} \, \mathrm{d}y_{i+1} \ldots \mathrm{d}y_k,
\end{aligned}
$$

which is equivalent to the representation in (8.2.17). □

For random vectors in $\mathbb{R}^k$ we define the Lévy distance by

$$
\begin{aligned}
L(X,Y) \;=\; \inf\{\varepsilon > 0;\; & P(X \in B_x) \le P(Y \in B_x(\varepsilon)) + \varepsilon, \\
& P(Y \in B_x) \le P(X \in B_x(\varepsilon)) + \varepsilon,\ \forall x \in \mathbb{R}^k\},
\end{aligned} \tag{8.2.19}
$$

where $B_x := \{y \in \mathbb{R}^k;\ y_i \le x_i,\ 1 \le i \le k\}$ and

$$B_x(\varepsilon) \ := \ \{y \in \mathbb{R}^k;\ \|y - B_x\| \le \varepsilon\}, \quad \|y\| = \left(\sum_{i=1}^k y_i^2\right)^{1/2};$$

note that $P(X \in B_x) = F_X(x)$. L metrizes the topology of weak convergence. If X has a bounded density u_X, then

$$\begin{aligned} \varrho(X,Y) &:= \sup |F_X(x) - F_Y(x)| \\ &\le \left(1 + \sum_{i=1}^k \sup_x u_{X_i}(x)\right) L(X,Y), \quad X = (X_1, \ldots, X_k). \end{aligned} \tag{8.2.20}$$

Next, we establish that $\vartheta_{s,p}$ convergence implies weak convergence by providing a lower bound of $\vartheta_{s,p}$ in terms of L.

Theorem 8.2.6 *Let* $s = 1, 2, \ldots, p \in [1, \infty]$, $r = s - 1 + \frac{1}{p}$. *Then*

$$\vartheta_{s,p}(X,Y) \ \ge \ a(s,k) L^{r+1}(X,Y), \tag{8.2.21}$$

where

$$a(s,k) \ := \ \frac{V_r}{2^{s+k} s!}, \quad V_r \ := \int_{\{\|x\| \le 1\}} (1 - \|x\|^2)^{r+1}\, dx. \tag{8.2.22}$$

Proof: Let $L(X,Y) > \varepsilon$. Then without loss of generality we can assume that for some $z = (z_1, \ldots, z_k)$,

$$P(X \in B_z) - P(Y \in B_z(\varepsilon)) \ > \ \varepsilon. \tag{8.2.23}$$

We define

$$g_r(x) \ = \ (1 - \|x\|^2)_+^{r+1}, \quad x \in \mathbb{R}^k \ (a_+ = \max(0, a)), \tag{8.2.24}$$

and "normalize" g_r by

$$\overline{g}_r(x) \ := \ \frac{g_r(x)}{V_r}. \tag{8.2.25}$$

Finally, we define the smoothed version of the indicator of B_z:

$$\begin{aligned} u_\varepsilon(x) &= \int_{\mathbb{R}^k} I\left\{x - \frac{\varepsilon}{2} y \in B_z\left(\frac{\varepsilon}{2}\right)\right\} \overline{g}_r(y)\, dy \\ &= \left(\frac{2}{\varepsilon}\right)^k \int_{\mathbb{R}^k} I\left\{\widetilde{y} \in B_z\left(\frac{\varepsilon}{2}\right)\right\} \overline{g}_r\left(\frac{2}{\varepsilon}(x - \widetilde{y})\right) d\widetilde{y}. \end{aligned} \tag{8.2.26}$$

Since $0 \le u_\varepsilon(x) \le \int\limits_{\mathbb{R}^k} \bar{g}_r(y)\,dy \le 1$, we have

$$|f_\varepsilon| \le 1, \quad \text{where } f_\varepsilon(x) := 2\,u_\varepsilon(x) - 1; \tag{8.2.27}$$

and furthermore,

$$f_\varepsilon(x) = \begin{cases} 1 & \text{if } x \in B_z, \\ 0 & \text{if } x \notin B_z(\varepsilon). \end{cases} \tag{8.2.28}$$

In fact, we have for $x \in B_z$,

$$\begin{aligned} u_\varepsilon(x) &= \int\limits_{\mathbb{R}^k} I\left\{x - \frac{\varepsilon}{2} y \in B_z\left(\frac{\varepsilon}{2}\right)\right\} \bar{g}_r(y)\,dy \\ &= \int\limits_{\mathbb{R}^k} I\left\{x - \frac{\varepsilon}{2} y \in B_z\left(\frac{\varepsilon}{2}\right), \|y\| \le 1\right\} \bar{g}_r(y)\,dy \\ &= \int\limits_{\mathbb{R}^k} \bar{g}_r(y)\,dy \;=\; 1. \end{aligned}$$

Similarly, for $x \notin B_z(\varepsilon)$, $u_\varepsilon(x) = 0$.

In the next step we establish bounds on the derivatives of f_ε. To that purpose let

$$L_s(f) \;=\; \sup_{1 \le i \le k} \operatorname*{ess\,sup}_{x \in \mathbb{R}^{k-1}} \|D^s_{i_1,\ldots,i_s} f\|_{q,i}(x). \tag{8.2.29}$$

Then

$$L_s(f_\varepsilon) \;\le\; \frac{2\varepsilon^{-r}}{a(s,k)}. \tag{8.2.30}$$

To show (8.2.30), observe that

$$\begin{aligned} &D^s_{i_1,\ldots,i_s} f_\varepsilon(x) \qquad\qquad (8.2.31) \\ &= 2\left(\frac{2}{\varepsilon}\right)^{k+s} \int\limits_{\mathbb{R}^k} I\left\{\widetilde{y} \in B_z\left(\frac{\varepsilon}{2}\right)\right\} D^s_{i_1,\ldots,i_s} \bar{g}_s\left(\frac{2}{\varepsilon}(x - \widetilde{y})\right) d\widetilde{y} \\ &= 2\left(\frac{2}{\varepsilon}\right)^{s} \int\limits_{\mathbb{R}^k} I\left\{x - \varepsilon y \in B_z\left(\frac{\varepsilon}{2}\right)\right\} D^s_{i_1,\ldots,i_s} \bar{g}_s(y)\,dy. \end{aligned}$$

By Minkowski's inequality, we get the following bound for the norm of the above quantity:

$$\|D^s_{i_1,\ldots,i_s} f_\varepsilon\|_{q,1}(x) \tag{8.2.32}$$

$$= \left[\int_{\mathbb{R}} |D^s_{i_1,\ldots,i_s} f_\varepsilon(x_1, x_2, \ldots, x_n)|^q \, \mathrm{d}x_1 \right]^{1/q}$$

$$= 2\left(\frac{2}{\varepsilon}\right)^s \left\{ \int_{\mathbb{R}} \left| \int_{\mathbb{R}^k} I\left\{x - \varepsilon y \in B_z\left(\frac{\varepsilon}{2}\right), \|y\| \le 1\right\} \right.\right.$$

$$\left.\left. \cdot I\{B_z(\varepsilon)\backslash B_z\} D^s_{i_1,\ldots,i_s} \overline{g}_s(y) \, \mathrm{d}y \right|^q \mathrm{d}x_1 \right\}^{1/q}$$

$$= 2\left(\frac{2}{\varepsilon}\right)^s \int_{\{\|y\|\le 1\}\cap(B_z(\varepsilon)\backslash B_z)} |D^s_{i_1,\ldots,i_s} \overline{g}_s(y)| \left\{ \int_{\mathbb{R}} I\{x - \varepsilon y \in B_z\left(\frac{\varepsilon}{2}\right) \mathrm{d}x_1 \right\}^{1/q} \mathrm{d}y$$

$$\le 2\left(\frac{2}{\varepsilon}\right)^s \int_{\{\|y\|\le 1\}\cap(B_z(\varepsilon)\backslash B_z)} |D^s_{i_1,\ldots,i_s} \overline{g}_s(y)|$$

$$\cdot \left\{ \int_{\mathbb{R}} I\{z_1 \le x_1 - \varepsilon y_1 \le z_1 + \varepsilon\} \, \mathrm{d}x_1 \right\}^{1/q} \mathrm{d}y.$$

In fact, the inequality is valid a.e. with respect to Lebesgue measure λ^{k-1}. The last integrals are estimated as follows:

$$\left\{ \int_{\mathbb{R}} I\{z_1 \le x_1 - \varepsilon y_1 \le z_1 + \varepsilon\} \, \mathrm{d}x_1 \right\}^{1/q} = \varepsilon^{1/q}, \tag{8.2.33}$$

and

$$\int_{\{\|y\|\le 1\}\cap(B_z(\varepsilon)\backslash B_z)} |D^s_{i_1,\ldots,i_s} \overline{g}_s(y)| \, \mathrm{d}y \tag{8.2.34}$$

$$= \frac{1}{V_r} \int_{\{\|y\|\le 1\}} I\{B_z(\varepsilon)\backslash B_z\} \left| D^{s-1}_{i_1,\ldots,i_{s-1}} s \left(1 - \sum_{i=1}^n y_i^2\right)_+^{s-1} 2y_{i_s} \right| \mathrm{d}y$$

$$= \frac{1}{V_r} \int_{\{\|y\|\le 1\}} I\{B_z(\varepsilon)\backslash B_z\} s! 2^s |y_{i_1} \cdots y_{i_s}| \, \mathrm{d}y$$

$$\le \frac{1}{V_r} s! 2^s \int_{-1}^{1} \cdots \int_{-1}^{1} |y_{i_1} \cdots y_{i_s}| \, \mathrm{d}y \le \frac{1}{V_r} s! 2^s \cdot 2^{k-s} = \frac{1}{V_r} s! 2^k.$$

Similarly, we can argue for any index $1 \le i \le k$, and thus (8.2.30) follows from (8.2.33) and (8.2.34).

From the inequality in (8.2.30) we obtain that given

$$f^*(x) := \frac{f_\varepsilon(x)}{L_s(f_\varepsilon)},$$

$$\begin{aligned} \vartheta_{s,p}(X,Y) &\geq E(f^*(X) - f^*(Y)) \\ &\geq \frac{\varepsilon^r}{2} a(s,k) E(f_\varepsilon(X) - f_\varepsilon(Y)). \end{aligned} \tag{8.2.35}$$

Applying (8.2.27), (8.2.28) we arrive at the following decomposition:

$$\begin{aligned} E(f_\varepsilon(X) - f_\varepsilon(Y)) &= \int_{\mathbb{R}^k} (f_\varepsilon(x)+1)(P_X - P_Y)(\mathrm{d}x) \\ &= \left(\int_{B_z} + \int_{B_z(\varepsilon)\setminus B_z} + \int_{\mathbb{R}^k \setminus B_z(\varepsilon)} \right) (f_\varepsilon(x)+1)(P_X - P_Y)(\mathrm{d}x) \\ &=: I_1 + I_2 + I_3, \end{aligned}$$

where

$$\begin{aligned} I_1 &= \int_{B_z} (f_\varepsilon(x)+1)(P_X - P_Y)(\mathrm{d}x) = 2(P_X - P_Y)(B_z); \\ I_2 &\geq -2\,P_Y(B_z(\varepsilon)\setminus B_z); \\ I_3 &= 0. \end{aligned}$$

Thus by (8.2.23),

$$I_1 + I_2 + I_3 \geq 2(P_X(B_z) - P_Y(B_z)) - 2(P_Y(B_z(\varepsilon)) - P_Y(B_z)) \geq 2\varepsilon.$$

From (8.2.35) we finally obtain $\vartheta_{s,p}(X,Y) \geq \varepsilon^{r+1} a(s,k)$. With $\varepsilon \to L(X,Y)$, this implies (8.2.21). □

Remark 8.2.7 (a) *Let us use the polar transformation*

$$\begin{aligned} x_1 &= \varrho \cos\vartheta_1 \cdots \cos\vartheta_{k-2}\cos\vartheta_{k-1}, \\ x_2 &= \varrho \cos\vartheta_1 \cdots \cos\vartheta_{k-2}\sin\vartheta_{k-1}, \\ x_3 &= \varrho \cos\vartheta_1 \cdots \sin\vartheta_{k-2}, \\ &\vdots \\ x_k &= \varrho \sin\vartheta_1, \end{aligned} \tag{8.2.36}$$

where $\varrho > 0$, $0 \le \vartheta_1 \le 2\pi$, $0 \le \vartheta_j \le \pi$, $2 \le j \le k-1$, *and*

$$\frac{\partial(x_1, \ldots, x_k)}{\partial(\varrho, \vartheta_1, \ldots, \vartheta_{k-1})} = \varrho^{k-1} D_k(\vartheta), \tag{8.2.37}$$

$$D_k(\vartheta) := \det \begin{pmatrix} \sin\vartheta_1 & \cdots & \sin\vartheta_{k-1} & \sin\vartheta_{k-1} \\ \sin\vartheta_1 & \cdots & \sin\vartheta_{k-2} & \cos\vartheta_{k-1} \\ \sin\vartheta_1 & \cdots & \cos\vartheta_{k-2} & 0 \\ \vdots & & & \\ \cos\vartheta_1 & 0 & \cdots & 0 \end{pmatrix}.$$

Then we have

$$\begin{aligned} V_r &= D_k \int_0^1 (1-\varrho^2)^{r+1} \varrho^{k-1}\, \mathrm{d}\varrho \qquad (8.2.38) \\ &= D_k \int_0^1 (1-\varrho^2)^{r+1} (\varrho^2)^{\frac{k-2}{2}}\, \mathrm{d}\varrho^2 \\ &= D_k \frac{1}{2} \frac{\Gamma(r+2)\Gamma(\frac{k}{2})}{\Gamma(r+1+\frac{k}{2})}, \end{aligned}$$

where

$$D_k := \int D_k(\vartheta)\, \mathrm{d}\vartheta.$$

(b) *Note that lower bound of* $\vartheta_{s,p}$ ($r = s - 1 + \frac{1}{p} \notin \mathbb{N}$) *in terms of the Prohorov metric exists. This follows from an example in Maejima and Rachev (1987) in the case* $k = 1$.

We next investigate smoothing inequalities, which play an important role in the proof of Berry–Esséen-type theorems. They are also of interest for the study of intrinsic properties of probability metrics.

Lemma 8.2.8 (a) *Let* Z *be independent of* X, Y, $\varepsilon > 0$, $r = s - 1 + \frac{1}{p}$. *Then*

$$\vartheta_{s,p}(X,Y) \le \vartheta_{s,p}(X+\varepsilon Z, Y+\varepsilon Z) + 2\frac{\Gamma(1+\frac{1}{p})}{\Gamma(1+r)} \varepsilon^r k E\|Z\|_1^r. \tag{8.2.39}$$

(b) *If* Z, W *are independent of* X, Y, *then*

$$\vartheta_{s,p}(X+Z, Y+Z) \le \vartheta_{s,p}(X,Y)\sigma(W,Z) + \vartheta_{s,p}(X+W, Y+W); \tag{8.2.40}$$

and moreover,

$$\vartheta_{s,p}(X+Z, Y+Z) \le \vartheta_{s,p}(W,Z)\sigma(X,Y) + \vartheta_{s,p}(X+W, Y+W), \tag{8.2.41}$$

where σ is the total variation distance.

Proof: (a) By the regularity of $\vartheta_{s,p}$ (cf. 8.2.1) we have

$$\begin{aligned}\vartheta_{s,p}(X,Y) &\le \vartheta_{s,p}(X+\varepsilon Z, Y+\varepsilon Z) + 2\vartheta_{s,p}(0,\varepsilon Z)\\ &\le \vartheta_{s,p}(X+\varepsilon Z, Y+\varepsilon Z) + 2\varepsilon^r\vartheta_{s,p}(0,Z).\end{aligned}$$

By (8.2.11) and (8.2.15),

$$\vartheta_{s,p}(0,Z) \le \frac{\Gamma(1+\alpha)}{\Gamma(1+r)} \sum_{1\le i_1,\dots,i_{s-1}\le k} E\sum_{i=1}^{k}|Z_i|^\alpha|Z_{i_1}\cdots Z_{i_{s-1}}|.$$

(b) For any $f\in G_{s,p}$ we have

$$\begin{aligned}&|E(f(X+Z)-f(Y+Z))| \qquad (8.2.42)\\ &\le |E(f(X+Z)-f(X+W))-E(f(Y+Z)-f(Y+W))|\\ &\quad +|E(f(X+W)-f(Y+W))|.\end{aligned}$$

If $f\in G_{s,p}$, then the translates $f_x(z) := f(x+z)$ are also in $G_{s,p}$, and therefore, the first term is estimated by conditioning on X (respectively Y):

$$\begin{aligned}&\left|\int E(f(x+Z)-f(x+W))\,\mathrm{d}P_X(x) - \int E(f(x+Z)-f(x+W))\,\mathrm{d}P_Y(x)\right|\\ &= \left|\int E(f(x+Z)-f(x+W))\,\mathrm{d}(P_X-P_Y)(x)\right| \qquad (8.2.43)\\ &\le \vartheta_{s,p}(Z,W)\sigma(X,Y).\end{aligned}$$

Indeed, the inequalities (8.2.42), (8.2.43) imply

$$\vartheta_{s,p}(X+Z,Y+Z) \le \vartheta_{s,p}(Z,W)\sigma(X,Y)+\vartheta_{s,p}(X+W,Y+W).$$

The other case is derived similarly. □

Lemma 8.2.9 *If Z is independent of X, Y and P_Z has a density p_Z having integrable $(s-1)$-fold derivatives*

$$C_{s,Z} := \sup_{1\le i_1,\dots,i_{s-1}\le k} \int_{\mathbb{R}^k} |D^{s-1}_{i_1,\dots,i_{s-1}} p_Z(x)|\,\mathrm{d}x < \infty, \qquad (8.2.44)$$

then

$$\vartheta_{1,p}(X+Z,Y+Z) \le C_{s,Z}\vartheta_{s,p}(X,Y). \qquad (8.2.45)$$

Proof: For any $f \in G_{1,p}$,

$$E(f(X+Z) - f(Y+Z)) = \int_{\mathbb{R}^k} f(x)\,\mathrm{d}(F_{X+Z} - F_{Y+Z})(x) \tag{8.2.46}$$
$$= \int_{\mathbb{R}^k} \left(\int_{\mathbb{R}^k} f(x) p_Z(x-z)\,\mathrm{d}(F_X - F_Y)(z) \right) \mathrm{d}x$$
$$= \int_{\mathbb{R}^k} f^*(z)\,\mathrm{d}(F_X - F_Y)(z),$$

where $f^*(z) = \int_{\mathbb{R}^k} f(x) p_Z(x-z)\,\mathrm{d}x$. From the Taylor expansion,

$$f^*(z) = f^*(0) + \sum_{j=1}^{s-1} \sum_{1 \le i_1,\ldots,i_j \le k} D^j_{i_1,\ldots,i_j} f^*(0) z_{i_1} \cdots z_{i_j} \tag{8.2.47}$$
$$+ \sum_{1 \le i_1,\ldots,i_s \le k} \int_0^1 \frac{(1-t)^{s-1}}{(s-1)!} D^s_{i_1,\ldots,i_s} f^*(tz) z_{i_1} \cdots z_{i_s}\,\mathrm{d}t.$$

Since $f \in G_{1,p}$, i.e.,

$$\left(\int_{\mathbb{R}} |D^1_i f(x_1, \ldots, x_i, \ldots, x_n)|^q\,\mathrm{d}x_i \right)^{1/q} \le 1 \quad \text{a.s.,} \tag{8.2.48}$$

we have the following bound for the qth norm of f^*-derivatives:

$$\left(\int_{\mathbb{R}} |D^s_{i_1,\ldots,i_s} f^*(z_1, \ldots, z_n)|^q\,\mathrm{d}z_i \right)^{1/q} \tag{8.2.49}$$
$$= \left(\int_{\mathbb{R}} \left| D^s_{i_1,\ldots,i_s} \int_{\mathbb{R}^k} f(x) p_Z(x-z)\,\mathrm{d}x \right|^q \mathrm{d}z_i \right)^{1/q}$$
$$= \left(\int_{\mathbb{R}} \left| D^1_{i_1} \int_{\mathbb{R}^k} f(x) D^{s-1}_{i_2,\ldots,i_s} p_Z(x-z)\,\mathrm{d}x \right|^q \mathrm{d}z_i \right)^{1/q}$$
$$= \left(\int_{\mathbb{R}} \left| D^1_{i_1} \int_{\mathbb{R}^k} f(x+z) D^{s-1}_{i_2,\ldots,i_s} p_Z(x)\,\mathrm{d}x \right|^q \mathrm{d}z_i \right)^{1/q}$$

$$= \left(\int_{\mathbb{R}} \left| \int_{\mathbb{R}^k} (D^1_{i_1} f(x+z))\,(D^{s-1}_{i_2,\dots,i_s} p_Z(x))\,\mathrm{d}x \right|^q \mathrm{d}z_i \right)^{1/q}$$

$$\le \int_{\mathbb{R}^k} \left\{ \int_{\mathbb{R}} \left|(D^1_{i_1} f(x+z))(D^{s-1}_{i_2,\dots,i_s} p_Z(x))\right|^q \mathrm{d}z_i \right\}^{1/q} \mathrm{d}x$$

$$= \int_{\mathbb{R}^k} \left|D^{s-1}_{i_2,\dots,i_s} p_Z(x)\right| \left\{ \int_{\mathbb{R}} \left|D^1_{i_1} f(x+z)\right|^q \mathrm{d}z_i \right\}^{1/q} \mathrm{d}x$$

$$\le \int_{\mathbb{R}^k} \left|D^{s-1}_{i_2,\dots,i_s} p_Z(x)\right| \mathrm{d}x \;=\; C_{s,Z} \quad \text{by (8.2.48)).}$$

Summarizing the results in (8.2.46), (8.2.47), and (8.2.48), we derive the desired inequality (8.2.45). □

As a consequence of Lemmas 8.2.8, 8.2.9 we next obtain an estimate between $\vartheta_{1,p}$ and $\vartheta_{s,p}$.

Theorem 8.2.10 *For every $s = 1, 2, \dots$, $p \in [1, \infty)$, $r := s - 1 + \frac{1}{p}$, and random vectors X, Y on $\mathbb{R}^k$ we have*

$$\vartheta_{1,p}(X,Y) \;\le\; A(s,p,k)\vartheta_{s,p}^{\frac{1}{pr}}(X,Y), \tag{8.2.50}$$

where

$$A(s,p,k) \;:=\; a^{\frac{1}{pr}} b^{\frac{s-1}{r}} (p(s-1))^{\frac{1}{pr}} \frac{r}{s-1} \tag{8.2.51}$$

and

$$a \;:=\; \frac{1}{\sqrt{s}} \left(\frac{2s}{\pi}\right)^{s/2}, \qquad b \;:=\; 2k \left(k\sqrt{\frac{2}{\pi}}\right)^{1/p}. \tag{8.2.52}$$

Proof: Recall the inequality (8.2.39). Then for any $\varepsilon > 0$ and any $Z \stackrel{d}{=} N(0,1)$ independent of X, Y, we have

$$\vartheta_{1,p}(X,Y) \;\le\; \vartheta_{1,p}(X+\varepsilon Z, Y+\varepsilon Z) + 2\varepsilon^{1/p} k\, E\|Z\|_1^{1/p}; \tag{8.2.53}$$

and furthermore,

$$E\|Z\|_1^{1/p} \;\le\; (E\|Z\|_1)^{1/p} \;=\; \left(\sum_{i=1}^{k} E|Z_i|\right)^{1/p} \;=\; \left(k\sqrt{\frac{2}{\pi}}\right)^{1/p}. \tag{8.2.54}$$

Now we apply Lemma 8.2.9 to get

$$\vartheta_{1,p}(X+\varepsilon Z, Y+\varepsilon Z) \;\le\; C_{s,\varepsilon Z}\vartheta_{s,p}(X,Y). \tag{8.2.55}$$

Next, assuming that Z is standard normal with independent components, we bound the constant in (8.2.55) as follows:

$$\begin{aligned} C_{s,\varepsilon Z} &= \sup_{i_1,\dots,i_{s-1}} \int_{\mathbb{R}^k} |D^{s-1}_{i_1,\dots,i_{s-1}} p_Z(\frac{x}{\varepsilon})| \varepsilon^{-s-n+1}\,\mathrm{d}x \qquad (8.2.56)\\ &= \varepsilon^{1-s} C_{s,Z} \;=\; \varepsilon^{1-s} C_{s,s^{-1/2}(Z_1+\cdots+Z_s)}. \end{aligned}$$

Here, $Z_1,\dots,Z_s$ are i.i.d. copies of Z, and thus (see Zolotarev (1977, 1983, 1986))

$$\begin{aligned} C_{s,\varepsilon Z} &= \varepsilon^{1-s} s^{-\frac{1}{2}(1-s)} C_{s,Z_1+\cdots+Z_s} \qquad (8.2.57)\\ &\le \varepsilon^{1-s} s^{\frac{s-1}{2}} (C_{1,Z})^s, \end{aligned}$$

where

$$\begin{aligned} C_{1,Z} &= \sup_{1\le i\le k} \int_{\mathbb{R}^k} |D^1_i p_Z(x)|\,\mathrm{d}x \\ &= \sup_i \int_{\mathbb{R}^k} \left(\frac{1}{\sqrt{2\pi}}\right)^n |x_i| e^{-\frac{x_i^2}{2}} \prod_{j\ne i} e^{-\frac{x_j^2}{2}}\,\mathrm{d}x \\ &= \sup_i \int_{\mathbb{R}} \frac{1}{\sqrt{2\pi}} |x_i| e^{-\frac{x_i^2}{2}}\,\mathrm{d}x_i \;=\; \sqrt{\frac{2}{\pi}}. \end{aligned}$$

Therefore, from (3.15), (3.17), (3.19) we obtain

$$\vartheta_{1,p}(X,Y) \;\le\; \varepsilon^{1-s}\frac{1}{\sqrt{s}}\left(\frac{2s}{\pi}\right)^{s/2} \vartheta_{s,p}(X,Y) + 2\varepsilon^{1/p}k\left(k\sqrt{\frac{2}{\pi}}\right)^{1/p}. \tag{8.2.58}$$

Define $\varphi(x) := ax^{1-s} + bx^{1/p}$,

$$a \;:=\; \frac{1}{\sqrt{s}}\left(\frac{2s}{\pi}\right)^{s/2}\vartheta_{s,p}(X,Y), \qquad b \;:=\; 2k\left(k\sqrt{\frac{2}{\pi}}\right)^{1/p}.$$

Minimizing φ with respect to x yields (8.2.50). □

As a consequence of the smoothing properties we shall establish a Berry–Esséen-type result, that will provide the right order estimate in the stable central limit theorem in terms of the metric $\vartheta_{1,1}$. Let (X_i) be an i.i.d.

sequence of random vectors in $\mathbb{R}^k$; let (Θ, Θ_i) be an i.i.d. sequence of symmetric α-stable distributed random vectors, i.e. $n^{-1/\alpha}(\Theta_1 + \cdots + \Theta_n) \stackrel{d}{=} \Theta$; and define

$$\vartheta_r := \vartheta_r(X_1, \Theta) := \sup_{h>0} h^{r-1} \vartheta_{1,1}(X_1 + h\Theta, \Theta_1 + h\Theta), \tag{8.2.59}$$

$$\sigma_r := \sigma_r(X_1, \Theta) := \sup_{h>0} h^r \, \sigma(X_1 + h\Theta, \Theta_1 + h\Theta), \tag{8.2.60}$$

$$\vartheta := \vartheta_{1,1}(X_1, \Theta), \quad \sigma := \sigma(X_1, \Theta), \quad \tau_r := \max\left(\vartheta, \sigma, \sigma_r^{\frac{1}{(r-\alpha)}}\right). \tag{8.2.61}$$

Theorem 8.2.11 *Suppose that* $1 < \alpha \le 2$, $\alpha < r$, *and* $\vartheta + \vartheta_r + \sigma + \sigma_r < \infty$. *Let* $Z_n = n^{-1/\alpha} \sum_{i=1}^n X_i$. *Then for some absolute constant* $C = C(k)$ *depending only on the dimension* k,

$$\vartheta_{1,1}(Z_n, \Theta) \le C\left(n^{1-\frac{r}{\alpha}} \vartheta_r + \tau_r \, n^{-1/\alpha}\right). \tag{8.2.62}$$

Proof: We shall use the notation $\vartheta(X,Y) = \vartheta_{1,1}(X,Y)$ during this proof. Note that by (8.2.10), $\vartheta(X,Y) = \overline{\zeta}_1(X,Y) = \zeta_1(X,Y)$. From the smoothing inequality (8.2.39) we obtain the following bound: For any $\varepsilon > 0$ with Θ_i, Θ independent and identically distributed,

$$\vartheta(Z_n, \Theta_1) \le \vartheta(Z_n + \varepsilon\Theta, \, \Theta_1 + \varepsilon\Theta) + C\varepsilon, \tag{8.2.63}$$

where $C := 2\,k\,E\|\Theta\|_1$. Our proof will be based on the Bergström convolution method (cf. Rachev (1991, Chapter 18) and the references therein). We start by making use of the triangle inequality:

$$\begin{aligned}
&\vartheta(Z_n + \varepsilon\Theta, \Theta_1 + \varepsilon\Theta) \qquad\qquad (8.2.64)\\
&\le \vartheta\left(Z_n + \varepsilon\Theta, \; \frac{\Theta_1 + X_2 + \cdots + X_n}{n^{1/\alpha}} + \varepsilon\Theta\right)\\
&\quad + \sum_{j=1}^m \vartheta\left(\frac{\Theta_1 + \cdots + \Theta_j + X_{j+1} + \cdots + X_n}{n^{1/\alpha}} + \varepsilon\Theta,\right.\\
&\qquad\qquad \left.\frac{\Theta_1 + \cdots + \Theta_{j+1} + X_{j+2} + \cdots + X_n}{n^{1/\alpha}} + \varepsilon\Theta\right)\\
&\quad + \vartheta\left(\frac{\Theta_1 + \cdots + \Theta_{m+1} + X_{m+2} + \cdots + X_n}{n^{1/\alpha}} + \varepsilon\Theta, \; \Theta_1 + \varepsilon\Theta\right)\\
&=: I_o + \sum_{j=1}^m I_j + I_{m+1}, \qquad m = [n/2].
\end{aligned}$$

Applying the smoothing property (8.2.41), we obtain

$$I_o \le \vartheta\left(\frac{X_2 + \cdots + X_n}{n^{1/\alpha}}, \; \frac{\Theta_2 + \cdots + \Theta_n}{n^{1/\alpha}}\right) \tag{8.2.65}$$

$$\cdot \sigma\left(n^{-1/\alpha}X_1 + \varepsilon\Theta,\ n^{-1/\alpha}\Theta_1 + \varepsilon\Theta\right)$$
$$+ \vartheta\left(\frac{X_1 + \Theta_2 + \cdots + \Theta_n}{n^{1/\alpha}} + \varepsilon\Theta,\ \frac{\Theta_1 + \cdots + \Theta_n}{n^{1/\alpha}} + \varepsilon\Theta\right).$$

Similarly, for $1 \le j \le m$, we have

$$\begin{aligned} I_j \;\le\; & \vartheta\left(\frac{X_{j+2} + \cdots + X_n}{n^{1/\alpha}},\ \frac{\Theta_{j+2} + \cdots + \Theta_n}{n^{1/\alpha}}\right) \\ & \cdot \sigma\left(\frac{\Theta_1 + \cdots + \Theta_j + X_{j+1}}{n^{1/\alpha}} + \varepsilon\Theta,\ \frac{\Theta_1 + \cdots + \Theta_{j+1}}{n^{1/\alpha}} + \varepsilon\Theta\right) \\ & + \vartheta\left(\frac{\Theta_1 + \cdots + \Theta_j + X_{j+1} + \Theta_{j+2} + \cdots + \Theta_n}{n^{1/\alpha}} + \varepsilon\Theta,\ \Theta_1 + \varepsilon\Theta\right). \end{aligned} \tag{8.2.66}$$

Summarizing the above inequalities, we get the bound

$$\vartheta(Z_n, \Theta_1) \;\le\; \sum_{j=1}^{5} \Delta_j, \tag{8.2.67}$$

where

$$\begin{aligned} \Delta_1 \;&:=\; \vartheta\left(\frac{X_2 + \cdots + X_n}{n^{1/\alpha}},\ \frac{\Theta_2 + \cdots + \Theta_n}{n^{1/\alpha}}\right) \sigma\left(\frac{X_1}{n^{1/\alpha}} + \varepsilon\Theta,\ \frac{\Theta_1}{n^{1/\alpha}} + \varepsilon\Theta\right), \\ \Delta_2 \;&:=\; \sum_{j=1}^{m} \vartheta\left(\frac{X_{j+2} + \cdots + X_n}{n^{1/\alpha}},\ \frac{\Theta_{j+2} + \cdots + \Theta_n}{n^{1/\alpha}}\right) \\ & \qquad \cdot \sigma\left(\frac{\Theta_1 + \cdots + \Theta_j + X_{j+1}}{n^{1/\alpha}} + \varepsilon\Theta,\ \frac{\Theta_1 + \cdots + \Theta_{j+1}}{n^{1/\alpha}} + \varepsilon\Theta\right), \\ \Delta_3 \;&:=\; (m+1)\vartheta\left(\frac{X_1 + \Theta_2 + \cdots + \Theta_n}{n^{1/\alpha}},\ \Theta\right), \\ \Delta_4 \;&:=\; \vartheta\left(\frac{\Theta_1 + \cdots + \Theta_{m+1} + X_{m+2} + \cdots + X_n}{n^{1/\alpha}} + \varepsilon\Theta,\ \Theta_1 + \varepsilon\Theta\right), \\ \Delta_5 \;&:=\; C\varepsilon \;=\; 2k\,E\|\Theta_1\|\varepsilon. \end{aligned}$$

To estimate the terms Δ_i, we introduce smoothed versions of the metrics ϑ, σ defined as follows: For $r > 1$,

$$\vartheta_r(X, Y) \;:=\; \sup_{h>0} h^{r-1}\vartheta(X + h\Theta, Y + h\Theta), \tag{8.2.68}$$

$$\sigma_r(X, Y) \;:=\; \sup_{h>0} h^{r}\sigma(X + h\Theta, Y + h\Theta) \tag{8.2.69}$$

(cf. also (8.2.59), (8.2.60)). It is easy to see that both ϑ_r, σ_r are ideal metrics of order r.

We first estimate Δ_3 and Δ_4 using the ideality of ϑ_r. In the rest of the proof, c stands for a general constant, which may be different at different places:

$$\begin{aligned}\Delta_3 &= (m+1)\vartheta\left(n^{-1/\alpha}X_1+\left(\frac{n-1}{n}\right)^{1/\alpha}\Theta,\, n^{-1/\alpha}\Theta_1+\left(\frac{n-1}{n}\right)^{1/\alpha}\Theta\right)\\ &\le (m+1)\left(\frac{n}{n-1}\right)^{\frac{r-1}{\alpha}}\vartheta_r\left(n^{-1/\alpha}X_1,\, n^{-1/\alpha}\Theta\right) \qquad (8.2.70)\\ &\le cn^{1-\frac{r}{\alpha}}\vartheta_r.\end{aligned}$$

Similarly,

$$\begin{aligned}\Delta_4 &\le \vartheta_r\left(\frac{X_{m+2}+\cdots+X_n}{n^{1/\alpha}},\frac{\Theta_{m+2}+\cdots+\Theta_n}{n^{1/\alpha}}\right)\left(\frac{n}{m+1}\right)^{\frac{r-1}{\alpha}} \qquad (8.2.71)\\ &\le cn^{1-\frac{r}{\alpha}}\vartheta_r.\end{aligned}$$

Define

$$\varepsilon := A\max\left(\sigma_1,\sigma_r^{\frac{1}{r-\alpha}}\right)n^{-1/\alpha}, \quad A>0. \qquad (8.2.72)$$

The proof continues by induction on n. The induction hypothesis states that for all $j<n$,

$$\vartheta\left(\frac{X_1+\cdots+X_j}{j^{1/\alpha}},\frac{\Theta_1+\cdots+\Theta_j}{j^{1/\alpha}}\right) \le B(\vartheta_r j^{1-\frac{r}{\alpha}}+\tau_r j^{-1/\alpha}). \qquad (8.2.73)$$

(For $n=1,\ldots,n_0$, n_0 fixed, (8.2.62) follows from $\tau_r\ge\vartheta$ and the ideality of $\vartheta_{1,1}$.) Then, for $\Delta_1=\Delta_1(n)$, we obtain

$$\begin{aligned}\Delta_1 &\le Bc(n^{1-\frac{r}{\alpha}}\vartheta_r+n^{-1/\alpha}\tau_r)\sigma(n^{-1/\alpha}X_1+\varepsilon\Theta,\, n^{-1/\alpha}\Theta_1+\varepsilon\Theta) \qquad (8.2.74)\\ &\le Bc(n^{1-\frac{r}{\alpha}}\vartheta_r+n^{-1/\alpha}\tau_r)\varepsilon^{-1}n^{-1/\alpha}\sigma_1\\ &\le \frac{1}{A}Bc\left(n^{1-\frac{r}{\alpha}}\vartheta_r+n^{-1/\alpha}\tau_r\right).\end{aligned}$$

Similarly, we estimate Δ_2:

$$\begin{aligned}\Delta_2 &\le cB(\vartheta_r(n-m-2)^{1-\frac{r}{\alpha}}+\tau_r(n-m-2)^{-1/\alpha}) \qquad (8.2.75)\\ &\quad\cdot\sum_{i=1}^{\infty}\sigma_r\left(n^{-1/\alpha}X_1,n^{-1/\alpha}\Theta\right)\left(\frac{j}{n}+\varepsilon^\alpha\right)^{-r/\alpha}\\ &\le cB\left(\vartheta_r n^{1-\frac{r}{\alpha}}+\tau_r n^{-1/\alpha}\right)\sum_{j=1}^{\infty}\frac{\sigma_r}{(j+\varepsilon^\alpha n)^{r/\alpha}}\\ &\le cB\left(\vartheta_r n^{1-\frac{r}{\alpha}}+\tau_r n^{-1/\alpha}\right)\frac{1}{A^{r-\alpha}}.\end{aligned}$$

From (8.2.70)–(8.2.75) we infer

$$\begin{aligned} \vartheta(Z_n,\Theta) &\le C_1\left(\frac{1}{A}+\frac{1}{A^{r-\alpha}}\right)B\left(\vartheta_r\, n^{1-\frac{r}{\alpha}}+\tau_r\, n^{-1/\alpha}\right) \\ &\quad + C_2\left(\vartheta_r\, n^{1-\frac{r}{\alpha}}+A\tau_r n^{-1/\alpha}\right). \end{aligned} \tag{8.2.76}$$

Choosing A big enough so that $C_1(\frac{1}{A}+\frac{1}{A^{r-\alpha}})<\frac{1}{2}$ and then B such that $B/2>C_2(1+A)$, we complete the proof. □

Remark 8.2.12 (a) *We note that the conditions concerning the domain of attraction of* Θ *are given solely in terms of the metrics appearing in the upper bounds.*

(b) *Since* Θ *has a density* p_Θ *with integrable derivatives, we can get, similarly to the proof of Lemma* 8.2.9, *that*

$$\sigma(X+\delta\Theta,Y+\delta\Theta) \le C\delta^{-1}\vartheta_{1,1}(X,Y), \tag{8.2.77}$$

where $C=C(\Theta)$. *This implies that for any* $0<\varepsilon<1$,

$$\begin{aligned} \sigma_r(X,Y) &= \sup_{h>0} h^r\sigma(X+h\Theta,Y+h\Theta) \\ &= \sup_{h>0} h^r\sigma\Big(X+h(1-\varepsilon)^{1/\alpha}\Theta_1+h\varepsilon^{1/\alpha}\Theta_2, \\ &\qquad\qquad Y+h(1-\varepsilon)^{1/\alpha}\Theta_1+h\varepsilon^{1/\alpha}\Theta_2\Big) \\ &\le \sup_{h>0} h^r\vartheta_{1,1}\left(X+h(1-\varepsilon)^{-1/\alpha}\Theta_1, Y+h(1-\varepsilon)^{1/\alpha}\Theta_2\right) \\ &\qquad\qquad \cdot C(\Theta)/(h(1-\varepsilon)^{1/\alpha}) \\ &= C(\Theta)(1-\varepsilon)^{(1-r)/\alpha}\varepsilon^{-1/\alpha}\vartheta_r(X,\Theta). \end{aligned}$$

The minimum in the right-hand side is attained for $\varepsilon=1/r$, *implying that*

$$\begin{aligned} \sigma_r(X,Y) &\le C(\Theta)r^{r/\alpha}(r-1)^{(1-r)/\alpha}\vartheta_r(X,\Theta) \\ &\le C(\Theta)2^{r/\alpha}\vartheta_r(X,\Theta). \end{aligned} \tag{8.2.78}$$

Similarly to the proof of Theorem 8.2.10, *relations* (8.2.77), (8.2.78) *allow us to replace* (8.2.62) *by a bound involving only* ϑ, ϑ_r:

$$\vartheta_{1,1}(Z_n,\Theta) \le C\left(n^{1-\frac{r}{\alpha}}\vartheta_r+\max\{\vartheta,\vartheta_r^{1/(r-\alpha)}\}n^{-1/\alpha}\right). \tag{8.2.79}$$

For $r\in\mathbb{N}$, ϑ_r, *the* r*-smoothed* ϑ *metric, can be estimated from above by the* ζ_r *metric:*

$$\vartheta_r \le c_r\zeta_r, \tag{8.2.80}$$

where c_r *depends on* $\sup_{\|z\|\le 1}\int |p_\Theta^{(r)}(y)(z)|\,dy$ *(cf. Zolotarev (1983, p. 294, property 6)). Also, for* $r \in \mathbb{N}$, ζ_r *is estimated from above by the* r*th difference pseudomoment* k_r, *defined by*

$$k_r(X,Y) = \sup\Big\{|E(f(X)-f(Y))|;\ f \text{ bounded},\ f:\mathbb{R}^k \to \mathbb{R}, \quad (8.2.81)$$
$$|f(x)-f(y)| \le \big\|x\|x\|^{r-1} - y\,\|y\|^{r-1}\big\|\Big\}$$

and

$$k_r(X,Y) \le 2^r v_r(X,Y) = 2^r \int \|x\|^r\,|P_X - P_Y|(\,dx), \qquad (8.2.82)$$

where v_r *is the absolute pseudomoment of order* r*. From* (8.2.79)– (8.2.82) *we obtain easy-to-check criteria for finiteness of the upper bounds. In particular, in the normal case* $\alpha = 2$, *we take* $r = 3$, *and so the finiteness of the third moments of* X_i *implies the Berry–Esséen result. In the case* $1 < \alpha < 2$ *we use the boundness of* k_r *in* (8.2.81).

(c) *In the case* $k = 1$, $\alpha = 2$ *(normal case, dimension one) the result of Theorem* 8.2.11 *is due to Zolotarev (1987), based on the proof of Senatov (1980).*

From part (b) of the above remark, it follows that we can replace the terms ϑ, σ, ϑ_r, σ_r in the upper bound in (8.2.62) by k_1 and k_r. Since k_r is topologically weaker than v_r, it is of interest to obtain alternative bounds for k_r. To this end let us recall the minimal ℓ_r-metric: for $r > 0$,

$$\ell_r(X,Y) := \inf\Big\{(E\|\widetilde{X} - \widetilde{Y}\|^r)^{(1/r)\wedge 1};\ \widetilde{X} \stackrel{d}{=} X,\ \widetilde{Y} \stackrel{d}{=} Y\Big\}. \qquad (8.2.83)$$

Then

$$\begin{aligned} k_r(X,Y) &= \inf\Big\{|E\|\widetilde{X}\|\widetilde{X}\|^{r-1} - \widetilde{Y}\|\widetilde{Y}\|^{r-1}|;\ \widetilde{X} \stackrel{d}{=} X,\ \widetilde{Y} \stackrel{d}{=} Y\Big\} \qquad (8.2.84)\\ &= \ell_1(X\|X\|^{r-1},\ Y\|Y|^{r-1})\\ &= \inf\Big\{E\|U - V\|;\ U \stackrel{d}{=} X\|X\|^{r-1},\ V \stackrel{d}{=} Y\|Y\|^{r-1}\Big\}\end{aligned}$$

(cf. Rachev and Rüschendorf (1990)). If X and Y have densities f_X and f_Y, respectively, then (cf. Rachev (1991, pp. 249–252)) we use that $k_1 = \ell_1$ to get the bound

$$\begin{aligned} k_r(X,Y) &\le \alpha_r(X,Y) \qquad (8.2.85)\\ &:= \int_{\mathbb{R}^k} \|x\|_1 \left| \int_0^1 t^{-k-1}\left(f_{X\|X\|^{r-1}}\left(\frac{x}{t}\right) - f_{Y\|Y\|^{r-1}}\left(\frac{x}{t}\right)\right) dt \right| dx.\end{aligned}$$

For some examples with an equality in (8.2.85), see Rachev (1991, p. 252). The densities of $X\|X\|^{r-1}$, $Y\|Y\|^{r-1}$ are obtainable from the transformation formula. In particular, this gives explicit bounds in the case $r = 1$, where the expression in (8.2.85) simplifies. The following upper bound for k_r, $r > 1$, will turn out to be useful.

Lemma 8.2.13 *If* $E\|X\|^r < \infty$, $E\|Y\|^r < \infty$, $r > 1$, *then*

$$k_r(X,Y) \leq c\ell_r(X,Y), \tag{8.2.86}$$

where the constant c *depends on the* r*th moments of* X, Y.

Proof: Let $r' := \frac{r}{r-1}$ and $U \stackrel{d}{=} X$, $V \stackrel{d}{=} Y$. Then

$$\begin{aligned} & E\left\|U\|U\|^{r-1} - V\|V\|^{r-1}\right\| \\ \leq\ & E\|U\|^{r-1}\|U - V\| + E\|V\| \left|\|U\|^{r-1} - \|V\|^{r-1}\right| \\ =:\ & I_1 + I_2. \end{aligned}$$

For I_1, I_2 we readily get the bounds $I_1 \leq (E\|U\|^r)^{1/r}(E\|U - V\|^r)^{1/r}$ and

$$I_2 \leq \begin{cases} (E\|V\|^r)^{1/r}(E\|U - V\|^r)^{1/r}, & 1 < r \leq 2, \\ (r-1)(E\|U - V\|^r)^{1/r} & \\ \quad \cdot \left((E\|V\|^r)^{1/r} + (E\|V\|^r)^{\frac{1}{r-1}}(E\|U\|^r)^{\frac{r-2}{r-1}}\right), & r > 2. \end{cases}$$

So $I_1 + I_2 \leq c(E\|U - V\|^r)^{1/r}$, and passing to the corresponding minimal metrics, we get $k_r(X,Y) \leq c\ell_r(X,Y)$, as required. □

In some examples one can determine k_r explicitly. Suppose that for some radial transformation

$$\phi : \mathbb{R}^k \to \mathbb{R}^k, \quad \phi(x) := \begin{cases} \frac{\alpha(\|x\|)}{\|x\|}x & \text{if } x \neq 0, \\ 0 & \text{if } x = 0, \end{cases}$$

with α monotonically nondecreasing, we have $Y \stackrel{d}{=} \phi(X)$. Examples of this relation include spherically invariant distributions and spherically equivalent distributions, as for example, the uniform distribution on a p-ball in $\mathbb{R}^k$ and the product of Weibull distributions (cf. Section 3.2). By (8.2.84), it is easy to see that the pair $(X, \phi(X))$ is an optimal coupling with respect to k_r, and so we obtain

$$\begin{aligned} k_r(X,Y) &= E\left\|X\|X\|^{r-1} - \phi(X)\|\phi(X)\|^{r-1}\right\| \\ &= E\left|\|X\|^r - \alpha(\|X\|)^r\right|. \end{aligned} \tag{8.2.87}$$

(A related explicit formula was derived for the ℓ_r distance in Section 3.2.) Note that α is determined by the equation $F_{\|Y\|}(y) = P(\alpha(\|X\|) \le y) = F_{\|X\|}(\alpha^{-1}(y))$, which in the case of $F_{\|X\|}$ continuous leads to $\alpha(t) = F_{\|Y\|}^{-1} \circ F_{\|X\|}(t)$.

We illustrate the above resutls, invoking again the stable limit theorem 8.2.11. Let Θ be a k-dimensional α-stable random vector with spectral measure m such that $\int_{\mathbb{R}^k} \|x\|^\alpha \, dm(x) < \infty$; i.e.,

$$E\exp\{i\langle t, \Theta\rangle\} = \exp\left\{-\frac{1}{2}\int_{\mathbb{R}^k} |\langle t, s\rangle|^\alpha \, dm(s)\right\}. \tag{8.2.88}$$

We apply the LePage representation for symmetric α-stable laws. Let

1. $(\widetilde{Y}_j)$ be an i.i.d. sequence of random vectors with distribution $m/|m|$ and let $Y_j := |m|^{1/\alpha}\widetilde{Y}_j$;
2. $(\widetilde{\eta}_j)$ be i.i.d. symmetric random variables with $\|\widetilde{\eta}_1\|_\alpha = (E|\widetilde{\eta}_1|^\alpha)^{1/\alpha} < \infty$ and let $\eta_j := \widetilde{\eta}_j / \|\widetilde{\eta}_j\|_\alpha$;
3. (Γ_j) be the sequence of successive times of jump of a standard Poisson process and assume that the three sequences are independent.

Then

$$\Theta \overset{d}{=} c_\alpha \sum_{j=1}^\infty \Gamma_j^{-1/\alpha} \eta_j Y_j, \tag{8.2.89}$$

where the constant c_α is determined by the tail behavior of the law of Θ. Without loss of generality we set $c_\alpha = 1$ (cf. Ledoux and Talagrand (1991, Section 5.1) and Samorodnitsky and Taqqu (1994)).

Suppose that the distribution of X has a similar representation in distribution

$$X \overset{d}{=} \sum_{j=1}^\infty \Gamma_j^{-1/\alpha} \eta_j^* Y_j^*, \tag{8.2.90}$$

where (Y_j^*), (η_j^*) are independent but not necessarily identically distributed. Recall the bound (8.2.80) to see that what we need is an estimate for $\zeta_r(X, \Theta)$ from above.

Proposition 8.2.14 *Let $r > \max\{1, \alpha\}$. Suppose that $Y_j, Y_j^*, \eta_j, \eta_j^*$ have finite rth moments with $\sup_j E|\eta_j^*|^r < \infty$. Then*

$$\zeta_r(X, \Theta) \le C \sup_{j \ge 1} \left(\ell_r(Y_j^*, Y_j) + \int_{-\infty}^\infty |x|^{r-1} |F_{\eta_j^*}(x) - F_{\eta_j}(x)| \, dx\right). \tag{8.2.91}$$

Proof: By the ideality of ζ_r,

$$\zeta_r(X,\Theta) \leq \sum_{j=1}^{\infty} E(\Gamma_j^{-r/\alpha})\zeta_r(\eta_j^* Y_j^*, \eta_j Y_j).$$

Since $r > \alpha$, and for $j > r/\alpha$,

$$E\Gamma_j^{-r/\alpha} = \frac{\Gamma(j - r/\alpha)}{\Gamma(j)} \sim j^{-r/\alpha},$$

the series $S_r = \sum_{j=1}^{\infty} E(\Gamma_j^{-r/\alpha})$ converges. Furthermore,

$$\begin{aligned}
\zeta_r(\eta_j^* Y_j^*, \eta_j Y_j) &\leq \zeta_r(\eta_j^* Y_j^*, \eta_j^* Y_j) + \zeta_r(\eta_j^* Y_j, \eta_j Y_j) \\
&\leq (E|\eta_j^*|^r)\zeta_r(Y_j^*, Y_j) + \sum_{i=1}^{k}(E|Y_{j,i}|^r)\zeta_r(\eta_j^*, \eta_j) \\
&\leq C\left(\zeta_r(Y_j^*, Y_j) + \zeta_r(\eta_j^*, \eta_j)\right),
\end{aligned}$$

where $Y_{j,i}$ is the ith compoment of Y_j. Since

$$\zeta_r(\eta_j^*, \eta_j) \leq \frac{1}{r!}k_r(\eta_j^*, \eta_j) = \frac{1}{(r-1)!}\int_{-\infty}^{\infty} |x|^{r-1}|F_{\eta_j^*}(x) - F_{\eta_j}(x)|\,dx,$$

we obtain (8.2.91) after applying the inequality $\zeta_r(Y_j^*, Y_j) \leq \frac{1}{r!}\kappa_r(Y_j^*, Y_j)$ and Lemma 8.2.13. □

Under the additional assumption

$$\sup_j E\|Y_j^*\|^r < \infty \tag{8.2.92}$$

we obtain (by the obvious bound $\ell_r(Y_j^*, Y_j) \leq (E\|Y_j^*\|^r)^{1/r} + (E\|Y_j\|^r)^{1/r}$) the finiteness of the upper bound in the limit theorem in (8.2.62). In this way we establish a stable limit theorem (with an estimate of the rate of convergence) for random vectors in the ℓ_r-neighborhood of a stable symmetric law in the sense of the LePage representation. For $r = 1$ we use the estimate in (8.2.85). For $r > 1$ (and in particular $r = 2$) explicit expressions for ℓ_r are known in several cases (cf. Section 3.2), for example for the distance $\ell_r(X,Y)$ between normal distributed random vectors X and Y, between uniform distributions on balls and multivariate normal, or Weibull, distributions, and between spherically equivalent distributions.

8.3 Application to Summability Methods and Compound Poisson Approximation

In this section we apply the $\vartheta_{s,p}$-metric (see (8.2.2)) to obtain rate of convergence results in stable limit theorems for multivariate summability methods, thus extending some results of Maejima (1985) in the real case. We also study the approximation of sums of independent random variables by compound Poisson distributions. Let $(X_n)_{n\geq 0}$ be an i.i.d. sequence of random vectors in $\mathbb{R}^k$ and consider the weighted sums

$$T(\lambda) := \sum_{j=0}^{\infty} c_j(\lambda) X_j, \quad c_j(\lambda) \geq 0, \tag{8.3.1}$$

where for $\lambda > 0$ or $\lambda \in \mathbb{N}$, $(c_j(\lambda)), j \geq 0$, is a summability method Some classical summability methods are

$$\text{“Césaro method”} \quad c_j(\lambda) = \begin{cases} \frac{1}{n+1}, & 0 \leq j \leq n, \\ 0, & \text{otherwise}; \end{cases} \tag{8.3.2}$$

$$\text{“Borel method”} \quad c_j(\lambda) = \frac{\lambda^j}{j!} e^{-\lambda}, \ \lambda > 0, \ j \in \mathbb{N}_0; \tag{8.3.3}$$

$$\text{“Euler method”} \quad c_j(\lambda) = \binom{n}{j} \lambda^j (1-\lambda)^{n-j}, \quad 0 \leq j \leq n, \ 0 < \lambda < 1; \tag{8.3.4}$$

$$\text{“Abel method”} \quad c_j(\lambda) = (1 - e^{-1/\lambda}) e^{-j/\lambda}, \ 0 \leq j < \infty; \tag{8.3.5}$$

$$\text{“random walk method”} \quad c_j(n) = P(S_n = j), \quad 0 \leq j < \infty, \tag{8.3.6}$$

where S_n is a random walk on the integers $\mathbb{N}_0$.

For a review and discussion of these methods in the univariate case we refer to Maejima (1985). Let $\Theta_{(\alpha)}$ denote a random vector with symmetric stable distribution of index α, $0 < \alpha \leq 2$. Recall (see Samorodnitsky and Taqqu (1994)) that for $0 < \alpha < 2$, $\Theta_{(\alpha)}$ is symmetric α-stable in $\mathbb{R}^k$ if and only if there exists a (unique) symmetric finite measure Γ on the unit sphere S_k such that

$$\begin{aligned} \varphi_{\Theta_{(\alpha)}}(t) &= E \exp\{i\langle t, \Theta_{(\alpha)}\rangle\} \\ &= \exp\left\{ -\int_{S_k} |(t,s)|^{\alpha} \Gamma(ds) \right\}, \quad t \in \mathbb{R}^k. \end{aligned} \tag{8.3.7}$$

Define then

$$d_\alpha(\lambda) = \left(\sum_{j=0}^{\infty} c_j(\lambda)^\alpha\right)^{1/\alpha}. \tag{8.3.8}$$

Theorem 8.3.1 *Let* $0 < \alpha < r = s - 1 + \frac{1}{p}$. *Then*

$$\vartheta_{s,p}\left(\frac{1}{d_\alpha(\lambda)} T(\lambda), \Theta_{(\alpha)}\right) \le R(\lambda)\vartheta_{s,p}, \tag{8.3.9}$$

where

$$R(\lambda) = \left(\sum_{j=0}^{\infty} c_j(\lambda)/d_\alpha(\lambda)\right)^r, \quad \vartheta_{s,p} := \vartheta_{s,p}(X_0, \Theta_{(\alpha)}). \tag{8.3.10}$$

Proof: Let (Θ_j) be an i.i.d. sequence with the same distribution as $\Theta_{(\alpha)}$. Let us show first that

$$\frac{1}{d_\alpha(\lambda)} \sum_{i=0}^{\infty} c_j(\lambda)\Theta_j \stackrel{d}{=} \Theta_{(\alpha)}. \tag{8.3.11}$$

Consider the characteristic function of the right-hand side quantity in (8.3.11):

$$\begin{aligned} E\, e^{i\left\langle \frac{1}{d_\alpha(\lambda)} \sum_{j=0}^{\infty} c_j(\lambda)\Theta_j, t\right\rangle} &= \prod_{j=0}^{\infty} E\, e^{i\left\langle \frac{c_j(\lambda)}{d_\alpha(\lambda)}\Theta_j, t\right\rangle} \qquad (8.3.12) \\ &= \prod_{j=0}^{\infty} \exp\left\{ - \int_{S_k} \left| \left\langle \frac{c_j(\lambda)}{d_\alpha(\lambda)} t, s\right\rangle \right|^\alpha \Gamma(\,ds) \right\} \\ &= \exp\left\{ - \sum_{j=0}^{\infty} \left(\frac{c_j(\lambda)}{d_\alpha(\lambda)}\right)^\alpha \int_{S_k} |\langle t, s\rangle|^\alpha \Gamma(\,ds) \right\} \\ &= \varphi_{\Theta_{(\alpha)}}(t). \end{aligned}$$

By Lemma 8.2.3, $\vartheta_{s,p}$ is ideal of order $r = s - 1 + \frac{1}{p} > \alpha$. Therefore,

$$\begin{aligned} &\vartheta_{s,p}\left(\frac{1}{d_\alpha(\lambda)} T(\lambda), \Theta_{(\alpha)}\right) \qquad (8.3.13) \\ &= \vartheta_{s,p}\left(\frac{1}{d_\alpha(\lambda)} \sum_{j=0}^{\infty} c_j(\lambda) X_j, \frac{1}{d_\alpha(\lambda)} \sum_{j=0}^{\infty} c_j(\lambda)\Theta_j\right) \end{aligned}$$

$$\leq \; d_\alpha(\lambda)^{-r} \sum_{j=0}^{\infty} c_j(\lambda)^r \, \vartheta_{s,p}(X_0, \Theta_{(\alpha)})$$
$$= \; R(\lambda)\vartheta_{s,p}(X_0, \Theta_{(\alpha)}).$$

□

Note that various upper bounds for $\vartheta_{s,p}$ were established in Section 8.2. In particular, if $r \in \mathbb{N}$, or if X_0 has a density, we have obtained upper bounds in terms of difference pseudomoments. Maejima (1985) showed that

$$R(\lambda) \;\leq\; c\lambda^{-(r-\alpha)/\alpha} \tag{8.3.14}$$

for the Césaro and Abel methods, and

$$R(\lambda) \;\leq\; c\lambda^{-(r-\alpha)/2\alpha} \tag{8.3.15}$$

for the random walk method (which includes the Euler method and the Borel method as particular cases).

In the Gaussian case, for $r = 3$ the metric $\vartheta_{s,p}$ in (8.3.9) is finite, provided that (i) if $\mathrm{Cov}\,(X_0) = I_k$, the unity matrix, and (ii) the components have finite third moments. Furthermore, the corresponding rate of convergence is $\lambda^{-1/2}$ for the Césaro and Abel methods and $\lambda^{-1/4}$ for the random walk method.

We complete this section with an application of the ideality properties of our metrics to the approximation of the distribution of sums of nonidentically distributed random vectors by a compound Poisson law. Let $X_1, \ldots, X_n$ be independent random vectors in $\mathbb{R}^k$ with distributions $P_1, \ldots, P_n$ of the form

$$P_i \;=\; (1-p_i)\delta_0 + p_i Q_i, \quad 0 \leq p_i \leq 1, \; 1 \leq i \leq n. \tag{8.3.16}$$

Here, δ_0 stands for the one point distribution at zero. We can write X_i in the form

$$X_i \;=\; C_i D_i, \quad 1 \leq i \leq n, \tag{8.3.17}$$

where C_i has distribution Q_i, D_i is $\mathcal{B}(1, p_i)$-distributed, and C_i, D_i are independent. We shall consider the approximation of

$$S^{\text{ind}} \;:=\; \sum_{i=1}^{n} X_i \tag{8.3.18}$$

by a multivariate compound Poisson distribution $\mathcal{P}(\mu, Q)$. $\mathcal{P}(\mu, Q)$ is defined as the distribution of

$$S^{\text{coll}} \;:=\; \sum_{i=1}^{N} Z_i, \tag{8.3.19}$$

where (Z_i) is an i.i.d. sequence; $P^{Z_i} = Q$, N is Poisson distributed with parameter μ, $P^N = \mathcal{P}(\mu)$; and N, (Z_i) are independent. The notation S^{ind}, S^{coll} is taken from risk theory. Recall that in the risk-theory framework in the "individual model" p_i is the probability of a claim C_i with distribution Q_i, corresponding to k different types of claims. S^{coll} denotes the approximation of S^{ind} by the "collective model"; we refer to the books of Gerber (1981) and Hipp and Michel (1990) for these and related notions.

The usual choice of Q, μ in risk theory is

$$\mu = \widetilde{\mu} := \sum_{i=1}^{n} p_i, \qquad Q = \widetilde{Q} := \sum_{i=1}^{n} \frac{p_i}{\mu} Q_i. \tag{8.3.20}$$

This leads to the following representation of S^{coll}:

$$S^{\text{coll}} = \sum_{i=1}^{n} S_i^{\text{coll}}, \tag{8.3.21}$$

where $S_i^{\text{coll}} \sim \mathcal{P}(p_i, Q_i)$ ($X \sim Q$ denoting that X has distribution Q) and $\{S_i^{\text{coll}}\}$ independent. Note that with this choice $\mu = \widetilde{\mu}$, $Q = \widetilde{Q}$, and moreover,

$$E\, S^{\text{ind}} = E\, S^{\text{coll}}, \tag{8.3.22}$$

if the expectations exist. If $\Sigma_i = \operatorname{Cov}(C_i)$, $\alpha_i = (\alpha_{i,1}, \ldots, \alpha_{i,k}) = EC_i$, then

$$\operatorname{Cov}\left(S^{\text{ind}}\right) = \sum_{i=1}^{n} p_i \Sigma_i + \sum_{i=1}^{n} p_i q_i\, \alpha_i^T\, \alpha_i, \tag{8.3.23}$$

while

$$\operatorname{Cov}\left(S^{\text{coll}}\right) = \sum_{i=1}^{n} p_i \Sigma_i + \sum_{i=1}^{n} p_i\, \alpha_i^T\, \alpha_i. \tag{8.3.24}$$

As a consequence we obtain the following majorization result:

$$\operatorname{Cov}\left(S^{\text{ind}}\right) <_L \operatorname{Cov}\left(S^{\text{coll}}\right) \tag{8.3.25}$$

in the sense of the Loewner ordering $<_L$.

In particular, $\vartheta_{s,p}\left(S^{\text{ind}}, S^{\text{coll}}\right) = \infty$ if $r = s - 1 + \frac{1}{p} \geq 3$.

In Rachev and Rüschendorf (1990) it is shown that a better choice of (μ, Q) is possible for $k = 1$ by appropriately chosen scale transformations. For an extension of this result to $k \geq 1$, define

$$\begin{aligned} \overline{\mu}_i &:= (1-p_i)\alpha_i, & \Gamma_i &:= (1-p_i)\Sigma_i, & \overline{\beta}_i &:= \frac{p_i}{1-p_i}; \\ \overline{\mu} &:= \sum_{i=1}^n \overline{\mu}_i, & \overline{Q} &:= \sum_{i=1}^n \frac{\overline{\beta}_i}{\overline{\beta}} \overline{Q}_i, & \overline{\beta} &:= \sum_{i=1}^n \overline{\beta}_i, \end{aligned} \tag{8.3.26}$$

where $\overline{Q}_i$ is a probability measure with mean $\overline{\mu}_i$ and covariance Γ_i. We approximate X_i by a compound Poisson distributed r.v., $S_i^{\text{coll}} \sim \mathcal{P}(\overline{\beta}_i, \overline{Q}_i)$. This leads to an approximation of S^{ind} by

$$S^{\text{coll}} \sim \mathcal{P}(\overline{\beta}, \overline{Q}). \tag{8.3.27}$$

From our construction, it follows that

$$\begin{aligned} EX_i &= p_i\alpha_i = E\,S^{\text{coll}}; \\ \text{Cov}\,(X_i) &= \Sigma_i = \text{Cov}\left(S_i^{\text{coll}}\right) = \overline{\beta}_i\Gamma_i + \beta_i\mu_i^T\mu_i. \end{aligned} \tag{8.3.28}$$

The "ideal" properties of the metric $\vartheta_{s,p}$ derived in Section 8.2 yield closeness between the "individual" and the "collective" models using the following bounds:

$$\vartheta_{s,p}\left(S^{\text{ind}}, S^{\text{coll}}\right) \leq \sum_{i=1}^n \vartheta_{s,p}\left(X_i, S_i^{\text{coll}}\right) \tag{8.3.29}$$

and

$$\vartheta_{1,p}\left(S^{\text{ind}}, S^{\text{coll}}\right) \leq A(s,p,k)\left(\sum_{i=1}^n \vartheta_{s,p}\left(X_i, S_i^{\text{coll}}\right)\right)^{\frac{1}{pr}}. \tag{8.3.30}$$

The constant $A(s,p,k)$ can be determined from (8.2.45). $\vartheta_{s,p}\left(X_i, S_i^{\text{coll}}\right)$ is estimated from above by the metric $\alpha_{s,p}$ (Theorem 8.2.5), which depends only on pseudo-difference moments. In particular, for $r = s - 1 + \frac{1}{p} = s$ and $b_i = E\|X_i\|_1^s$,

$$\vartheta_{s,p}(X_i, S_i^{\text{coll}}) \leq p_i b_i \frac{1}{(s-1)!} \quad \text{(cf. (8.2.11))}. \tag{8.3.31}$$

Define the normalizations

$$\widetilde{X}_i = X_i - EX_i, \quad \widetilde{Y}_i = S_i^{\text{coll}} - E\,S_i^{\text{coll}}. \tag{8.3.32}$$

Consider the i.i.d case; we shall establish an estimate similar to that in (8.3.29) but without the dependence on n in the upper bound.

Theorem 8.3.2 *Suppose* (X_n) *is an i.i.d. sequence with* $\alpha = EC_i$, $\Sigma = \mathrm{Cov}\,(C_i)$, $p = P(D_i = 1)$. *Then*

$$\vartheta_{1,1}\left(S^{\mathrm{ind}}, S^{\mathrm{coll}}\right) \le C(\vartheta_r + \widetilde{\vartheta}_r + \tau_r + \widetilde{\tau}_r), \tag{8.3.33}$$

where $\vartheta_r = \vartheta_r(\widetilde{X}_1, \Theta)$, $\widetilde{\vartheta}_r = \vartheta_r(\widetilde{Y}_1, \Theta)$, $\tau_r = \tau_r(\widetilde{X}_1, \Theta)$, $\widetilde{\tau}_r = \tau_r(\widetilde{Y}_1, \Theta)$, C *is as defined in* (8.2.62), *and* Θ *is an* $N(0, \Sigma)$*-distributed r.v.*

Proof: The ideality of $\vartheta_{1,1}$ and the triangle inequality yield

$$\begin{aligned}\vartheta_{1,1}\left(S^{\mathrm{ind}}, S^{\mathrm{coll}}\right) &= \vartheta_{1,1}\left(\sum_{i=1}^n X_i, \sum_{i=1}^n S_i^{\mathrm{coll}}\right) \\ &= \vartheta_{1,1}\left(\sum_{i=1}^n \widetilde{X}_i, \sum_{i=1}^n \widetilde{Y}_i\right) \\ &\le \sqrt{n}\vartheta_{1,1}\left(\frac{1}{\sqrt{n}}\sum_{i=1}^n \widetilde{X}_i, \Theta\right) + \sqrt{n}\vartheta_{1,1}\left(\frac{1}{\sqrt{n}}\sum_{i=1}^n \widetilde{Y}_i, \Theta\right),\end{aligned}$$

where Θ is normally distributed with mean zero and covariance Σ. The proof of Theorem 8.2.11 (in the case $\alpha = 2$ with Σ being the identity matrix) extends to the general case with the same constant, thus implying (8.3.33). □

8.4 Application to Operator-Stable Limit Theorems: Statement of the Results and Auxiliary Lemmas

In this section we generalize some of the results from Section 8.2, studying the rate of convergence problem for a more general summation scheme for random vectors. Namely, suppose that the $\mathbb{R}^d$-valued random vector θ is strictly operator-stable in the sense that $\widehat{\mu}$, the characteristic function of θ, satisfies $\widehat{\mu}(z)^t = \widehat{\mu}(t^{B^*}z)$ for every $t > 0$, for some invertible linear operator B on $\mathbb{R}^d$. Suppose also that for the i.i.d. random vectors $\{X_i\}$ in $\mathbb{R}^d$, $n^{-B}\sum_{i=1}^n X_i \xrightarrow{w} \theta$. In this and the next section we study the rate of convergence of this operator-stable limit theorem in terms of several probability metrics including the Kantorovich metric and L_p-minimal versions, see (2.6.2). The results in this and thc ncxt section are due to Maejima and Rachev (1996).

We start with some definitions and notation related to operator-stable limit theorems. A probability distribution μ on $\mathbb{R}^d$ is said to be *full* if μ is not concentrated on a proper hyperplane in $\mathbb{R}^d$. A full distribution μ on $\mathbb{R}^d$ is called *operator-stable* if there exists an invertible linear operator B on $\mathbb{R}^d$ and a function $b : (0, \infty) \to \mathbb{R}^d$ such that for all $t > 0$,

$$\widehat{\mu}(z)^t = \widehat{\mu}(t^{B^*} z)e^{ib(t)}, \quad \text{for all } z \in \mathbb{R}^d. \tag{8.4.1}$$

Here $\widehat{\mu}$ is the characteristic function of μ, B^* is the adjoint operator of B, and $t^A = \exp\{(\ln t)A\} = \sum_{k=0}^{\infty}(k!)^{-1}(\ln t)^k A^k$. The distribution μ is called strictly operator-stable if we can choose $b(t) \equiv 0$. In this section, we always assume that μ is a full strictly operator-stable distribution on $\mathbb{R}^d$. Sharpe (1969) showed that if 1 is not in the spectrum of B, then the operator-stable law can be centered so as to become strictly operator-stable. Thus, the assumption of strict operator-stability is not so restrictive.

The invertible linear operator B in (8.4.1) is called an exponent of μ. When μ is operator-stable with an exponent B, μ may satisfy (8.4.1) for other B's; i.e., the exponent of μ is not necessarily unique. Further, we fix the value of the exponent B and denote by θ the random vector in $\mathbb{R}^d$ having the full strictly operator-stable distribution μ with this fixed B. It is known that every eigenvalue of B has its real part not less than $\frac{1}{2}$ (see Sharpe (1969)).

Recall that for a given sequence $X_1, X_2, \ldots$ of i.i.d. random vectors in $\mathbb{R}^d$ for which the normalized sum converges to θ, namely,

$$n^{-B}\sum_{i=1}^{n} X_i \xrightarrow{w} \theta, \tag{8.4.2}$$

we say that $\{X_i\}$ belongs to the *domain of normal attraction of* μ. As in the previous sections, we will be interested in the rate of convergence of (8.4.2).

Remark 8.4.1 *Some of the results in this section can be extended to Banach space–valued random variables. Also, our arguments can be used for similar rate of convergence problems of the* max-*operator-stable limit theorem.*

We start with some notation. Let $\|\cdot\|_0$ be the usual Euclidean norm of $\mathbb{R}^d$ and let $S(\mu)$ be the symmetry group associated with μ, that is, the group of all invertible linear operators A on $\mathbb{R}^d$ such that for some $a \in \mathbb{R}^d$, $\widehat{\mu}(z) = \widehat{\mu}(A^* z)e^{ia}$. Since by assumption μ is full, $S(\mu)$ is compact, and thus there exists a Haar probability H on $S(\mu)$. We introduce the following norm $\|\cdot\|$, which depends on the particular operator-stable law μ but not on the choice of exponent:

$$\|x\| = \int_{S(\mu)}\int_0^1 \|gt^B x\|_0 \frac{dt}{t}\, dH(g) \tag{8.4.3}$$

(see Hudson et al. (1986) and Hahn et al. (1989)). It has the following properties:

(i) $\|\cdot\|$ does not depend on the choice of the exponent B.

(ii) The map $t \mapsto \|t^B x\|$ is strictly increasing on $(0, \infty)$ for $x \neq 0$.

Define the norm of the linear operator A on $\mathbb{R}^d$ in the usual way by $\|A\| = \sup_{\|x\|=1} \|Ax\|$. Then property (ii) implies

(iii) The map $t \mapsto \|t^B\|$ is strictly increasing on $(0, \infty)$; i.e., $t \mapsto \|t^{-B}\| = \|(t^{-1})^B\|$ is strictly decreasing on $(0, \infty)$.

Further, we will need estimates of the growth rate of $R(t) = \|t^B x\|$. Meerschaert (1989) showed that for every x, the function $R_0(t) = \|t^B x\|_0$ varies regularly with index between λ_B and Λ_B, where λ_B and Λ_B are the minimum and the maximum of the real parts of the eigenvalues of B, respectively. Clearly, for every norm $\|\cdot\|$ on $\mathbb{R}^d$, the function $R(t) = \|t^B x\|$ will be of the same order as the regularly varying function $R_0(t)$. In particular, for any $\eta > 0$, there exists $t_0 > 0$ such that for any $t > t_0$,

$$t^{\lambda_B - \eta}\|x\| \;<\; \|t^B x\| \;<\; t^{\Lambda_B + \eta}\|x\|, \tag{8.4.4}$$

and

$$t^{-\Lambda_B - \eta}\|x\| \;<\; \|t^{-B} x\| \;<\; t^{-\lambda_B + \eta}\|x\|. \tag{8.4.5}$$

Let $\mathcal{X}(\mathbb{R}^d)$ be the class of all random vectors in $\mathbb{R}^d$, and ϱ the Kolmogorov metric in $\mathcal{X}(\mathbb{R}^d)$,

$$\varrho(X, Y) \;:=\; \sup_{x \in \mathbb{R}^d} |P(X \leq x) - P(Y \leq x)|. \tag{8.4.6}$$

Here, and throughout this section, $x \leq y$ or $x < y$, x, $y \in \mathbb{R}^d$, means component-wise inequality. Also, all the probability metrics μ that we shall use are in fact metrics in the space of probability laws: We write $\mu(X, Y)$ instead of $\mu(P_X, P_Y)$ only for the sake of simplicity, where P_X, P_Y stand for the probability distributions of X, Y, respectively. Next, we define a *uniform metric depending on the exponent* B,

$$\varrho^*(X, Y) \;:=\; \sup_{t>0} \varrho(t^B X,\, t^B Y). \tag{8.4.7}$$

This metric plays a crucial role in our approach to the rate of convergence problem (8.4.2).

Let Var be the total variation distance in $\mathcal{X}(\mathbb{R}^d)$,

$$\begin{aligned}\mathrm{Var}(X,Y) &:= 2 \sup_{A\in\mathcal{B}(\mathbb{R}^d)} |P(X\in A) - P(Y\in A)| \qquad (8.4.8)\\ &= \int_{\mathbb{R}^d} |P_X - P_Y|(\,\mathrm{d}x)\\ &= \sup\{|Ef(X) - Ef(Y)|;\ f:\mathbb{R}^d\to\mathbb{R}, \text{ continuous},\\ &\qquad\qquad |f(x)|\le 1 \text{ for all } x\in\mathbb{R}^d\}.\end{aligned}$$

Remark 8.4.2 *It is not difficult to check that ϱ^* is topologically "between" ϱ and* Var*; that is,*

$$\varrho \overset{\text{top}}{\prec} \varrho^* \overset{\text{top}}{\prec} \mathrm{Var}\,.$$

Here we use the standard notation $\mu \overset{\text{top}}{\prec} \nu$, meaning that ν-convergence implies μ-convergence, but the inverse is not generally valid.

Remark 8.4.3 *Our aim is to present a general approach to the rate of convergence problems associated with* (8.4.2) *that is designed to work for different metrics in terms of which we want to obtain estimates of the rate of convergence. We start with uniform-type metrics* $(\varrho, \varrho^*, \mathrm{Var})$*, and then we will proceed with Kantorovich-type minimal distances.*

For $r > 0$, define a convolution-type metric associated to Var:

$$\mu_r(X,Y) := \sup_{t>0} \|t^B\|^{-r}\ \mathrm{Var}\ (t^BX+\theta,\ t^BY+\theta). \qquad (8.4.9)$$

Here and in what follows, the notation $X_1 + X_2$ means the sum of two *independent* random vectors X_1 and X_2. We shall first list our results and then prove them, extending the general method we have outlined in Section 8.1.

Theorem 8.4.4 *Let θ be a full strictly operator-stable random vector in $\mathbb{R}^d$, and B an exponent of θ. Let $r > \frac{\Lambda_B}{\lambda_B^2}(\ge \frac{1}{\lambda_B})$ and take p such that $\frac{1}{\lambda_B} < p < \frac{\lambda_B}{\Lambda_B} r$. Let $\{X_i\}_{i=1}^\infty$ be a sequence of i.i.d. random vectors in $\mathbb{R}^d$ with*

$$\varrho^* = \varrho^*(X_1,\theta), \quad \mu_1 = \mu_1(X_1,\theta), \quad \mu_r = \mu_r(X_1,\theta), \quad (8.4.10)$$

satisfying the moment-type condition

$$\tau_r = \tau_r(X_1,\theta) := \max\left\{\varrho^*, \mu_1, \mu_r^{\frac{1}{r-p}}\right\} < \infty. \qquad (8.4.11)$$

Then for some absolute constant $K = K(d, B, r, p) > 0$,

$$\begin{aligned} \varrho\left(n^{-B}\sum_{i=1}^{n} X_i, \theta\right) &\leq \varrho^*\left(n^{-B}\sum_{i=1}^{n} X_i, \theta\right) \\ &\leq K\left(n\|n^{-B}\|^r \mu_r + \|n^{-B}\|\tau_r\right) \quad \textit{for all } n \geq 1. \end{aligned}$$

Remark 8.4.5 *In our theorem, we do not explicitly assume that* $\{X_i\}$ *belongs to the domain of normal attraction of* θ. *However, since* $\lambda_B > \frac{1}{2}$ *and* $r\lambda_B > 1$,

$$n\|n^{-B}\|^r \mu_r + \|n^{-B}\|\tau_r \ \rightarrow\ 0 \quad \textit{as} \quad n\ \rightarrow\ \infty,$$

because of (8.4.5). *Consequently, conditions* (8.4.10) *and* (8.4.11) *are sufficient for* $\{X_i\}$ *to be in the domain of normal attraction of* θ. *As to the decreasing rate of* $\|n^{-B}\|$, *by* (8.4.5), *for every* $\eta > 0$, *there exists* n_0 *such that* $\|n^{-B}\| \leq n^{-\lambda_B+\eta}$ *for every* $n \geq n_0$. *However, we also see that for any* $\eta > 0$, $\|n^{-B}\| \leq Mn^{-\lambda_B+\eta}$ *for all* $n \geq 1$, *where* $M = \sup_{t\geq 1} \|t^{-B+(\lambda_B-\eta)I}\|$ $(< \infty)$. *Note, however, that rate of convergence theorems typically describe only a relatively small subset of that domain of attraction.*

Letting $B = \frac{1}{\alpha}I$, $0 < \alpha \leq 2$, we have the following:

Corollary 8.4.6 *Let* θ *be a strictly* α*-stable random vector with index* $0 < \alpha \leq 2$. *Let* $\alpha < p < r$ *and* $\{X_i\}$ *be a sequence of i.i.d. random vectors in* $\mathbb{R}^d$ *satisfying* $\tau_r < \infty$. *Then for some absolute constant* $K = K(d, \alpha, p) > 0$,

$$\varrho^*\left(n^{-1/\alpha}\sum_{i=1}^{n} X_i, \theta\right) \leq K\left(n^{1-\frac{r}{\alpha}}\mu_r + n^{-\frac{1}{\alpha}}\tau_r\right) \quad \textit{for all } n \geq 1. \quad (8.4.12)$$

Resnick and Greenwood (1979) studied the limit theorem for (α_1, α_2)-stable laws, which corresponds to the operator-stable limit theorem with exponent

$$B = \begin{pmatrix} 1/\alpha_1 & 0 \\ 0 & 1/\alpha_2 \end{pmatrix}.$$

Theorem 8.4.4 provides a bound for the rate of convergence in this particular case.

Corollary 8.4.7 *Let* $\theta = (\theta^{(1)}, \theta^{(2)})$ *be a strictly* (α_1, α_2)*-stable bivariate vector,* $0 < \alpha_1 \leq \alpha_2 \leq 2$. *Let* $r > \frac{\alpha_2^2}{\alpha_1}(\geq \alpha_2)$ *and take* p *such that* $\alpha_2 <$

$p < \frac{\alpha_1}{\alpha_2} r$. *Let* $\{X_i = (X_i^{(1)}, X_i^{(2)})\}_{i \geq 1}$ *be a sequence of i.i.d. random vectors satisfying* $\tau_r < \infty$. *Then for all* $n \geq 1$,

$$\varrho^* \left(\left(n^{-1/\alpha_1} \sum_{i=1}^{n} X_i^{(1)}, \ n^{-1/\alpha_2} \sum_{i=1}^{n} X_i^{(2)} \right), \theta \right) \leq K \left(n^{1-r/\alpha_1} \mu_r + n^{-1/\alpha_1} \tau_r \right).$$

We next state our results on the rates of convergence in another type of uniform metric: the total variation distance Var and the uniform distance between characteristic functions. Let

$$\begin{aligned} r &> \tfrac{1}{\lambda_B}, & b &= \tfrac{5}{4} \left\| 2^{-B} \right\|^r \left\| \left(\tfrac{2}{5}\right)^{-B} \right\|^r, \\ c &= \left\| 2^B \right\|^r + \left\| 3^B \right\|^r, & a &= \tfrac{1}{bc} \left\| 2^{-B} \right\|^r, \end{aligned} \tag{8.4.13}$$

and

$$M = \sup_{x \geq 1} \left\| x^{\frac{1}{r} I - B} \right\|^r \ (< \infty). \tag{8.4.14}$$

Theorem 8.4.8 *Let* $\{X_i\}_{i=1}^{\infty}$ *be a sequence of i.i.d. random vectors in* $\mathbb{R}^d$ *satisfying*

$$\nu_r = \nu_r(X_1, \theta) := \max\{\mathrm{Var}(X_1, \theta), \mu_r\} \leq \frac{a}{M}. \tag{8.4.15}$$

Then

$$\begin{aligned} \mathrm{Var}\left(n^{-B} \sum_{i=1}^{n} X_i, \ \theta \right) &\leq cn \left\| n^{-B} \right\|^r \nu_r \qquad (8.4.16) \\ &\leq \frac{1}{bM} \left\| 2^{-B} \right\|^r n \left\| n^{-B} \right\|^r \quad \textit{for all } n \geq 1. \end{aligned}$$

It would be interesting to have a version of this theorem without condition (8.4.15). Our next theorem concerns the rate of convergence of a third type of uniform metric χ that lies "between" ϱ and Var,

$$\varrho \overset{\text{top}}{\prec} \chi \overset{\text{top}}{\prec} \mathrm{Var},$$

namely, the uniform distance between characteristic functions:

$$\chi(X, Y) := \sup_{s \in \mathbb{R}^d} |\phi_X(s) - \phi_Y(s)|, \quad \phi_X(s) := E\left[e^{i\langle s, X \rangle} \right], \tag{8.4.17}$$

where $\langle \cdot, \cdot \rangle$ is the inner product in $\mathbb{R}^d$. The corresponding "t^B-uniform" (recall the definition of ϱ^* (8.4.7)) and "smoothed" versions of χ are defined by

$$\chi^*(X, Y) := \sup_{t > 0} \chi(t^B X, \ t^B Y) \tag{8.4.18}$$

and

$$\chi_r(X,Y) := \sup_{t>0} \left\|t^B\right\|^{-r} \chi^*\left(t^B X + \theta,\ t^B Y + \theta\right). \tag{8.4.19}$$

Theorem 8.4.9 *Let $\{X_i\}_{i=1}^{\infty}$ be a sequence of i.i.d. random vectors in $\mathbb{R}^d$ satisfying*

$$\nu_r^* = \nu_r^*(X_1,\theta) := \max\{\chi^*(X_1,\theta), \chi_r(X_1,\theta)\} \le \frac{a}{M}.$$

Then for all $n \ge 1$,

$$\chi^*\left(n^{-B}\sum_{i=1}^{n} X_i,\ \theta\right) \le cn \left\|n^{-B}\right\|^r \nu_r^* \le \frac{1}{bM}\left\|2^{-B}\right\|^r n \left\|n^{-B}\right\|^r.$$

Let us now denote the density of the random vector X by $p_X(x)$ (when it exists) and define the fourth type of uniform metric

$$d(X,Y) := \operatorname*{ess\,sup}_{x\in\mathbb{R}^d} |p_X(x) - p_Y(x)|. \tag{8.4.20}$$

This is "topologically" the strongest:

$$\varrho \overset{\text{top}}{\prec} \chi \overset{\text{top}}{\prec} \mathrm{Var} \overset{\text{top}}{\prec} d. \tag{8.4.21}$$

Let

$$K(d,B) := \max_{2\le x\le 3}\ \max_{1\le i,j\le d} \left|\left(x^B\right)_{ij}\right| \quad (<\infty),$$

where A_{ij} is the (i,j) component of the matrix A, and put

$$C(d,B) = d!\, K(d,B)^d. \tag{8.4.22}$$

Let d_r be the smoothed version of d,

$$d_r(X,Y) := \sup_{t>0} \left\|t^B\right\|^{-r} d\left(t^B X + \theta,\ t^B Y + \theta\right). \tag{8.4.23}$$

Applying Theorem 8.4.8, we obtain the following rate of convergence bound in the local central limit theorem for operator-stable random vectors.

Theorem 8.4.10 *Suppose X_1 has a density. Let*

$$A = \max\left\{C(d,B)\left(\left\|2^B\right\|^r + \frac{6}{5}\left\|3^B\right\|^r\right),\ 1\right\}$$

and

$$D \ge C(d,B)\left\|3^B\right\|^r.$$

If

$$T_r = T_r(X_1,\theta) := \max\{d(X_1,\theta),\ d_r(X_1,\theta)\} < \infty$$

and

$$\nu_r \leq \min\left\{\frac{a}{M}, \frac{1}{McD}\right\},$$

then

$$d\left(n^{-B}\sum_{i=1}^{n} X_i,\ \theta\right) \leq An\|n^{-B}\|^r T_r \quad \textit{for all } n \geq 1.$$

Remark 8.4.11 *Operator-stable random vectors have bounded densities (Hudson (1980)).*

The rest of our rate of convergence results are concerned with the minimal L_p-metrics, $0 \leq p \leq \infty$, and in particular with ℓ_1, the Kantorovich metric, $\ell_1 = \widehat{L}_1$; see Section 8.1. Recall that the total variation distance Var is in fact the minimal L_0-metric. Recall the definition of the L_p-compound metric: For any $X, Y \in \mathcal{X}(\mathbb{R}^d)$,

$$L_p(X,Y) := \{E[\|X-Y\|^p]\}^{\min(1,1/p)}, \quad 0 < p < \infty, \tag{8.4.24}$$

$$L_0(X,Y) := E[I[X \neq Y]] = P(X \neq Y), \tag{8.4.25}$$

$$L_\infty(X,Y) := \operatorname{ess\,sup} \|X-Y\| = \inf\{\varepsilon > 0;\ P(\|X-Y\| > \varepsilon) = 0\}, \tag{8.4.26}$$

where $I[A]$ is the indicator function of a set A. As always in this book, we assume that all random vectors $X \in \mathcal{X}(\mathbb{R}^d)$ are defined on a nonatomic probability space $(\Omega, \mathcal{A}, P)$; in this way the space of all joint laws $P_{X,Y}$ coincides with the space of all probability measures on $\mathbb{R}^{2d}$. The L_p-minimal metric for $0 \leq p \leq \infty$ was defined in (8.2.23):

$$\ell_p(X,Y) = \widehat{L}_p(X,Y) = \widehat{L}_p(P_X,P_Y) := \inf L_p(\widetilde{X},\widetilde{Y}), \tag{8.4.27}$$

where the infimum is taken over all $P_{\widetilde{X},\widetilde{Y}}$ with fixed marginals $\widetilde{X} \overset{d}{\sim} X$, $\widetilde{Y} \overset{d}{\sim} Y$.

Remark 8.4.12 *For every $p \in [0,\infty]$ fixed, we shall be interested in the rate of convergence of $\widehat{L}_p\left(n^{-B}\sum_{i=1}^n X_i,\ \theta\right) \to 0$. As a consequence, we shall derive the rate of convergence results in terms of the Prohorov metric*

$$\pi(X,Y) = \inf\{\varepsilon; P(X \in A) \leq P(Y \in A^\varepsilon) + \varepsilon \ \textit{for all } A \in \mathcal{B}(\mathbb{R}^d)\}, \tag{8.4.28}$$

where $A^\varepsilon := \{x;\ \|x - A\| \leq \varepsilon\}$.

Case $p = 0$. *For $p = 0$, $\widehat{L}_0 = \frac{1}{2}\mathrm{Var}$, and so Theorem 8.4.8 gives the desired bound for the rate of convergence.*

Case $0 < p \le 1$. *Suppose first that B, the exponent of θ, satisfies*

$$\left(\frac{1}{\lambda_B} \le\right) \frac{\Lambda_B}{\lambda_B^2} < p \le 1. \tag{8.4.29}$$

Then by (8.4.5), $n\|n^{-B}\|^p \to 0$ as $n \to \infty$.

Theorem 8.4.13 *Suppose $0 < p \le 1$ and (8.4.29) holds. Let $X, X_1, X_2, \dots$ be a sequence of i.i.d. random vectors satisfying*

$$\widehat{L}_p := \widehat{L}_p(X, \theta) < \infty. \tag{8.4.30}$$

Then

$$\widehat{L}_p\left(n^{-B}\sum_{i=1}^n X_i,\ \theta\right) \le n\left\|n^{-B}\right\|^p \widehat{L}_p,$$

and furthermore,

$$\pi\left(n^{-B}\sum_{i=1}^n X_i,\ \theta\right) \le n^{\frac{1}{p+1}}\left\|n^{-B}\right\|^{\frac{p}{p+1}} \widehat{L}_p^{\frac{1}{p+1}}.$$

In the case where (8.4.29) is not satisfied, we shall prove a result similar to that in Theorem 8.4.10. Define the convolution-type metric associated with $\widehat{L}_p$: For $r > 0$,

$$\widehat{L}_{p,r}(X,Y) := \sup_{t>0} \|t^B\|^{-r}\widehat{L}_p(t^B X + \theta,\ t^B Y + \theta). \tag{8.4.31}$$

Theorem 8.4.14 *Let $0 < p \le 1$ and $X, X_1, X_2, \dots$ be a sequence of i.i.d. random vectors in $\mathbb{R}^d$. Let $r > \frac{1}{\lambda_b}$,*

$$A = \max\left\{\left\|2^{-B}\right\|^p\left(\left\|2^B\right\|^r + \frac{6}{5}\left\|3^B\right\|^r\right), 1\right\}$$

and

$$D \ge \left\|2^{-B}\right\|^p \left\|3^B\right\|^r.$$

Suppose

$$R_{p,r} = R_{p,r}(X,\theta) := \max\{\widehat{L}_p(X,\theta),\ \widehat{L}_{p,r}(X,\theta)\} < \infty$$

and

$$\nu_r \;\le\; \min\left\{\frac{a}{M},\ \frac{1}{McD}\right\},$$

where a, c, *and* M *are defined in* (8.4.13) *and* (8.4.14). *Then, for all* $n \ge 1$,

$$\left\{\pi\left(n^{-B}\sum_{i=1}^{n} X_i,\ \theta\right)\right\}^{p+1} \;\le\; \widehat{L}_p\left(n^{-B}\sum_{i=1}^{n} X_i,\ \theta\right) \le\; An\|n^{-B}\|^r R_{p,r}.$$

In the next theorem we shall relax the assumption $\nu_r \le \min\left\{\frac{a}{M},\ \frac{1}{McD}\right\}$ at the cost of losing a little of the order of convergence $n\|n^{-B}\|^r$. The next result has a form resembling Theorem 8.4.4.

Theorem 8.4.15 *Let* $0 < p \le 1$. *Let* $r > \frac{\Lambda_B}{\lambda_B^2} (\ge \frac{1}{\lambda_B})$, *and take* q *such that* $\frac{1}{\lambda_B} < q < \frac{\lambda_B}{\Lambda_B} r$. *Let* $\{X_i\}_{i=1}^{\infty}$ *be a sequence of i.i.d. random vectors in* $\mathbb{R}^d$ *satisfying*

$$\widehat{L}_{p,r} \;=\; \widehat{L}_{p,r}(X_1,\theta) \;<\; \infty$$

and

$$Q_{p,r} \;=\; Q_{p,r}(X_1,\theta) \;:=\; \max\left\{\widehat{L}_p, \mu_1, \mu_r^{\frac{1}{r-q}}\right\} \;<\; \infty,$$

where $\widehat{L}_p = \widehat{L}_p(X_1,\theta)$, $\mu_1 = \mu_1(X_1,\theta)$, *and* $\mu_r = \mu_r(X_1,\theta)$. *Then for some absolute constant* $K = K(d, B, r, p, q) > 0$,

$$\widehat{L}_p\left(n^{-B}\sum_{i=1}^{n} X_i,\ \theta\right) \;\le\; K\left(n\left\|n^{-B}\right\|^r \widehat{L}_{p,r} + \left\|n^{-B}\right\|^p Q_{p,r}\right),$$

for all $n \ge 1$.

Case 1< p ≤2. For $1 < p \le 2$, we use a completely different method in the rate of convergence problem, which relies on the minimality property of $\widehat{L}_p$ and was introduced in Rachev and Rüschendorf (1992).

Theorem 8.4.16 *Suppose* $1 < p \le 2$ *and* $p > \frac{\Lambda_B}{\lambda_B^2}$ $(\ge \frac{1}{\lambda_B})$. *Let* X, X_1, $X_2, \ldots$ *be a sequence of i.i.d. random vectors satisfying*

$$\widehat{L}_p \;=\; \widehat{L}_p(X,\theta) \;<\; \infty \quad \textit{and} \quad E[X-\theta] \;=\; 0.$$

Then there exists $C_r > 0$ *such that for all* $n \ge 1$,

$$\widehat{L}_p\left(n^{-B}\sum_{i=1}^{n} X_i,\ \theta\right) \;\le\; C_p n^{\frac{1}{p}} \left\|n^{-B}\right\| \widehat{L}_p,$$

and moreover, the right-hand side vanishes as $n \to \infty$. Furthermore,

$$\pi\left(n^{-B}\sum_{i=1}^{n} X_i,\ \theta\right) \ \le\ C_p^{\frac{p}{p+1}} n^{\frac{1}{p+1}} \left\|n^{-B}\right\|^{\frac{p}{p+1}} \widehat{L}_p^{\frac{p}{p+1}}.$$

Corollary 8.4.17 *Let $\theta = (\theta^{(1)}, \theta^{(2)})$ be a strictly (α_1, α_2)-stable bivariate vector, $0 < \alpha_1 \le \alpha_2 \le 2$. Let $2 \ge p > \frac{\alpha_2^2}{\alpha_1} (\ge \alpha_2)$. Let $\{X_i = (X_i^{(1)}, X_i^{(2)})\}_{i\ge 1}$ be a sequence of i.i.d. random vectors satisfying $\widehat{L}_p = \widehat{L}_p(X_1, \theta) < \infty$, and if $p > 1$, we additionally assume that $E[X_1 - \theta] = 0$. Then for all $n \ge 1$,*

$$\begin{aligned}
&\pi\left((n^{-1/\alpha_1}\sum_{i=1}^{n} X_i^{(1)}, n^{-1/\alpha_2}\sum_{i=1}^{n} X_i^{(2)}),\ \theta\right) \\
&\le\ \widehat{L}_p\left(\left(n^{-1/\alpha_1}\sum_{i=1}^{n} X_i^{(1)}, n^{-1/\alpha_2}\sum_{i=1}^{n} X_i^{(2)}\right),\ \theta\right)^{\frac{\max(1,p)}{p+1}} \\
&\le\ \begin{cases} \left(n^{1-\frac{p}{\alpha_1}}\widehat{L}_p\right)^{\frac{1}{p+1}} & \text{for } 0 < p \le 1, \\ \left(C_p n^{\frac{1}{p}-\frac{1}{\alpha_1}}\widehat{L}_p\right)^{\frac{p}{p+1}} & \text{for } 1 < p \le 2. \end{cases}
\end{aligned}$$

Remark 8.4.18 *Our approach based on the use of "ideality" of $\widehat{L}_p$ can be extended to bound the distance between the maxima $M_X(n) := n^{-B}\bigvee_{k=1}^{n} (\sum_{i=1}^{k} X_i)$ and $M_\theta(n) := n^{-B}\bigvee_{k=1}^{n}(\sum_{i=1}^{k}\theta_i)$. (Here $\bigvee_{k=1}^{n}$ stands for the componentwise maximum, and $\{\theta_i\}$ are i.i.d. copies of θ.) Also, we can compare $m_X(n) := n^{-B}\bigwedge_{k=1}^{n}(\sum_{i=1}^{k} X_i)$ with $m_\theta(n) := n^{-B}\bigwedge_{k=1}^{n}(\sum_{i=1}^{k}\theta_i)$ (where $\bigwedge_{k=1}^{n}$ stands for the componentwise minimum) and $a_X(n) := n^{-B}\bigvee_{k=1}^{n}\|\sum_{i=1}^{k} X_i\|$ with $a_\theta(n) := n^{-B}\bigvee_{k=1}^{n}\|\sum_{i=1}^{k}\theta_i\|$.*

Theorem 8.4.19 *Let $1 < p \le 2$ and θ be a full strictly operator-stable random vector with exponent B such that*

$$n^2\|n^{-B}\|^p \to 0 \quad \text{as} \quad n \to \infty.$$

Let $X, X_1, X_2, \ldots$ be i.i.d. random vectors in $\mathbb{R}^d$ with $E[X - \theta] = 0$ and such that $L_p(X, \theta) < \infty$. Then there exists $C_{p,d} > 0$ such that for every $n \ge 1$,

$$\begin{aligned}
&\max\left\{\widehat{L}_p\left(M_X(n), M_\theta(n)\right),\ \widehat{L}_p\left(m_X(n), m_\theta(n)\right), \widehat{L}_p\left(a_X(n), a_\theta(n)\right)\right\} \\
&\le\ C_{p,d}\, n^{2/p}\left\|n^{-B}\right\|\widehat{L}_p(X, \theta).
\end{aligned}$$

Remark 8.4.20 *Note that $a_X(n)$ and $a_\theta(n)$ are positive random variables, and therefore*

$$\widehat{L}_p(a_X(n), a_\theta(n)) = \left(\int_0^1 \left| F^{-1}_{a_X(n)}(t) - F^{-1}_{a_\theta(n)}(t) \right|^p \mathrm{d}t \right)^{1/p},$$

where F_X^{-1} is the generalized inverse of F_X; cf. Theorem 3.1.2. Also, from the above bound we can get the rate of convergence for π by making use of the bound $\pi \le L_p^{\frac{p}{p+1}}$.

Let us compare Theorem 8.4.16 with a similar result on the rate of convergence in terms of the Zolotarev metric ζ_r, $r > 0$; see (8.2.1).

Theorem 8.4.21 *Let $X, X_1, X_2, \ldots$ be i.i.d. random vectors in $\mathbb{R}^d$, and r a positive constant satisfying the conditions*

$$\zeta_r := \zeta_r(X, \theta) < \infty \quad \text{and} \quad n\|n^{-B}\|^r \to 0 \text{ as } n \to \infty.$$

Then for every $n \ge 1$,

$$\zeta_r\left(n^{-B} \sum_{i=1}^n X_i,\ \theta \right) \le n\|n^{-B}\|^r \zeta_r,$$

and for some $C_r > 0$,

$$\pi\left(n^{-B} \sum_{i=1}^n X_i,\ \theta \right) \le C_r^{\frac{1}{r+1}} n^{\frac{1}{r+1}} \left\|n^{-B}\right\|^{\frac{r}{r+1}} \zeta_r^{\frac{1}{r+1}}.$$

It is known that if r is an integer, then ζ_r on the right-hand sides of the above bounds can be estimated by the *rth difference pseudomoment* κ_r from above (see Zolotarev (1993)). Namely, if all mixed moments of order less than or equal to $r-1$ for X and Y agree, then

$$\zeta_r(X, Y) \le \frac{1}{r!} \kappa_r(X, Y), \quad r \in \mathbb{N}, \tag{8.4.32}$$

where κ_r is *rth difference pseudomoment*

$$\kappa_r(X, Y) = \sup\Big\{ |E[f(X) - f(Y)]|;\ f : \mathbb{R}^d \to \mathbb{R}, \tag{8.4.33}$$
$$|f(x) - f(y)| \le \left\| \|x\|^{r-1} x - \|y\|^{r-1} y \right\| \text{ for all } x, y \in \mathbb{R}^d \Big\}.$$

For arbitrary $r > 0$, ζ_r is bounded from above by the absolute pseudomoment, namely, if all mixed moments of X and Y of order less than or equal to m ($r = m + \alpha$, $m \in \mathbb{N}$, $\alpha \in (1, 2]$) agree, then

$$\zeta_r \le \frac{\Gamma(1+\alpha)}{\Gamma(1+r)} \xi_r, \tag{8.4.34}$$

where ξ_r is the *rth absolute pseudomoment*

$$\xi_r(X,Y) := \int_{\mathbb{R}^d} \|x\|^r |P_X - P_Y|(\mathrm{d}x). \tag{8.4.35}$$

Let us now compare the rate of convergence in Theorem 8.4.21 with that in Theorem 8.4.16 for $r = p \in (1,2]$. Recall that (8.3.32) is true only for $r \in \mathbb{N}$, and the known estimates for ζ_r from above by κ_r (r being noninteger) involve $E[\|X\|^r]$ and $E[\|Y\|^r]$. However, for *any* $p \geq 1$,

$$\widehat{L}_p^p(X,Y) \leq 2^p \kappa_p(X,Y) \leq 2^p \xi_p(X,Y). \tag{8.4.36}$$

Therefore, the restriction $\widehat{L}_p(X,\theta) < \infty$ in Theorem 8.4.16 is preferable to $\zeta_r(X,\theta) < \infty$ in Theorem 8.4.21. On the other hand, the estimate for $\zeta_r(n^{-B}\sum_{i=1}^n X_i,\ \theta)$ holds for any $r > 0$ and provides us with the exact order of convergence (as $n \to \infty$) under the assumption $\zeta_r(X_1,\theta) < \infty$.

Case 2$< p \leq \infty$.

Theorem 8.4.22 *T8.4.21 Let θ be a full strictly operator-stable random vector that does not have a Gaussian component, or equivalently, whose exponent B satisfies*

$$n^{1/2}\|n^{-B}\| \to 0 \quad as \quad n \to \infty.$$

Let $X, X_1, X_2, \ldots$ be i.i.d. random vectors such that $\widehat{L}_p(X,\theta) < \infty$. Then for some $C(d,p) > 0$,

$$\widehat{L}_p\left(n^{-B}\sum_{i=1}^n X_i,\ \theta\right) \leq C(d,p) n^{1/2}\|n^{-B}\|\widehat{L}_p(X,\theta).$$

Before starting with the proof of our theorems, we introduce a notion of ideality for a probability metric, designed for problem (8.4.2).

Definition 8.4.23 *A metric $\zeta : \mathcal{X}(\mathbb{R}^d) \times \mathcal{X}(\mathbb{R}^d) \to [0,\ \infty)$ is called operator-ideal of order $r \geq 0$ if*

(i) (homogeneity) $\zeta(a^B X, a^B Y) \leq \|a^B\|^r \zeta(X,Y)$ *for any* $a > 0$,

and

(ii) (regularity) $\zeta(X+Z, Y+Z) \leq \zeta(X,Y)$ *for any Z independent of X and Y.*

We next show a few lemmas needed for the proof of our main results.

Lemma 8.4.24 *ϱ, κ_r, and ℓ are regular (that is, (ii) holds); ϱ^*, Var, and χ^* are operator-ideal of order $r = 0$; and $\varrho \le \varrho^* \le \frac{1}{2}\mathrm{Var}, \chi \le \chi^* \le \mathrm{Var}$.*

This follows directly from the definitions of the metrics.

Lemma 8.4.25 *$\mu_r, \chi_r, d_r, \widehat{L}_{p,r}$, and ζ_r are operator-ideal of order $r > 0$. $\widehat{L}_p$ is operator-ideal of order $p \wedge 1$.*

Proof: We first show the operator-ideality of μ_r. For any $a > 0$,

$$\begin{aligned}
\mu_r(a^B X,\ a^B Y) &= \sup_{t>0} \left\|t^B\right\|^{-r} \mathrm{Var}((ta)^B X + \theta,\ (ta)^B Y + \theta) \\
&= \sup_{t>0} \left\|\left(\frac{t}{a}\right)^B\right\|^{-r} \mathrm{Var}\left(\left(\frac{t}{a}a\right)^B X + \theta,\ \left(\frac{t}{a}a\right)^B Y + \theta\right) \\
&\le \|a^B\|^r \sup_{t>0} \left\|t^B\right\|^{-r} \mathrm{Var}(t^B X + \theta,\ t^B Y + \theta),
\end{aligned}$$

since

$$\begin{aligned}
\left\|t^B\right\| \left\|a^B\right\|^{-1} &= \left\|t^B a^{-B} a^B\right\| \left\|a^B\right\|^{-1} \le \left\|t^B a^{-B}\right\| \\
&= \|a^B\|^r \mu_r(X, Y),
\end{aligned}$$

which shows the homogeneity of μ_r of order $r > 0$. We also have

$$\begin{aligned}
&\mu_r(X + Z, Y + Z) \\
&= \sup_{t>0} \|t^B\|^{-r} \mathrm{Var}(t^B(X + Z) + \theta,\ t^B(Y + Z) + \theta) \\
&\le \sup_{t>0} \|t^B\|^{-r} \mathrm{Var}(t^B X + \theta,\ t^B Y + \theta) \\
&= \mu_r(X, Y),
\end{aligned}$$

since $t^B Z$ is independent of $t^B X$ and θ, and Var is regular. This demonstrates the regularity of μ_r. One can check the ideality of χ_r, d_r, and $\widehat{L}_{p,r}$ in a similar fashion.

We next show the operator-ideality of ζ_r. The regularity of ζ_r is known. As for the homogeneity, we have

$$\begin{aligned}
\zeta_r(a^B X, a^B Y) = \sup\Big\{&|E[f(a^B X) - f(a^B Y)]|; \\
&\|f^{(r-1)}(x) - f^{(r-1)}(y)\| \le \|x - y\|\Big\}. \qquad (8.4.37)
\end{aligned}$$

Let $f_a(x) := f(a^B x)$. Then

$$f_a^{(r-1)}(x)(h)^{(r-1)} = f^{(r-1)}\left(a^B x\right)\left(a^B h\right)^{(r-1)},$$

implying that

$$\left\|f_a^{(r-1)}(x) - f_a^{(r-1)}(y)\right\| \;\le\; \left\|a^B\right\|^{r-1} \left\|f^{(r-1)}\left(a^B x\right) - f^{(r-1)}\left(a^B y\right)\right\|.$$

Then the side condition in (8.4.37),

$$\left\|f^{(r-1)}(x) - f^{(r-1)}(y)\right\| \;\le\; \|x-y\|,$$

results in

$$\left\|f_a^{(r-1)}(x) - f_a^{(r-1)}(y)\right\| \;\le\; \left\|a^B\right\|^{r-1} \left\|a^B x - a^B y\right\| \;\le\; \left\|a^B\right\|^{r} \|x-y\|.$$

Consequently, by (8.4.37),

$$\begin{aligned}
&\zeta_r\left(a^B X, a^B Y\right)\\
&\quad\le\; \sup\left\{|E[f_a(X) - f_a(Y)]|\,;\ \|f_a^{(r-1)}(x) - f_a^{(r-1)}(y)\| \le \|a^B\|^r \|x-y\|\right\}\\
&\quad=\; \|a^B\|^r \zeta_r(X,Y),
\end{aligned}$$

which shows the regularity of ζ_r. □

Lemma 8.4.26 *If Z, $W \in \mathcal{X}(\mathbb{R}^d)$ are independent of X, $Y \in \mathcal{X}(\mathbb{R}^d)$, then*

$$\varrho^*(X+Z,\ Y+Z) \;\le\; \varrho^*(Z,W)\,\mathrm{Var}(X,Y) + \varrho^*(X+W,\ Y+W)$$

and

$$\varrho^*(X+Z,\ Y+Z) \;\le\; \varrho^*(X,Y)\,\mathrm{Var}(Z,W) + \varrho^*(X+W,\ Y+W).$$

Lemma 8.4.27 *If Z, $W \in \mathcal{X}(\mathbb{R}^d)$ are independent of X, $Y \in \mathcal{X}(\mathbb{R}^d)$, then*

$$\mathrm{Var}(X+Z,\ Y+Z) \;\le\; \mathrm{Var}(Z,W)\,\mathrm{Var}(X,Y) + \mathrm{Var}(X+W,\ Y+W)$$

and

$$\chi^*(X+Z,\ Y+Z) \;\le\; \chi^*(Z,W)\chi^*(X,Y) + \chi^*(X+W,\ Y+W).$$

Lemma 8.4.28 *If Z, $W \in \mathcal{X}(\mathbb{R}^d)$ are independent of X, $Y \in \mathcal{X}(\mathbb{R}^d)$, then*

$$\begin{aligned}
d(X+Z,\ Y+Z) &\;\le\; d(Z,W)\,\mathrm{Var}(X,Y) + d(X+W,\ Y+W),\\
d(X+Z,\ Y+Z) &\;\le\; d(X,Y)\,\mathrm{Var}(Z,W) + d(X+W,\ Y+W),
\end{aligned}$$

and for $0 < p \le 1$ both inequalities hold with d replaced by $\widehat{L}_p$.

Proof: The proofs are very similar to those in Lemma 8.1.15; cf. also Lemma 2 in Senatov (1980) or Lemmas 14.3.3 and 14.3.6 in Rachev (1991). We shall demonstrate only the proof of the smoothing inequality for $\widehat{L}_p$, $0 < p \leq 1$. We use the dual representation for $\widehat{L}_p$:

$$\begin{aligned}&\widehat{L}_p(X+Z, Y+Z)\\ &\quad= \sup_{f\in \mathrm{Lip}_b(p)} |E[f(X+Z)-f(Y+Z)]|\end{aligned}$$

(recall that $\mathrm{Lip}_b(p)$ consists of all bounded continuous functions on $\mathbb{R}^d$ satisfying $|f(x)-f(y)| \leq \|x-y\|^p$ for all $x, y \in \mathbb{R}^d$)

$$\begin{aligned}
&= \sup_{f\in \mathrm{Lip}_b(p)} \left| \int P_Z(\mathrm{d}z)(E[f(X+z)] - E[f(Y+z)]) \right| \\
&\leq \sup_{f\in \mathrm{Lip}_b(p)} \left| \int (P_Z - P_W)(\mathrm{d}z)\,(E[f(X+z)] - E[f(Y+z)]) \right| \\
&\quad + \sup_{f\in \mathrm{Lip}_b(p)} \left| \int P_W(\mathrm{d}z)\,(E[f(X+z)] - E[f(Y+z)]) \right| \\
&\leq \int |P_Z - P_W|(\mathrm{d}z) \sup_{f\in \mathrm{Lip}_b(p)} |(E[f(X+z)] - E[f(Y+z)])| \\
&\quad + L_p(X+W, Y+W) \\
&= \mathrm{Var}(Z,W)L_p(X,Y) + L_p(X+W, Y+W),
\end{aligned}$$

as desired. □

Lemma 8.4.29 (Smoothing inequalities for ϱ^* and $\widehat{L}_p$, $0<p\leq 1$) *Let θ and θ_1 be independent random vectors in $\mathbb{R}^d$ having the same full strictly operator-stable distribution with exponent B. Then for any $X \in \mathcal{X}(\mathbb{R}^d)$ independent of θ_1 and $\delta > 0$,*

$$\varrho^*(X,\theta) \;\leq\; C_1\varrho^*(X+\delta^B\theta_1,\ \theta+\delta^B\theta_1) + C_2\|\delta^B\|,$$

and for $0<p\leq 1$, if $E[\|\theta\|^p] < \infty$,

$$\widehat{L}_p(X,\theta) \;\leq\; L_p(X+\delta^B\theta_1,\ \theta+\delta^B\theta_1) + 2\|\delta^B\|^p E[\|\theta\|^p].$$

Here and in what follows, the C_i's are absolute constants depending only on d and B, unless stated otherwise explicitly.

Proof: Fix $\varepsilon \in (0,1)$ and choose $\widetilde{X} = t_\varepsilon^B X$, $\widetilde{\theta} = t_\varepsilon^B\theta$ for some $t_\varepsilon > 0$ such that $\varrho^*(X,\theta) \leq \varrho(\widetilde{X},\widetilde{\theta}) + \varepsilon$. We first show the inequality

$$\varrho(\widetilde{X},\widetilde{\theta}) \;\leq\; C_1\varrho(\widetilde{X}+\delta^B\widetilde{\theta}_1,\ \widetilde{\theta}+\delta^B\widetilde{\theta}_1) + C_2\|\delta^B\| \tag{8.4.38}$$

for $\widetilde{\theta}_1 \overset{d}{\sim} \widetilde{\theta}$. Since $\varrho(\widetilde{X}, \widetilde{\theta}) = \varrho(c\widetilde{X}, c\widetilde{\theta})$ for any $c > 0$, we can shrink $\widetilde{X}$, $\widetilde{\theta}$, and $\widetilde{\theta}_1$ without any loss of generality. So we assume

$$P(\|\widetilde{\theta}\| < 1) \;>\; \frac{2}{3}. \tag{8.4.39}$$

For brevity we shall delete the "$\sim$" from now on. Let $\theta^{(i)} \in \mathbb{R}$, $i = 1, \dots, d$, be the ith component of the operator-stable random vector $\theta \in \mathbb{R}^d$. Then for each $i = 1, \dots, d$, $\theta^{(i)}$ has a bounded density; that is, for some $M < \infty$,

$$\sup_{x \in \mathbb{R}} \left| \frac{\mathrm{d}}{dx} P(\theta^{(i)} \le x) \right| \;=:\; \sup_{x \in \mathbb{R}} |p_{\theta^{(i)}}(x)| \;\le\; M \quad \text{for all } i.$$

(See Hudson (1976).) The idea of the following proof is taken from Lemma 12.1 in Bhattacharya and Rao (1976).

First consider the case

$$\begin{aligned} \varrho \;:=\; \varrho(X, \theta) \;&=\; \sup_{x \in \mathbb{R}^d} |P(X \le x) - P(\theta \le x)| \\ &=\; -\inf_{x \in \mathbb{R}^d} \left(P(X \le x) - P(\theta \le x)\right). \end{aligned} \tag{8.4.40}$$

Given $\eta \in (0,\, \varrho)$, there exists $x_0 \in \mathbb{R}^d$ such that

$$P(X \le x_0) - P(\theta \le x_0) \;<\; -\varrho + \eta. \tag{8.4.41}$$

We then have

$$\begin{aligned} I &:= P(X + \delta^B \theta_1 \le x_0 - \|\delta^B\| e) - P(\theta + \delta^B \theta_1 \le x_0 - \|\delta^B\| e) \\ &= \int_{\mathbb{R}^d} \left[P(X + z \le x_0 - \|\delta^B\| e) - P(\theta + z \le x_0 - \|\delta^B\| e)\right] P(\delta^B \theta_1 \in \mathrm{d}z) \\ &= \int_E + \int_{E^c}, \end{aligned}$$

where

$$E \;:=\; \{z \in \mathbb{R}^d \;\; -\|\delta^B\| e < z < \|\delta^B\| e\} \quad \text{and} \quad e = (1, 1, \dots, 1)^t \in \mathbb{R}^d.$$

Then estimating both integrals in the representation for I, we get

$$\begin{aligned} I \;\le\; & \int_E \left[P(X \le x_0) - P(\theta \le x_0 - z - \|\delta^B\| e)\right] P(\delta^B \theta_1 \in \mathrm{d}z) \\ & \quad + \varrho P(\delta^B \theta_1 \in E^c). \end{aligned} \tag{8.4.42}$$

To estimate the last term observe that

$$\beta \;:=\; P(\delta^B\theta_1 \in E) \;\geq\; P(\|\delta^B\|\,\|\theta_1\| < \|\delta^B\|) \;>\; \frac{2}{3}, \tag{8.4.43}$$

by (8.4.39). On the other hand, denoting the distribution function of θ by $F(x)$, $x = (x^{(1)}, x^{(2)}, \ldots, x^{(d)})^t \in \mathbb{R}^d$, and $\varepsilon = (\varepsilon^{(1)}, \varepsilon^{(2)}, \ldots, \varepsilon^{(d)})^t \in \mathbb{R}^d$, we have

$$\begin{aligned}
&|P(\theta \leq x + \varepsilon) - P(\theta \leq x)| \\
&\leq \sum_{i=1}^{d} \Big| F(x^{(1)}, \ldots, x^{(i-1)}, x^{(i)} + \varepsilon^{(i)}, \ldots, x^{(d)} + \varepsilon^{(d)}) \\
&\qquad\qquad - F(x^{(1)}, \ldots, x^{(i)}, x^{(i+1)} + \varepsilon^{(i+1)}, \ldots, x^{(d)} + \varepsilon^{(d)}) \Big| \\
&\leq \sum_{i=1}^{d} P(\theta^{(i)} \in I_i).
\end{aligned}$$

Here $I_i := (x^{(i)}, x^{(i)} + \varepsilon^{(i)}]$ or $:= (x^{(i)} + \varepsilon^{(i)} x^{(i)}]$ depending on the sign of $\varepsilon^{(i)}$. Therefore,

$$\sum_{i=1}^{d} P(\theta^{(i)} \in I_i) \;\leq\; \sum_{i=1}^{d} |\varepsilon^{(i)}| \sup_{x \in \mathbb{R}} |p_{\theta^{(i)}}(x)| \;\leq\; M\|\varepsilon\|_1,$$

where $\|\cdot\|_1$ is the L_1-norm. Hence,

$$-P(\theta \leq x + \varepsilon) \;\leq\; -P(\theta \leq x) + M\|\varepsilon\|_1. \tag{8.4.44}$$

Thus we have, by (8.4.41), (8.4.42), and (8.4.44) with $\varepsilon = -z - \|\delta^B\|e$, that

$$\begin{aligned}
I \;&\leq\; \int_E \left[P(X \leq x_0) - P(\theta \leq x_0) + M(\|z\|_1 + d\|\delta^B\|)\right] P(\delta^B\theta_1 \in \mathrm{d}z) \\
&\qquad + \varrho P(\delta^B\theta_1 \in E^c) \\
&\leq\; \int_E \left[-\varrho + \eta + M(\|z\|_1 + d\|\delta^B\|)\right] P(\delta^B\theta_1 \in \mathrm{d}z) + \varrho P(\delta^B\theta_1 \in E^c).
\end{aligned}$$

Since $\|z\|_1 \leq d\|\delta^B\|$ on E, it follows that

$$\begin{aligned}
I \;&\leq\; (-\varrho + \eta + 2Md\|\delta^B\|)P(\delta^B\theta_1 \in E) + \varrho P(\delta^B\theta_1 \in E^c) \\
&\leq\; \varrho(1 - 2\beta) + \eta + 2Md\|\delta^B\|.
\end{aligned}$$

Consequently,

$$(2\beta - 1)\varrho \;\leq\; \varrho(X + \delta^B\theta_1,\; \theta + \delta^B\theta_1) + 2Md\|\delta^B\| + \eta.$$

Since η can be taken arbitrarily small, we have

$$\varrho \le \frac{1}{2\beta - 1}\left(\varrho(X + \delta^B\theta_1,\ \theta + \delta^B\theta_1) + 2Md\|\delta^B\|\right).$$

Since $\beta > \frac{2}{3}$ by (8.4.43), $\varrho \le 3\left(\varrho(X + \delta^B\theta_1,\ \theta + \delta^B\theta_1) + 2Md\|\delta^B\|\right)$. This proves the inequality $\varrho(X, \theta) \le C_1\varrho(X + \delta^B\theta_1, \theta + \delta^B\theta_1) + C_2\|\delta^B\|$ with $C_1 = 3$ and $C_2 = 6Md$, provided that (8.4.40) holds.

If, on the other hand, $\varrho = \sup_{x\in\mathbb{R}^d}\left(P(X \le x) - P(\theta \le x)\right)$, then given $\eta \in (0,\ \varrho)$, there exists x_0 such that

$$P(X \le x_0) - P(\theta \le x_0) \ > \ \varrho - \eta.$$

Then we similarly have

$$\begin{aligned}
&P(X + \delta^B\theta_1 \le x_0 + \|\delta^B\|e) - P(\theta + \delta^B\theta_1 \le x_0 + \|\delta^B\|e)\\
&\quad = \int_{\mathbb{R}^d} \left[P(X + z \le x_0 + \|\delta^B\|e) - P(\theta + z \le x_0 + \|\delta^B\|e)\right] P(\delta^B\theta_1 \in \mathrm{d}z)\\
&\quad = \int_E + \int_{E^c}\\
&\quad \ge \int_E \left[P(X \le x_0) - P(\theta \le x_0 - z + \|\delta^B\|e)\right] P(\delta^B\theta_1 \in \mathrm{d}z)\\
&\qquad\quad - \varrho P(\delta^B\theta_1 \in E^c)\\
&\quad \ge \int_E \left[P(X \le x_0) - P(\theta \le x_0) - M(\|z\|_1 + d\|\delta^B\|)\right] P(\delta^B\theta_1 \in \mathrm{d}z)\\
&\qquad\quad - \varrho P(\theta^B\theta_1 \in E^c)\\
&\quad \ge \ \varrho(2\beta - 1) - \eta - 2Md\|\delta^B\|.
\end{aligned}$$

Hence $(2\beta - 1)\varrho \le \varrho(X + \delta^B\theta_1,\ \theta + \delta^B\theta_1) + 2Md\|\delta^B\| + \eta$, so that $\varrho \le 3\left(\varrho(X + \delta^B\theta_1,\ \theta + \delta^B\theta_1) + 2Md\|\delta^B\|\right)$. This completes the proof of the inequality (8.4.28), and reintroducing the symbol "$\widetilde{\ }$", we write

$$\varrho(\widetilde{X}, \widetilde{\theta}) \ \le \ C_1\varrho(\widetilde{X} + \delta^B\widetilde{\theta}_1, \widetilde{\theta} + \delta^B\widetilde{\theta}_1) + C_2\|\delta^B\|.$$

Therefore, by the definition of $\varrho^*, \widetilde{X}$, and $\widetilde{\theta}$,

$$\begin{aligned}
\varrho^*(X, \theta) \ &\le \ \varrho(\widetilde{X}, \widetilde{\theta}) + \varepsilon\\
&\le \ C_1\varrho(\widetilde{X} + \delta^B\widetilde{\theta}_1,\ \widetilde{\theta} + \delta^B\widetilde{\theta}_1) + C_2\|\delta^B\| + \varepsilon\\
&\le \ C_1\varrho(t_\varepsilon^B(X + \delta^B\theta_1),\ t_\varepsilon^B(\theta + \delta^B\theta_1)) + C_2\|\delta^B\| + \varepsilon\\
&\le \ C_1\varrho^*(X + \delta^B\theta_1,\ \theta + \delta^B\theta_1) + C_2\|\delta^B\| + \varepsilon,
\end{aligned}$$

which yields the first smoothing inequality of Lemma 8.4.29.

Now let us show an inequality of a similar type for $\widehat{L}_p(0 < p \leq 1)$. By the triangle inequality and the regularity of $\widehat{L}_p$,

$$\begin{aligned}\widehat{L}_p(X,\theta) &\leq \widehat{L}_p(X,\ X+\delta^B\theta_1)+\widehat{L}_p(X+\delta^B\theta_1,\ \theta+\delta^B\theta_1)+\widehat{L}_p(\theta,\ \theta+\delta^B\theta_1)\\ &\leq \widehat{L}_p(X+\delta^B\theta_1,\ \theta+\delta^B\theta_1)+2\widehat{L}_p(0,\ \delta^B\theta_1).\end{aligned}$$

From the definition of $\widehat{L}_p$ as a minimal metric with respect to the L_p-metric, it follows that

$$\widehat{L}_p(0,\ \delta^B\theta) = E[\|\delta^B\theta\|^p] \leq \|\delta^B\|^p E[\|\theta\|^p],$$

which completes the proof of Lemma 8.4.29. □

The proof of the next two lemmas is obvious.

Lemma 8.4.30 *For any $a > 0$ and $r > 0$,*

$$\begin{aligned}\mathrm{Var}(a^BX+\theta,\ a^BY+\theta) &\leq \|a^B\|^r\mu_r(X,Y),\\ \chi^*(a^BX+\theta,\ a^BY+\theta) &\leq \|a^B\|^r\chi_r(X,Y),\\ d(a^BX+\theta,\ a^BY+\theta) &\leq \|a^B\|^r d_r(X,Y),\end{aligned}$$

and for $0 < p \leq 1$,

$$\widehat{L}_p(a^BX+\theta,\ a^BY+\theta) \leq \|a^B\|^r\widehat{L}_{p,r}(X,Y),$$

where $X,\ Y \in \mathcal{X}(\mathbb{R}^d)$ are independent of θ.

Lemma 8.4.31 *Let $\mathrm{Aut}(\mathbb{R}^d)$ be the set of all invertible linear operators (automorphisms). Then, for any $A \in \mathrm{Aut}(\mathbb{R}^d)$,*

$$d(AX,\ AY) = |J_{A^{-1}}|d(X,Y),$$

where J_A is the Jacobian of the matrix A.

Lemma 8.4.32 *For $x \in [2,3]$,*

$$|J_{x^B}| \leq C(d,B),$$

where $C(d,B)$ is defined in (8.4.22).

Proof: Note that

$$|J_{x^B}| = |\det\ x^B|.$$

If A is $d \times d$ matrix, then

$$|\det A| \le d! \, |\max_{1 \le i,j \le d} A_{ij}|^d,$$

which proves the lemma. □

Lemma 8.4.33 *Let θ be a full strictly operator-stable random vector in $\mathbb{R}^d$ with exponent B. Then for any two independent copies θ_1 and θ_2 of θ and for any t, $s > 0$,*

$$t^B\theta_1 + s^B\theta_2 \overset{d}{\sim} (t+s)^B\theta.$$

Proof: By (8.4.1) with $b(t) \equiv 0$,

$$\begin{aligned}
E\left[e^{i\langle z, t^B\theta_1 + s^B\theta_2\rangle}\right] &= E\left[e^{i\langle z, t^B\theta_1\rangle}\right] E\left[e^{i\langle z, s^B\theta_2\rangle}\right] \\
&= \widehat{\mu}\left(t^{B^*} z\right) \widehat{\mu}\left(s^{B^*} z\right) = \widehat{\mu}(z)^t \, \widehat{\mu}(z)^s \\
&= \widehat{\mu}(z)^{t+s} = \widehat{\mu}\left((t+s)^{B^*} z\right) \\
&= E\left[e^{i\langle z, (t+s)^B\theta\rangle}\right].
\end{aligned}$$

□

The following lemmas are proved in Sections 2.5 and 2.6. Further, $C_b(\mathbb{R}^d)$ stands for the set of all bounded continuous functions on $\mathbb{R}^d$.

Lemma 8.4.34 (Duality theorems for $\widehat{L}_p$) *$\widehat{L}_p$ admits the following representation:*

(i) For $p = 0$,

$$\begin{aligned}
\widehat{L}_0(X,Y) &= \sup\{|E[f(X) - f(Y)]|;\ f \in C_b(\mathbb{R}^d) \\
&\qquad \textit{such that } |f(x) - f(y)| \le 1 \textit{ for all } x, y \in \mathbb{R}^d\} \\
&= \frac{1}{2}\ \mathrm{Var}(X,Y).
\end{aligned}$$

(ii) For $0 < p \le 1$,

$$\begin{aligned}
\widehat{L}_p(X,Y) &= \sup\{|E[f(X) - f(Y)]|;\ f \in C_b(\mathbb{R}^d) \\
&\qquad \textit{such that } |f(x) - f(y)| \le \|x - y\|^p \textit{ for all } x, y \in \mathbb{R}^d\}.
\end{aligned}$$

(iii) For $1 < p < \infty$,

$$\widehat{L}_p(X,Y) = \sup\{E[f(X)] - E[g(Y)];\ f,g \in C_b(\mathbb{R}^d) \text{ such that } |f(x) - g(y)| \leq \|x-y\|^p \text{ for all } x,y \in \mathbb{R}^d\}.$$

(iv) For $p = \infty$,

$$\widehat{L}_\infty(X,Y) = \inf\{\varepsilon > 0;\ P(X \in A) \leq P(Y \in A^\varepsilon) \text{ for all } A \in \mathcal{B}(\mathbb{R}^d)\}.$$

Lemma 8.4.35 ($\widehat{L}_p$-convergence) (i) *For any $0 \leq p \leq \infty$, $\widehat{L}_p$-convergence implies weak convergence. Moreover, if π is the Prohorov metric, then*

$$\pi \leq \widehat{L}_p^{\frac{1}{p+1}} \quad \text{for all } 0 \leq p \leq 1$$

and

$$\pi \leq \widehat{L}_p^{\frac{p}{p+1}} \quad \text{for all } 1 \leq p \leq \infty.$$

(ii) *Let $0 < p < \infty$ and $E[\|X_n\|^p] + E[\|X\|^p] < \infty$. Then*

$$\widehat{L}_p(X_n, X) \to 0$$

if and only if

$$X_n \xrightarrow{w} X \quad \text{and} \quad E[\|X_n\|^p] \to E[\|X\|^p].$$

Lemma 8.4.36 (Explicit representations for $\widehat{L}_p$ in $\mathcal{X}(\mathbb{R})$) *For $d = 1$, $1 \leq p \leq \infty$,*

$$\widehat{L}_p(X,Y) = \left(\int_0^1 |F_X^{-1}(t) - F_Y^{-1}(t)|^p dt\right)^{1/p}, \quad 1 \leq p < \infty,$$

$$\widehat{L}_\infty(X,Y) = \sup_{0 \leq t \leq 1} |F_X^{-1}(t) - F_Y^{-1}(t)|.$$

Lemma 8.4.37 (Upper bounds for $\widehat{L}_p$) *For $1 \leq p < \infty$,*

$$2^{-p+1}\widehat{L}_p \leq \kappa_p \leq \xi_p,$$

where κ_p (resp. ξ_p) is the pth difference (resp. absolute) pseudomoment.

8.5 Application to Operator-Stable Limit Theorems: Proofs of the Rate of Convergence Results

In this section we give the proofs of the rate of convergence results stated in Section 8.4.

Proof of Theorem 8.4.4: All probability metrics for the random vectors in this section are defined by their marginal distributions and are consequently independent of their joint distributions. So, without loss of generality, we assume that $\{X_i\}$ and θ are independent of each other.

Let $\{\theta_i\}$ be independent copies of θ and assume that $\{\theta_i\}$ are independent of $\{X_i\}$ and θ. Then by the definition of θ (or by (8.4.1) with $b(t) \equiv 0$),

$$n^{-B}\sum_{i=1}^{n}\theta_i \overset{d}{\sim} \theta \quad \text{for any } n = 1, 2, \dots . \tag{8.5.1}$$

Now, by Lemma 8.4.29 and (8.5.1), for any $\delta > 0$,

$$\begin{aligned}
&\varrho^*\left(n^{-B}\sum_{i=1}^{n}X_i,\, \theta\right) &(8.5.2)\\
&\le C_1\varrho^*\left(n^{-B}\sum_{i=1}^{n}X_i + \delta^B\theta_1,\, \theta + \delta^B\theta_1\right) + C_2\|\delta^B\|\\
&= C_1\varrho^*\left(n^{-B}\sum_{i=1}^{n}X_i + \delta^B\theta,\, n^{-B}\sum_{i=1}^{n}\theta_i + \delta^B\theta\right) + C_2\|\delta^B\|.
\end{aligned}$$

Furthermore, by the triangle inequality,

$$\begin{aligned}
&\varrho^*\left(n^{-B}\sum_{i=1}^{n}X_i + \delta^B\theta, n^{-B}\sum_{i=1}^{n}\theta_i + \delta^B\theta\right)\\
&\le \varrho^*\left(n^{-B}\sum_{i=1}^{n}X_i + \delta^B\theta, n^{-B}\theta_1 + n^{-B}\sum_{i=2}^{n}X_i + \delta^B\theta\right)\\
&\quad + \sum_{j=1}^{m}\varrho^*\Bigg(n^{-B}\left(\sum_{i=1}^{j}\theta_i + \sum_{i=j+1}^{n}X_i\right) + \delta^B\theta,\\
&\qquad\qquad n^{-B}\left(\sum_{i=1}^{j+1}\theta_i + \sum_{i=j+2}^{n}X_i\right) + \delta^B\theta\Bigg)\\
&\quad + \varrho^*\left(n^{-B}\left(\sum_{i=1}^{m+1}\theta_i + \sum_{i=m+2}^{n}X_i\right) + \delta^B\theta, n^{-B}\sum_{i=1}^{n}\theta_i + \delta^B\theta\right),
\end{aligned}$$

where $m = [\frac{n}{2}]$, $n \geq 5$. By Lemma 8.4.26, the above is

$$\begin{aligned}
&\leq \varrho^* \left(n^{-B} \sum_{i=2}^{n} X_i, n^{-B} \sum_{i=2}^{n} \theta_i \right) \mathrm{Var}\left(n^{-B} X_1 + \delta^B \theta, n^{-B} \theta_1 + \delta^B \theta \right) \\
&\quad + \varrho^* \left(n^{-B} (X_1 + \sum_{i=2}^{n} \theta_i) + \delta^B \theta, n^{-B} \sum_{i=1}^{n} \theta_i + \delta^B \theta \right) \\
&\quad + \sum_{j=1}^{m} \left[\varrho^* \left(n^{-B} \sum_{i=j+2}^{n} X_i, n^{-B} \sum_{i=j+2}^{n} \theta_i \right) \right. \\
&\qquad \times \mathrm{Var} \left(n^{-B} \left(\sum_{i=i}^{j} \theta_i + X_{j+1} \right) + \delta^B \theta, n^{-B} \sum_{i=1}^{j+1} \theta_i + \delta^B \theta \right) \\
&\quad \left. + \varrho^* \left(n^{-B} (\sum_{i=1}^{j} \theta_i + X_{j+1} + \sum_{i=j+2}^{n} \theta_i) + \delta^B \theta, n^{-B} \sum_{i=1}^{n} \theta_i + \delta^B \theta \right) \right] \\
&\quad + \varrho^* \left(n^{-B} \left(\sum_{i=1}^{m+1} \theta_i + \sum_{i=m+2}^{n} X_i \right) + \delta^B \theta, n^{-B} \sum_{i=1}^{n} \theta_i + \delta^B \theta \right) \\
&=: \sum_{k=1}^{4} \Delta_k.
\end{aligned}$$

Here, the summands Δ_k are defined as follows:

$$\begin{aligned}
\Delta_1 &= \varrho^* \left(n^{-B} \sum_{i=2}^{n} X_i, n^{-B} \sum_{i=2}^{n} \theta_i \right) \mathrm{Var}\left(n^{-B} X_1 + \delta^B \theta, n^{-B} \theta_1 + \delta^B \theta \right), \\
\Delta_2 &= \sum_{j=1}^{m} \varrho^* \left(n^{-B} \sum_{i=j+2}^{n} X_i, n^{-B} \sum_{i=j+2}^{n} \theta_i \right) \\
&\qquad \times \mathrm{Var} \left(n^{-B} \left(\sum_{i=1}^{j} \theta_i + X_{j+1} \right) + \delta^B \theta, n^{-B} \sum_{i=1}^{j+1} \theta_i + \delta^B \theta \right), \\
\Delta_3 &= (m+1) \varrho^* \left(n^{-B} \left(X_1 + \sum_{i=2}^{n} \theta_i \right) + \delta^B \theta, \theta_1 + \delta^B \theta \right), \\
\Delta_4 &= \varrho^* \left(n^{-B} \left(\sum_{i=1}^{m+1} \theta_i + \sum_{i=m+2}^{n} X_i \right) + \delta^B \theta, n^{-B} \sum_{i=1}^{n} \theta_i + \delta^B \theta \right).
\end{aligned}$$

Next, by (8.5.2),

$$\varrho^* \left(n^{-B} \sum_{i=1}^{n} X_i, \theta \right) \leq C_1 \sum_{k=1}^{4} \Delta_k + \Delta_5, \tag{8.5.3}$$

with

$$\Delta_5 = C_2\|\delta^B\|.$$

We shall estimate each Δ_k separately.

(I) Estimate for Δ_3. We have

$$\begin{aligned}\Delta_3 &= (m+1)\varrho^*\Bigg(n^{-B}X_1 + n^{-B}(n-1)^B(n-1)^{-B}\sum_{i=2}^n \theta_i + \delta^B\theta, \\ &\qquad n^{-B}\theta_1 + n^{-B}(n-1)^B(n-1)^{-B}\sum_{i=2}^n \theta_i + \delta^B\theta\Bigg) \\ &= (m+1)\varrho^*\Big(n^{-B}X_1 + n^{-B}(n-1)^B\theta_2 + \delta^B\theta, \\ &\qquad n^{-B}\theta_1 + n^{-B}(n-1)^B\theta_2 + \delta^B\theta\Big)\end{aligned}$$

[by (8.5.1)]

$$\leq (m+1)\varrho^*\left(n^{-B}X_1 + n^{-B}(n-1)^B\theta_2, n^{-B}\theta_1 + n^{-B}(n-1)^B\theta_2\right)$$

[by the regularity of ϱ^*]

$$\leq \frac{1}{2}(m+1)\,\mathrm{Var}\left(n^{-B}X_1 + n^{-B}(n-1)^B\theta, n^{-B}\theta_1 + n^{-B}(n-1)^B\theta\right)$$

[since $\varrho^* \leq \frac{1}{2}\,\mathrm{Var}$]

$$\leq \frac{1}{2}(m+1)\,\mathrm{Var}\left((n-1)^{-B}X_1 + \theta, (n-1)^{-B}\theta_1 + \theta\right)$$

[by the homogeneity of Var]. Thus, we have

$$\Delta_3 \leq \frac{1}{2}n\left\|(n-1)^{-B}\right\|^r \mu_r(X_1,\theta_1), \tag{8.5.4}$$

invoking Lemma 8.4.30. Now, using the fact that $t \mapsto \|t^B\|$ is strictly increasing on $(0,\infty)$, we have

$$\begin{aligned}\|(n-1)^{-B}\| &\leq \left\|\left(\frac{n-1}{n}\right)^{-B}\right\| \|n^{-B}\| \\ &\leq \left\|\left(\frac{1}{2}\right)^{-B}\right\| \|n^{-B}\| = \|2^B\|\,\|n^{-B}\|.\end{aligned} \tag{8.5.5}$$

Thus it follows from (8.5.4) and (8.5.5) that

$$\Delta_3 \leq C_3 n\left\|n^{-B}\right\|^r \mu_r, \tag{8.5.6}$$

with $C_3 = \frac{1}{2}\|2^B\|^r$.

(II) Estimate for Δ_4.

Similarly, we have

$$\begin{aligned}
\Delta_4 &= \varrho^*\Bigg(n^{-B}(m+1)^B(m+1)^{-B}\sum_{i=1}^{m+1}\theta_i + n^{-B}\sum_{i=m+2}^{n} X_i + \delta^B\theta, \\
&\qquad n^{-B}(m+1)^B(m+1)^{-B}\sum_{i=1}^{m+1}\theta_i + n^{-B}\sum_{i=m+2}^{n}\theta_i + \delta^B\theta\Bigg) \\
&\leq \varrho^*\Bigg(n^{-B}(m+1)^B\theta_1 + n^{-B}\sum_{i=m+2}^{n} X_i, \\
&\qquad n^{-B}(m+1)^B\theta_1 + n^{-B}\sum_{i=m+2}^{n}\theta_i\Bigg) \\
&\text{[by (8.5.1) and the regularity of } \varrho^*] \\
&\leq \frac{1}{2}\operatorname{Var}\Bigg(n^{-B}(m+1)^B\theta_1 + n^{-B}\sum_{i=m+2}^{n} X_i, \\
&\qquad n^{-B}(m+1)^B\theta_1 + n^{-B}\sum_{i=m+2}^{n}\theta_i\Bigg) \\
&\leq \frac{1}{2}\operatorname{Var}\Bigg((m+1)^{-B}\sum_{i=m+2}^{n} X_i + \theta, (m+1)^{-B}\sum_{i=m+2}^{n}\theta_i + \theta\Bigg) \\
&\leq \frac{1}{2}\|(m+1)^{-B}\|^r\mu_r\Bigg(\sum_{i=m+2}^{n} X_i, \sum_{i=m+2}^{n}\theta_i\Bigg) \\
&\text{[by Lemma 8.4.30]} \\
&\leq \frac{1}{2}\left\|(m+1)^{-B}\right\|^r(n-m-1)\mu_r(X_1,\theta_1)
\end{aligned}$$

by the triangle inequality and the repeated use of the regularity of μ_r. Finally, by $\left\|\left(\frac{n}{m+1}\right)^{-B}\right\| \leq \left\|\left(\frac{1}{2}\right)^{-B}\right\| = \|2^B\|$, we have

$$\Delta_4 \leq C_3 n \left\|n^{-B}\right\|^r \mu_r. \tag{8.5.7}$$

We prove the theorem by induction. For $n = 1$, the theorem is valid because

$$\varrho^*(X_1,\theta) \leq \tau_r(X_1,\theta).$$

For $n = 2, 3$, and 4, the estimates are similar to those for $n \geq 5$, the case we are going to prove. However, the absolute constants in the bounds for $n = 2, 3$, and 4 have smaller values.

In the following we assume $n \geq 5$. Assume that for all $j < n$,

$$\varrho^* \left(j^{-B} \sum_{i=1}^{j} X_i, \theta \right) \leq K \left(j \|j^{-B}\|^r \mu_r + \|j^{-B}\| \tau_r \right).$$

Since we have already estimated Δ_3 and Δ_4 independently of the induction hypothesis, we shall estimate only Δ_1, Δ_2, and Δ_5. To this end, take $\delta > 0$ such that $n\delta = \varepsilon^{-1}$ for some $\varepsilon > 0$, where ε will be suitably chosen later.

(III) A bound for Δ_1.

By the definition of ϱ^*,

$$\begin{aligned}
&\varrho^* \left(n^{-B} \sum_{i=2}^{n} X_i, n^{-B} \sum_{i=2}^{n} \theta_i \right) \\
&= \sup_{t>0} \varrho \left(t^B n^{-B} \sum_{i=2}^{n} X_i, t^B n^{-B} \sum_{i=2}^{n} \theta_i \right) \\
&= \sup_{t>0} \varrho \left(t^B (\frac{n-1}{n})^B (n-1)^{-B} \sum_{i=2}^{n} X_i, t^B (\frac{n-1}{n})^B (n-1)^{-B} \sum_{i=2}^{n} \theta_i \right) \\
&\leq \sup_{u>0} \varrho \left(u^B (n-1)^{-B} \sum_{i=2}^{n} X_i, u^B (n-1)^{-B} \sum_{i=2}^{n} \theta_i \right) \\
&= \varrho^* \left((n-1)^{-B} \sum_{i=2}^{n} X_i, (n-1)^{-B} \sum_{i=2}^{n} \theta_i \right) \\
&\leq K \left((n-1) \|(n-1)^{-B}\|^r \mu_r + \|(n-1)^{-B}\| \tau_r \right)
\end{aligned}$$

by the induction hypothesis. Furthermore,

$$\begin{aligned}
\operatorname{Var}\left(n^{-B} X_1 + \delta^B \theta, n^{-B} \theta_1 + \delta^B \theta \right) &\leq \operatorname{Var}\left((n\delta)^{-B} X_1 + \theta, (n\delta)^{-B} \theta_1 + \theta \right) \\
&\leq \|(n\delta)^{-B}\| \mu_1(X_1, \theta_1)
\end{aligned}$$

by the homogeneity of Var of order 0 and Lemma 8.4.30. Thus, we have

$$\begin{aligned}
\Delta_1 &\leq K \left((n-1) \left\|(n-1)^{-B}\right\|^r \mu_r + \left\|(n-1)^{-B}\right\| \tau_r \right) \left\|(n\delta)^{-B}\right\| \mu_1 \qquad (8.5.8) \\
&\leq K C_4 \left(n \left\|n^{-B}\right\|^r \mu_r + \left\|n^{-B}\right\| \tau_r \right) \left\|(n\delta)^{-B}\right\| \mu_1 \\
&\leq K C_4 \left(n \left\|n^{-B}\right\|^r \mu_r + \left\|n^{-B}\right\| \tau_r \right) \left\|\varepsilon^B\right\| \tau_r,
\end{aligned}$$

with $C_4 = \max\{\|2^B\|^r, \|2^B\|\}$.

(IV) A bound for Δ_2.

We assume $n \geq 5$. Then, as in the case for Δ_3, we have, for $j \leq m = [\frac{n}{2}]$,

$$\begin{aligned}
&\varrho^* \left(n^{-B} \sum_{i=j+2}^{n} X_i, n^{-B} \sum_{i=j+2}^{n} \theta_i \right) \\
&\leq \varrho^* \left((n-j-1)^{-B} \sum_{i=j+2}^{n} X_i, (n-j-1)^{-B} \sum_{i=j+2}^{n} \theta_i \right) \\
&\leq K \left((n-j-1) \|(n-j-1)^{-B}\|^r \mu_r + \|(n-j-1)^{-B}\| \tau_r \right).
\end{aligned}$$

Also, we have

$$\begin{aligned}
&\mathrm{Var} \left(n^{-B} (\sum_{i=1}^{j} \theta_i + X_{j+1}) + \delta^B \theta, n^{-B} \sum_{i=1}^{j+1} \theta_i + \delta^B \theta \right) \\
&\leq \mathrm{Var} \left(X_{j+1} + \sum_{i=1}^{j} \theta_i + (n\delta)^B \theta, \theta_{j+1} + \sum_{i=1}^{j} \theta_i + (n\delta)^B \theta \right) \\
&= \mathrm{Var} \left(X_{j+1} + j^B \theta_1 + (n\delta)^B \theta, \theta_{j+1} + j^B \theta_1 + (n\delta)^B \theta \right) \\
&= \mathrm{Var} \left(X_1 + (j+n\delta)^B \theta, \theta_1 + (j+n\delta)^B \theta \right) \quad \text{(by Lemma 8.4.33)} \\
&\leq \mathrm{Var} \left((j+n\delta)^{-B} X_1 + \theta, (j+n\delta)^{-B} \theta_1 + \theta \right) \\
&\qquad \text{(by the homogeneity of Var of order 0,)} \\
&\leq \|(j+n\delta)^{-B}\|^r \mu_r(X_1, \theta) \quad \text{by Lemma 8.4.30. Thus we have}
\end{aligned}$$

$$\begin{aligned}
\Delta_2 \leq \sum_{j=1}^{m} K \Big(&(n-j-1) \|(n-j-1)^{-B}\|^r \mu_r \\
&+ \|(n-j-1)^{-B}\| \tau_r \Big) \|(j+n\delta)^{-B}\|^r \mu_r.
\end{aligned} \tag{8.5.9}$$

Now, for $1 \leq j \leq m = [\frac{n}{2}]$, $n \geq 5$, we have $\frac{n-j-1}{n} \geq \frac{n-m-1}{n} \geq \frac{1}{4}$. Hence

$$\|(n-j-1)^{-B}\| \leq \|4^B\| \, \|n^{-B}\|, \tag{8.5.10}$$

and so by (8.5.9) and (8.5.10),

$$\Delta_2 \leq K C_5 \left(n \|n^{-B}\|^r \mu_r + \|n^{-B}\| \tau_r \right) \sum_{j=1}^{\infty} \|(j+n\delta)^{-B}\|^r \mu_r, \tag{8.5.11}$$

where $C_5 = \max\{\|4^B\|^r, \|4^B\|\}$. Furthermore,

$$\sum_{j=1}^{\infty} \|(j+n\delta)^{-B}\|^r \leq \int_0^{\infty} \|(x+n\delta)^{-B}\|^r \, dx \tag{8.5.12}$$

$$= \int_{n\delta}^{\infty} \|y^{-B}\|^r \, \mathrm{d}y \;\leq\; (n\delta)\|(n\delta)^{-B}\|^r \int_{1}^{\infty} \|z^{-B}\|^r \, \mathrm{d}z.$$

Recall that $r > \Lambda_B/\lambda_B^2 \geq 1/\lambda_B$. Take q such that $1/\lambda_B < q < r$. Then, by (8.4.5), $\|x^{\frac{1}{q}I-B}\| \to 0$ as $x \to \infty$, and hence $M_1 := \sup_{x\geq 1} \|x^{\frac{1}{q}I-B}\| < \infty$. Thus for $z \geq 1$, $\|z^{-B}\|^r \leq M_1^r z^{-r/q}$, and hence

$$\int_{1}^{\infty} \|z^{-B}\|^r \, \mathrm{d}z \;\leq\; M_1 \int_{1}^{\infty} z^{-r/q} \, \mathrm{d}z \;=:\; C_6 \;<\; \infty. \tag{8.5.13}$$

It also follows from the assumption for p ($p > 1/\lambda_B$) that for $z \geq 1$, $z^{\frac{1}{p}} \leq M_2\|z^{-B}\|^{-1} \leq M_2\|z^B\|$ since $\|z^B\| \, \|z^{-B}\| \geq 1$, where $M_2 = \sup_{x\geq 1}\left\|x^{\frac{1}{p}I-B}\right\|$. Thus if $n\delta \geq 1$, then

$$n\delta \;\leq\; M_2^p \|(n\delta)^B\|^p. \tag{8.5.14}$$

Finally, we have by (8.5.11)–(8.5.14) and (8.4.5) that if $n\delta = \varepsilon^{-1}$ (where $\varepsilon > 0$ will be taken small), then

$$\begin{aligned}
\Delta_2 &\leq KC_5\left(n\|n^{-B}\|^r\mu_r + \|n^{-B}\|\tau_r\right) C_6(n\delta)\left\|(n\delta)^{-B}\right\|^r \mu_r \qquad (8.5.15)\\
&\leq KC_5C_6\left(n\|n^{-B}\|^r\mu_r + \|n^{-B}\|\tau_r\right) M_2^p \left\|\varepsilon^{-B}\right\|^p \left\|\varepsilon^{B}\right\|^r \mu_r\\
&\leq KC_5C_6M_2^p\left(n\left\|n^{-B}\right\|^r \mu_r + \left\|n^{-B}\right\|\tau_r\right)\left\|\varepsilon^{-B}\right\|^p \left\|\varepsilon^{B}\right\|^r \tau_r^{r-p}.
\end{aligned}$$

(V) A bound for Δ_5.

We have

$$\Delta_5 \;\leq\; C_2\|n^{-B}\| \, \|\varepsilon^{-B}\| \;=\; C_2\|n^{-B}\|\tau_r\|\varepsilon^{-B}\|\tau_r^{-1}. \tag{8.5.16}$$

Altogether we have from (8.5.3), (8.5.6), (8.5.7), (8.5.8), (8.5.15), and (8.5.16) that

$$\begin{aligned}
&\varrho\left(n^{-B}\sum_{i=1}^{n} X_i, \theta\right) \qquad (8.5.17)\\
&\leq 2C_1C_3n\|n^{-B}\|^r\mu_r + KC_1C_4\left\|\varepsilon^B\right\|\tau_r\left(n\left\|n^{-B}\right\|^r\mu_r + \left\|n^{-B}\right\|\tau_r\right)\\
&\quad + KC_1C_5C_6M_2^p\left\|\varepsilon^{-B}\right\|^p\left\|\varepsilon^{B}\right\|^r\tau_r^{r-p}\left(n\left\|n^{-B}\right\|^r\mu_r + \left\|n^{-B}\right\|\tau_r\right)\\
&\quad + C_2\left\|\varepsilon^{-B}\right\|\tau_r^{-1}\left\|n^{-B}\right\|\tau_r\\
&\leq \Big\{K\left(C_1C_4\left\|\tau_r\varepsilon^B\right\| + C_1C_5C_6M_2^p\left\|(\tau_r\varepsilon^B)^{-1}\right\|^p\left\|\tau_r\varepsilon^B\right\|^r\right)\\
&\quad + \left(2C_1C_3 + C_2\left\|(\tau_r\varepsilon^B)^{-1}\right\|\right)\Big\}\left(n\left\|n^{-B}\right\|^r\mu_r + \left\|n^{-B}\right\|\tau_r\right).
\end{aligned}$$

We first show that

$$\lim_{t\to 0} \left\|t^{-B}\right\|^p \left\|t^B\right\|^r = 0. \tag{8.5.18}$$

It follows from (8.4.4) and (8.4.5) that for any $\eta > 0$ and for some small $t_0 > 0$, $\left\|t^{-B}\right\| \le t^{-\Lambda_B - \eta}$, $t < t_0$, and $\|t^B\| \le t^{\lambda_B - \eta}$, $t < t_0$. Thus for $t < t_0$,

$$\left\|t^{-B}\right\|^p \left\|t^B\right\|^r \le t^{-(\Lambda_B+\eta)p+(\lambda_B-\eta)r} \le t^{-\Lambda_B p+\lambda_B r-\eta(p+r)},$$

where by the restrictions on r and p, $-\Lambda_B p + \lambda_B r > 0$. Thus, taking $\eta > 0$ sufficiently small, we get (8.5.18). Also, of course, $\lim_{t\to 0} \|t^B\| = 0$. Therefore, we can find a sufficiently small $\varepsilon > 0$ such that the matrix $\tau_r \varepsilon^B$ satisfies

$$C_1C_4 \left\|\tau_r\varepsilon^B\right\| + C_1C_5C_6M_2^p \left\|\left(\tau_r\varepsilon^B\right)^{-1}\right\|^p \left\|\tau_r\varepsilon^B\right\|^r \le \frac{1}{2}. \tag{8.5.19}$$

Then choose K such that

$$2C_1C_3 + C_2 \left\|(\tau_r\varepsilon^B)^{-1}\right\| \le \frac{K}{2}. \tag{8.5.20}$$

Finally, we obtain from (8.5.17), (8.5.19), and (8.5.20) that

$$\varrho\left(n^{-B}\sum_{i=1}^n X_i, \theta\right) \le K\left(n\left\|n^{-B}\right\|^r \mu_r + \left\|n^{-B}\right\| \tau_r\right). \qquad \square$$

Proof of Theorems 8.4.8 and 8.4.9: By the same reasoning as that mentioned at the beginning of this section, we can assume that $\{X_i\}$ and θ are independent of each other and that the $\{\theta_i\}$ are independent copies of θ and are independent of the $\{X_i\}$ and θ. We prove the theorem by induction. For $n = 1$, the assertion is trivial.

For $n = 2$, we have

$$\operatorname{Var}\left(2^{-B}(X_1 + X_2),\, \theta\right) = \operatorname{Var}\left(2^{-B}(X_1 + X_2),\, 2^{-B}(\theta_1 + \theta_2)\right) \text{ [by (8.5.1)]}$$

$$\le \operatorname{Var}(X_1 + X_2,\, \theta_1 + \theta_2) \text{ [by the homogeneity of Var of order } r = 0]$$

$$\le \operatorname{Var}(X_1 + X_2,\, X_1 + \theta_2) + \operatorname{Var}(X_1 + \theta_2,\, \theta_1 + \theta_2)$$

[by the triangle inequality]

$$\le \operatorname{Var}(X_2,\, \theta_2) + \operatorname{Var}(X_1,\, \theta_1) \text{ [by the regularity of Var]}$$

$$= 2\operatorname{Var}(X_1,\, \theta) \le 2\nu_r \le 2c\left\|2^{-B}\right\|^r \nu_r,$$

since

$$c\left\|2^{-B}\right\|^r = \left(\left\|2^B\right\|^r + \left\|3^B\right\|^r\right)\left\|2^{-B}\right\|^r \geq \left\|2^B\right\|^r \left\|2^{-B}\right\|^r \geq 1.$$

For $n = 3$, we similarly have

$$\operatorname{Var}\left(3^{-B}(X_1 + X_2 + X_3), \theta\right) \leq 3c\left\|3^{-B}\right\|^r \nu_r.$$

Now suppose that for all $j < n$,

$$\operatorname{Var}\left(j^{-B}\sum_{i=1}^{j} X_i, \theta\right) \leq cj\left\|j^{-B}\right\|^r \nu_r. \tag{8.5.21}$$

Then for any $j < n$,

$$\operatorname{Var}\left(j^{-B}\sum_{i=1}^{j} X_i, \theta\right) \leq cM\nu_r \leq ca = \frac{1}{b}\left\|2^{-B}\right\|^r \tag{8.5.22}$$

by our assumptions.

For any integer $n \geq 4$ and $m = [\frac{n}{2}]$, we have

$$\operatorname{Var}\left(n^{-B}\sum_{i=1}^{n} X_i, \theta\right) \tag{8.5.23}$$

$$= \operatorname{Var}\left(n^{-B}\sum_{i=1}^{n} X_i, n^{-B}\sum_{i=1}^{n}\theta_i\right)$$

$$\leq \operatorname{Var}\left(n^{-B}\sum_{i=1}^{n} X_i, n^{-B}\left(\sum_{i=1}^{m}\theta_i + \sum_{i=m+1}^{n} X_i\right)\right)$$

$$+ \operatorname{Var}\left(n^{-B}(\sum_{i=1}^{m}\theta_i + \sum_{i=m+1}^{n} X_i), n^{-B}\sum_{i=1}^{n}\theta_i\right)$$

[by the triangle inequality]

$$\leq \operatorname{Var}\left(n^{-B}\sum_{i=m+1}^{n} X_i, n^{-B}\sum_{i=m+1}^{n}\theta_i\right)\operatorname{Var}\left(n^{-B}\sum_{i=1}^{m} X_i, n^{-B}\sum_{i=1}^{m}\theta_i\right)$$

$$+ \operatorname{Var}\left(n^{-B}\left(\sum_{i=1}^{m} X_i + \sum_{i=m+1}^{n}\theta_i\right), n^{-B}\sum_{i=1}^{n}\theta_i\right)$$

$$+ \operatorname{Var}\left(n^{-B}(\sum_{i=1}^{m}\theta_i + \sum_{i=m+1}^{n} X_i), n^{-B}\sum_{i=1}^{n}\theta_i\right)$$

[by Lemma 8.4.26]

$$=: \quad I_1 + I_2 + I_3.$$

By the induction hypotheses (8.5.21) and (8.5.22),

$$\begin{aligned}
I_1 &= \operatorname{Var}\left(\left(\frac{n}{n-m}\right)^{-B}(n-m)^{-B}\sum_{i=m+1}^{n} X_i,\ \left(\frac{n}{n-m}\right)^{-B}\theta_1\right) \\
&\quad \times \operatorname{Var}\left(\left(\frac{n}{m}\right)^{-B} m^{-B}\sum_{i=m+1}^{n} X_i,\ \left(\frac{n}{m}\right)^{-B}\theta_1\right) \\
&\le \operatorname{Var}\left((n-m)^{-B}\sum_{i=m+1}^{n} X_i,\ \theta_1\right)\operatorname{Var}\left(m^{-B}\sum_{i=1}^{m} X_i,\ \theta_1\right) \\
&\le \frac{1}{b}\left\|2^{-B}\right\|^r cm\left\|m^{-B}\right\|^r \nu_r.
\end{aligned}$$

Note that $m = [\frac{n}{2}] \ge \frac{2}{5}n$ for $n \ge 4$. Hence

$$\left\|m^{-B}\right\|^r \le \frac{n}{2}\left\|\left(\frac{2}{5}n\right)^{-B}\right\|^r \le \frac{n}{2}\left\|\left(\frac{2}{5}\right)^{-B}\right\|^r \left\|n^{-B}\right\|^r.$$

Thus

$$I_1 \le \frac{1}{b}\left\|2^{-B}\right\|^r c\frac{1}{2}\left\|\left(\frac{2}{5}\right)^{-B}\right\|^r n\left\|n^{-B}\right\|^r \nu_r = \frac{2}{5}cn\left\|n^{-B}\right\|^r \nu_r \tag{8.5.24}$$

by the definition of b. To estimate I_2, observe that

$$\begin{aligned}
I_2 &= \operatorname{Var}\left(n^{-B}\sum_{i=1}^{m} X_i + \left(\frac{n}{n-m}\right)^{-B}(n-m)^{-B}\sum_{i=m+1}^{n}\theta_i,\right. \\
&\qquad\qquad \left. n^{-B}\sum_{i=1}^{m}\theta_i + \left(\frac{n}{n-m}\right)^{-B}(n-m)^{-B}\sum_{i=m+1}^{n}\theta_i\right) \\
&\le \operatorname{Var}\left((n-m)^{-B}\sum_{i=1}^{m} X_i + \theta,\ (n-m)^{-B}\sum_{i=1}^{m}\theta_i + \theta\right).
\end{aligned}$$

Then we have

$$I_2 \le \left\|(n-m)^{-B}\right\|^r \mu_r\left(\sum_{i=1}^{m} X_i,\ \sum_{i=1}^{m}\theta_i\right) \quad \text{[by Lemma 8.4.30]}$$

$\le \left\|(n-m)^{-B}\right\|^r m\mu_r(X_1,\ \theta)$ [by the triangle inequality and the repeated use of the regularity of μ_r]

$$\le \frac{1}{2}\left\|\left(\frac{1}{2}\right)^{-B}\right\|^r n\left\|n^{-B}\right\|^r \mu_r \text{ since } \frac{n-m}{n} \ge \frac{1}{2}. \text{ Hence}$$

$$I_2 \;\leq\; \frac{1}{2}\left\|2^B\right\|^r n \left\|n^{-B}\right\|^r \nu_r. \tag{8.5.25}$$

As to I_3, we have

$$\begin{aligned} I_3 &= \operatorname{Var}\left(n^{-B}\sum_{i=m+1}^{n} X_i + n^{-B}m^B\theta,\, n^{-B}\sum_{i=m+1}^{n}\theta_i + n^{-B}m^B\theta\right) \\ &\leq \operatorname{Var}\left(m^{-B}\sum_{i=m+1}^{n} X_i + \theta,\, m^{-B}\sum_{i=m+1}^{n}\theta_i + \theta\right) \qquad (8.5.26) \\ &\leq \left\|m^{-B}\right\|^r (n-m)\mu_r(X_1,\theta) \;\leq\; \left\|n^{-B}\right\|^r \left\|\left(\frac{n}{m}\right)^B\right\|^r (n-m)\mu_r \\ &\leq \frac{3}{5}\left\|3^B\right\|^r n \left\|n^{-B}\right\|^r \nu_r, \end{aligned}$$

since $\frac{n}{m} \leq 3$ for $n \geq 4$ and $n - m \leq \frac{3}{5}n$.

Altogether, we have from (8.5.23)–(8.5.26),

$$\begin{aligned} \operatorname{Var}\left(n^{-B}\sum_{i=1}^{n} X_i,\, \theta\right) &\leq \left(\frac{2}{5}c + \frac{1}{2}\|2^B\|^r + \frac{3}{5}\|3^B\|^r\right) n \left\|n^{-B}\right\|^r \nu_r \\ &\leq cn\|n^{-B}\|^r\nu_r. \end{aligned}$$

This completes the proof of Theorem 8.4.8. The proof of Theorem 8.4.9 is similar and is therefore omitted. □

Proof of Theorem 8.4.10: Again, we assume that the $\{X_i\}$ and θ are independent of each other. Let $\{\theta_i\}$ be independent copies of θ, and assume that the $\{\theta_i\}$ are independent of $\{X_i\}$ and θ.

We prove the theorem by induction. For $n = 1$,

$$d(X_1,\theta) \;\leq\; Ad(X_1,\theta) \;\leq\; AT_r.$$

For $n = 2$, we have by Lemma 8.4.31, the regularity of d, and the triangle inequality,

$$\begin{aligned} d\left(2^{-B}(X_1+X_2),\,\theta\right) &= d\left(2^{-B}(X_1+X_2),\, 2^{-B}(\theta_1+\theta_2)\right) \\ &= \left|J_{2^B}\right| d(X_1+X_2,\,\theta_1+\theta_2) \;\leq\; 2\left|J_{2^B}\right| d(X_1,\theta) \\ &\leq 2C(d,B)T_r \;\text{[by Lemma 8.4.32]} \\ &\leq 2A\|2^{-B}\|^r T_r, \end{aligned}$$

since $A\left\|2^{-B}\right\|^r \;\geq\; C(d,B)\left\|2^B\right\|^r\left\|2^{-B}\right\|^r \;\geq\; C(d,B)$. Similarly, we have for $n = 3$,

$$d\left(3^{-B}(X_1+X_2+X_3),\,\theta\right) \;\leq\; 3C(d,B)T_r \;\leq\; 3A\left\|3^{-B}\right\|^r T_r,$$

since $A\left\|3^{-B}\right\|^r \geq C(d,B)\left\|3^B\right\|^r\left\|3^{-B}\right\|^r \geq C(d,B)$.

To prove the theorem by induction, assume for all $j<n$ that

$$d\left(j^{-B}\sum_{i=1}^{j}X_j,\,\theta\right) \leq Aj\left\|j^{-B}\right\|^r T_r.$$

For any $n\geq 4$ and $m=[\frac{n}{2}]$, we have by Lemma 8.4.28,

$$\begin{aligned}
&d\left(n^{-B}\sum_{i=1}^{n}X_i,\,\theta\right) \qquad (8.5.27)\\
&\leq\; d\left(n^{-B}\sum_{i=1}^{n}X_i,\,n^{-B}\left(\sum_{i=1}^{m}\theta_i+\sum_{i=m+1}^{n}X_i\right)\right)\\
&\quad +d\left(n^{-B}\left(\sum_{i=1}^{m}\theta_i+\sum_{i=m+1}^{n}X_i\right),\,n^{-B}\sum_{i=1}^{n}\theta_i\right)\\
&\leq\; d\left(n^{-B}\sum_{i=1}^{m}X_i,\,n^{-B}\sum_{i=1}^{m}\theta_i\right)\mathrm{Var}\left(n^{-B}\sum_{i=m+1}^{n}X_i,\,n^{-B}\sum_{i=m+1}^{n}\theta_i\right)\\
&\quad +d\left(n^{-B}\left(\sum_{i=1}^{m}X_i+\sum_{i=m+1}^{n}\theta_i\right),\,n^{-B}\sum_{i=1}^{n}\theta_i\right)\\
&\quad +d\left(n^{-B}\left(\sum_{i=1}^{m}\theta_i+\sum_{i=m+1}^{n}X_i\right),\,n^{-B}\sum_{i=1}^{n}\theta_i\right)\\
&=:\; I_1+I_2+I_3.
\end{aligned}$$

By Lemma 8.4.31,

$$\begin{aligned}
I_1 &\leq d\left(\left(\frac{n}{m}\right)^{-B}m^{-B}\sum_{i=1}^{m}X_i,\,\left(\frac{n}{m}\right)^{-B}\theta\right)\mathrm{Var}\left((n-m)^{-B}\sum_{i=m+1}^{n}X_i,\,\theta\right)\\
&= \left|J_{(\frac{n}{m})^B}\right| d\left(m^{-B}\sum_{i=1}^{m}X_i,\,\theta\right)\mathrm{Var}\left((n-m)^{-B}\sum_{i=m+1}^{n}X_i,\,\theta\right). \qquad (8.5.28)
\end{aligned}$$

By the induction hypothesis and Lemma 8.4.32,

$$\begin{aligned}
\left|J_{(\frac{n}{m})^B}\right| d\left(m^{-B}\sum_{i=1}^{m}X_i,\,\theta\right) &\leq C(d,B)Am\left\|m^{-B}\right\|^r T_r \qquad (8.5.29)\\
&\leq C(d,B)A\frac{n}{2}\left\|n^{-B}\right\|^r\left\|\left(\frac{m}{n}\right)^{-B}\right\|^r T_r\\
&\leq C(d,B)A\frac{1}{2}\left\|3^B\right\|^r n\left\|n^{-B}\right\|^r T_r.
\end{aligned}$$

On the other hand, since $\nu_r \leq \frac{a}{M}$, by Theorem 8.4.8,

$$\begin{aligned}\operatorname{Var}\left((n-m)^{-B}\sum_{i=m+1}^{n} X_i,\, \theta\right) &\leq c(n-m)\left\|(n-m)^{-B}\right\|^r \nu_r \qquad (8.5.30)\\ &= c\left\|(n-m)^{\frac{1}{r}I-B}\right\|^r \nu_r\\ &\leq cM\nu_r \leq \frac{1}{D},\end{aligned}$$

where we have used $\nu_r \leq \frac{1}{McD}$. Therefore, we have, by (8.5.28)–(8.5.30),

$$I_1 \leq C(d,B)A\frac{1}{2}\left\|3^B\right\|^r \frac{1}{D} n\left\|n^{-B}\right\|^r T_r \leq \frac{1}{2}An\left\|n^{-B}\right\|^r T_r, \qquad (8.5.31)$$

since $D \geq C(d,\, B)\left\|3^B\right\|^r$. Similarly, for the estimate of I_2,

$$\begin{aligned}I_2 &= d\left(n^{-B}\sum_{i=1}^{m} X_i + \left(\frac{n}{n-m}\right)^{-B}\theta,\, n^{-B}\sum_{i=1}^{m}\theta_i + \left(\frac{n}{n-m}\right)^{-B}\theta\right)\\ &= \left|J_{(\frac{n}{n-m})^B}\right| d\left((n-m)^{-B}\sum_{i=1}^{m} X_i + \theta,\, (n-m)^{-B}\sum_{i=1}^{m}\theta_i + \theta\right)\\ &\qquad\qquad\qquad\qquad\qquad\qquad \text{[by Lemma 8.4.31]}\\ &\leq C(d,B)\|(n-m)^{-B}\|^r d_r\left(\sum_{i=1}^{m} X_i,\, \sum_{i=1}^{m}\theta_i\right) \quad \text{[by Lemma 8.4.32]}\\ &\leq C(d,B)\left\|(n-m)^{-B}\right\|^r m d_r(X_1,\, \theta)\end{aligned}$$

by the triangle inequality and the repeated use of the regularity of d_r.

Hence

$$I_2 \leq C(d,B)\frac{1}{2}\left\|2^B\right\|^r n\left\|n^{-B}\right\|^r T_r. \qquad (8.5.32)$$

Finally, we have

$$\begin{aligned}I_3 &= d\left(\left(\frac{n}{m}\right)^{-B}\theta + n^{-B}\sum_{i=m+1}^{n} X_i,\, \left(\frac{n}{m}\right)^{-B}\theta + n^{-B}\sum_{i=m+1}^{n}\theta_i\right) \qquad (8.5.33)\\ &\leq C(d,B)d\left(m^{-B}\sum_{i=m+1}^{n} X_i + \theta,\, m^{-B}\sum_{i=m+1}^{n}\theta_i + \theta\right)\\ &\leq C(d,B)\left\|m^{-B}\right\|^r (n-m)d_r(X_1,\, \theta)\\ &\leq C(d,B)\frac{3}{5}\left\|3^B\right\|^r n\left\|n^{-B}\right\|^r T_r.\end{aligned}$$

Combining the estimates for I_1, I_2, and I_3, we finally obtain from (8.5.27) and (8.5.31)–(8.5.33) that

$$\begin{aligned} d\left(n^{-B}\sum_{i=1}^{n} X_i,\, \theta\right) &\le I_1 + I_2 + I_3 \\ &\le \left(\frac{1}{2}A + \frac{1}{2}C(d,B)\left\|2^B\right\|^r + \frac{3}{5}C(d,B)\left\|3^B\right\|^r\right) n \left\|n^{-B}\right\|^r T_r \\ &\le An\|n^{-B}\|^r T_r \end{aligned}$$

by the definition of A. □

Proof of Theorem 8.4.13: We apply the "minimality" property of the $\widehat{L}_p$-metric :

$$\widehat{L}_p\left(n^{-B}\sum_{i=1}^{n} X_i,\, \theta\right) \le \inf L_p\left(n^{-B}\sum_{i=1}^{n} \widetilde{X}_i,\, n^{-B}\sum_{i=1}^{n}\widetilde{\theta}_i\right), \quad (8.5.34)$$

where the infimum is taken over all independent identically distributed pairs $(\widetilde{X}_i, \widetilde{\theta}_i) \overset{d}{\sim} (\widetilde{X}, \widetilde{\theta})$ with fixed marginals $\widetilde{X} \overset{d}{\sim} X$ and $\widetilde{\theta} \overset{d}{\sim} \theta$. The right-hand side in (8.5.34) is less than or equal to

$$\inf\left\{\left\|n^{-B}\right\|^p n\mathcal{L}_p(\widetilde{X},\widetilde{\theta});\ \widetilde{X} \overset{d}{\sim} X,\, \widetilde{\theta} \overset{d}{\sim} \theta\right\} = n\left\|n^{-B}\right\|^p \widehat{L}_p(X,\theta).$$

The bound for π follows from Lemma 8.4.33. □

Proof of Theorem 8.4.14: The proof is similar to that of Theorem 8.4.10 and is therefore omitted. □

Proof of Theorem 8.4.15: The proof resembles that of Theorem 8.4.4 with the replacement of the smoothing inequality for ϱ^* by that for $\widehat{L}_p$ in Lemma 8.4.27 and hence is omitted. □

Proof of Theorem 8.4.16: Using the Marcinkiewicz–Zygmund inequality (see for example Kawata (1972, Theorem 13.6.1)), if $1 < p \le 2$ and $\{\xi_i\}_{i\ge 1}$ are independent random vectors with $E[\xi] = 0$, then

$$E\left[\left\|\sum_{i=1}^{n}\xi_i\right\|^p\right] \le C_p \sum_{i=1}^{n} E[\|\xi_i\|^p], \quad (8.5.35)$$

for some $C_p > 0$. Take $\xi_i = n^{-B}(\widetilde{X}_i - \widetilde{\theta}_i)$, where $\{(\widetilde{X}_i - \widetilde{\theta}_i)\}_{i\ge 1}$ are i.i.d. pairs, $\widetilde{X}_i \overset{d}{\sim} X$ and $\widetilde{\theta}_i \overset{d}{\sim} \theta$. Thus we get

$$E\left[\left\|n^{-B}\sum_{i=1}^{n}(\widetilde{X}_i - \widetilde{\theta}_i)\right\|^p\right] \le C_p \sum_{i=1}^{n} E\left[\left\|n^{-B}(\widetilde{X}_i - \widetilde{\theta}_i)\right\|^p\right]$$

$$= C_p n \left\|n^{-B}\right\|^p E\left[\left\|\widetilde{X} - \widetilde{\theta}\right\|^p\right].$$

Passing to the minimal metrics gives the necessary inequality. Finally, note that $p > \frac{\Lambda_B}{\lambda_B^2}$ implies $n\|n^{-B}\|^p \to 0$ as $n \to \infty$. The bound for the Prohorov metric π comes from the inequality $\pi \leq \widehat{L}_p^{\frac{p}{p+1}}$ for $p \geq 1$. □

Proof of Theorem 8.4.19: We use the minimality property of $\widehat{L}_p$ to get

$$\begin{aligned}
&\left\{\widehat{L}_p(M_X(n), M_\theta(n))\right\}^p \\
&= \left\{\widehat{L}_p\left(n^{-B}\bigvee_{k=1}^{n}\left(\sum_{i=1}^{k}X_i\right), n^{-B}\bigvee_{k=1}^{n}\left(\sum_{i=1}^{k}\theta_i\right)\right)\right\}^p \\
&\leq E\left[\left\|n^{-B}\bigvee_{k=1}^{n}\left(\sum_{i=1}^{k}X_i\right) - n^{-B}\bigvee_{k=1}^{n}\left(\sum_{i=1}^{k}\theta_i\right)\right\|^p\right] \\
&\leq \|n^{-B}\|^p E\left[\left\|\bigvee_{k=1}^{n}\left(\sum_{i=1}^{k}X_i\right) - \bigvee_{k=1}^{n}\left(\sum_{i=1}^{k}\theta_i\right)\right\|^p\right] \\
&\leq C'_{p,d}\|n^{-B}\|^p E\left[\bigvee_{k=1}^{n}\left\|\sum_{i=1}^{k}X_i - \sum_{i=1}^{k}\theta_i\right\|_0^p\right]
\end{aligned}$$

(here we have used $\|\cdot\| \leq \{C'_{p,d}\}^{1/p}\|\cdot\|_0$ for some $C'_{p,d} > 0$ and $\|\bigvee x_k - \bigvee y_k\|_0 \leq \bigvee \|x_k - y_k\|_0$)

$$\leq C'_{p,d}\|n^{-B}\|^p \sum_{k=1}^{n} E\left[\left\|\sum_{i=1}^{k}(X_i - \theta_i)\right\|_0^p\right].$$

Since $E[X - \theta] = 0$ and $1 < p \leq 2$, by (8.5.35) again the above is, for some other constant $C_{p,d} > 0$,

$$\begin{aligned}
&\leq \left\|n^{-B}\right\|^p C_{p,d}\sum_{k=1}^{n}\sum_{i=1}^{k} E\left[\|X - \theta\|^p\right] \\
&= C_{p,d}\frac{n(n+1)}{2}\left\|n^{-B}\right\|^p E\left[\|X - \theta\|^p\right].
\end{aligned}$$

Passing to the minimal metric gives the necessary bound for $\widehat{L}_p(M_X(n), M_\theta(n))$.

The same argument leads to the bound for $\widehat{L}_p(m_X(n), m_\theta(n))$. Further,

$$\begin{aligned}
\left\{\widehat{L}_p(a_X(n), a_\theta(n)\right\}^p &\leq \|n^{-B}\|^p E\left[\left\|\bigvee_{k=1}^n \left\|\sum_{i=1}^k X_i\right\| - \bigvee_{k=1}^n \left\|\sum_{i=1}^k \theta_i\right\|\right\|^p\right] \\
&\leq \|n^{-B}\|^p E\left[\bigvee_{k=1}^n \left\|\left\|\sum_{i=1}^k X_i\right\| - \left\|\sum_{i=1}^k \theta_i\right\|\right\|^p\right] \\
&\leq \|n^{-B}\|^p E\left[\bigvee_{k=1}^n \left\|\sum_{i=1}^k X_i - \sum_{i=1}^k \theta_i\right\|^p\right] \\
&\leq C_p \frac{n(n+1)}{2} \left\|n^{-B}\right\|^p E\left[\|X-\theta\|^p\right]
\end{aligned}$$

as before. Combining our estimates, we complete the proof of the theorem. □

Proof of Theorem 8.4.21: From the definition of the Zolotarev metric ζ_r and its ideality of order r, we get the first bound:

$$\begin{aligned}
\zeta_r\left(n^{-B}\sum_{i=1}^n X_i, \theta\right) &\leq \zeta_r\left(n^{-B}\sum_{i=1}^n X_i, n^{-B}\sum_{i=1}^n \theta_i\right) \\
&\leq \left\|n^{-B}\right\|^r \zeta_r\left(\sum_{i=1}^n X_i, \sum_{i=1}^n \theta_i\right) \leq n\left\|n^{-B}\right\|^r \zeta_r\left(X_1, \theta\right).
\end{aligned}$$

Applying the universal bound for the Prohorov metric π by ζ_r,

$$\pi^{r+1} \leq C_r^{r+1}\zeta_r$$

on $\mathcal{X}(\mathbb{R}^d)$ for some $C_r > 0$ (cf. Zolotarev (1983)), we obtain the final estimate. □

Proof of Theorem 8.4.22: Let $\xi_i = n^{-B}(\widetilde{X}_i - \widetilde{\theta}_i)$, where $\widetilde{X}_i \overset{d}{\sim} X$ and $\widetilde{\theta}_i \overset{d}{\sim} \theta$, and $(\widetilde{X}_i, \widetilde{\theta}_i)$ are i.i.d. Then by the Rosenthal inequality (see, for example, Araujo and Giné (1980, p. 205)), for $2 < p < \infty$,

$$\left\{E\left[\left\|\sum_{i=1}^n \xi_i\right\|^p\right]\right\}^{1/p} \leq C(d,p)\max\left\{\left(\sum_{i=1}^n E[\|\xi_i\|^p]\right)^{1/p}, \left(\sum_{i=1}^n E[\|\xi_i\|^2]\right)^{1/2}\right\},$$

for some $C(d,p) > 0$. Then

$$L_p\left(n^{-B}\sum_{i=1}^n \widetilde{X}_i, n^{-B}\sum_{i=1}^n \widetilde{\theta}_i\right)$$

$$
\begin{aligned}
&\leq C(d,p)\max\left\{\left(\sum_{i=1}^{n} E[\|n^{-B}(\widetilde{X}_i-\widetilde{\theta}_i)\|^p]\right)^{1/p},\right.\\
&\qquad\qquad\left.\left(\sum_{i=1}^{n} E[\|n^{-B}(\widetilde{X}_i-\widetilde{\theta}_i)\|^2]\right)^{1/2}\right\}\\
&\leq C(d,p)\max\left\{\left(n\|n^{-B}\|^p E[\|\widetilde{X}-\widetilde{\theta}\|^p]\right)^{1/p}, \left(n\|n^{-B}\|^2 E[\|\widetilde{X}-\widetilde{\theta}\|^2]\right)^{1/2}\right\}\\
&\leq C(d,p)\max\left\{n^{1/p}\left\|n^{-B}\right\| L_p(\widetilde{X},\widetilde{\theta}), n^{1/2}\left\|n^{-B}\right\| L_2(\widetilde{X},\widetilde{\theta})\right\}\\
&\leq C(d,p) n^{1/2}\left\|n^{-B}\right\| L_p(\widetilde{X},\widetilde{\theta}),
\end{aligned}
$$

since $L_2 \leq L_p$ and $n^{1/p} \leq n^{1/2}$ for $p > 2$. The case $p = \infty$ is similar. □

In Theorems 8.4.4, 8.4.8, 8.4.9, 8.4.10, and 8.4.14 we have assumed the finiteness of the metrics μ_r, χ_r, d_r, and $\widehat{L}_{p,r}$. Since $\chi_r \leq \mu_r$, the natural question is how we can assure the finiteness of μ_r, d_r and $\widehat{L}_{p,r}$, which may not be easily checked just by a direct use of the definitions. The rest of this section is devoted to the construction of upper bounds for μ_r, d_r, and $\widehat{L}_{p,r}$, where the metrics used in the upper bounds are more familiar distances in the literature.

We shall construct bounds from above for μ_r, d_p, and $L_{p,r}$, using the Zolotarev ζ_r-metric. Define the following probability metrics: For $X, Y \in \mathcal{X}(\mathbb{R}^d)$,

$$
\begin{aligned}
\overline{\mu}_r(X,Y) &= \sup_{T\in \mathrm{Aut}(\mathbb{R}^d)} \|T\|^{-r}\,\mathrm{Var}(TX+\theta,\ TY+\theta),\\
\overline{d}_r(X,Y) &= \sup_{T\in \mathrm{Aut}(\mathbb{R}^d)} \|T\|^{-r} d(TX+\theta,\ TY+\theta),
\end{aligned}
$$

and

$$
\widehat{\widehat{L}}_{p,r}(X,Y) = \sup_{T\in \mathrm{Aut}(\mathbb{R}^d)} \|T\|^{-r} L_p(TX+\theta,\ TY+\theta).
$$

Clearly, $\mu_r \leq \overline{\mu}_r$, $d_r \leq \overline{d}_r$, $\widehat{L}_{p,r} \leq \widehat{\widehat{L}}_{p,r}$. In the next two theorems we are going to estimate $\overline{\mu}_r, \overline{d}_r$, and $\widehat{\widehat{L}}_{p,r}$ from above by ζ_r. Let $p_\theta(x)$ be the density function of the strictly operator-stable random vector $\theta \in \mathbb{R}^d$. For $m \in \mathbb{N}$ let

$$
C_m(\theta) := \sup_{\|h\|=1} \int_{\mathbb{R}^d} |p_\theta^{(m)}(x)(h)^m|\,dx
$$

and

$$D_m(\theta) := \sup_{x\in\mathbb{R}^d}\sup_{\|h\|=1} |p_\theta^{(m)}(x)(h)^m|.$$

Theorem 8.5.1 (i) *For* $m \in \mathbb{N}$,

$$\overline{\mu}_m(X,Y) \le C_m(\theta)\zeta_m(X,Y).$$

(ii) *If* $r = m+p, m \in \mathbb{N}, 0 < p \le 1,$ *then*

$$\widehat{\overline{L}}_{p,r}(X,Y) \le C_m(\theta)\zeta_r(X,Y).$$

Theorem 8.5.2

$$d_m(X,Y) \le D_m(\theta)\zeta_m(X,Y), \quad m = 1,2,\ldots .$$

Proof of Theorem 8.5.1: (i) For any X and Y,

$$\begin{aligned}\mathrm{Var}(X+\theta, Y+\theta) &= \sup_{A\in\mathcal{B}(\mathbb{R}^d)} |P(X+\theta\in A) - P(Y+\theta\in A)| \quad (8.5.36)\\ &= \sup\{|E[f(X+\theta) - f(Y+\theta)]|;\ f\in\mathcal{F},\ \|f\|_\infty \le 1\}\\ &= \sup\{|E[g(X) - g(Y)]|;\ f\in\mathcal{F},\ \|f\|_\infty \le 1\},\end{aligned}$$

where $g(x) := E[f(x+\theta)]$. Since $p_\theta(x)$ is differentiable infinitely many times (see Hudson (1980)),

$$g(x) = \int_{\mathbb{R}^d} f(x+y)p_\theta(y)\,\mathrm{d}y = \int_{\mathbb{R}^d} f(z)p_\theta(z-x)\,\mathrm{d}z$$

has derivatives of every order, and furthermore,

$$\begin{aligned}|g^{(m)}(x)(h)^m| &= \left|\int_{\mathbb{R}^d} f(z)p_\theta^{(m)}(z-x)(h)^m\,\mathrm{d}z\right|\\ &= \left|\int_{\mathbb{R}^d} f(x+y)p_\theta^{(m)}(y)(h)^m\,\mathrm{d}y\right|.\end{aligned}$$

Since for f with $\|f\|_\infty \le 1$, $\sup_{x\in\mathbb{R}^d}\sup_{\|h\|=1} |g^{(m)}(x)(h)^m| \le C_m(\theta)$, we have

$$\|g^{(m-1)}(x) - g^{(m-1)}(y)\| \le C_m(\theta)\|x-y\|. \quad (8.5.37)$$

Hence by (8.5.36) and (8.5.37),

$$\begin{aligned}
\operatorname{Var}(X+\theta, Y+\theta) \\
&\leq \sup\Big\{|E\,[g(X)-g(Y)]|\,;\ \Big\|g^{(m-1)}(x)-g^{(m-1)}(y)\Big\| \\
&\qquad\qquad \leq C_m(\theta)\,\|x-y\|\Big\}
\end{aligned}$$

and

$$\begin{aligned}
\overline{\mu}_m(X,Y) &= \sup_{T\in\operatorname{Aut}(\mathbb{R}^d)} \|T\|^{-m}\operatorname{Var}(TX+\theta,\ TY+\theta) \qquad (8.5.38)\\
&\leq \sup_T \|T\|^{-m}\sup\Big\{\,|E\,[g(TX)-g(TY)]|\,; \\
&\qquad \Big\|g^{(m-1)}(x)-g^{(m-1)}(y)\Big\| \leq C_m(\theta)\,\|x-y\|\,\Big\}.
\end{aligned}$$

Let $g_T(x) := g(Tx)$. Then

$$g_T^{(m-1)}(x)(h)^{m-1} \;=\; g^{(m-1)}(Tx)(Th)^{m-1} \quad \text{for any } x,h\in\mathbb{R}^d,$$

implying that

$$\Big\|g_T^{(m-1)}(x)-g_T^{(m-1)}(y)\Big\| \;\leq\; \|T\|^{m-1}\Big\|g^{(m-1)}(Tx)-g^{(m-1)}(Ty)\Big\|.$$

Then the side condition in (8.5.38),

$$\Big\|g^{(m-1)}(x)-g^{(m-1)}(y)\Big\| \;\leq\; C_m(\theta)\|x-y\|,$$

results in

$$\begin{aligned}
\Big\|g_T^{(m-1)}(x)-g_T^{(m-1)}(y)\Big\| &\leq C_m(\theta)\|T\|^{m-1}\|Tx-Ty\| \\
&\leq C_m(\theta)\|T\|^{m}\|x-y\|.
\end{aligned}$$

Consequently, by (8.5.38),

$$\begin{aligned}
\overline{\mu}_m(X,Y) &\leq \sup_T \|T\|^{-m}\sup\Big\{|E[g_T(X)-g_T(Y)]|; \\
&\qquad \|g_T^{(m-1)}(x)-g_T^{(m-1)}(y)\| \leq C_m(\theta)\|T\|^m\|x-y\|\Big\} \\
&= C_m(\theta)\zeta_m(X,Y),
\end{aligned}$$

as desired.

(ii) Let us now prove a similar bound for $\widehat{\overline{L}}_{p,r}$. Let $r=m+p$, $m\in\mathbb{N}$, $0<p\leq 1$. Then by Lemma 8.4.34 (ii),

$$\begin{aligned}
\widehat{L}_p(X+\theta,\ Y+\theta) &= \sup\{|E\,[f(X+\theta)-f(Y+\theta)]|\,;\ f\in \operatorname{Lip}_b(p)\} \\
&= \sup\{|E\,[g(X)-g(Y)]|\,;\ f\in \operatorname{Lip}_b(p)\}\,,
\end{aligned}$$

where $g(x) := E[f(x+\theta)]$. Since $p_\theta(x)$ is differentiable infinitely many times, the function $g(x) = \int\limits_{\mathbb{R}^d} f(x+y)p_\theta(y)\,\mathrm{d}y = \int\limits_{\mathbb{R}^d} f(z)p_\theta(z-x)\,\mathrm{d}z$ has derivatives of all orders, and for $m \in \mathbb{N}, r = m+p$,

$$|g^{(m)}(x)(h)^m| = \left| \int\limits_{\mathbb{R}^d} f(x+y)p_\theta^{(m)}(y)(h)^m\,\mathrm{d}y \right|.$$

By the requirement for f,

$$\begin{aligned}
\left\|g^{(m)}(x) - g^{(m)}(y)\right\| &= \sup_{\|h\|=1} \left|g^{(m)}(x)(h)^m - g^{(m)}(y)(h)^m\right| \\
&= \sup_{\|h\|=1} \left| \int\limits_{\mathbb{R}^d} [f(x+z) - f(y+z)]\, p_\theta^{(m)}(z)(h)^m\,\mathrm{d}z \right| \\
&\leq \sup_{\|h\|=1} \int\limits_{\mathbb{R}^d} \|x-y\|^p \left|p_\theta^{(m)}(z)(h)^m\right|\,\mathrm{d}z \\
&\leq \|x-y\|^p C_m(\theta).
\end{aligned}$$

Therefore $\widehat{L}_p(X+\theta, Y+\theta) \leq \sup\Big\{ |E\left[g(X) - g(Y)\right]|;\ \|g^{(m)}(x) - g^{(m)}(y)\| \leq C_m(\theta)\|x-y\|^p \text{ for any } x, y \in \mathbb{R}^d \Big\}$.

Next consider $\widehat{\widehat{L}}_p(X,Y)$:

$$\begin{aligned}
\widehat{\widehat{L}}_{p,r}(X,Y) &= \sup_{T\in \mathrm{Aut}(\mathbb{R}^d)} \|T\|^{-r} L_p(TX+\theta,\ TY+\theta) \\
&\leq \sup_T \|T\|^{-r} \sup\Big\{|E[g(TX) - g(TY)]|; \\
&\qquad \left\|g^{(m)}(x) - g^{(m)}(y)\right\| \leq C_r(\theta)\|x-y\|^p\Big\}.
\end{aligned}$$

Let $g_T(x) := g(Tx)$. Then for all $x, h \in \mathbb{R}^d$,
$g_T^{(m)}(x)(h)^m = g^{(m)}(Tx)(Th)^m$ for any $x, h \in \mathbb{R}^d$, implying that

$$\left\|g_T^{(m)}(x) - g_T^{(m)}(y)\right\| = \|T\|^m \left\|g^{(m)}(Tx) - g^{(m)}(Ty)\right\|.$$

Applying $\|g^{(m)}(x) - g^{(m)}(y)\| \leq C_m(\theta)\|x-y\|^p$, we get that

$$\begin{aligned}
\left\|g_T^{(m)}(x) - g_T^{(m)}(y)\right\| &\leq C_m(\theta)\|T\|^m\|Tx - Ty\|^p \\
&\leq C_m(\theta)\|T\|^{m+p}\|x-y\|^p,
\end{aligned}$$

and $m+p=r$ by assumption. Similarly, we get

$$\begin{aligned}\widehat{\overline{L}}_{p,r}(X,Y) &\le \sup_T \|T\|^{-r} \sup\{|E[g_T(X)-g_T(Y)]|;\\ &\qquad \left\|g_T^{(m)}(x)-g_T^{(m)}(y)\right\| \le C_m(\theta)\|T\|^r\|x-y\|^p\Big\}\\ &\le C_m(\theta)\zeta_r(X,Y).\end{aligned}$$

□

Proof of Theorem 8.5.2: We have

$$\begin{aligned}&d_m(X,Y) &(8.5.39)\\ &= \sup_{T\in \mathrm{Aut}(\mathbb{R}^d)} \|T\|^{-m} d(TX+\theta,\ TY+\theta)\\ &= \sup_T \|T\|^{-m} \sup_{x\in\mathbb{R}^d} |p_{TX+\theta}(x)-p_{TY+\theta}(x)|\\ &= \sup_T \|T\|^{-m} \sup_{x\in\mathbb{R}} \left|\int_{\mathbb{R}^d} p_\theta(x-y)[P(TX\in \mathrm{d}y)-P(TY\in\mathrm{d}y)]\right|.\end{aligned}$$

$$\text{Let}\quad g(y) = p_\theta(x-y). \tag{8.5.40}$$

Then

$$\sup_{y\in\mathbb{R}^d}\sup_{x\in\mathbb{R}^d}\sup_{\|h\|=1}\left|g^{(m)}(y)(h)^m\right| \le \sup_{y,x,h}\left|p_\theta^{(m)}(x-y)(h)^m\right| = D_m(\theta).$$

$$\text{Hence}\quad \left\|g^{(m-1)}(x)-g^{(m-1)}(y)\right\| \le D_m(\theta)\|x-y\|. \tag{8.5.41}$$

From by (8.5.39)–(8.5.41) we have

$$\begin{aligned}d_m(X,Y) &\le \sup_T \|T\|^{-m}\sup\Big\{|E[g(TX)-g(TY)]|;\\ &\qquad \left\|g^{(m-1)}(x)-g^{(m-1)}(y)\right\| \le D_m(\theta)\|x-y\|\Big\}.\end{aligned}$$

This upper bound is the same as that of $\overline{\mu}_m(X,\,Y)$ in (8.5.38) if $C_m(\theta)$ is replaced by $D_m(\theta)$. Therefore the proof of Theorem 8.5.1 also implies that

$$d_m(X,Y) \le D_m(\theta)\zeta_m(X,Y).$$

□

The next question is the finiteness of $C_m(\theta)$ and $D_m(\theta)$.

Theorem 8.5.3 *We have $D_m(\theta)<\infty,\ \ m=1,2,\ldots,$ and if $\Lambda_B<1$, then $C_m(\theta)<\infty,\ \ m=1,2,\ldots.$*

Proof: We first show the finiteness of $D_m(\theta)$. Let $\widehat{\mu}(z), z \in \mathbb{R}^d$, be the characteristic function of θ. As was shown in Hudson (1980), for some $c > 0$,

$$|\widehat{\mu}(z)| \leq \exp\{-c\|z\|^{1/\|B\|}\} \quad \text{for every } z \text{ with } \|z\| > 1. \tag{8.5.42}$$

Hence, for every $m = 1, 2, \ldots$, $\int_{\mathbb{R}^d} \|z\|^m |\widehat{\mu}(z)|\, dz < \infty$, implying the existence of $p_\theta^{(m)}(x)$, and furthermore the finiteness of $D_m(\theta)$.

To prove the finiteness of $C_m(\theta)$, we assume $\Lambda_B < 1$, which implies $E[\|\theta\|] < \infty$. (See Hudson et al. (1988).) We start with Carlson's inequality for one-variable functions f. If $\widehat{f}$, the Fourier transform of f, is in $L^2(\mathbb{R})$, and $\widehat{f}'$ exists and is in $L^2(\mathbb{R})$, then

$$\left(\int_{-\infty}^{\infty} |f(x)|\, dx\right)^4 \leq K \left(\int_{-\infty}^{\infty} \widehat{f}(z)^2\, dz\right) \left(\int_{-\infty}^{\infty} \widehat{f}'(z)^2\, dz\right). \tag{8.5.43}$$

A version of this inequality for several variable functions f is, for each $h \in \mathbb{R}^d$,

$$\left(\int_{\mathbb{R}^d} |f(x)|\, dx\right)^4 \leq K \left(\int_{\mathbb{R}^d} \widehat{f}(z)^2\, dz\right) \left(\int_{\mathbb{R}^d} (D\widehat{f}(z)h)^2\, dz\right), \tag{8.5.44}$$

where $D\widehat{f}(z)$ is the gradient (row) vector of $\widehat{f}(z)$. The proof of (8.5.44) can be carried out in the same manner as for (8.5.43), so we omit it.

Since we are assuming $E[\|\theta\|] < \infty$, $\widehat{\mu}(z)$ is differentiable. Fix $h \in \mathbb{R}^d$ and apply (8.5.44) to $f(x) := p_\theta^{(m)}(x)(h)^m, \quad x \in \mathbb{R}^d$.

Then $\widehat{f}(z) = \int_{\mathbb{R}^d} e^{i\langle z,x\rangle} p_\theta^{(m)}(x)(h)^m\, dx$, and

$$D\widehat{f}(z)h = i\int_{\mathbb{R}^d} \langle x, h\rangle e^{i\langle z,x\rangle} p_\theta^{(m)}(x)(h)^m\, dx.$$

Thus $\int_{\mathbb{R}^d} \widehat{f}(z)^2\, dz \leq \int_{\mathbb{R}^d} \|z\|^{2m} |\widehat{\mu}(z)|^2\, dz < \infty$ by (8.5.42), and

$$\sup_{\|h\|=1} \int_{\mathbb{R}^d} \left(D\widehat{f}(z)h\right)^2 dz \leq \int_{\mathbb{R}^d} \|z\|^{2m} |D\widehat{\mu}(z)|^2\, dz =: I.$$

So it remains to show the finiteness of I. We recall that the characteristic function $\widehat{\mu}(z)$ is given by

$$\widehat{\mu}(z) = \exp\Bigg\{ i\langle z, c\rangle + \langle z, Az\rangle$$

$$+ \int_S \gamma(\mathrm{d}x) \int_0^\infty \left[e^{i\langle z, s^B x\rangle} - 1 - i\langle z, s^B x\rangle I_Q(s^B x) \right] \frac{1}{s^2} \, \mathrm{d}s \Big\}.$$

Here $z \in \mathbb{R}^d$, $c \in \mathbb{R}^d$, A is a nonnegative definite symmetric matrix, $S = \{x \in \mathbb{R}^d; \|x\| = 1 \text{ and } \|t^B x\| > 1 \text{ for all } t > 1\}$, $Q = \{x \in \mathbb{R}^d; \|x\| \leq 1\}$, γ is a probability measure on S. We write $\widehat{\mu}(z) = e^{\psi(z)}$. Since $\Lambda_B < 1$,

$$M_1 \;:=\; \int_S \gamma(\mathrm{d}x) \int_{\{\|s^B x\| > 1\}} \frac{\|s^B x\|}{s^2} \, \mathrm{d}s \;<\; \infty. \tag{8.5.45}$$

Note that if $\gamma(S) > 0$, the non-Gaussian part exists, and the restriction of B to the support of the measure γ (we shall call it B again) satisfies $\lambda_B > \frac{1}{2}$. (See Hudson and Mason (1981).) Hence we also have

$$M_2 \;:=\; \int_S \gamma(\mathrm{d}x) \int_{\{\|s^B x\| \leq 1\}} \frac{\|s^B x\|^2}{s^2} \, \mathrm{d}s \;<\; \infty. \tag{8.5.46}$$

We have, for $h \in \mathbb{R}^d$, $D\widehat{\mu}(z)h = \widehat{\mu}(z) D\psi(z)h$.
Let $z = (z_1, \ldots, z_d)^t$, $c = (c_1, \ldots, c_d)^t$, $A = (a_{ij})$, $s^B x = \left((s^B x)_1, \ldots, (s^B x)_d \right)^t$, and $h = (h_1, \ldots, h_d)^t$. Then

$$\begin{aligned}\tfrac{\partial}{\partial z_j} \psi(z) \;&=\; i c_j + 2(Az)_j \\ &\quad + \int_S \gamma(\mathrm{d}x) \int_0^\infty \left[i \left(s^B x\right)_j e^{i\langle z, s^B x\rangle} - i \left(s^B x\right)_j I_Q\left(s^B x\right) \right] \frac{1}{s^2} \, \mathrm{d}s.\end{aligned}$$

Thus

$$\begin{aligned}\left| \tfrac{\partial}{\partial z_j} \psi(z) \right| \;&\leq\; |c_j| + 2\|Az\| + \int_S \gamma(\mathrm{d}x) \int_0^\infty \left| \left(s^B x\right)_j \right| \left| e^{i<z, s^B x>} - I_Q(s^B x) \right| \frac{1}{s^2} \, \mathrm{d}s \\ &\leq\; \|c\| + 2\|Az\| + \int_S \gamma(\mathrm{d}x) \int_{\{\|s^B x\| > 1\}} \|s^B x\| \, \frac{1}{s^2} \, \mathrm{d}s \\ &\quad + \int_S \gamma(\mathrm{d}x) \int_{\{\|s^B x\| \leq 1\}} \|s^B x\| \left| e^{i<z, s^B x>} - 1 \right| \frac{1}{s^2} \, \mathrm{d}s \\ &\leq\; \|c\| + 2\|Az\| + M_1 + M_2 \|z\|,\end{aligned}$$

where M_1 and M_2 are finite by (8.5.45) and (8.5.46). We thus finally have

$$|D\psi(z)h| \;=\; \left| \sum_{j=1}^d \tfrac{\partial}{\partial z_j} \psi(z) h_j \right|$$

$$\begin{aligned} &\le d\|h\|\left(\|c\| + 2\|A\|\|z\| + M_1 + M_2\|z\|\right) \\ &\le C_1 + C_2\|z\|, \end{aligned}$$

and

$$|D\widehat{\mu}(z)h| \le (C_1 + C_2\|z\|)\,|\widehat{\mu}(z)|.$$

Hence by (8.5.42) we conclude that

$$I = \int_{\mathbb{R}^d} \|z\|^{2m} |D\widehat{\mu}(z)h|^2 \, dz < \infty.$$

The proof of Theorem 8.5.3 is now complete. □

The final question is the finiteness of $\zeta_m(X_1, \theta)$. As we have noted in (8.4.32),

$$\zeta_m(X_1, \theta) \le \frac{1}{m!}\kappa_m(X_1, \theta), \quad m = 1, 2, \ldots,$$

where $\kappa_m(X_1, \theta)$ is the difference pseudomoment, namely

$$\kappa_m(X_1, \theta) = \sup\left\{|E[f(X_1) - f(\theta)]|;\ |f(x) - f(y)| \le d_m(x, y) \text{ for any } x, y \in \mathbb{R}^d\right\},$$

where $d_m(x, y) = \left\| x\,\|x\|^{m-1} - y\,\|y\|^{m-1} \right\|$.

It would be difficult to check conditions implying $\kappa_m(X_1, \theta) < \infty$. Instead we give an example for laws of X_1 with $\zeta_m(X_1, \theta) < \infty$. The idea is to use the series representation for θ (compare Remark 8.1.5).

Let $\{W_j\}_{j=1}^{\infty}$ be a sequence of i.i.d. random variables taking their values in $S = \{x \in \mathbb{R}^d; \|x\| = 1\}$ with a common probability distribution λ on S and $\Gamma_j = \delta_1 + \cdots + \delta_j$. Here $\{\delta_j\}$ is a sequence of independent exponentially distributed random variables with $E[\delta_1] = 1$ that are independent of $\{W_j\}$. It is known that the series

$$\sum_{j=1}^{\infty} \left[\Gamma_j^{-B} W_j - E\left[\Gamma_j^{-B} I[\Gamma_j \ge 1]\right] E[W_j]\right]$$

converges almost surely and is distributed as an operator-stable random vector with exponent B.

Suppose $E[W_j] = 0$ for all j, and set

$$\theta \stackrel{d}{=} \sum_{j=1}^{\infty} \Gamma_j^{-B} W_j.$$

Let $r > 1/\lambda_B$ and $j_0 = [r\lambda_B]$. Then it is easy to show that if $j > j_0$, then $E[\|\Gamma_j^{-B}\|^r] < \infty$.

Consider another sequence of independent random variables $\{V_j\}$ on S, which are independent of $\{\Gamma_j\}$, such that

$$V_j \stackrel{d}{=} W_j \quad \text{for } j \le j_0,$$

and for $j > j_0$, $\{V_j\}$ are arbitrary but not identically distributed random variables on S. Define

$$X \stackrel{d}{=} \sum_{j=1}^{\infty} \Gamma_j^{-B} V_j,$$

assuming that the series converges almost surely.

Theorem 8.5.4 *Suppose that all mixed moments of order less than or equal to $r-1$ coincide; that is,*

$$E\left[V_1^{\alpha_1}\cdots V_p^{\alpha_p} - W_1^{\alpha_1}\cdots W_p^{\alpha_p}\right] = 0 \tag{8.5.47}$$

for any $\alpha_i \ge 0$ with $\sum_{i=1}^p \alpha_i \le r-1$. Then

$$\zeta_r(X,\theta) < \infty.$$

Proof: By the operator-ideality of ζ_r of order r and the triangle inequality,

$$\begin{aligned}
\zeta_r(X,\theta) &= \zeta_r\left(\sum_j \Gamma_j^{-B} V_j, \sum_j \Gamma_j^{-B} W_j\right) \\
&\le \sum_{j=1}^{\infty} \zeta_r(\Gamma_j^{-B} V_j, \Gamma_j^{-B} W_j) = \sum_{j=j_0+1}^{\infty} \zeta_r(\Gamma_j^{-B} V_j, \Gamma_j^{-B} W_j)
\end{aligned}$$

(by Zolotarev (1983, Property 4 on p. 293))

$$\begin{aligned}
&\le \sum_{j=j_0+1}^{\infty} E[\|\Gamma_j^{-B}\|^r]\zeta_r(V_j, W_j) \\
&\le \sup_j \zeta_r(V_j, W_j) \sum_{j=j_0+1}^{\infty} E\left[\|\Gamma_j^{-B}\|^r\right].
\end{aligned}$$

Since $r\lambda_B > 1$, the final series converges. This is shown as follows. Note that for any $\varepsilon > 0$, there exists $C > 0$ such that

$$\|x^{-B}\| < Cx^{-(\Lambda_B+\varepsilon)}, \quad 0 < x \le 1,$$

and

$$\|x^{-B}\| < Cx^{-(\lambda_B - \varepsilon)}, \quad x > 1.$$

Thus we have for $j > j_0 = [r\lambda_B]$,

$$\begin{aligned} E[\|\Gamma_j^{-B}\|^r] &= E[\|\Gamma_j^{-B}\|^r I[\Gamma_j \le 1]] + E[\|\Gamma_j^{-B}\|^r I[\Gamma_j > 1]] \\ &\le E[\Gamma_j^{-r(\Lambda_B+\varepsilon)}] + E[\Gamma_j^{-r(\lambda_B-\varepsilon)}]. \end{aligned}$$

On the other hand, if $j > p$, then

$$E[\Gamma_j^{-p}] = \frac{\Gamma(j-p)}{\Gamma(j)} \sim j^{-p}.$$

Hence

$$E[\|\Gamma_j^{-B}\|^r] \le C(j^{-r(\Lambda+\varepsilon)} + j^{-r(\lambda_B+\varepsilon)}) \le Cj^{-r(\lambda_B+\varepsilon)})$$

for large j, because $\Lambda_B \ge \lambda_B$. This shows the convergence of the series in question.

Also, as we have noted in (8.4.34), under (8.5.47),

$$\zeta_r(V_j, W_j) \le \frac{\Gamma(1+\alpha)}{\Gamma(1+r)} \xi_r(V_j, W_j),$$

where $r = m + \alpha, m \in \mathbb{N}, \alpha \in (1, 2]$. In our case, V_j and W_j take their values on the unit sphere, and therefore,

$$\xi_r(V_j, W_j) = \operatorname{Var}(V_j, W_j) \le 1.$$

This concludes the proof of the theorem. □

8.6 Ideal Metrics in the Problem of Rounding

It is a widely accepted fact that sums of rounded proportions often fail to add to 1. In their pioneering works, Mosteller, Youtz, and Zahn (1967) and Diaconis and Freedman (1979) assessed the probability that a vector of conventionally rounded percentages adds to 100. The conventional rule picks the midpoint of each interval as the threshold for rounding. However, the goal of rounding to maximize the probability that the sum of roundings is the rounding of the sum may well not be a "significant" question: Instead, it seems that the goal of rounding to obtain a distribution as much like the original one as possible is more fundamental.

Suppose, for example, that $q_1, \ldots, q_s$ are s independent identically $[0,1]$-uniformly distributed random numbers and that each is to be rounded to either $0, \frac{1}{2}$, or 1. The usual method is to let $x = 0$ if $0 \le q < \frac{1}{4}, x = \frac{1}{2}$ if $\frac{1}{4} \le q < \frac{3}{4}$, and $x = 1$ if $\frac{3}{4} \le q \le 1$. Then $Ex = Eq = \frac{1}{2}$, and $\operatorname{Var} x = \frac{1}{8} \ne \operatorname{Var} q = \frac{1}{12}$. On the other hand, if instead of rounding "at" $\frac{1}{4}$ and $\frac{3}{4}$ one rounds at $\frac{1}{6}$ and $\frac{5}{6}$, that is, $x^* = 0$ if $0 \le q < \frac{1}{6}$, $x^* = \frac{1}{2}$ if $\frac{1}{6} \le q < \frac{5}{6}$, and $x^* = 1$ if $\frac{5}{6} \le q \le 1$, then $Ex^* = Eq = \frac{1}{2}$ and $\operatorname{Var} x^* = \operatorname{Var} q = \frac{1}{12}$. The importance of the deviations between the sums of the resulting roundings $x_S = x_1 + \cdots + x_s$ and $x_S^* = x_1^* + \cdots + x_s^*$ from $q_S = q_1 + \cdots + q_s$ may be seen by comparing the differences

$$\triangle_s = \sup_{a<b} |P(a < q_S < b) - P(a < x_S < b)|$$

and

$$\triangle_s^* = \sup_{a<b} |P(a < q_S < b) - P(a < x_S^* < b)|,$$

Then, by the central limit theorem,

$$\lim_{s \to \infty} \triangle_s \ge 0.049, \qquad \text{whereas } \lim_{s\to\infty} \triangle_s^* = 0,$$

showing that at least by this specific criterion, conventional rounding is not the best choice.

In this section the Diaconis and Freedman results are extended, and the motivation for changing the nature of the goal of rounding is given.[(2)] We define and study the properties of optimal roundings in terms of ideal metrics. We start with some definitions from the theory of rounding (see Balinski and Young (1982)).

A *vector problem* is a pair (p, h) where $p = (p_j)$ is a vector of real numbers, $j \in N = \{1, \ldots, n\}$, and h is a real number as well. Given any positive real $t > 0$, a rule of $(1/t)$-*rounding* is a mapping ϱ_t such that

$$\varrho_t(p, h) \in \{x = (x_j);\ x_j = k_j/t,\ k_j \text{ integer}, j \in N\}. \tag{8.6.1}$$

In the sequel we write $x = \varrho_t(p)$, since h remains fixed and it is generally the case that $h = \sum_j p_j$.

Our interest concerns problems with $p_N := \sum_N p_j = h$; for example, p_js are probabilities and $h = 1$.[(3)] This motivates two immediate questions: (a) Given a rule ϱ_t what is the chance that $x_N := \sum_N x_j = h$ for $x = \varrho_t(p)$? (b) What specific rule ϱ_t maximizes this chance? So from now on, it is assumed that problem (p, h) satisfies $p_N = h$.

[(2)] The results in this section are due to Balinski and Rachev (1993).

[(3)] There is no loss of generality in considering $h = 1$.

The conventional rule of $(1/t)$-rounding, $x = \varrho_t(p)$, with x_i equal to p_j rounded to the nearest $1/t$, was first discussed by Mosteller, Youtz, and Zahn (1967) for the vector problem $(p, 1)$. They computed the chance that $x_N = 1$ with this rule of rounding for several different probability models generating p. Diaconis and Freedman (1979) assessed the limiting probability that $x_N = 1$ as $t \to \infty$ under the assumption that p is absolutely continuously distributed on the simplex

$$S_n := \{p = (p_j) \geq 0, j \in N;\ p_N = 1\}. \tag{8.6.2}$$

The MYZ-rule of rounding (see Mosteller, Youtz, and Zahn (1967) and Diaconis and Freedman (1979)) is defined by

$$x_j := [p_j]_t^{1/2} = k/t \quad \text{if} \quad k - \frac{1}{2} < tp_j \leq k + \frac{1}{2}, k \text{ an integer.} \tag{8.6.3}$$

The MYZ-rule is only one example of an infinite class of rules of rounding first discussed by Balinski and Young (1978) called *divisor rules of* $(1/t)$-*rounding based on* d, ϱ_t^d, defined by

$$x_j := [p_j]_t^d := k/t \quad \text{if} \quad d(k-1) < tp_j \leq d(k), k \text{ an integer,} \tag{8.6.4}$$

where $d(h) \in [k, k+1]$, the *divisor criterion*, is any real-valued function from the integers into the closed interval k to $k+1$. It is the "threshold" for rounding: Above the threshold round up, at or below it round down. The divisor rules arose as a characterization of rules satisfying certain desirable properties in the context of apportionment problems; see Balinski and Young (1982). The best known and most discussed among them are the following (for integer k):

$$\text{Adams (or round up):} \quad d(k) = k; \tag{8.6.5}$$

$$\text{Dean (or harmonic mean):} \quad d(k) = k(k+1)\Big/\left(k + \frac{1}{2}\right); \tag{8.6.6}$$

$$\text{Hill (or geometric mean):} \quad d(k) = \sqrt{k(k+1)}; \tag{8.6.7}$$

$$\text{Webster (or arithmetic mean, the MYZ-rule):} \quad d(k) = k + \frac{1}{2}; \tag{8.6.8}$$

$$\text{Jefferson (or round down):} \quad d(k) = k + 1. \tag{8.6.9}$$

Theorem 8.6.1 (Diaconis and Freedman (1979)) *Suppose p is absolutely continuously distributed on the simplex and ϱ_t^d is based on $d(k) = k+C, C \in [0,1]$. Then, as $t \to \infty$,*

$$P(x_N = 1) \to P(C - 1 \leq V_1 + \cdots + V_{n-1} \leq C), \tag{8.6.10}$$

where the V_i's are independent and uniformly distributed on $[-C, 1+C]$.

A *stationary divisor rule based on* C is a divisor rule based on a criterion $d(k) = k + C, 0 \leq C \leq 1$. A K*-stationary divisor* ϱ_t^K *based on* $(C_0, \dots, C_{K-1}, C), 0 \leq C \leq 1, 0 \leq C_i \leq 1$ for all i, is a divisor rule based on the divisor criterion

$$d(k) = \begin{cases} k + C_k & \text{if } 0 \leq k \leq K-1, \\ k + C & \text{otherwise.} \end{cases} \tag{8.6.11}$$

From Theorem 8.6.1 it follows that among the stationary divisor rules (or 0-stationary ones) the MYZ-rule, or Webster's rule ϱ_t^w, maximizes the probability that in the limit as $t \to \infty$ the sum of the roundings is 1. The following theorem strengthens Diaconis–Freedman's result.

Theorem 8.6.2 *Suppose* p *is uniformly distributed on the simplex* S_n. *Then the maximum of the limiting probability* $\lim_{t\to x} P(x_N = 1)$ *over the set of all* K*-stationary divisor rules is attained with* $C = \frac{1}{2}$ *and* $C_0, \dots, C_{K-1}$ *arbitrary.*

The proof is similar to that in Theorem 8.6.1; for details we refer to Balinski and Rachev (1993). This theorem remains valid under the weaker assumption that p is absolutely continuous on the simplex. However, the result suggests that maximizing the limiting probability in (8.6.10) is not in fact a reasonable objective. The rate of convergence in the best case,

$$\begin{aligned} &\lim_{t\to x} P(x_N = 1) \qquad\qquad (8.6.12) \\ &= P\left(-\frac{1}{2} < \sum_1^{n-1} V_i \leq \frac{1}{2};\ V_i \text{ i.i.d. uniform on } \left(-\frac{1}{2}, \frac{1}{2}\right)\right) \\ &\approx \sqrt{\frac{6}{\pi(n-1)}} + o(n^{-1/2}), \end{aligned}$$

can be very slow. Indeed, for every n and every t it is possible to choose an absolutely continuous distribution $\mu_{n,t}$ on S_n such that $P(x_N = 1)$, whereas the right-hand side of (8.6.10) is less than 1 for $n \geq 3$ and converges to 0 as n grows. To construct such a distribution choose $p_1, \dots, p_{n-1}$ to be independent identically $(0, \delta)$-uniformly distributed random variables with $\delta \in (0, \frac{1}{2m})$ and $p_n = 1 - p_1 - \cdots - p_{n-1}$. This determines an absolutely continuous distribution on S_n such that the Webster rule of rounding gives $x_i = 0$ for $i = 1, \dots, n-1$ and $x_n = 1$, with probability equal to 1.

Perhaps a more serious limitation of this approach is the fact that a "best" K-stationary rule—"best" in the sense of maximizing the limit in (8.6.10), an example of which is the Webster rule ϱ_t^w—may not satisfy a natural "continuity" property: If two problems $(p, 1)$ and $(p^*, 1)$ come from distributions that are "close," then the distributions of their roundings should be close as well. Letting $\varrho\mu$ represent the distribution of the roundings under rule ϱ when the original data have the distribution μ, "continuity" means that

$$\mu \approx \mu^* \text{ implies } \varrho\mu \approx \varrho\mu^*. \tag{8.6.13}$$

For example, the continuity property may fail if μ is an absolutely continuous distribution on S_n and μ^* is a discrete distribution on S_n. Specifically, take μ^* to be the point distribution $p_1 = \cdots = p_{n-1} = 0, p_n = 1$, and $\mu^{(i)}$ to be absolutely continuous distributions satisfying $\nu_1(\mu^{(i)}, \mu^*) \to 0$ as $i \to \infty$, with ν_1 a distance in the space of distributions on S_n that metrizes the weak convergence. Given a similar distance $\nu_2(\varrho_t^w \mu, \varrho_t^w \mu^*)$ in the space of the distributions of the roundings on the lattice $\{1, 1 \pm 1/t, \ldots\}^n$, the continuity property says that

$$\text{as } i \to \infty, \quad \nu_1(\mu^{(i)}, \mu^*) \to 0 \quad \text{implies} \quad \lim_{t \to x} \nu_2(\varrho_t^w, \mu^{(i)}, \varrho_t^w \mu^*) \quad \to \quad 0.$$

However, in our example the first sequence goes to 0, but the second is strictly positive. To see this, recall that if $\varrho_t^w \mu^{(i)}$ converges weakly to $\varrho_t^w \mu^*$, then the distribution of $f(x^{(i)})$ converges to that of $f(x^*)$ for any continuous function f on R^n, and in particular, the distribution of $t(x_N^{(i)} - 1)$ converges to that of $t(x_N^* - 1)$. But, by construction $t(x_N^* - 1) = 0$, whereas for any i, t, we have $(x_N^{(i)} - 1) \to V_1 + \cdots + V_{n-t}$ as $t \to \infty$, and the last term is not 0 with probability 1. In contrast with Theorem 8.6.1, where the result is independent of the particular absolutely continuous distribution that obtains, the above remarks show that constructing a reasonable rule of rounding should depend on the information that is available concerning the distribution μ.

We next consider the vector problem of rounding as a special case of the *matrix problem*: $(p_{ij}), i \in N = \{1, \ldots, n\}, j \in S = \{1, \ldots, s\}$, made up of s observations $p_j = (p_{1j}, \ldots, p_{nj})^T, j \in S$ (the columns of the matrix) of the random vector $p = (p_1, \ldots, p_n)^T$, consisting of independent nonidentically distributed random variables.

The ith row of data $p_i = (p_{i1}, \ldots, p_{is})$ consists of s independent observations of p_i. For simplicity, set $q := (q_j)_{j \in S}$, where $q_j = (p_{ij})$, the vector problem with i.i.d. random variables. It is assumed throughout that q_1 has a continuous distribution.

Suppose $x = (x_j)_{j \in S}$ is obtained by some $1/t$-rule of rounding; that is, $x_j = k_j/t$, where the k_j are integer-valued i.i.d. random variables, and that $q \geq 0$, so the x_j's take values on the lattice $\{0, 1/t, 2/t, \ldots\}$. The question of interest is, What is the deviation between the sum of the observations $q_s = q_1 + \cdots + q_s$ and the sum of the roundings $x_s = x_1 + \cdots + x_s$. The answer will, of course, depend upon the distribution of the q_j's, the way by which the deviation is measured, and the rule of rounding that is used.

A rule of rounding $x^* = \varrho_t^*(q)$ is *optimal* with respect to the metric μ over a class of rules $\mathcal{R}$ if for any q,

$$\mu(q_s, x_s^*) = \min_{\varrho_t}\{\mu(q_s, x_s);\ x = \varrho_t(q), \varrho_t \in \mathcal{R}\} \tag{8.6.14}$$

and

$$\mu\left(\frac{1}{s}q_s, \frac{1}{s}x_s^*\right) \to 0 \quad \text{as } s \to \infty. \tag{8.6.15}$$

Roughly speaking, optimality asks that the deviation between the sum of the observations and the sum of the roundings should be as small as possible and that the deviation between their respective sample means should go to 0 as the number of observations grows.

Suppose $X = (X_1, \ldots, X_n)$ and $Y = (Y_1, \ldots, Y_n)$ are random vectors each with i.i.d. components. A variety of probability metrics μ have been proposed to measure the deviation between two such distributions (see Rachev (1991) and the previous Sections 8.1–8.5). Probably the best known is the Kolmogorov, or uniform, distance ϱ,

$$\varrho(X, Y) := \sup_{-\infty < x < \infty} |F_X(x) - F_Y(x)|,$$

where F_Z is the distribution function of Z. Others include the Kantorovich metric, Dall'Aglio's extension of the Kantorovich metric, and the Lévy metric. But for these metrics there exists no optimal rule of rounding because it is impossible to meet condition (8.6.15). The only type of metric that seems to be able to meet this condition is an *ideal metric of order* $r > 0$ (see Section 8.1) that satisfies

$$\mu\left(c\sum_1^n X_j, c\sum_1^n Y_j\right) \le c^r \sum_1^n \mu(X_j, Y_j) \quad \text{for any } c > 0. \tag{8.6.16}$$

The class of all ideal metrics has not been characterized; indeed, only a few have been identified. An example of an ideal metric of order $r = 1 + (1/p) \in (1, 2)$, $p \ge 1$, is

$$\theta_r(X, Y) = \sup\{|E(f(X) - f(Y))|;\ f \in \mathcal{F}_r\}. \tag{8.6.17}$$

Here $\mathcal{F}_r$ is the set of all functions f whose second derivative f'' has a bounded q-norm: $\|f''|_q = (f|f''|^q)^{1/q} \le 1$, with $(1/p) + (1/q) = 1$ (see Maejima and Rachev (1987)). It is easy to check that

$$\theta_r\left(\frac{1}{s}q_s, \frac{1}{s}x_s\right) = s^{-r}\theta_r(q_s, x_s) \le s^{-r}\sum_1^s \theta_r(q_j, x_j) = s^{1-r}\theta_r(q_1, x_1).$$

Thus

$$\theta_r\left(\frac{1}{s}q_s, \frac{1}{s}x_s\right) \;=\; O(s^{-1/p}) \text{ as } s \to \infty \text{ whenever } \theta_r(q_1, x_1) \;<\; \infty. \tag{8.6.18}$$

Notice that by the definition of θ_r,

$$\theta_r(X, Y) \;<\; \infty \quad \text{implies} \quad E(X - Y) \;=\; 0.$$

In fact, $\theta_r(X,Y) \geq \sup_{a>0} |E(aX - aY)| = +\infty$ if $E(X - Y) \neq 0$. Thus, a *necessary* condition for $x^* = \varrho_t^x(q)$ to be an optimal stationary rule with respect to θ_r is the equality of the first moments of q_1 and of its $(1/t)$-rounding:

$$Eq_1 \;=\; Ex_1^* \quad \text{or} \quad Eq_1 \;=\; \frac{1}{t}\sum_0^\infty P(q_1 > d(k)/t). \tag{8.6.19}$$

This condition is *sufficient* under the mild assumption that

$$Eq_1^r \;<\; \infty. \tag{8.6.20}$$

In fact, (8.6.19) implies

$$\theta_r(q_s, x_s^*) \;\leq\; s\frac{1}{\Gamma(r-1)}\kappa_r(q_1, r_1),$$

where $\kappa_r(X,Y) = r\int\limits_0^\infty x^{r-1}|F_X(x) - F_Y(x)|\,dx$ is the Kantorovich rth pseudomoment, and by (8.6.20),

$$\kappa_r(q_1, x_1^*) \;\leq\; Eq_1^r + E\left(q_1 + \frac{1}{t}\right)^r \;<\; \infty.$$

Summarizing, this yields the following theorem.

Theorem 8.6.3 *Suppose that the vector $q = (q_1, \ldots, q_s)$ consists of i.i.d. random variables, and Eq_1^r is finite for some $r \in (1,2)$. Then*

$$\theta_r(q_s, x_s) \;=\; \infty, \qquad \theta_r\left(\frac{1}{s}q_s, \frac{1}{s}x_s\right) \;=\; \infty$$

for any rule of $(1/t)$-rounding $x = \varrho_t(q)$ with $Eq_1 \neq Ex_1$. However, if $Eq_1 = Ex_1^*$ for some stationary rule $x^* = \varrho_t^*(q)$, then ϱ_t^* is an optimal rule with respect to μ over the class of all stationary rules. Moreover,

$$\theta_r\left(\frac{1}{s}q_s, \frac{1}{s}x_s^*\right) \;\leq\; s^{1-r}\frac{1}{\Gamma(r-1)}\kappa_r(q_1, x_1^*) \;=\; O(s^{1-r}).$$

Corollary 8.6.4 *If in Theorem* 8.6.3 ϱ_t *is stationary, then it is optimal with respect to* θ_r *if and only if*

$$tEq_1 = \sum_0^\infty P(tq_1 > k + C). \tag{8.6.21}$$

Equation (8.6.21) always has a solution C that is unique for any $t > 0$ under the condition that $F_{q_1}(x)$ is strictly increasing. Thus, an optimal stationary rule with respect to θ_r exists and is unique over the set of stationary rules.

Example 8.6.5 *Suppose* q_1 *is uniform on the interval* $(0, 1)$. *Then* (8.6.21) *becomes*

$$\frac{1}{2} = \frac{1}{t}\sum_0^{[t-C]}\left(1 - \frac{k+C}{t}\right).$$

If $t \in \mathbb{N} = \{1, 2, \ldots\}$, *then this is equivalent to*

$$\frac{1}{2} = \frac{1}{t}\sum_0^{t-1}\left(1 - \frac{k+C}{t}\right),$$

whose solution is $C(t) = \frac{1}{2}$, *so the Webster rule is optimal. However, if* $t \in N + \frac{1}{2}$, *then the solution is* $C(t) = (t - \frac{1}{4})/(2t+1) < \frac{1}{2}$, *so the Webster rule is only asymptotically optimal.*

The set of stationary rules are clearly a very restrictive class within the class of all divisor rules. Moreover, the rate of convergence of (8.6.15) for θ_r can be very slow, as can be seen from Theorem 8.6.3, where $1 - r \in (-1, 0)$. Indeed, simple examples show that the order of convergence $O(s^{1-r})$ is exact. Given two optimal rounding rules with respect to some μ, $x^* = \varrho_t^*(q)$ and $x' = \varrho_t'(q)$, ϱ_t^* is *preferred to* ϱ_t' if $\mu((1/s)q_s, (1/s)x_s^*) \to 0$ at a faster rate than $\mu((1/s)q_s, (1/s)x_s') \to 0$ as $s \to \infty$. And ϱ_t^* is *optimal of order* $\lambda > 0$ with respect to μ over a class $\mathcal{R}$ of rules if for any q it is optimal, and

$$\mu\left(\frac{1}{s}q_s, \frac{1}{s}x_s^*\right) \to O(s^{-\lambda}) \quad \text{as } s \to \infty. \tag{8.6.22}$$

Theorem 8.6.3 tells us that there exists an optimal stationary rule ϱ_t^* of order $\lambda = r - 1$ with respect to θ_r if the rth moment of q_1 exists and is finite. Is it possible that an ideal metric μ other than θ_r would determine an optimal stationary rule different from ϱ_t^*? The answer is negative for all "nonpathological" metrics.

Corollary 8.6.6 *Suppose μ is an ideal metric of order $r > 1$ such that the law of large numbers holds with respect to μ; that is,*

$$\mu\left(\frac{X_1 + \cdots + X_n}{n}, EX_1\right) \to 0 \quad \text{as} \quad n \to \infty \tag{8.6.23}$$

for any nonnegative i.i.d. X_i's with finite EX_1. Then there is a unique stationary rule of $(1/t)$-rounding that is optimal of order $r-1$, and it is determined by the solution C to (8.6.21).

Proof: Suppose two different stationary rules $x^* = \varrho_t^*(q)$ and $x' = \varrho_t'(q)$ are optimal of orders λ^* and λ' (where λ^* and λ' may be different). Then

$$\begin{aligned}\mu(Ex_1^*, Ex_1') \leq\ & \mu\left(Ex_1^*, \frac{1}{s}x_s^*\right) + \mu\left(\frac{1}{s}x_s^*, \frac{1}{s}q_s\right)\\ & + \mu\left(\frac{1}{s}q_s, \frac{1}{s}x_s'\right) + \mu\left(\frac{1}{s}x_s', Ex_1'\right).\end{aligned}$$

However, the right-hand side goes to 0 as $s \to \infty$ by (8.6.16) and (8.6.23), and therefore $Ex_1^* = Ex_1'$. Since the rules are stationary, they are both determined by (8.6.4), and so they are the same. □

To obtain faster rates of convergence one must consider a wider class of divisor rules than the stationary ones. It is our objective to show that if one extends the analysis to K-stationary rules, then one can find an optimal rule of order $\lambda = K+1$. To do this it is necessary to generalize the definition of θ_r (since by the previous definition $r \in (1,2)$), which can be done as follows:

$$\theta_r(X,Y) = \sup\{|E(f(X) - f(Y))|;\ f \in \mathcal{F}_r\}, \tag{8.6.24}$$

where $r = r_0 + 1/p > 0, r_0 \geq 0$ an integer, and $p \geq 1$. Furthermore, in (8.6.24), $\mathcal{F}_r$ is the set of all functions f with (r_0+1)-derivative $f^{(r_0+1)}$ satisfying $\|f^{(r_0+1)}\|_q = [\int |f^{(r_0+1)}|^q]^{1/q} \leq 1$, with $(1/p) + (1/q) = 1$. As before, it may be checked that

$$\theta_r\left(\frac{1}{s}q_s, \frac{1}{s}x_s\right) \leq s^{1-r}\theta_r(q_1, x_1),$$

so that (8.6.18) holds, provided that $\theta_r(q_1, x_1) < \infty$. In addition,

$$\theta_r(q_1, x_1) < \infty \quad \text{implies} \quad E(q_1^k - x_1^k) = 0 \quad \text{for } 0 < k \leq r_0. \tag{8.6.25}$$

Conversely, if $Eq_1^r < \infty$, then

$$E(q_1^k - x_1^k) = 0 \quad \text{for} \quad 0 < k \leq r_0 \quad \text{implies } \theta_r(q_1, x_1) < \infty. \tag{8.6.26}$$

In fact, $\theta_r(q_1, x_1)$ can be bounded by

$$\theta_r(q_1, x_1) \leq C'_r K \, \kappa_r(q_1, x_1) \leq C''_r E q_1^r < \infty, \tag{8.6.27}$$

where C'_r and C''_r are constants, and κ_r is the pseudomoment of order r. It is possible to say more. $\theta_r(q_1, x_1) \to 0$ means that the distributions of q_1 and of x_1 "θ-merge"; that is, $L(q_1, x_1) \to 0$ and $E(q_1^r - x_1^r) \to 0$, where L is the Lévy metric between the distributions of q_1 and x_1 (for merging, the Lévy metric, and related concepts see D'Aristotile, Diaconis, and Freedman (1988) and Rachev (1991)).

A K-stationary rule $x^* = \varrho_t^*(q)$ of order $\lambda = r-1$ is *optimal* with respect to the metric θ_r over the class of K-stationary rules if for any q,

$$\theta_r(q_s, x_s^*) \;=\; \min_{\varrho_t}\{\theta_r(q_s, x_s);\;\; x = \varrho_t(q), \varrho_t \; K\text{-stationary}\}, \tag{8.6.28}$$

and furthermore, the rate of θ_r-merging is

$$\theta_r\left(\frac{1}{s}q_s, \frac{1}{s}x_s^*\right) \;=\; O(s^{1-r}) \quad \text{as } s \to \infty \tag{8.6.29}$$

whenever $Eq_1^r < \infty$. In fact, (8.6.25) through (8.6.27) imply that if for a rule $x = \varrho_t(q)$ the moment conditions $E(q_1^k - x_1^k) = 0$ do not hold for some $k = 0, \ldots, r_0$, then $\theta_r(q_1, x_1) = \infty$. So (8.6.28) fails for $s = 1$, and thus ϱ_t is not optimal. On the other hand, if these moment conditions do hold for all k for some rule $x^* = \varrho_t^*(q)$, then by (8.6.27) and the ideality of $\theta_r, \theta_r(q_s, x_s^*) \leq sC'_r Eq < \infty$ for any fixed s, so (8.6.28) is fulfilled. This proves

Theorem 8.6.7 *Suppose $Eq_1^r < \infty$ with $r = K+1+1/p$. Then $x^* = \varrho_t^*(q)$ is an optimal K-stationary rule of order $\lambda = r - 1$ with respect to the metric θ_r if and only if the thresholds $C_0, \ldots, C_{K-1}, C = C_K = C_{K+1} \ldots$ are chosen such that*

$$E(q_1^j - x_1^{*j}) = 0 \quad \text{for } j = 1, \ldots, K+1. \tag{8.6.30}$$

This means that the $K+1$ thresholds must be chosen such that the following system of equations is satisfied

$$Eq_1^j \;=\; \sum_{k=1}^{\infty} \left(\frac{k}{t}\right)^j P(k - 1 + C_{k-1} < tq_1 < k + C_k), \tag{8.6.31}$$
$$j = 1, \ldots, K+1,$$

where $C_k = C_K$ for $k \geq K$.

It has been observed above that when q_1 is uniform over the interval $(0, 1)$, t is an integer, and $K = 0$, then the optimal rule is Webster's. But if $K > 0$, there is *no* stationary rule that meets (8.6.31) even in the uniform case.

Example 8.6.8 *Suppose q_1 is uniform over the interval $(0, 1)$, t is an integer, and $K = 1$.*[4] *Then (8.6.31) yields the solution $C_0 = \frac{1}{3}, C = (3t-2)/(6t-6)$, so the optimal 1-stationary rule is determined by $d(0) = \frac{1}{3}, d(k) = k + (3t-2)/(6t-6)$ for $k \neq 0$.*

Suppose $t = 2$. Then q_1 is rounded to either $0, \frac{1}{2}$, or 1 as follows: $x_1^* = 0$ if $0 < q_1 \leq \frac{1}{6}, x_1^* = \frac{1}{2}$ if $\frac{1}{6} < q_1 \leq \frac{5}{6}$, and $x_1^* = 1$ if $\frac{5}{6} < q_1 \leq 1$. The first two moments of q_1 and x_1^* agree: $Eq_1 = \frac{1}{2} = Ex_1^*, \operatorname{Var} q_1 = \frac{1}{12} = \operatorname{Var} x_1^*$. (Indeed, the third moments also agree: $Eq_1^3 = E(x_1^*)^3 = \frac{1}{4}$.)

The Webster rule, on the other hand, rounds q_1 as follows: $x_1^w = 0$ if $0 < q_1 \leq \frac{1}{4}, x_1^w = \frac{1}{2}$ if $\frac{1}{4} < q \leq \frac{3}{4}$, and $x_1^w = 1$ if $\frac{3}{4} < q_1 \leq 1$. The first moments of q_1 and x_1^w agree but not the second: $Eq_1 = \frac{1}{2} = Ex_1^w, \operatorname{Var} q_1 = \frac{1}{12} < \frac{1}{8} = \operatorname{Var} x_1^w$.

Since the first two moments of q_1 and x_1^* are equal for the optimal 1-stationary rule, the central limit theorem applies, and for the Kolmogorov metric $\varrho(q_s, x_s^*) = \sup\{|P(q_s \leq x) - P(x_s^* \leq x)|;\ x \in \mathbb{R}\}$ we have

$$\varrho(q_s, x_s^*) \quad = \quad \varrho\left[\frac{q_s - s/2}{\sqrt{12s}}, \frac{x_s^* - s/2}{\sqrt{12s}}\right] \quad \approx \quad O(s^{-1/2}) \quad \text{as} \quad s \to \infty.$$

In contrast, the Webster rule gives

$$\begin{aligned}\varrho(q_s, x_s^w) \quad &= \quad \varrho\left[\frac{q_s - s/2}{\sqrt{12s}}, \frac{x_s^w - s/2}{\sqrt{12s}}\right] \quad = \quad \varrho\left[\frac{q_s - s/2}{\sqrt{12s}}, \frac{x_s^w - s/2}{\sqrt{8s}}\sqrt{\frac{8}{12}}\right] \\ &\to \quad \varrho\left(N_{(0,1)}, N_{(0,\sqrt{2/3})}\right) \quad > \quad 0 \quad \text{as } s \to \infty,\end{aligned}$$

where $N_{(m,\sigma)}$ is the normal distribution with mean m and standard deviation σ.

In terms of the ideal metric θ_3 there is even better evidence for the advantage of the optimal 1-stationary rule over the standard Webster rounding. For

$$\theta_3\left(\frac{1}{s}q_s, \frac{1}{s}x_s^*\right) \quad \leq \quad s^{-2}\theta_3(q_1, x_1^*) \quad \leq \quad \text{constant } s^{-2},$$

the last inequality is due to the optimality of x_1^*, whereas

$$\theta_3\left(\frac{1}{s}q_s, \frac{1}{s}x_s^*\right) \quad = \quad +\infty \quad \text{for any } s.$$

---

[4] For other examples including the case $K = 2$, we refer to Balinski and Rachev (1993).

For more results concerning the vector problem with more i.i.d. observations and the problem of rounding tables we refer to Balinski and Rachev (1993).

9

Mass Transportation Problems and Recursive Stochastic Equations

In this chapter we use the regularity properties of metrics and distances defined via mass transportation problems in order to investigate the asymptotic behavior of stochastic algorithms and recursive equations. The recursive structure allows us to apply fixed-point and approximation techniques to the space of probability measures supplied with adapted probability metrics in order to describe the limiting behavior of various algorithms.

9.1 Recursive Algorithms and Contraction of Transformations

Several different approaches to the asymptotic analysis of algorithms have been given in the literature. Interesting results have been obtained by the transformation method, the method of branching processes, the method based on stochastic approximations, the martingale method, and others. The analysis of algorithms is an important application of stochastics in computer science and poses challenging questions and problems. It has led to some new developments also in stochastics.

Based on the properties of minimal metrics as introduced in Chapter 8, a promising new method for asymptotic analysis has recently been introduced. Rösler (1991) gave an asymptotic analysis of the quicksort algorithm based on the minimal ℓ_p-metric. His proof has been extended in several papers by Rachev and Rüschendorf to a general "contraction method" with a wide range of possible applications. A series of examples and further developments of the method have been found in some recent work.

The contraction method (in its basic form) uses the following sequence of steps:

1. Find the correct normalization of the algorithms. (Typically by studying the first moments or tails.)

2. Determine the recursion for the normalized algorithm.

3. Determine the limiting form of the normalized algorithms. The limiting equation typically is defined via a transformation T on the set of probability measures.

4. Choose an ideal metric μ such that T has good contraction properties with respect to μ. This ideal metric has to reflect the structure of the algorithm. It also has to have good bounds in terms of interpretable other metrics and has to allow one to estimate bounds (in terms of moments usually). As a consequence one obtains

5. The conjectured limiting distribution is the unique fixed point of T. Finally, one should ensure that the recursion is stable enough so that the contraction in the limit can be pulled over to establish contraction properties of the recursion itself for $n \to \infty$. This is the technically most involved step in the analysis.

6. Establish convergence of the algorithms to the fixed point.

Applications of this method have been given to several sorting algorithms, to the communication resolution interval (CRI) algorithm, to generalized branching-type algorithms, to bootstrap estimators, iterated function systems, learning algorithms, and others. For several examples modifications of this method have been considered. There are examples where the contraction factors converge to one. In several cases there is a trivial limiting recursion that gives no clue to a possible limit distribution. Also, logarithmic normalizations and convergence rates have to be handled by special considerations. We begin with a discussion of contraction properties of transformations T on the set of probability distributions on a basis space U.

Stochastic algorithms are typically directly described by the iterates of a transformation T on the set of all probability distributions on the basic space U, or else they are asymptotically closely related to the iterations (and thereby to the fixed points) of a transformation T that describes the limiting equation. They can, for example, differ from iterations by a stochastic sequence converging to zero. Examples for recursive algorithms that are asymptotically related to iterations of transformations T are studied in Section 9.2.

Consider a contraction transformation $T : M^1(U) \to M^1(U)$, where $M^1(U)$ is the set of probability measures on U supplied with probability metrics as in Chapter 8. Applying the fixed-point theorems for complete, separable metric spaces, we can infer the convergence of the iterates $(T^n F), F \in M^1(U)$, to a fixed point of T. Some of the following examples serve to describe the influence of the choice of the metrics, while others indicate the range of applicability to different fields.

(a) Consider at first a transformation of the form

$$TF \stackrel{d}{=} \sum_{i=1}^{N} a_i(\tau) Y_i + C(\tau). \tag{9.1.1}$$

Here, for $F \in M^1(U)$, $(Y_i)_{1 \leq i \leq N}$ is an i.i.d. sequence with distribution F. Furthermore, $C(\tau), a_i(\tau), \tau$ are real random variables independent of (Y_i), and finally, TF is the law of $\sum_{i=1}^{N} a_i(\tau) Y_i + C(\tau)$.

Consider Zolotarev's ideal metric ζ_r of order $r > 0$: For $r = m + \alpha$, $m \in \mathbb{N}$, and $0 < \alpha \leq 1$,

$$\begin{aligned} &\zeta_r(X,Y) \\ &:= \sup\left\{ |E(f(X) - f(Y))|;\ |f^{(m)}(x) - f^{(m)}(y)| \leq \|x - y\|^\alpha \right\}. \end{aligned} \tag{9.1.2}$$

Here $f^{(m)}(x)$ denotes the Fréchet derivative of order m, and $\|\cdot\|$ is a norm on U. Next, suppose that $F, G \in M^1(U)$, where $(U, \|\cdot\|)$ is a Banach space.

Proposition 9.1.1

$$\zeta_r(TF, TG) \leq \left(\sum_{i=1}^{N} E|a_i(\tau)|^r \right) \zeta_r(F, G). \tag{9.1.3}$$

Proof: The proof of (9.1.3) uses the ideality properties of ζ_r; that is,

$$\text{(i) } \zeta_r(X + Z, Y + Z) \leq \zeta_r(X,Y) \quad \text{for } Z \text{ independent of } X, Y, \tag{9.1.4}$$

and

$$\text{(ii) } \zeta_r(cX, cY) = |c|^r \zeta_r(X,Y), \quad \text{for all } c \in \mathbb{R}. \tag{9.1.5}$$

Let $Y_i \stackrel{d}{=} F$, $Z_i \stackrel{d}{=} G$ be independent r.v.s. Then, with $r = m + \alpha$,

$$\begin{aligned} \zeta_r(TF, TG) = \sup\Big\{ \Big| Ef\left(\sum a_i(\tau) Y_i + C(\tau)\right) - Ef\left(\sum a_i(\tau) Z_i + C(\tau)\right) \Big|;& \\ |f^{(m)}(x) - f^{(m)}(y)| \leq |x - y|^\alpha \Big\}& \end{aligned}$$

$$
\begin{aligned}
&\leq \int \zeta_r\left(\sum a_i(t)Y_i + C(t), \sum a_i(t)Z_i + C(t)\right) \mathrm{d}P^\tau(t) \\
&\leq \int \zeta_r\left(\sum a_i(t)Y_i, \sum a_i(t)Z_i\right) \mathrm{d}P^\tau(t) \\
&\leq \sum \int |a_i(t)|^r \,\mathrm{d}P^\tau(t)\zeta_r(Y_i, Z_i) \;=\; \sum E|a_i(\tau)|^r \zeta_r(F,G),
\end{aligned}
$$

which proves (9.1.3). □

For the property to hold it suffices to require that

$$
\sum E|a_i(\tau)|^r \;<\; 1 \qquad \text{and} \qquad \zeta_r(F,G) \;<\; \infty. \tag{9.1.6}
$$

In some cases, the last condition can be established by making use of the inequality

$$
\zeta_r \;\leq\; \frac{\Gamma(1+\alpha)}{\Gamma(1+r)} v_r, \tag{9.1.7}
$$

where

$$
v_r(X,Y) \;:=\; \int \|x\|^r \,\mathrm{d}|P_X - P_Y|(x)
$$

is the absolute pseudomoment of order r. For random vectors X and Y, and $m \leq r < m+1$, (9.1.7) requires that all moments X and Y of order $\leq m$ coincide.

Recall that the minimal L_p-metrics ℓ_p are ideal of order $r = \min(1,p)$.

Proposition 9.1.2 *For $F, G \in M^1(U)$,*

$$
\ell_p(TF,TG) \;\leq\; \left(\sum_{i=1}^N E|a_i(\tau)|^r\right) \ell_p(F,G). \tag{9.1.8}
$$

Proof: Let $Y_i \stackrel{d}{=} F, Z_i \stackrel{d}{=} G, 1 \le i \le N$, be independent pairs of random variables with laws F, G, and $L_p(Y_i, Z_i) = \ell_p(F,G)$. Then

$$\begin{aligned}
\ell_p(TF,TG) &= \ell_p\left(\sum a_i(\tau)Y_i + C(\tau), \sum a_i(\tau)Z_i + C(\tau)\right) \\
&\le \left\|\sum a_i(\tau)Y_i - \sum a_i(\tau)Z_i\right\|_p \\
&\le \sum_{i=1}^N E|a_i(\tau)|^r \|Y_i - Z_i\|_p \\
&= \left(\sum_{i=1}^N E|a_i(\tau)|^r\right) \ell_p(F,G).
\end{aligned}$$

□

So, the contraction property will hold if

$$\sum_{i=1}^N E|a_i(\tau)|^r < 1 \quad \text{and} \quad \ell_p(F,G) < \infty. \tag{9.1.9}$$

Under additional assumptions we may improve (9.1.9). Let U be a Hilbert space and let F, G have identical first moments. Then the following is a refinement of Proposition 9.1.2.

Proposition 9.1.3

$$\ell_2(TF,TG) \le \left(\sum_{i=1}^N E|a_i(\tau)|^2\right)^{1/2} \ell_2(F,G). \tag{9.1.10}$$

Proof: With Y_i, Z_i as in the proof of (9.1.8), we have

$$\begin{aligned}
\ell_2^2(TF,TG) &\le \left\|\sum a_i(\tau)(Y_i - Z_i)\right\|_2^2 \\
&= \sum_{i=1}^N E|a_i(\tau)|^2 \|Y_i - Z_i\|_2^2 \\
&= \left(\sum E|a_i(\tau)|^2\right) \ell_2^2(F,G).
\end{aligned}$$

□

If the Banach space U is of type p, $1 \le p \le 2$, and F, G have identical first moments (more precisely, $E(Y - Z) = 0$ for $Y \stackrel{d}{=} F, Z \stackrel{d}{=} G$), then for $1 \le p \le 2$,

$$\ell_p(TF,TG) \le B_p^{1/p} \left(\sum_{i=1}^N E|a_i(\tau)|^p\right)^{1/p} \ell_p(F,G). \tag{9.1.11}$$

Here B_p is the constant arising in the Woyczinski inequality (cf. Rachev and Rüschendorf (1992a)). For $U = L^p(\mu)$ ($L^p(\mu)$ is the space of all r.v.s X with finite $\int |X|^p \, d\mu$), one can choose the constants $B_1 = 1, B_p = 18p^{3/2}/(p-1)^{1/2}$ for $1 < p \leq 2$. The proof of (9.1.11) is similar to that of (9.1.10), but in (9.1.11) we use the Woyczinski inequality instead of the Hilbert space structure. If the underlying space is Euclidean, we can derive similar contraction properties with respect to other metrics defined in Chapter 8.

Example 9.1.4 *Let $N = 2$, let τ be uniformly distributed on (0,1), and $a_1(\tau) = \tau, a_2(\tau) = 1 - \tau$. Then the contraction factor α with respect to the ℓ_p-metric is given in the following list:*

$$\begin{aligned} \ell_1\text{-metric}: \ \alpha &= E\tau + E(1-\tau) = 1, \quad \text{i.e., "no contraction"}; \qquad (9.1.12) \\ \ell_2\text{-metric}: \ \alpha &= (E\tau^2 + E(1-\tau)^2)^{1/2} = \sqrt{2/3}; \\ \zeta_2\text{-metric}: \ \alpha &= E\tau^2 + E(1-\tau)^2 = 2/3; \\ \zeta_3\text{-metric}: \ \alpha &= 1/2. \end{aligned}$$

Clearly, if for a probability metric μ on $M^1(U)$ the contraction factor is $\alpha < 1, \mu(TF, TG) \leq \alpha\mu(F, G)$, then

$$\mu(T^{n+1}F, T^nF) \leq \alpha^n \mu(TF, F); \qquad (9.1.13)$$

i.e., one obtains an exponential convergence rate to a fixed point. In the example above we consider the recursion $X_{n+1} \stackrel{d}{=} \tau_n X_n + (1 - \tau_n)\overline{X}_n + C(\tau_n), \tau_n \stackrel{d}{=} \tau$, where $\overline{X}_n \stackrel{d}{=} X_n$, and $\tau_n, X_n, \overline{X}_n$ are independent. The corresponding fixed point equation is $X \stackrel{d}{=} \tau X + (1-\tau)\overline{X} + C(\tau)$. So under the condition of equal first moments, the convergence rate is $(2/3)^n$ for the "ideal" metric ζ_2, in comparison to $(2/3)^{n/2}$ for the ℓ_2-metric.

If $a_1(\tau) = \sqrt{\tau}, a_2(\tau) = \sqrt{1-\tau}$, then with respect to the ℓ_2-metric the contraction factor is $\alpha = E\tau + E(1-\tau) = 1$; i.e., there is no contraction. The same "no contraction" property is valid for ζ_2. For ζ_3 the contraction factor is $\alpha = E\tau^{3/2} + E(1-\tau)^{3/2} = \frac{4}{5} < 1$; so the contraction property holds if $\zeta_3(F, G) < \infty$.

(b) We next consider the transformation

$$TF \stackrel{d}{=} \max_{1 \leq i \leq N} \{a_i(\tau) Y_i\}. \qquad (9.1.14)$$

For $U = \mathbb{R}^k, k = 1, 2, \ldots, \infty$, $F \in M^1(U)$, let $(Y_i)_{1 \leq i \leq N}, \tau$ be as in (a). Let $a_i(\tau) \geq 0$ and consider

$$TF \stackrel{d}{=} \max_{1 \leq i \leq N} \{a_i(\tau) Y_i\}. \qquad (9.1.15)$$

We shall study the contraction property of T by making use of the weighted uniform metric ϱ_r:

$$\varrho_r(X,Y) = \sup_{x\in\mathbb{R}^k} M(x)^r |F_X(x) - F_Y(x)|, \tag{9.1.16}$$

where $M(x) := \min_{i\le k} |x_i|$. In the next proposition we use the fact that ϱ_r is an ideal metric of order r with respect to the maxima of i.i.d. r.v.s.

Proposition 9.1.5

$$\varrho_r(TF,TG) \le \left(\sum E(a_i(\tau))^r\right) \varrho_r(F,G). \tag{9.1.17}$$

Proof: Let (Z_i) be i.i.d. and $Z_1 \stackrel{d}{=} G$. Then using the max-ideality of ϱ_r, we have

$$\begin{aligned}
\varrho_r(TF,TG) &= \varrho_r\left(\max_{i\le N}\{a_i(\tau)Y_i\}, \max_{i\le N}\{a_i(\tau)Z_i\}\right)\\
&= \sup_x |x|^r \left|\int (F_{\max\{a_i(t)Y_i\}}(x) - F_{\max\{a_i(t)Z_i\}}(x))\,\mathrm{d}P^\tau(t)\right|\\
&\le \int \varrho_r(\max\{a_i(t)Y_i\}, \max\{a_i(t)Z_i\})\,\mathrm{d}P^\tau(t)\\
&\le \sum_i \int a_i(t)^r\,\mathrm{d}P^\tau(t)\varrho_r(Y_i,Z_i) = \left(\sum_i Ea_i(\tau)^r\right)\varrho_r(Y_1,Z_1).
\end{aligned}$$

□

For more general maxima we can again use the ℓ_p-metrics. Let $U = L^\lambda(\mu)$, $1 \le \lambda < \infty$. For $F \in M^1(U)$ and $a_i(\tau) \ge 0$ consider

$$TF \stackrel{d}{=} \max_{i\le i\le N} a_i(\tau)Y_i + C(\tau). \tag{9.1.18}$$

Here (Y_i) are i.i.d., $Y_1 \stackrel{d}{=} F, \tau$ is independent of (Y_i), and $C(\tau)$ has values in U. For any p, λ we have

$$\ell_p(TF,TG) \le \left(\sum_{i=1}^N Ea_i(\tau)^r\right)\ell_p(F,G), \quad r = \min(1,p). \tag{9.1.19}$$

For $1 \le p \le \lambda < \infty$ we have the following improvement:

Proposition 9.1.6 *If $1 \le p \le \lambda < \infty$, then*

$$\ell_p(TF,TG) \le \sum_{i=1}^N (Ea_i(\tau)^p)^{1/p}\ell_p(F,G). \tag{9.1.20}$$

Proof: Let $Y_i \stackrel{d}{=} F$, $Z_i \stackrel{d}{=} G$ satisfy $\|Y_i - Z_i\|_{\lambda,\mu}^p = \ell_p(F,G)^p$, where $\|X\|_{\lambda,\mu} = (\int |X(t)|^\lambda \, \mathrm{d}\mu(t))^{1/\lambda}$. Then

$$\begin{aligned}
\ell_p(TF,TG) &\leq (E\|\max a_i(\tau)Y_i - \max a_i(\tau)Z_i\|_{\lambda,\mu}^p)^{1/p} \\
&= \left(\int E\left(\int |\max a_i(t)Y_i(s) - \max a_i(t)Z_i(s)|^\lambda \, \mathrm{d}\mu(s)\right)^{p/\lambda} \mathrm{d}P^\tau(t)\right)^{1/p} \\
&\leq \left(\int E\left(\sum_i \int a_i(t)^\lambda |Y_i(s) - Z_i(s)|^\lambda \, \mathrm{d}\mu(s)\right)^{p/\lambda} \mathrm{d}P^\tau(t)\right)^{1/p} \\
&\leq \left(\int \sum_i E\left[\int a_i(t)^\lambda |Y_i(s) - Z_i(s)|^\lambda \, \mathrm{d}\mu(s)\right]^{p/\lambda} \mathrm{d}P^\tau(t)\right)^{1/p} \\
&\qquad\qquad (\text{since } p/\lambda \leq 1) \\
&= \left(\sum_i E a_i(\tau) E\|Y_i - Z_i\|_{\lambda,\mu}^p\right)^{1/p} = \left(\sum_i E a_i(\tau)^p\right)^{1/p} \ell_p(F,G).
\end{aligned}$$

□

(c) *Bootstrap Estimators*

For a separable Banach space U and $F \in M^1(U)$, let $\mu(F) = \int x \, \mathrm{d}F(x)$, $\mu_n(F) = \frac{1}{n}\sum_{i=1}^n X_i$, where (X_i) are i.i.d., $X_i \stackrel{d}{=} F$, and F_n is the empirical measure of $X_1, \ldots, X_n$. For $p > 0$ denote by Γ_p the class of distributions with finite pth moment. From the strong law of large numbers, for any $F \in \Gamma_p$ we then obtain (cf. Chapter 8)

$$\ell_p(F_n, F) \to 0 \text{ a.s.} \quad \text{and} \quad E\ell_p(F_n, F) \to 0. \tag{9.1.21}$$

Let now $X_1^*, \ldots, X_m^*$ be a bootstrap sample; i.e., the (X_i^*) are i.i.d., $X_1^* \stackrel{d}{=} F_n$ (conditionally on $X_1, \ldots, X_n$), and $F_{n,m}^*$ is the empirical distribution of $X_1^*, \ldots, X_m^*, m = m(n)$. The condition $\ell_p(F_{n,m}^*, F) \to 0$ a.s. (conditionally on X) is equivalent to the joint convergence

$$\begin{aligned}
\frac{1}{m}\sum_{i=1}^m f(X_i^*) &\to \int f \, \mathrm{d}F \text{ a.s.}, \quad f \in C_b(U), \\
\frac{1}{m}\sum_{i=1}^m d^p(X_i^*, a) &\to \int d^p(x,y) \, \mathrm{d}F(x) \text{ a.s.}
\end{aligned}$$

(cf. Chapter 8), representing a special form of the SLLN for real-valued r.v.s. In the case $(U,d) = (\mathbb{R}^r, \|\cdot\|), p > 1$, we are able to obtain a rate of convergence for the bootstrap approximation. Let $\gamma = kr/[(k-r)(k-2)]$, $k > r$, $k > 2$, and $\int \|x\|^\gamma F(dx) < \infty$. Then

$$E\ell_p^p(F_n, F) \ \le \ C(r,k,p)n^{-(1-\frac{1}{p})/k} \tag{9.1.22}$$

(cf. Rachev (1984d, pp. 667–668)), and thus

$$\begin{aligned} E\ell_p^p(F_{n,m}^*, F) \ &\le \ 2^p E E_{X_1,\dots,X_n} \ell_p^p(F_{n,m}^*, F_n) + 2^p C(r,k,p) n^{-(1-\frac{1}{p})/k} \\ &\le \ C^*(r,k,p)\left(m^{-(1-\frac{1}{p})/k} + n^{-(1-\frac{1}{p})/k}\right). \end{aligned} \tag{9.1.23}$$

If, however, the (X_i) are in the domain of an α-stable distribution, then it is more natural to choose the bootstrap estimator from a distribution $\widetilde{F}_n$ which has a tail behavior similar to F. Let then $\widetilde{X}_1, \dots, \widetilde{X}_n$ be a bootstrap sample with adapted tail behavior such that

$$\ell_p(\widetilde{F}_n, F) \ \to \ 0. \tag{9.1.24}$$

Consider $\widetilde{V}_n = \frac{1}{n^{1/\alpha}} \sum_{i=1}^n (\widetilde{X}_i - E_{\widetilde{F}_n} \widetilde{X}_i) \overset{d}{=} T_n(\widetilde{F}_n)$ $(1 \le \alpha \le 2)$ as a bootstrap estimator of $V_n = \frac{1}{n^{1/\alpha}} \sum_{i=1}^n (X_i - E_F X_i) \overset{d}{=} T_n(F)$. Then for a Banach space of type $p, 1 \le p \le 2$ and $1 \le \alpha \le p$, it follows from (9.1.11) that

$$\begin{aligned} \ell_p(\widetilde{V}_n, V_n) \ &\le \ B_p n^{\frac{1}{p}-\frac{1}{\alpha}} \ell_p(\widetilde{X}_1 - E\widetilde{X}_1, X_1 - E_F X_1) \\ &\le \ B_p n^{\frac{1}{p}-\frac{1}{\alpha}} (\ell_p(\widetilde{F}_n, F) + |E\widetilde{X}_1 - E_F X_1|) \ \to \ 0. \end{aligned} \tag{9.1.25}$$

Since $\ell_p(V_n, Y_{(\alpha)}) \to 0$, where $Y_{(\alpha)}$ an α-stable r.v., then in the case $\ell_p(X_1, Y_{(\alpha)}) < \infty$, it follows from the bound in (9.1.25) that

$$\ell_p(\widetilde{V}_n, Y_{(\alpha)}) \ \to \ 0. \tag{9.1.26}$$

Moreover, the rate of convergence is of order $o(n^{1/p-1/\alpha})$. In the case $p = 2$ and $F \in \Gamma_2$, the condition (9.1.24) is satisfied for $\widetilde{F}_n = F_n^*$; for Euclidean spaces this case has been considered by Bickel and Freedman (1981). Their investigation of more general functionals on the set of empirical measures can also be extended to the setting we described.

(d) *Transformation by Markov Kernels, Image Encoding*

Let (U,d) be a separable metric space and let $w_i : U \to U, 1 \le i \le N$, be mappings satisfying

$$d(w_i x, w_i y) \ \le \ s_i\, d(x,y). \tag{9.1.27}$$

Given a probability distribution $(p_i)_{1\le i\le N}$ define the Markov kernel

$$K(x,\cdot) = \sum_{i=1}^{N} p_i \varepsilon_{w_i(x)}, \quad N \le \infty. \tag{9.1.28}$$

The implied transformation on $M^1(U)$ is denoted by

$$TF = KF, \quad \text{where } KF(A) = \int K(x,A)F(\mathrm{d}x). \tag{9.1.29}$$

Let now $\mathrm{Lip}(U)$ be the set of Lipschitz functions, $|f(x)-f(y)| \le d(x,y)$ for all $x,y \in U$. Then for $Kf(x) = \sum p_i f \circ w_i(x)$, we have

$$\begin{aligned} |Kf(x) - Kf(y)| &\le \sum p_i |f\circ w_i(x) - f\circ w_i(y)| \\ &\le \sum p_i d(w_i(x), w_i(y)) \le \left(\sum p_i s_i\right) d(x,y). \end{aligned} \tag{9.1.30}$$

Let us look at the contraction properties for the mapping T with respect to the Kantorovich metric

$$\mu_L(F,G) = \sup\{|E(f(X) - f(Y))|; X \overset{d}{=} F, Y \overset{d}{=} G, f \in \mathrm{Lip}(U)\}. \tag{9.1.31}$$

We have then

$$\begin{aligned} \mu_L(TF,TG) &= \sup\{|E(f(K(X,\cdot) - f(K(Y,\cdot)))|;\ f\in \mathrm{Lip}(U)\} \\ &= \sup\{|E(Kf(X) - Kf(Y))|;\ f \in \mathrm{Lip}(U)\} \\ &\le \left(\sum p_i s_i\right) \sup\{|E(g(X) - g(Y))|;\ g\in \mathrm{Lip}(U)\} \\ &= \left(\sum p_i s_i\right) \mu_L(F,G). \end{aligned} \tag{9.1.32}$$

If $\sum p_i s_i < 1$, then T is a contractive mapping. By the Kantorovich–Rubinstein theorem μ_L coincides with the minimal L_1-metric, and therefore,

$$\ell_1(TF,TG) \le \left(\sum p_i s_i\right) \ell_1(F,G). \tag{9.1.33}$$

Moreover, for any $p > 0$, we can extend this result as follows.

Proposition 9.1.7

$$\ell_p(TF,TG) \le \left(\sum p_i s_i^p\right)^{1/p\wedge 1} \ell_p(F,G). \tag{9.1.34}$$

Proof: Suppose $X \overset{d}{=} F, Y \overset{d}{=} G$ satisfy

$$\ell_p(F,G) = \begin{cases} (E\,d(X,Y)^p)^{1/p}, & \text{if } p \ge 1, \\ E\,d(X,Y)^p, & \text{if } p < 1. \end{cases}$$

Take $\widetilde{I}$ to be a random variable with values in $\{1, 2, \dots, N\}$ and distribution (p_i) that is independent of X, Y. Then for $1 \leq p$,

$$\begin{aligned}
\ell_p^p(TF, TG) &\leq E\, d(w_{\widetilde{I}}(X),\, w_{\widetilde{I}}(Y))^p \qquad (9.1.35)\\
&= \sum_{i=1}^{N} E(d(w_i(X), w_i(Y))^p\, I(\widetilde{I} = i))\\
&= \sum_{i=1}^{N} p_i\, E\, d(w_i(X), w_i(Y))^p\\
&\leq \left(\sum_{i=1}^{N} p_i s_i^p\right) E\, d(X, Y) \;=\; \left(\sum p_i s_i^p\right) \ell_p^p(F, G).
\end{aligned}$$

The proof for the case $0 < p < 1$ is similar. □

Remark 9.1.8 *Another proof of* (9.1.32) *can be given via the dual representation of ℓ_p. Indeed, for $0 < p < \infty, p' = 1 \vee p$,*

$$\begin{aligned}
\ell_p^{p'}(F, G) &= \sup\left\{ \int f\, \mathrm{d}F + \int g\, \mathrm{d}G;\ f, g \text{ bounded continuous}, \right. \qquad (9.1.36)\\
&\qquad \left. f(x) + g(y) \leq d^p(x, y), \quad \forall x, y \in U \right\}.
\end{aligned}$$

Therefore,

$$\begin{aligned}
&\ell_p^{p'}(TF, TG) \qquad (9.1.37)\\
&= \sup\{E\, f(K(X, \cdot)) + E\, g(K(Y, \cdot));\ f(x) + g(y) \leq d^p(x, y)\},
\end{aligned}$$

where $X \stackrel{d}{=} F, Y \stackrel{d}{=} G$. Since

$$Kf(x) + Kg(y) \;\leq\; \sum p_i d^p(w_i(x), w_i(y)) \;\leq\; \left(\sum p_i s_i^p\right) d^p(x, y),$$

we obtain

$$\begin{aligned}
\ell_p^{p'}(TF, TG) &= \sup\{E\, Kf(X) + E\, Kg(Y)), f(x) + g(y) \leq d^p(x, y)\}\\
&\leq \left(\sum p_i s_i^p\right) \ell_p^{p'}(F, G). \qquad (9.1.38)
\end{aligned}$$

Remark 9.1.9 *Hutchinson (1981) was the first to prove convergence with respect to the metric μ_L in the case $s_i \leq 1$. Barnsley and Elton (1988) used the above Markov chain to "construct images" by so-called iterated* *function systems (IFS). They established the existence of a unique attractive invariant measure* μ *under the assumption that*

$$\prod_{i=1}^{N} d(w_i(x), w_i(y))^{p_i} \leq r\, d(x,y), \quad r < 1. \tag{9.1.39}$$

The above inequality is indeed implied by the condition (9.1.27) *with*

$$\prod_{i=1}^{N} s_i^{p_i} < 1. \tag{9.1.40}$$

In the case of affine maps on $\mathbb{R}^k$ we can improve the arguments in the following way (see Proposition 9.1.10 below, cf. also Burton and Rösler (1995)). Define

$$TF \stackrel{d}{=} AX + b, \tag{9.1.41}$$

where A is a random matrix, b a random vector, (A, b) independent of X, and $X \stackrel{d}{=} F$.

Consider the operator norm of the expected product $EA^T A$,

$$\|EA^T A\| := \sup_{\substack{x \in \mathbb{R}^k \\ x \neq 0}} \frac{\|(EA^T A)x\|}{\|x\|}. \tag{9.1.42}$$

Then

$$\|EA^T A\| = \sup_{x \neq 0} \frac{E\langle Ax, Ax\rangle}{\|x\|^2} = \|E\langle A, A\rangle\|, \tag{9.1.43}$$

where the right-hand side is the L_2-norm of $E\langle A, A\rangle$.

Proposition 9.1.10 *Assume that* $\|\,\|b\|\,\|_2 < \infty$. *Then*

$$\ell_2(T\mu, T\nu) \leq \sqrt{\|EA^T A\|}\, \ell_2(\mu, \nu) \tag{9.1.44}$$

for any $\mu, \nu \in M^1(\mathbb{R}^k)$ *with finite second moments.*

Proof: Let Y, Z be random vectors with distributions μ, ν and $\ell_2(\mu,\nu) = (E\|Y-Z\|^2)^{1/2}$, where Y, Z are independent of (A, b). Then

$$\begin{aligned} \ell_2(T\mu, T\nu) &\leq \| \|AY - AZ\| \|_2 \\ &\leq \sqrt{E\langle A(Y-Z), A(Y-Z)\rangle} \\ &= \sqrt{E\langle Y-Z, E(A^TA)(Y-Z)\rangle} \\ &\leq \sqrt{\|EA^TA\|}\sqrt{E\langle Y-Z, Y-Z\rangle} \\ &= \sqrt{\|EA^TA\|}\,\ell_2(\mu,\nu). \end{aligned} \tag{9.1.45}$$

□

Notice that the estimate from above defined in (9.1.45) is an improvement (in the case $p = 2$) over the general estimate

$$\| \|AX\| \|_p \leq \| \|A\| \cdot \|X\| \|_p \leq \| \|A\| \|_p \cdot \| \|X\| \|_p. \tag{9.1.46}$$

In fact, the above general bound requires the stronger condition $\| \|A\| \|_p < 1$ to yield the contraction property.

(e) *Environmental Processes*

Let (Y_i, Z_i) be a sequence of i.i.d. pairs of r.v.s with values in $U \times \mathbb{R}$, where U is a separable Banach space. Define a sequence of r.v.s (S_n) by

$$S_{n+1} = (Y_n + S_n)Z_n, \quad S_0 \geq 0. \tag{9.1.47}$$

This kind of process has found several applications in environmental modeling and has been studied intensively. If we write $\tau_n = (Y_n, Z_n)$, and $a(\tau_n) = Z_n, C(\tau_n) = Y_nZ_n$, then

$$S_{n+1} = a(\tau_n)S_n + C(\tau_n), \tag{9.1.48}$$

so we have a special case of (9.1.1). Under the condition that $E|a(\tau)|^r < 1$, the operator $TS \stackrel{d}{=} a(\tau)S + C(\tau)$ is contractive. Therefore, (S_n) converges (with respect to some ideal metric of order r such as ζ_r, for example) to a fixed point, i.e., a solution of

$$S \stackrel{d}{=} (Y + S)Z. \tag{9.1.49}$$

Numerous properties of the solutions of the above equation have been studied in the literature; see, for example, Rachev and Samorodnitsky (1995) and Rachev and Rüschendorf (1995).

9.2 Convergence of Recursive Algorithms

In this section we apply the contraction properties established in Section 9.1 to study limits for recursive algorithms. We shall use the "method of probability metrics." The main idea of this method is to transform the recursive equations in such a way that with respect to a suitable metric we can derive contraction properties in the limit; i.e., we consider decompositions $X_n = \widetilde{Y}_n + \widetilde{W}_n$ such that $(\widetilde{Y}_n)$ has contraction properties and $\widetilde{W}_n$ converges to zero. This idea will be demonstrated in various examples.

The approach is natural from the following point of view. If $S_n = \sum_{i=1}^n Y_i$ is a sum of independent (centered) random variables and $X_n = n^{-1/\alpha} S_n$ is the normalized sum, then X_n satisfies the following simple recursion:

$$X_{n+1} = \left(\frac{n}{n+1}\right)^{-1/\alpha} X_n + (n+1)^{-1/\alpha} Y_{n+1}. \tag{9.2.1}$$

Thus the central limit theorem can be considered as the limit theorem of this simple (stochastic) recursion. The form of the recursion corresponding to the strong law of large numbers is even simpler.

9.2.1 *Learning Algorithm*

Let $Y_1, Y_2, \ldots$ be an i.i.d. sequence of r.v.s with values in a separable Banach space with first moment μ. Define the following recursive sequence: Let X_1 be arbitrary with finite first moment, and let

$$X_{n+1} = \frac{n}{n+1} X_n + \frac{1}{n+1} Y_{n+1}. \tag{9.2.2}$$

X_n can be viewed as an easy recursive algorithm designed to "learn" about the unknown theoretical mean μ given the sample $(Y_1, \ldots, Y_n)$.

Proposition 9.2.1 $\zeta_r(X_n, \mu) \to 0$ if $\zeta_r(X_1, \mu) < \infty$.

Proof: Let $\ell_n = EX_n$.

Claim 1: $\ell_n \to \mu$.

For the proof of Claim 1 note that from (9.2.2), we obtain

$$\begin{aligned} \ell_{n+1} &= \frac{n}{n+1}\ell_n + \frac{1}{n+1}\mu \\ &= \frac{n-1}{n+1}\ell_{n-1} + \frac{2}{n+1}\mu \\ &= \frac{1}{n+1}\ell_1 + \frac{n}{n+1}\mu, \end{aligned} \tag{9.2.3}$$

where the last step follows from the inductive argument. This implies Claim 1.

Define next

$$Z_n = X_n - \ell_n, \qquad W_n = Y_n - \mu. \tag{9.2.4}$$

Then,

$$\begin{aligned}
Z_{n+1} + \ell_{n+1} &= \frac{n}{n+1}(Z_n + \ell_n) + \frac{1}{n+1}(W_{n+1} + \mu), \quad \text{(by (9.2.2))} \\
Z_{n+1} &= Z_n \frac{n}{n+1} + \frac{1}{n+1} W_{n+1} - \left(\ell_{n+1} - \ell_n \frac{n}{n+1} - \mu \frac{1}{n+1}\right) \\
&= Z_n \frac{n}{n+1} + W_{n+1} \frac{1}{n+1} \quad \text{(by (9.2.3))}. \qquad (9.2.5)
\end{aligned}$$

Now let μ_r be an ideal metric of order r, $1 < r < 2$, and $b_n = \mu_r(Z_n, 0)$ (for example we can choose $\mu_r = \zeta_r$).

Claim 2. $\mu_r(Z_n, 0) \to 0$ if $a = \mu_r(W_1, 0) < \infty$.

For the proof of this claim note that

$$\begin{aligned}
b_{n+1} &= \mu_r(Z_{n+1}, 0) = \mu_r\left(Z_n \frac{n}{n+1} + W_{n+1} \frac{1}{n+1}, 0\right) \\
&\leq \mu_r\left(\frac{n}{n+1} Z_n, 0\right) + \mu_r\left(\frac{1}{n+1} W_{n+1}, 0\right) \\
&\qquad \text{(since } Z_n \text{ is independent of } W_{n+1}\text{)} \\
&= \left(\frac{n}{n+1}\right)^r \mu_r(Z_n, 0) + \left(\frac{1}{n+1}\right)^r \mu_r(W_{n+1}, 0) \\
&= \left(\frac{n}{n+1}\right)^r b_n + \left(\frac{1}{n+1}\right)^r a.
\end{aligned}$$

Therefore,

$$\begin{aligned}
b_{n+1} &\leq \left(\frac{n}{n+1}\right)^r \left[\left(\frac{n-1}{n}\right)^r b_{n-1} + \left(\frac{1}{n}\right)^r a\right] + \left(\frac{1}{n+1}\right)^r_a \\
&= \left(\frac{n-1}{n+1}\right)^r b_{n-1} + 2\left(\frac{1}{n+1}\right)^r a \\
&\leq \left(\frac{1}{n+1}\right)^r b_1 + n\left(\frac{1}{n+1}\right)^r a. \qquad (9.2.6)
\end{aligned}$$

Since $1 < r$, it follows that $b_n \to 0$.

In particular, for $\mu_r = \zeta_r$, we obtain from Claim 1 that

$$\zeta_r(X_n, \mu) \;\to\; 0 \;\text{ if }\; \zeta_r(X_1, \mu) \;<\; \infty. \tag{9.2.7}$$

□

For the case of Euclidean spaces the condition $\zeta_r(Y_1, \mu) < \infty$ is satisfied if Y_1 has a finite absolute rth moment, $r > 1$. Therefore, under the assumption of a finite rth moment we obtain convergence of X_n to μ. The sequence (X_n) provides a simple example of a "learning algorithm" (for μ). Its convergence to μ in the real case can also be obtained as an application of the Robbins–Siegmund lemma (cf. Robbins and Siegmund (1971)) under the stronger assumption of a finite second moment. In this simple example we can, of course, directly prove the convergence of X_n to μ under the assumption of a finite first moment. The arguments above illustrate the general idea behind the method of probability metrics and show that in this simple case the method of probability metrics works with weaker assumptions than the method of stochastic approximation based on the Robbins–Siegmund lemma. Some further simple examples of the Robbins–Monroe-type recursion $X_{n+1} = f_n(X_n, Y_{n+1})$ can be treated similarly. Note that our method only needs a metric ideal of order $r > 1$ such that $\mu_r(X_n - \ell_n, 0) \to 0$ implies that $X_n - \ell_n \to 0$ in distribution. The ℓ_p metric will not work in this example, since its degree of ideality is only $r = \min(1, p)$.

9.2.2 Branching-Type Recursion

Consider the following recursive sequence (L_n):

$$L_0 \;\equiv\; 1, \qquad L_n \;\stackrel{d}{=}\; \sum_{i=1}^{K} X_i L_{n-1}^{(i)} + Y. \tag{9.2.8}$$

Here $L_{n-1}^{(i)}$ are i.i.d. copies of L_{n-1}, (X_i) is a real random sequence, K is a random number in $\mathbb{N}_0$, and Y is a random "immigration" such that $K, \{(X_i), Y\}, (L_{n-1}^{(i)})$ are independent. As usual, $\stackrel{d}{=}$ denotes equality in distribution. (9.2.8) induces a transformation T on M^1, the set of probability distributions on $(\mathbb{R}^1, \mathcal{B}^1)$. This is achieved by letting $T(\mu)$ be the distribution of $\sum_{i=1}^K X_i Z_i + Y$, where the (Z_i) are i.i.d. μ-distributed r.v.s, and moreover, $(Z_i), \{(X_i), Y\}, K$ are independent.

Some special cases of those transformation and recursion have been studied intensively in the literature. If $X_i \equiv 1$, then (9.2.8) describes a Galton–Watson process with immigration Y with the number of descendants of a parent described by K. The recursion (9.2.8) can be viewed as a branching process with random multiplicative weights. The special case where K is constant, $Y = 0$, and (X_i) are i.i.d. and nonnegative was introduced by Mandelbrot (1974) in his analysis of the Yaglom–Kolmogorov turbulence model. This case has been also studied by Kahane and Peyrière (1976) and Guivarch (1990), who considered the question of nontrivial fixed points of T, the existence of moments of the fixed points, and the convergence of (L_n). For $X_i \equiv K^{-1/\alpha}$, the solutions of the fixed-point equation $Z \stackrel{d}{=} \sum_{i=1}^{K} K^{-1/\alpha} Z_i$ are Paretian stable distributions (if $Z_i \geq 0$). For that reason the solutions are called semistable in Guivarch (1990). In this section we will be mainly interested in the case of multipliers X_i and solutions Z_i with moments of order ≥ 2. While the analysis of Kahane and Peyrière (1976) is based on an associated martingale, Guivarch (1990) uses a more elementary martingale property together with a conjugation relation and moment-type estimates for the L_p-distance, $0 < p < 1$.

Motivated by some problems in infinite particle systems, Holley and Liggett (1981) and Durrett and Liggett (1983) considered a smoothing transformation with (X_i) that are not not necessarily independent and assume that $X_i \geq 0$, K constant, and $Y = 0$. In Durrett and Liggett (1983) a complete analysis of the case is given. In particular, a necessary and sufficient condition for the existence and characterization of (all) fixed points as well as a general sufficient condition for convergence was derived, as well as a generalization of the result of Kahane and Peyrière on the existence of moments. The method of Durrett and Liggett is based on an associated branching random walk.

The use of contraction properties of minimal L_p-metrics in this section allows us to obtain quantitative approximation results for the recursion (9.2.8). Under moment assumptions used in this section, the recursion converges to the limiting distribution exponentially fast. This is demonstrated by simulations for several examples. Also, it is possible to remove the assumption of nonnegativity, to deal with a random number K, and to add immigration Y. This allows us to include applications to branching processes as well as to study the development of the total mass in the construction of multifractal measures (cf., for example, Arbeiter (1991)). For details we refer to Cramer and Rüschendorf (1996b).

(a) *Branching-Type Recursion with Multiplicative Weights*

In this section we shall study the recursion (9.2.8) allowing for dependent multipliers X_i but setting the immigration $Y \equiv 0$. In other words, we consider the recursion

$$L_0 \equiv 1, \qquad L_n \stackrel{d}{=} \sum_{i=1}^{K} X_i L_{n-1}^{(i)}, \tag{9.2.9}$$

where $\left(L_{n-1}^{(i)}\right)$ are i.i.d. copies of L_{n-1}, (X_i) is a square integrable real random sequence, K is a random number in $\mathbb{N}_0$, and $K, (X_i), \left(L_{n-1}^{(i)}\right)$ are independent r.v.s.

To determine the correct normalization of (L_n) we first consider the first moments of (L_n). Set $\ell_n := EL_n$, $c := E\left(\sum_{i=1}^K X_i\right)$, $v_n := \mathrm{Var}(L_n)$, $a := E\left(\sum_{i=1}^K X_i^2\right)$, and $b := \mathrm{Var}\left(\sum_{i=1}^K X_i\right)$.

Proposition 9.2.2 *Let $\ell_0 = 1$, $\ell_n = c^n$. Suppose that $b > 0, c \neq 0, a \neq c^2$. Then*

$$v_n = bc^{2n-2}\,\frac{1-(\frac{a}{c^2})^n}{1-\frac{a}{c^2}}, \quad n \geq 1, \; v_0 = 0. \tag{9.2.10}$$

If $a = c^2 \neq 0$, then $v_n = nba^{n-1}$.

Proof: Using the independence assumption in (9.2.9) and the conditional expectations, we obtain

$$\begin{aligned}
\ell_n &= E\left(E\left(\sum_{i=1}^K X_i L_{n-1}^{(i)} | K\right)\right) = E\left(\sum_{i=1}^K EX_i L_{n-1}^{(i)}\right) \\
&= E\left(\sum_{i=1}^K EX_i\right)\ell_{n-1} = c\,\ell_{n-1};
\end{aligned}$$

i.e., $\ell_n = c^n$. Similarly,

$$\begin{aligned}
v_n &= EL_n^2 - (EL_n)^2 \\
&= E\left[E\left(\left(\sum_{i=1}^K X_i L_{n-1}^{(i)}\right)^2 \Bigg| K\right)\right] - c^2\ell_{n-1}^2 \\
&= E\left[\sum_{i=1}^K E\left(X_i L_{n-1}^{(i)}\right)^2 + \sum_{i\neq j} E\left(X_i X_j L_{n-1}^{(i)} L_{n-1}^{(j)}\right)\right] - c^2\ell_{n-1}^2 \\
&= E\left[EL_{n-1}^2 \sum_{i=1}^K EX_i^2 + \ell_{n-1}^2 \sum_{i\neq j} E(X_i X_j)\right] - c^2\ell_{n-1}^2 \\
&= E\left[\sum_{i=1}^K EX_i^2\,(\mathrm{Var}\,L_{n-1} + \ell_{n-1}^2) + \ell_{n-1}^2 \sum_{i\neq j} E(X_i X_j)\right] - c^2\ell_{n-1}^2 \\
&= E\left(\sum_{i=1}^K X_i^2\right) v_{n-1} + \mathrm{Var}\left(\sum_{i=1}^K X_i\right)\ell_{n-1}^2 \\
&= a\,v_{n-1} + b\,c^{2(n-1)} = b\sum_{k=0}^{n-1} a^k c^{2(n-1-k)}
\end{aligned}$$

$$= \begin{cases} b\, c^{2n-2}\, \frac{1-(\frac{a}{c^2})^n}{1-\frac{a}{c^2}} = bc^{2n}\, \frac{1-(\frac{a}{c^2})^n}{c^2-a}, & \text{if} \quad a \neq c^2 \neq 0, \\ nba^{n-1}, & \text{if} \quad a = c^2. \end{cases}$$

□

In the case $b = 0$, we have $v_n = 0$ for all n. Therefore, we consider only the case $b > 0$.

From (9.2.10) we obtain that for $a < c^2$, $\sqrt{v_n}$ is of the same order as ℓ_n. This makes it possible to use a simple normalization by ℓ_n. Define for $c \neq 0$,

$$\widetilde{L}_n := L_n/c^n. \tag{9.2.11}$$

Then $E\widetilde{L}_n = 1$, and $\mathrm{Var}(\widetilde{L}_n) \to \frac{b}{c^2-a}$. Moreover, $\widetilde{L}_n$ satisfies the modified recursion

$$\widetilde{L}_n \stackrel{d}{=} \frac{1}{c}\sum_{i=1}^{K} X_i \widetilde{L}_{n-1}^{(i)}, \tag{9.2.12}$$

where $\widetilde{L}_{n-1}^{(i)} := \frac{L_{n-1}^{(i)}}{c^{n-1}}$. Define D_2 to be the set of distributions on $(\mathbb{R}^1, \mathcal{B}^1)$ with finite second moments and first moment equal to one. Next, define the mapping $T : D_2 \to D_2$ by

$$T(G) = \mathcal{L}\left(\frac{1}{c}\sum_{i=1}^{K} X_i Z_i\right), \tag{9.2.13}$$

where the (Z_i) are i.i.d. random variables with distribution G, and such that (X_i), (Z_i), K are independent r.v.s. Let ℓ_2 denote the minimal L_2-metric on D_2:

$$\begin{aligned} \ell_2(\mu,\nu) &= \inf\left\{\left(E(V-W)^2\right)^{1/2};\ V \stackrel{d}{=} \mu, W \stackrel{d}{=} \nu\right\} \\ &= \left(\int_0^1 \left(F^{-1}(u) - G^{-1}(u)\right)^2 du\right)^{1/2}. \end{aligned} \tag{9.2.14}$$

Here F, G are the distribution functions of μ, ν respectively. If $a < c^2$, then T is a contraction with respect to ℓ_2.

Proposition 9.2.3 *Assume that $a < c^2$. Then for $F, G \in D_2$,*

$$\ell_2(T(F), T(G)) \leq \sqrt{\frac{a}{c^2}}\, \ell_2(F, G). \tag{9.2.15}$$

Proof: Let the r.v.s $U^{(i)} \overset{d}{=} F, V^{(i)} \overset{d}{=} G, i \in \mathbb{N}$, be choosen on $(\Omega, \mathcal{A}, P)$ in such a way that $\|U^{(i)} - V^{(i)}\|_2 = \ell_2(F,G)$; for all i and K, (X_i), $(U^{(1)}, V^{(1)})$, $(U^{(2)}, V^{(2)}), \ldots$ are all assumed to be independent. Then

$$\begin{aligned}
\ell_2^2(T(F),T(G)) &\leq \left\| \frac{1}{c}\sum_{i=1}^K X_i U^{(i)} - \frac{1}{c}\sum_{i=1}^K X_i V^{(i)} \right\|_2^2 \\
&= \frac{1}{c^2} E\left(E\left[\left(\sum_{i=1}^K X_i U^{(i)} - \sum_{i=1}^K X_i V^{(i)} \right)^2 \middle| K \right] \right) \\
&= \frac{1}{c^2} E\left[\sum_{i=1}^K E\left(X_i^2 \left(U^{(i)} - V^{(i)} \right)^2 |K \right) \right. \\
&\qquad \left. + \sum_{i \neq j} E\left[X_i \left(U^{(i)} - V^{(i)} \right) X_j \left(U^{(j)} - V^{(j)} \right) |K \right] \right] \\
&= \frac{1}{c^2} E\left(\sum_{i=1}^K E X_i^2 E\left(U^{(i)} - V^{(i)} \right)^2 \right) \\
&= \frac{a}{c^2}\, \ell_2^2(F,G).
\end{aligned}$$

□

As a consequence of Proposition 9.2.3 it follows that T has exactly one fixed point in D_2 with variance equal to $b/(c^2 - a)$. The fixed-point equation is given in terms of the independent random variables $Z, Z_i \in D_2, Z_i \overset{d}{=} Z, (Z_i)$ as follows:

$$Z \overset{d}{=} \frac{1}{c}\sum_{i=1}^K X_i Z_i. \tag{9.2.16}$$

As a corollary we obtain

Theorem 9.2.4 *If* $a = E\left(\sum_{i=1}^K X_i^2\right) < c^2$, *then*

$$\ell_2(\widetilde{L}_n, Z) \leq \left(\frac{a}{c^2}\right)^{n/2} \frac{\sqrt{b}}{\sqrt{c^2 - a}}. \tag{9.2.17}$$

In particular, $\widetilde{L}_n$ *converges in distribution to* Z.

Proposition 9.2.5 *If* K *is constant and* $E\left(\sum_{i=1}^K |X_j|^k\right) < c^k$ *for all* $2 \leq k \leq h$, *then* $E|Z|^h < \infty$.

Proof: $\widetilde{L}_n$ can be equivalently represented by Y_n in the following form:

$$Y_0 = 1, \quad Y_n = \frac{1}{c^n} \sum_{(j_1,\ldots,j_n)\in\{1,\ldots K\}^n} \prod_{k=1}^{n} X_{j_1,\ldots,j_k},$$

where $(X_{j_1,\ldots,j_{k-1},1},\ldots,X_{j_1,\ldots,j_{k-1},K}) \stackrel{d}{=} (X_1,\ldots,X_K)$ (cf. Guivarch (1990)). Moreover, (Y_n) is a martingale, and therefore $|Y_n|^k$ is a submartingale. Representing the Y_n in the recursive way $Y_n = \frac{1}{c}\sum_{j=1}^{K} X_j Y_{n-1}^{(j)}$, where $Y_{n-1}^{(j)}$ are independent copies of Y_{n-1}, we have

$$\begin{aligned} c^k\, E|Y_n|^k &\le \left(E\sum_{j=1}^{K}|X_j|^k\right) E|Y_{n-1}|^k \\ &+ \sum_{\substack{k_1+\cdots+k_K=k \\ k_i\le k-1}} \binom{k}{k_1,\ldots,k_K} E\prod_{j=1}^{K}|X_j|^{k_j}\prod_{j=1}^{K}E|Y_{n-1}|^{k_j}. \end{aligned}$$

We can infer from Theorem 9.2.4 that $E|Y_n|^k$ is uniformly bounded for $k \le 2$. By induction over $k \le h$, we see that the lower-order terms in the above equation are uniformly bounded, say by a constant C. Since $E|Y_n|^k \ge E|Y_{n-1}|^k$, we obtain

$$E|Y_n|^k\left[c^k - E\left(\sum_{i=1}^{K}|X_j|^k\right)\right] \le C.$$

Therefore, the assumptions of this proposition ensure that $E|Y_n|^k$ is uniformly bounded for all $k \le h$. The submartingale convergence theorem now yields the existence of an integrable almost sure limit of $|Y_n|^h$. Since $\widetilde{L}_n \stackrel{d}{=} Y_n$, the weak limit Z of $\widetilde{L}_n$ is absolutely h-integrable. □

We can also obtain a "stability" result for the stationary equation (9.2.16). This will be achieved in terms of the ℓ_p metrics defined as in (9.2.14) with 2 replaced by p. Suppose we want to approximate the solution S of the equation

$$S \stackrel{d}{=} \sum_{i=1}^{K} X_i S_i \tag{9.2.18}$$

by the solution of the "approximate" equation

$$S^* \stackrel{d}{=} \sum_{i=1}^{K} X_i^* S_i^*.$$

Here we assume without loss of generality that $c = 1$ and consider the case of independent sequences $(X_i), (X_i^*)$ so that the pairs $(X_i), (S_i)$ and $(X_i^*), (S_i^*)$ are independent, and K is constant.

Proposition 9.2.6 *If K is constant, $\sum_{i=1}^K \ell_p(X_i, X_i^*) < \varepsilon$, and $\sum_{i=1}^K ||X_i||_p < 1$, then*

$$\ell_p(S, S^*) \leq \frac{\varepsilon ||S^*||_p}{1 - \sum_{i=1}^K ||X_i||_p}. \tag{9.2.19}$$

Proof: From the definition of S, S^*,

$$\begin{aligned}
\ell_p(S, S^*) &= \ell_p\left(\sum_{i=1}^K X_i S_i, \sum_{i=1}^K X_i^* S_i^*\right) \\
&\leq \sum_{i=1}^K \ell_p(X_i S_i, X_i^* S_i^*) \\
&\leq \sum_{i=1}^K (\ell_p(X_i S_i, X_i S_i^*) + \ell_p(X_i S_i^*, X_i^* S_i^*) \\
&\leq \left(\sum_{i=1}^K ||X_i||_p\right) \ell_p(S, S^*) + ||S^*||_p \cdot \varepsilon.
\end{aligned}$$

This implies that

$$\ell_p(S, S^*) \leq \frac{\varepsilon ||S^*||_p}{1 - \sum_{i=1}^K ||X_i||_p}.$$

□

A similar idea for establishing robustness of equations can be found in Rachev (1991, Chapter 19.3). For the case of a random K we replace Proposition 9.2.6 by the following one.

Proposition 9.2.7 *If $E\left(\sum_{i=1}^K (X_i - X_i^*)^2\right) \leq \varepsilon^2, EX_i = EX_i^*$, and $a = E\left(\sum_{i=1}^K X_i^2\right) < 1$, then*

$$\ell_2(S, S^*) \leq \frac{\varepsilon}{1 - \sqrt{a}} ||S^*||_2. \tag{9.2.20}$$

Proof: By the triangle inequality and the independence assumption and the assumption $EX_i = EX_i^*$,

$$\ell_2(S, S^*) = \ell_2\left(\sum_{i=1}^K X_i S_i, \sum_{i=1}^K X_i^* S_i^*\right)$$

$$
\begin{aligned}
&\leq \ell_2\left(\sum_{i=1}^{K} X_i^* S_i^*, \sum_{i=1}^{K} X_i S_i^*\right) + \ell_2\left(\sum_{i=1}^{K} X_i S_i^*, \sum_{i=1}^{K} X_i S_i\right) \\
&\leq \left(E\left(\sum_{i=1}^{K} X_i^2\right)\right)^{1/2} \ell_2(S, S^*) + ||S^*||_2 \left(E\sum_{i=1}^{K}(X_i - X_i^*)^2\right)^{1/2} \\
&= \sqrt{a}\, \ell_2(S, S^*) + \varepsilon||S^*||_2.
\end{aligned}
$$

Therefore,

$$
\ell_2(S, S^*) \leq \frac{\varepsilon||S^*||_2}{1 - \sqrt{a}}.
$$

□

Remark 9.2.8 *In the case of constant K and nonnegative X_i, Durrett and Liggett (1983) proved that the stationary solution Z of* (9.2.16) *has moments of order β if and only if*

$$
v(\beta) = \log\left(\frac{1}{c^\beta}\sum_{i=1}^{K} EX_i^\beta\right) < 0. \tag{9.2.21}
$$

For $\beta = 2$, (9.2.4) *is equivalent to the condition $a < c^2$ used in Proposition* 9.2.3. *In this sense this condition is sharp when using ℓ_2-distances. Guivarch (1990) has shown how to relax the second-moment assumption.*

Remark 9.2.9 *For the normalized recursion* (9.2.12) *with (X_i) i.i.d. r.v.s, K being a constant (we assume without loss of generality that $c = 1$), we can use the form*

$$
\widetilde{L}_0 = 1, \quad \widetilde{L}_n = \sum_{(j_1,\dots,j_n)\in\{1,\dots,K\}^n} \prod_{k=1}^{n} X_{j_1,\dots,j_k}, \tag{9.2.22}
$$

where $(X_{j_1,\dots,j_k})$ are independent and distributed as X_1; i.e., $\widetilde{L}_n$ is a sum over product weights in the complete K-ary tree; cf. the proof of Proposition 9.2.5. *For nonnegative multipliers X_i we also can consider functionals of the type*

$$
M_n = \max_{P_n} \prod_{k=1}^{n} X_{j_1,\dots,j_k}, \tag{9.2.23}
$$

where the maximum is taken over all paths of length n. Taking logarithms,

$$
-\ln M_n = -\max_{P_n} \sum_{k=1}^{n} \ln\left(X_{j_1,\dots,j_k}\right) = \min_{P_n} \sum_{k=1}^{n} \left(-\ln\left(X_{j_1,\dots,j_k}\right)\right),
$$

and applying Kingman's subadditive ergodic theorem yields that for some constant β,

$$\frac{1}{n} \log M_n \quad \to \quad \beta \text{ a.s.} \tag{9.2.24}$$

This shows that in some sense the max product weight is not larger in order of magnitude than the average product weight. In some cases, the constant β is explicitly known, for example, for $X_i \stackrel{d}{=} U[0,1]$, $\beta \approx -0.23196$ (cf. Mahmoud (1992, p. 165)).

Remark 9.2.10 *In some cases explicit solutions of* (9.2.16) *are known.*

(1) *If K is constant and $\frac{1}{c}X_i \stackrel{d}{=} \beta(\frac{a}{K}, a - \frac{a}{K})$ is beta distributed, then $Z \stackrel{d}{=} \Gamma(a, \beta)$ is gamma distributed (cf. Guivarch (1990)).*

(2) *Suppose that $Z_1 \stackrel{d}{=} \frac{1}{K}\sum_{i=1}^K X_i Z_i$, (Y_i) are i.i.d. r.v.s, $\overline{X} \stackrel{d}{=} X_1$, and $Y_1 \stackrel{d}{=} X_1 Z_1$ holds. Then $Y_1 \stackrel{d}{=} \frac{1}{K}\sum_{i=1}^K Y_i \overline{X}$. Conversely, if $Y_1 \stackrel{d}{=} \frac{1}{K}\sum_{i=1}^K Y_i X_1$ and (X_i) are i.i.d. r.v.s, then $Z_i \stackrel{d}{=} \frac{1}{K}\sum_{j=1}^K Y_j$. The sequence (Z_i) solves the equation $Z_1 \stackrel{d}{=} \frac{1}{K}\sum_{i=1}^K X_i Z_i$ (cf. Durrett and Liggett (1983)).*

(3) *Suppose (Z_i) solves $Z_1 \stackrel{d}{=} \sum_{i=1}^K X_i Z_i, X_i \geq 0$. Then $Y_i = Z_i^{1/\vartheta} W_i$, where $0 < \vartheta \leq 2$ and W_i are stable r.v.s with index ϑ, satisfy*

$$\sum_{i=1}^K X_i^{1/\vartheta} Y_i \stackrel{d}{=} Y_1. \tag{9.2.25}$$

To prove (9.2.25), *observe that*

$$\sum_{i=1}^K X_i^{1/\vartheta} Z_i^{1/\vartheta} W_i \stackrel{d}{=} \left(\sum_{i=1}^K X_i Z_i\right)^{1/\vartheta} W_1 \stackrel{d}{=} Z_1^{1/\vartheta} W_1 = Y_1.$$

This interesting transformation property is used in Guivarch (1990) to reduce the case with moments of X_i of higher order to the case of moments of lower order.

(4) *If $\sum_{i=1}^K X_i^2 \equiv c^2 \neq 0$, then the normally distributed r.v.s $Z \stackrel{d}{=} Z_i \stackrel{d}{=} \mathcal{N}(0, \sigma^2)$ satisfy* (9.2.16).

(5) *If Z solves* (9.2.16) *and $\overline{Z}$ is an independent copy of Z, then $Z^* := Z - \overline{Z}$ solves*

$$Z^* \stackrel{d}{=} \frac{1}{c} \sum_{i=1}^K X_i^* Z_i^*.$$

Here $X_i^* = \tau_i X_i$, and the τ_i are arbitrary random signs. In particular, if $K = 2$, and the r.v.s $X_i^* \stackrel{d}{=} U[-1, 1]$ are independent, then (9.2.16) *is solved by $Z^* := Z - \overline{Z}$, where $Z \stackrel{d}{=} \Gamma(2, \frac{1}{2}, 0)$.*

Remark 9.2.11 *The following simulations (Figures* 9.1 *and* 9.2*) of $\widetilde{L}_n$, with $K = 2$, X_1, X_2 independent r.v.s, $X_1 \stackrel{d}{=} X_2 \stackrel{d}{=} U[0, 1]$, $X_i \stackrel{d}{=} \beta(2, 2)$, show good approximation of the empirical d.f. by the theoretical gamma distribution.*

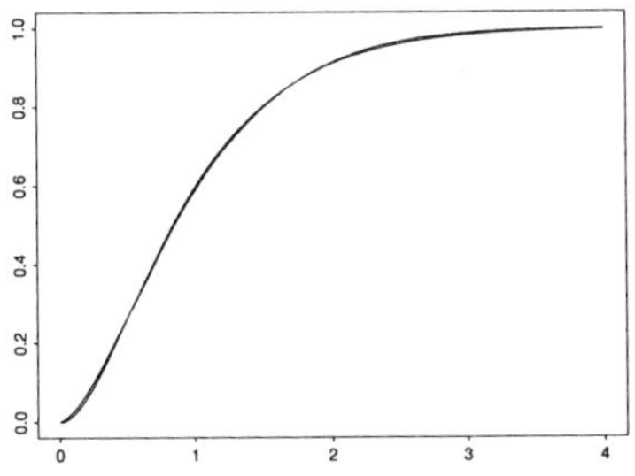

FIGURE 9.1. Empirical d.f., $X_1 \stackrel{d}{=} U[0, 1]$, $n = 10$, theoretical Gamma $\Gamma(2, \frac{1}{2}, 0)$

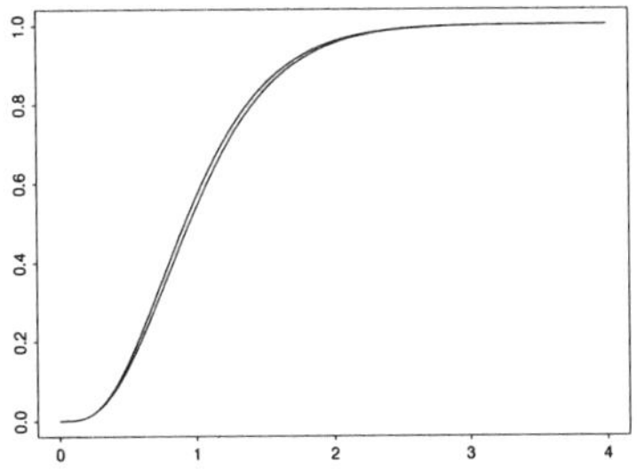

FIGURE 9.2. Empirical d.f., $X_1 \stackrel{d}{=} \beta(2, 2)$, $n = 8$, theoretical Gamma $\Gamma(4, \frac{1}{4}, 0)$

Remark 9.2.12 *In the case $K = 2$, X_1, X_2 independent r.v.s, $X_1 \stackrel{d}{=} X_2 \stackrel{d}{=} U\left[-\frac{1}{8}, \frac{9}{8}\right]$, no explicit solution of* (9.2.16) *is known. Nevertheless, the following simulation (Figure* 9.3*) shows that $\widetilde{L}_n$ converges very fast to the fixed point of* (9.2.16)*. The empirical distribution functions of $\widetilde{L}_{10}$ and $\widetilde{L}_{12}$ can hardly be distinguished. Therefore, they may be regarded as the limiting distribution function. The empirical distribution function of $\widetilde{L}_6$ is already very close to that limit (cf. Figure* 9.3*).*

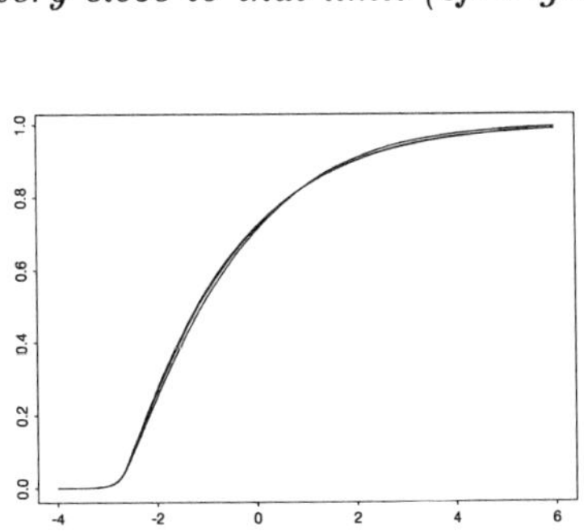

FIGURE 9.3. Empirical d.f. of $\widetilde{L}_n$ for $n = 6$, 10, and 12, $X_1 \stackrel{d}{=} U\left[-\frac{1}{8}, \frac{9}{8}\right]$

Remark 9.2.13 (Branching processes) *Equation (9.2.9) includes the Galton–Watson process as special case. A Galton–Watson process is defined by the recursion*

$$Z_0 = 1, \quad Z_{n+1} = \sum_{k=1}^{Z_n} X_k^n, \tag{9.2.26}$$

where $X_k^n \stackrel{d}{=} X$ are i.i.d. r.v.s, $n \in \mathbb{N}_0$. Define $K \stackrel{d}{=} X$ and $X_i \equiv 1$. Then

$$L_n \stackrel{d}{=} Z_n \tag{9.2.27}$$

for all n. This equality can be checked by induction on n. In fact, take first $Z_0 = L_0 = 1$. If $Z_k \stackrel{d}{=} L_k$ for $k \le n$, then

$$\begin{aligned} Z_{n+1} &\stackrel{d}{=} \sum_{k=l}^{L_n} X_k^n \stackrel{d}{=} \sum_{i=1}^{K} \sum_{k=1+\sum_{j=1}^{i-1} L_{n-1}^{(j)}}^{\sum_{j=1}^{i} L_{n-1}^{(j)}} X_k^n \\ &\stackrel{d}{=} \sum_{i=1}^{K} \left(\sum_{k=1}^{L_{n-1}} X_k^n \right)^{(i)} \stackrel{d}{=} \sum_{i=1}^{K} \left(\sum_{k=1}^{Z_{n-1}^{(i)}} X_k^n \right)^{(i)} = \sum_{i=1}^{K} Z_n^{(i)} \\ &\stackrel{d}{=} \sum_{i=1}^{K} L_n^{(i)} \stackrel{d}{=} L_{n+1}. \end{aligned}$$

The assumption $a < c^2$ is equivalent to the condition $EX > 1$.

From (9.2.27) we can derive explicit stationary distributions and even the extinction probabilities in some cases. If, for example, X is geometrically distributed, $P(X = k) = p(1-p)^k, k \in \mathbb{N}_0$, then $c = EX = \frac{1-p}{p} > 1$ if $p < \frac{1}{2}$ and $\operatorname{Var}(X) = \frac{1-p}{p^2}$. The normalized Galton–Watson process $\frac{Z_n}{\sqrt{\operatorname{Var}(Z_n)}}$ converges to a (unique) solution of the fixed-point equation

$$Z \stackrel{d}{=} \frac{1}{EX} \sum_{i=1}^{X} Z_i, \quad EZ = \sqrt{\frac{EX(EX-1)}{\operatorname{Var}(X)}}. \tag{9.2.28}$$

The extinction probability is easily seen to be $\frac{p}{1-p}$. For the normalized continuous part an equation identical to (9.2.28) (but with different variances) is also valid. It is well known that this equation is solved by the geometric stable distribution of order 1, i.e., the exponential distribution. This finally implies

$$Z \stackrel{d}{=} \frac{p}{1-p}\delta_0 + \frac{1-2p}{1-p} \exp\left(\frac{\sqrt{1-2p}}{1-p} \right), \tag{9.2.29}$$

since $EZ = \sqrt{1-2p}, EZ^2 = 2(1-p)$.

(b) *A Random Immigration Term*

In this section we admit an additional immigration term; i.e., we consider the recursion

$$L_n \stackrel{d}{=} \sum_{i=1}^{K} X_i L_{n-1}^{(i)} + Y, \tag{9.2.30}$$

where $\{(X_i), Y\}$, K, $L_{n-1}^{(1)}$, $L_{n-2}^{(2)}, \ldots$ are independent r.v.s. The analysis of (9.2.30) is essentially simplified if we assume $\ell_0 := EL_0$, $v_0 := \mathrm{Var}(L_0)$,

$$\ell_0 = \frac{EY}{1-c} \text{ (if } c \neq 1\text{)}, \quad v_0 = \frac{\mathrm{Var}(\ell_0 \sum_{i=1}^K X_i + Y)}{1-a}, \quad \text{where } a < 1. \tag{9.2.31}$$

If $c = 1$, then $EY = 0$ and ℓ_0 is arbitrary.

Lemma 9.2.14 *Under the assumption* (9.2.31),

$$\ell_n = EL_n = \ell_0, \quad v_n = \mathrm{Var}(L_n) = v_0, \quad \textit{for all } n \in \mathbb{N}. \tag{9.2.32}$$

Proof: From (9.2.31), $\ell_n = c\,\ell_{n-1} + EY = \frac{EY}{1-c} = \ell_{n-1}$,

$$\begin{aligned}
v_n &= \mathrm{Var}(L_n) = EL_n^2 - \ell_n^2 \\
&= E\left[\sum_{i=1}^{K} E\left(X_i^2 \left(L_{n-1}^{(i)}\right)^2 |K\right) + \sum_{i \neq j} E\left(X_i X_j L_{n-1}^{(i)} L_{n-1}^{(j)} |K\right)\right. \\
&\quad \left. + E(Y^2|K) + 2E\left(\sum_{i=1}^{K} Y X_i L_{n-1}^{(i)} |K\right)\right] - \ell_0^2 \\
&= a(v_{n-1} + \ell_0^2) + \left(\sum_{i \neq j} EX_i X_j\right) \ell_0^2 \\
&\quad + EY^2 + \ell_0 2E\left(\sum_{i=1}^{K} Y X_i\right) - \ell_0^2 \\
&= av_{n-1} + \mathrm{Var}\left(\ell_0 \sum_{i=1}^{K} X_i + Y\right) = v_{n-1}.
\end{aligned}$$

□

Condition (9.2.31) is fulfilled for a two-point distribution of L_0. Indeed, it allows us to use the method in the proof of section 9.2.2(a). A change of the initial condition leads to the necessity to change the method of proof and leads to a great variety of different cases to be considered. We therefore restrict ourselves to (9.2.31) in this section.

As in section (a), we introduce the operator

$$T : M(\ell_0, v_0) \;\to\; M(\ell_0, v_0), \quad T(G) \;=\; \mathcal{L}\left(\sum_{i=1}^{K} X_i V_i + Y\right). \tag{9.2.33}$$

Here $M(\ell_0, v_0)$ is the set of distributions with mean ℓ_0 and variance v_0, (V_i) are i.i.d. r.v.s, and the random quantities $V_1 \stackrel{d}{=} G, (V_i), \{(X_i), Y\}$, K are independent.

Similarly as in Proposition 9.2.3, we obtain the contractive inequality

$$\ell_2(T(F), T(G)) \;\le\; \sqrt{a}\, \ell_2(F, G). \tag{9.2.34}$$

This implies the convergence of L_n to a unique fixed point for the mapping T in $M(\ell_0, v_0)$ with respect to the ℓ_2-metric. The contraction factor is $\sqrt{a}$. A sharper result (i.e., a smaller contraction factor) is obtained by the use of the Zolotarev metric ζ_r instead of ℓ_2. Recall the definition for ζ_r (cf. (9.1.2)):

$$\zeta_r(F, G) \;=\; \sup\{|E(f(X) - f(Y))|;\ |f^{(m)}(x) - f^{(m)}(y)| \le |x - y|^\alpha\} \tag{9.2.35}$$

for $r = m + \alpha, m \in \mathbb{N}_0, 0 < \alpha \le 1$.

Proposition 9.2.15

$$\zeta_r(T(F), T(G)) \;\le\; E\left(\sum_{i=1}^{K} |X_i|^r\right) \zeta_r(F, G). \tag{9.2.36}$$

Proof: Recall that ζ_r is an ideal metric of order r with respect to summation; i.e.,

$$\zeta_r(X + Z, Y + Z) \;\le\; \zeta_r(X, Y)$$

for Z independent of X, Y, and moreover,

$$\zeta_r(cX, cY) \;=\; |c|^r \zeta_r(X, Y).$$

Then, for $(Z_i), (W_i)$ being i.i.d. r.v.s distributed according to F, G, we have

$$\begin{aligned}
\zeta_r(TF, TG) \;&=\; \sup\left\{\left| Ef\left(\sum_{i=1}^{K} X_i Z_i + Y\right) - Ef\left(\sum_{i=1}^{K} X_i W_i + Y\right)\right|;\right.\\
&\qquad\qquad \left. |f^{(m)}(X) - f^{(m)}(Y)| \le |x - y|^2 \right\}\\
&\le\; \int \zeta_r\left(\sum_{i=1}^{K} x_i Z_i + y, \sum_{i=1}^{K} x_i W_i + y\right) dP^{(X,Y,K)} \ (x, y, k)
\end{aligned}$$

$$\leq \int \sum_{i=1}^{k} |x_i|^r \zeta_r(Z_i, W_i) \, \mathrm{d}P^{(X,Y,K)} \ (x, y, k)$$
$$= E\left(\sum_{i=1}^{K} |X_i|^r\right) \zeta_r(F, G). \qquad \square$$

Note that for the recursion defined by T the first two moments are matched. Therefore, we can apply (ζ_r) with $r \leq 3$ and obtain as a corollary the following theorem.

Theorem 9.2.16 *Suppose either $c \neq 1$ and $\ell_0 = \frac{EY}{1-c}$ or $c = 1$. Suppose also that $EY = 0$ and $v_0 = \frac{\mathrm{Var}(\ell_0 \sum_{i=1}^{K} X_i + Y)}{1-a}$ for $a < 1$. Then for $0 < r \leq 3$, the inequality*

$$a_r := E \sum_{i=1}^{K} |X_i|^r < 1$$

implies

$$\zeta_r(L_n, Z) \leq \frac{a_r^n}{1 - a_r} \zeta_r(L_0, L_1) < \infty,$$

where Z is a fixed point of T in $M(\ell_0, v_0)$. In particular, L_n converges in distribution to Z.

Therefore, in the case with immigration we also obtain an exponential rate of convergence. As a consequence, after a few iterations, the limiting distribution is already well approximated.

Consider the following example: $L_0 \stackrel{d}{=} \frac{1}{10}\,\delta_{-5} + \frac{2}{5}\,\delta_0 + \frac{1}{2}\,\delta_2$, $K = 2$, X_1, X_2 independent, $X_1 \stackrel{d}{=} X_2 \stackrel{d}{=} U\left[-\frac{1}{2}, 1\right]$, and $Y \stackrel{d}{=} \frac{17}{32}\,\delta_{-1} + \frac{5}{64}\,\delta_0 + \frac{25}{64}\,\delta_2$.

In this situation the assumptions of Theorem 9.2.16 are fulfilled. The fast convergence is confirmed by the closeness of the empirical distribution functions of L_6 and L_8 in the simulation described in Figure 9.4.

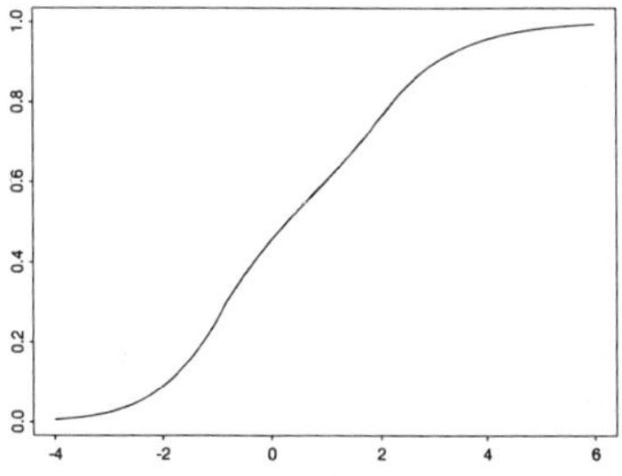

FIGURE 9.4. Empirical distribution functions for L_6 and L_8; the difference between the two curves is hardly visible. (Figs. 9.1–9.4: M. Cramer, L. Rüschendorf; *Annales de l'Institut Henri Poincaré*, 6, 1996; © from Gauthier-Villars Editeur)

9.2.3 Limiting Distribution of the Collision Resolution Interval

In this section we apply the method of probability metrics to investigate the contraction properties of stochastic algorithms arising in random-access communication protocols. The results are due to Feldman, Rachev, and Rüschendorf (1994); see also Rachev and Rüschendorf (1995).

The Capetanakis–Tsybakov–Mikhailov (CTM) protocol is one of the most elegant solutions to the classical multiple-access problem, in which a large population of users share a single communication channel. Throughput of this protocol is close to the throughput of the slotted Aloha protocol. The CTM protocol, unlike the classical "slotted Aloha," is inherently stable. The "tree splitting protocols," of which the CTM protocol is an example, pose some interesting mathematical problems and have been the subject of intensive study.

We briefly review the definition of the CTM protocol; see Bertsekas and Gallager (1987). Time is divided into slots of equal duration. During each slot, one of the following events occurs:

1. The slot is wasted because no one transmits.
2. Exactly one user transmits a message, in which case the message is successfully received.
3. The slot is wasted because two or more users transmit, interfering with each other. This is called a collision.

At the end of each slot, every user knows which of these three events occurred (this is sometimes called "trinary feedback").

When a collision occurs, all users involved (those that transmitted during the slot) divide themselves into two groups on a random basis. Each user performs the equivalent of an independent coin toss in order to make its decision; p is the probability that a user selects the first group. Users in the first group retransmit their messages during the slot following the one in which the collision occurred; users in the second group defer their retransmissions until all users in the first group have successfully transmitted their messages. If one of these groups contains more than one user, another collision will occur, in which case this group divides in the same way. Collisions are resolved on a last-come first-served (LCFS) basis; i.e., the most recent collision is resolved before any prior collisions.

We assume that new messages are generated according to a Poisson process with aggregate rate λ. Users who have transmitted a message that collided do not generate any new messages until their messages have been transmitted; however, since only a finite number of users are involved in any collision, the rate λ remains constant when the total user population is infinite.

Denote by L_n the number of slots required for resolution of a collision between n users. L_n includes the slot in which the initial collision occurred, plus the times for the two groups of users to transmit their messages. It is easily seen that the following stochastic recursion holds:

$$L_n \stackrel{d}{=} 1 + L_{I_n+X} + \widetilde{L}_{n-I_n+Y}, \quad n \geq 2, \tag{9.2.37}$$

with initial conditions $L_0 = L_1 = 1$. Here $I_n \stackrel{d}{=} B(n,p)$ is the number of users who retransmit immediately, X is the number of new arrivals in the collision slot, and Y is the number of new arrivals during the slot in which the deferred retransmissions occur. Moreover, $L_n \stackrel{d}{=} \widetilde{L}_n$, and the random quantities $X, Y, (L_n)_{n\geq 0}$, $(\widetilde{L}_n)_{n\geq 0}$ are assumed to be mutually independent. For real systems, the total number of users sharing a multiple-access channel might be as large as 10^3 or 10^4, but the number n of users involved in any collision would be a small fraction of this. Fayolle et al. (1985) showed that $\lim_{n\to\infty} EL_n/n$ exists if

$$\log p/\log(1-p) \text{ is irrational;} \tag{9.2.38}$$

otherwise, EL_n/n oscillates around a certain value. In a subsequent paper, Fayolle et al. (1986) proved the linearity of the variance of L_n under (9.2.38) and the finiteness of all moments of L_n. Confirming a conjecture of Massey (1981), Regnier and Jacquet (1989) proved that the variance of L_n is not linear for $I_n \stackrel{d}{=} B(n,p)$, $p = 1/2$, and $X = Y = 0$. In Jacquet and Regnier (1988) and Regnier and Jacquet (1989) the asymptotic normality of the standardized sequence $\{L_n\}$ (for $X = Y = 0$ or both Poisson) was established.

In this section we examine the asymptotic normality of the law of L_n without the specific assumptions on the distribution type of I_n, X, and Y, provided that the variance of L_n is asymptotically linear. In the second part of the section we numerically investigate the influence of nonlinearity in the case $I_n \stackrel{d}{=} B(n,p)$, $X = Y = 0$, and $p = \frac{1}{2}$. It turns out that $E\,L_n/n$ and $(\operatorname{Var} L_n)/n$ increase monotonically with n until n reaches a large value ($n = 39,488$). After that, the linearity breaks down, in agreement with the theoretical results. We consider the simple normalization

$$Y_n = (L_n - \ell_n)/\sqrt{n}, \qquad \text{when} \quad \ell_n = E\,L_n. \tag{9.2.39}$$

The main theoretical result indicates that normality holds if the variance behaves linearly and the numbers of retransmissions are not concentrated too much in the extremes. In this sense the result can be considered as a stability result for the asymptotic distribution.

This idea of stability is confirmed by simulations for some cases of immigration in part (b) of this section. In the numerical study we detect the theoretically predicted instability but only for extremely large n and with a practically negligible order of magnitude. Our simulation study confirms the stability in the standard model concerning dependence on p. Moreover, a simulation study of the empirical d.f. of Y_n confirms the normality for $10^2 \le n \le 10^4$. The "instability" of $E\,L_n/n$ and $\operatorname{Var} L_n/N$ and hence the fluctuation of the limit distribution of Y_n arises for very large values of n, $n \gg 10^4$. The order of magnitude of the instability is seen from our numerical results and simulation study to be extremely small (but existent; in accordance with the theoretical results) and can be neglected from the practical point of view. This has the valuable consequence that in practical applications one can use just simple linear normalizations as in (9.2.39) and the normal approximation also for n "moderately" large, $10^2 \le n \le 10^4$.

(a) *Asymptotic Normality of the Law of* L_n

In this section the asymptotic normality of Y_n is shown under the following assumptions: For some $r \in (2,3]$,

(a) $E\,X^{r/2} + E\,Y^{r/2} < \infty$ and $\dfrac{I_n}{n} \xrightarrow{L^r} p \in (0,1)$;

(b) $\sigma_n^2 = (\operatorname{Var} L_n)/n \to \sigma^2$;

(c) $\sup_n E|Y_n|^r < \infty$ for some $r \in (2,3]$.

Conditions (b), (c) amount to the correctness of the normalization in (9.2.39). Condition (a) implies that the subgroups are not allowed to be extremely large or small. Note that the number of retransmitting users I_n is not necessarily binomial in our assumptions. This allows us, for example, to consider departures from independence in the protocol. Regnier and Jacquet (1989) showed that (a), (b), and (c) hold for $I_n \stackrel{d}{=} B(n,p)$, (9.2.38), and $X = Y = 0$. More generally, one can allow $X \stackrel{d}{=} Y \stackrel{d}{=} \operatorname{Pois}(\lambda)$.

Theorem 9.2.17 *Under (a), (b), and (c), the distribution of* Y_n *is asymptotically* $N(0,\sigma^2)$.

Proof: From the definitions of L_n, Y_n, and (9.2.37), (9.2.39),

$$
\begin{aligned}
Y_n \stackrel{d}{=} & \left(\frac{I_n + X}{n}\right)^{1/2} Y_{I_n+X} \\
& + \left(\frac{n - I_n + Y}{n}\right)^{1/2} \widetilde{Y}_{n-I_n+Y} + C_n(I_n, X, Y).
\end{aligned}
\tag{9.2.40}
$$

Here $\widetilde{Y}_n$ is an independent copy of Y_n, and

$$C_n(k,m,\widetilde{m}) \;:=\; n^{-1/2}(1+\ell_{k+m}+\ell_{n-k+\widetilde{m}}-\ell_n).$$

Define a sequence of normal $N(0,\sigma_n^2)$-distributed independent r.v.s Z_n that are independent of (I_n), X, and Y, and let

$$Z_n^* \;=\; \left(\frac{I_n+X}{n}\right)^{1/2} Z_{I_n+X} + \left(\frac{n-I_n+Y}{n}\right)^{1/2} \widetilde{Z}_{n-I_n+Y} + C_n(I_n,X,Y),$$

where $\widetilde{Z}_n$ is an independent version of Z_n. Z_n^* is an accompanying sequence to Y_n. Let μ_r be one of the following ideal metrics of order $r>0$:

$$\begin{aligned}
\mu_r^{(1)}(X,Y) &= \sup\left\{|E(f(X)-f(Y))|;\ \|f^{(s)}\|_{\widetilde{q}}\le 1\right\}\\
\text{with}\qquad & r \;=\; s+1/\widetilde{p},\ \ s\in\mathbb{N},\ \ \widetilde{p}\in[1,\infty],\ \ \frac{1}{\widetilde{p}}+\frac{1}{\widetilde{q}}=1,\\
\mu_r^{(2)}(X,Y) &= \sup_{t\in\mathbb{R}}|t|^{-r}|E\,e^{it\,X}-E\,e^{it\,Y}|;\\
\mu_r^{(3)}(X,Y) &= \sup_{h\in\mathbb{R}}|h|^r \sup_{A\in\mathcal{B}(\mathbb{R})}|P(X+hN\in A)-P(Y+hN\in A)|,
\end{aligned}$$

where N is a standard normal r.v. independent of X and Y.

Claim 1. (μ_r-closeness of Z_n^* and Y_n) *Set* $a_n = \mu_r(Z_n,Y_n)$ *and suppose* $a \;:=\; \sup_n a_n \;<\; \infty$. *Then*

$$\sup_n \mu_r^{(i)}(Z_n^*,Y_n) \le a[p^{r/2}+(1-p)^{r/2}]. \tag{9.2.41}$$

For $\mu_r=\mu_r^{(i)}$ $(i=1,2,3)$,

$$\begin{aligned}
&\mu_r(Z_n^*,Y_n)\\
&\quad\le \sum_{k,m,\widetilde{m}} P(I_n=k,\ X=m,\ Y=\widetilde{m})\\
&\qquad\quad \mu_r\Bigg(\sqrt{\frac{k+m}{n}}\,Z_{k+m}+\sqrt{\frac{n-k+\widetilde{m}}{n}}\,\widetilde{Z}_{n-k+\widetilde{m}}+c_n(k,m,\widetilde{m}),\\
&\qquad\qquad \sqrt{\frac{k+m}{n}}\,Y_{k+m}+\sqrt{\frac{n-k+\widetilde{m}}{n}}\,\widetilde{Y}_{n-k+\widetilde{m}}+c_n(k,m,\widetilde{m})\Bigg)\\
&\quad\le \sum_{k,m,\widetilde{m}} P(I_n=k,X=m,Y=\widetilde{m})a\left[\left(\frac{k+m}{n}\right)^{r/2}+\left(\frac{n-k+\widetilde{m}}{n}\right)^{r/2}\right]\\
&\quad= a\,E\left[\left(\frac{I_n+X}{n}\right)^{r/2}+\left(\frac{n-I_n+Y}{n}\right)^{r/2}\right].
\end{aligned}$$

Using assumption (a), the right-hand side of the above inequality converges to $a[p^{r/2} + (1-p)^{r/2}]$.

Claim 2. (Condition (9.2.41) holds)

$$a \;\le\; C \sup_n (E|Y_n|^r + E|Z_n|^r) \;<\; \infty. \tag{9.2.42}$$

(Throughout this section, C stands for an absolute constant that can have different values in different places.) For the proof, note that for $i = 1, 2$, or 3,

$$\mu_r^{(i)}(X,Y) \;\le\; C(E|X|^r + E|Y|^r) \;<\; \infty,$$

provided that $E(X^j - Y^j) = 0$ for $j = 1, 2$ (see, for example, Rachev (1991, Chapters 14, 15)). Thus (9.2.42) holds.

Claim 3. (Asymptotic normality of Z_n^*)

For $n \to \infty$, $b_n = \mu_r(Z_n, Z_n^*) \to 0$.

We consider the case $\mu_r = \mu_r^{(1)}$ only. Let κ_r be the rth pseudomoment,

$$\kappa_r(X,Y) \;=\; r \int_{\mathbb{R}} |x|^{r-1} |F_X(x) - F_Y(x)|\, dx.$$

Then, since the mean and variance of Z_n matched those of Z_n^* ($\mu_r(Z_n^*, Y_n) < \infty$ implies $E((Z_n^*)^j - Y_n^j) = 0$, $j = 1, 2$), it follows that $b_n \le C\, \kappa_r(Z_n, Z_n^*)$.

Recall that $(Z_n)_{n\ge 1}$ is independent of $(I_n)_{n\ge 1}$ and X, Y. Let N_0 denote a standard normal r.v. independent of (I_n) and X, Y. Consequently,

$$\begin{aligned}
Z_n^* &= \sqrt{\frac{I_n + X}{n}}\, Z_{I_n+X} + \sqrt{\frac{n - I_n + Y}{n}}\, \widetilde{Z}_{n-I_n+Y} + C_n(I_n, X, Y) \\
&\overset{d}{=} \left(\frac{I_n + X}{n}\sigma^2_{I_n+X} + \frac{n - I_n + Y}{n}\sigma^2_{n-I_n+Y}\right)^{1/2} N_0 + C_n(I_n, X, Y) \\
&=: \eta_n\, N_0 + C_n(I_n, X, Y).
\end{aligned}$$

From assumptions (a), (b) we get the convergence of η_n in probability:

$$\eta_n \;\xrightarrow{P}\; (p\sigma^2 + (1-p)\sigma^2)^{1/2} \;=\; \sigma.$$

Since $Z_n^* = \eta_n\, N_0 + C_n(I_n, X, Y)$ has the same mean and variance as $Z_n \overset{d}{=} \sigma_n\, N_0$, then

$$\begin{aligned}
\sigma_n^2 &= E(\eta_n\, N_0 + C_n(I_n, X, Y))^2 \\
&= E\,\eta_n^2 + E(C_n(I_n, X, Y))^2.
\end{aligned}$$

As $\eta_n \xrightarrow{L^2} \sigma$, we conclude that $C_n(I_n, X, Y) \xrightarrow{P} 0$. This implies that $b_n = \mu_r(Z_n, Z_n^*) \to 0$, as desired in Claim 3.

With $a_n = \mu_r(Z_n, Y_n) \leq \mu_r(Z_n^*, Y_n) + b_n$ and $\overline{a} = \overline{\lim}\, a_n$ we finally obtain from claims 1–3 the following result:

Claim 4. $\overline{a} = 0$.

To prove the claim, choose $n_0 = n_0(\varepsilon)$ ($\varepsilon > 0$) such that $a_k \leq \overline{a} + \varepsilon$ for $k > n_0$. Then for $n \geq n_0$, as in the proof of Claim 1, we have

$$\begin{aligned} a_n &\leq \mu_r(Z_n^*, Y_n) + b_n \\ &\leq \left(\sum_{k=0}^{n_0-1} + \sum_{k=n-n_0}^{n} \right) P(I_n = k) \sup_{\substack{0 \leq k \leq n_0 - 1, \\ n - n_0 \leq k \leq n}} (a_{k+X} + a_{n-k+Y}) \\ &\qquad \times E\left[\left(\frac{k+X}{n} \right)^{r/2} + \left(\frac{n-k+Y}{n} \right)^{r/2} \right] \\ &\quad + \sum_{k=n_0}^{n-n_0-1} P(I_n = k) \\ &\qquad \times E\left[\left(\frac{k+X}{n} \right)^{r/2} (\overline{a} + \varepsilon) + \left(\frac{n-k+Y}{n} \right)^{r/2} (\overline{a} + \varepsilon) \right] + b_n. \end{aligned}$$

Recall Claim 2, $a = \sup_n a_n < \infty$, and thus as $n \to \infty$,

$$\begin{aligned} \overline{a} &\leq \limsup_n \left(\sum_{k=0}^{n_0-1} + \sum_{k=n-n_0}^{n} \right) P(I_n = k) 2a\, E(X^{r/2} + Y^{r/2}) \\ &\quad + (\overline{a} + \varepsilon)(p^{r/2} + (1-p)^{r/2}) + \limsup b_n \\ &= 0 + (\overline{a} + \varepsilon)(p^{r/2} + (1-p)^{r/2}) + 0. \end{aligned}$$

Since $r > 2$, we have $p^{r/2} + (1-p)^{r/2} < 1$, which implies that $\overline{a} = 0$, and thus the proof of the theorem is complete, since μ_r-convergence implies weak convergence. □

Remark 9.2.18 *Theorem 9.2.17 shows a remarkable stability of the central limit theorem for L_n. It says that the central limit theorem can be expected if the variance behaves approximately linearly and that it is even true under protocols that are not based on a binomial number of retransmitting users. In concrete examples it is not easy to obtain the asymptotic behavior of the first moments. Our method of proof separates this problem and establishes a general structural stability property concerning the asymptotic distribution. This should be of some interest for the application of the algorithm, too.*

This stability is not clear or expected from the methods that established the central limit theorem up to now in some very special cases.

(b) *Numerical Results*

In the first part of this section we study numerically the extent of nonlinearity of $EL_n, \operatorname{Var} L_n$ in the special case of (9.2.37) where $X = Y = 0, I_n \stackrel{d}{=} B(n,p), \log p/\log(1-p)$ rational.

Initial investigation of the behavior of the mean ℓ_n of L_n at $p = 0.5$ failed to show the predicted instability of ℓ_n/n. The normalized value ℓ_n/n seemed to converge rapidly, reaching a value of about 2885 for $n = 2400$, and showing no variation out to 7 decimal places with further increase in n. The increments $\ell_n/n - \ell_{n-1}/(n-1)$ were observed always to be positive, another indication of convergence. At $n = 38,488$, a negative increment appears, and subsequently, values of the increment oscillate in a sinusoidal fashion, with a peak magnitude of about 1×10^{-10}. The behavior of the increments is shown graphically in Figure 9.5 on a logarithmic scale.

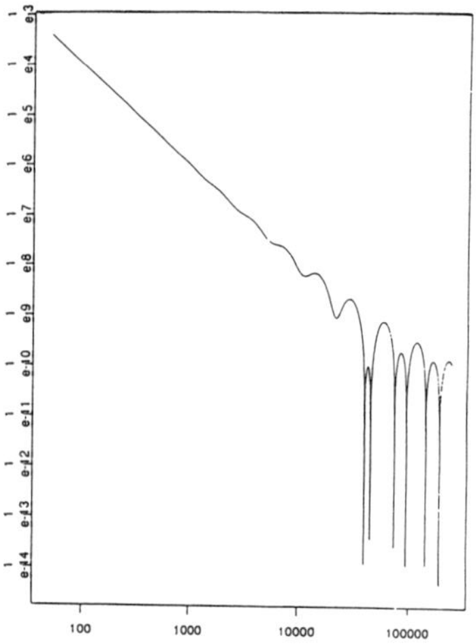

FIGURE 9.5. Increments of ℓ_n/n, $p = 0.5$, $n =$ number of users who initially collide

Based on recursions for the first moments, the numerical results for evaluation of $\ell_n/n, \Delta(\ell_n/n) := \frac{\ell_n}{n} - \frac{\ell_{n-1}}{n-1}, \operatorname{Var}_n := \operatorname{Var}(L_n/\sqrt{n})$, and $\Delta(\operatorname{Var}_n) := \operatorname{Var}_n - \operatorname{Var}_{n-1}$ are shown in Table 9.1.

On the other hand, a change in the initial conditions disturbs the value of ℓ_n/n and Var_n (see Table 9.2).

Table 9.1 confirms the stability of $\frac{\ell_n}{n} \approx 2.88, \operatorname{Var}_n \approx 3.38$ for moderate $n \in (10^2, 10^4)$ and $p = 0.5$. Slight perturbation of p around 0.5 does not change the overall stability of $\frac{\ell_n}{n}$ for practically relevant n; see Figure 9.6.

Summarizing the numerical findings, it appears that for reasonably large $n \geq 100$ and $p = 0.5$ the nonlinearity of ℓ_n/n and Var_n is not observed

TABLE 9.1. numerical results $p = 0.5$, $L_0 = L_1 = 1$

n	ℓ_n/n	$\Delta(\ell_n/n)$	Var_n	$\Delta(\mathrm{Var}_n)$
2	2.5000D+00	1.5000D+00	4.0000D+00	4.000D+00
3	2.5556D+00	5.5556D−02	3.2593D+00	−7.4074D−01
4	2.6310D+00	7.5397D−02	3.3832D+00	1.2396D−01
5	2.6838D+00	5.2857D−02	3.3875D+00	4.2812D−03
10	2.7853D+00	1.0985D−02	3.3832D+00	1.1672D−04
100	2.8754D+00	1.0113D−04	3.3834D+00	9.1046D−07
500	2.8834D+00	4.0528D−06	3.3834D+00	−8.5624D−08
1000	2.8844D+00	1.0224D−06	3.3834D+00	−4.1963D−08
5000	2.8852D+00	3.4639D−08	3.3834D+00	2.1844D−08
10000	2.8853D+00	7.3428D−09	3.3835D+00	−4.1539D−07

TABLE 9.2. $p = 0.5$, $L_0 = L_1 = 0$

n	ℓ_n/n	$\Delta(\ell_n/n)$	Var_n	$\Delta(\mathrm{Var}_n)$
2	1.0000E+00	1.000E+00	1.000E+00	1.000E+00
3	1.1111E+00	1.1111E−01	8.1481E−01	−1.8519E−01
4	1.1905E+00	7.9365E−02	8.4580E−01	3.0990E−02
5	1.2419E+00	5.1429E−02	8.4688E−01	1.0703E−03
10	1.3427E+00	1.1048E−02	8.4579E−01	2.9179E−05
100	1.4327E+00	1.0107E−04	8.4586E−01	2.2762E−07
500	1.4407E+00	4.0304E−06	8.4586E−01	−2.1503E−08
1000	1.4417E+00	1.0117E−06	8.4586E−01	−1.2471E−08

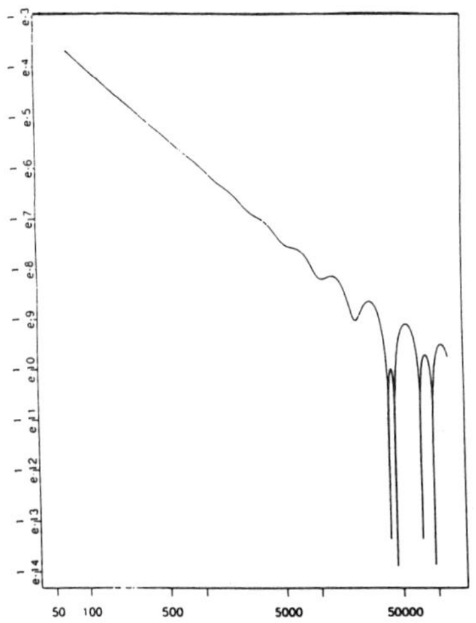

FIGURE 9.6. $|\Delta(\ell_n/n)|$, $p = 0.499$

in a practically relevant magnitude. Also, in this range of values of n the behavior of ℓ_n and Var_n is stable with respect to p. The following simulations (Figures 9.7, 9.8, and 9.9) show a good agreement with the normal approximation for $n \geq 100$.

For $n = 20$ or $n = 30$ the normal fit is no longer good. Further simulation results indicate stability with respect to the value of p.

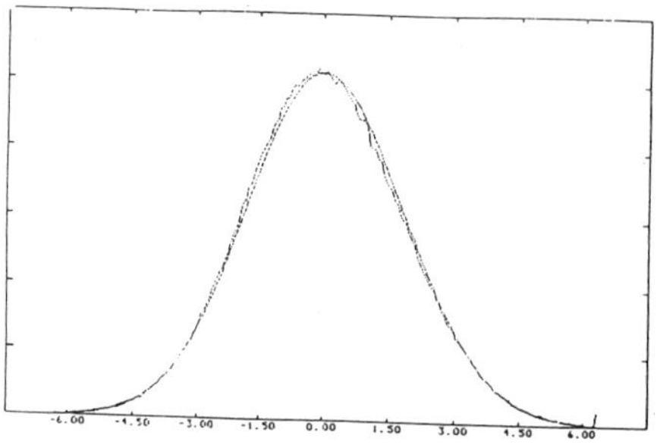

FIGURE 9.7. Simulation curve for $Y_n = (L_n - \ell_n)/\sigma_n$ for $n = 1000$, $p = 0.5$, $L_0 = L_1 = 1$, based on 936,725 trials, and the fitted normal curve with mean zero and variance 3.3834 as given in Table 9.1

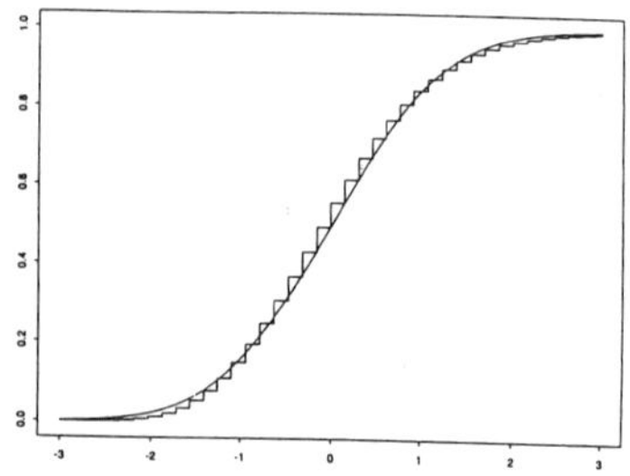

FIGURE 9.8. Normal fit to empirical d.f. with $n = 50, p = 0.5$

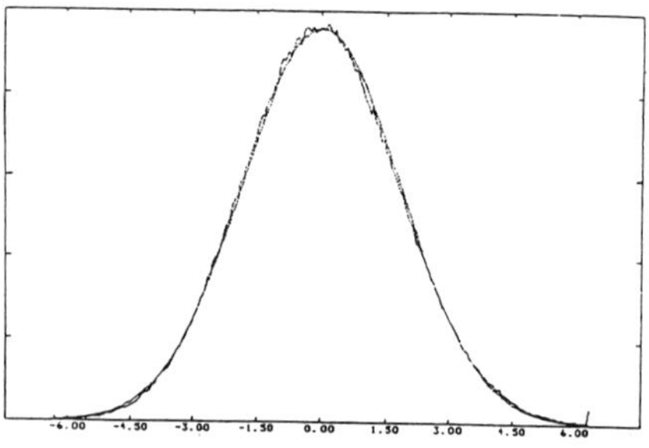

FIGURE 9.9. Normal fit with $\sigma^2 = 3.3874$ to the simulated Y_n's; $n = 1000$, $p = 0.49$, $L_0 = L_1 = 1$ based on 697,675 trials

In the final simulations (Figures 9.10, 9.11) we consider the case with nonzero immigrations X, Y in a symmetric and a nonsymmetric case with masses in 0,1,2.

These examples confirm the general robustness idea that asymptotic normality is approximatively valid if the variances behave approximately lin-

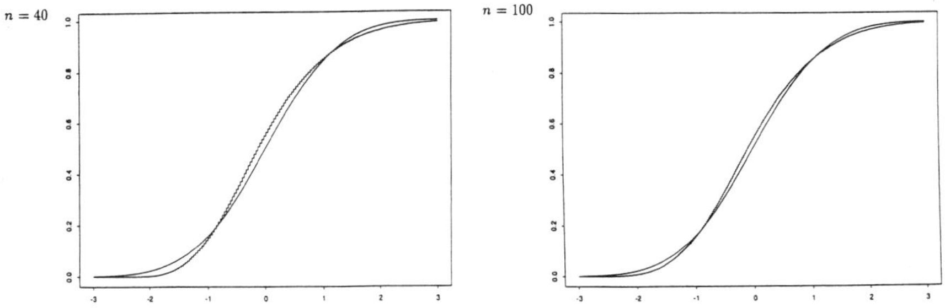

FIGURE 9.10. Normal fit for $n = 40/100$ and $X \sim \frac{3}{4}\delta_0 + \frac{1}{8}\delta_1 + \frac{1}{8}\delta_2, Y \sim \delta_0$, (nonsymmetric case)

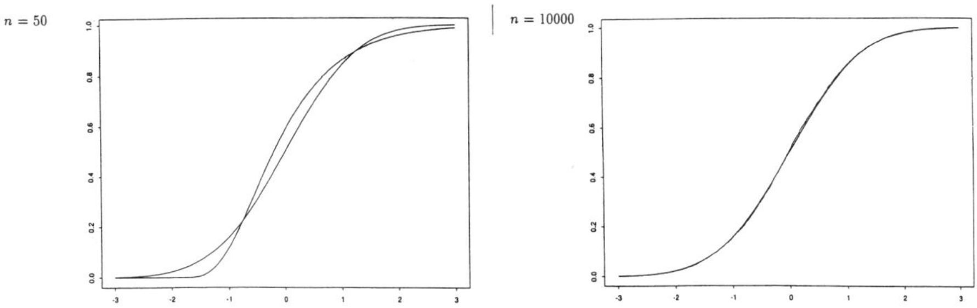

FIGURE 9.11. Normal fit for $n = 50/10000$ and $X \sim Y \sim \delta_0 \frac{3}{16}\delta_1 + \frac{1}{16}\delta_2$ (symmetric case). (Figs. 9.5–9.11): P. Feldman, S.T. Rachev, L. Rüschendorf; *Statistica Neerlandica*, 51:1, 1997; © from Blackwell Publishers)

early (which is observed in these examples empirically).

9.2.4 Quicksort

The quicksort algorithm, which was introduced by C.A.A. Hoare in 1961–1962, represents a standard sorting procedure in computer systems. From a list of n arbitrary (but different) real numbers it selects an element x randomly. Then the remaining numbers are divided into two groups, the group of numbers smaller and that of numbers larger than x. The same procedure is applied to each of these groups if they contain more than one element. The algorithm ends with a sorted list of the original numbers.

If L_n denotes the number of comparisons in the quicksort algorithm on its way to ordering n elements $x_1, \ldots, x_n$, then L_n satisfies the following recursive equation:

$$L_n \stackrel{d}{=} n - 1 + L_{I_n} + \overline{L}_{n-I_n}, \quad L_0 = L_1 = 0, \quad L_2 = 1. \tag{9.2.43}$$

Here I_n, $n - I_n$ are the sizes of the subgroups, and they are assumed to be uniformly distributed on $\{0, \ldots, n-1\}$. The expectation $\ell_n = EL_n$ satisfies then the resursion

$$\ell_n = n - 1 + \sum_{i=0}^{n-1} P(I_n = i)(\ell_i + \ell_{n-i}),$$

and therefore,

$$\frac{\ell_n}{n+1} = \frac{\ell_{n-1}}{n} + \frac{2(n-1)}{n(n+1)} = \sum_{i=1}^{n+1} \frac{1}{i} + \frac{2}{n+1} - 4.$$

This yields

$$\ell_n = 2n \ln n + n(2\gamma - 4) + 2 \ln n + 2\gamma + 1 + O(n^{-1} \ln n), \tag{9.2.44}$$

where $\gamma \approx 0.5772$ is the Euler constant. Similarly, for $v_n = \mathrm{Var}(L_n)$, we have

$$v_n = \left(7 - \frac{2}{3}\pi^2\right) n^2 + o(n^2). \tag{9.2.45}$$

The normalized random sequence

$$Y_n = \frac{L_n - \ell_n}{n} \tag{9.2.46}$$

satisfies the recursion

$$Y_n \stackrel{d}{=} \frac{I_n}{n} Y_{I_n} + \frac{n - I_n}{n} \overline{Y}_{n-I_n} + C_n(I_n), \tag{9.2.47}$$

where

$$C_n(i) = \frac{n-1}{n} E(L_i + L_{n-i} - L_n).$$

As $n \to \infty$, $\frac{I_n}{n}$ converges to some random variable τ that is uniformly distributed on $[0, 1]$. Moreover, $C_n(I_n) = C_n(n\frac{I_n}{n})$ can be uniformly approximated as follows:

$$\sup_{x \in (0,1)} |C_n(\lceil nx \rceil) - C(x)| \le \frac{6}{n} n \ln n + O(n^{-1}). \tag{9.2.48}$$

Here $C(x) = 2x\log x + 2(1-x)\log(1-x) + 1$, and $\lceil x \rceil$ is the smallest integer larger than or equal to x (cf. Rösler (1991)).

As a result we obtain the limiting equation

$$Y \stackrel{d}{=} \tau Y + (1-\tau)\overline{Y} + C(\tau). \tag{9.2.49}$$

In particular, it yields recursive formulas for the moments of Y. Using as an accompanying sequence $\widetilde{Y}_n := \tau Y + (1-\tau)\overline{Y} + C_n(I_n)$, Rösler (1991) established the convergence of Y_n to Y for the ℓ_p-metrics. From Proposition 9.1.3 there exists a unique solution Y (in distribution) of the fixed-point equation (9.2.49).

Theorem 9.2.19 *Let* Y *denote the solution of* (9.2.49). *Then*

$$\ell_p(Y_n, Y) \ \rightarrow \ 0. \tag{9.2.50}$$

The simulation result described on Figure 9.12 shows that the density of Y is very well approximated by a lognormal distribution (cf. Cramer (1996)). The maximal deviation of the fitted lognormal density and the smoothed empirical density is about 0.004.

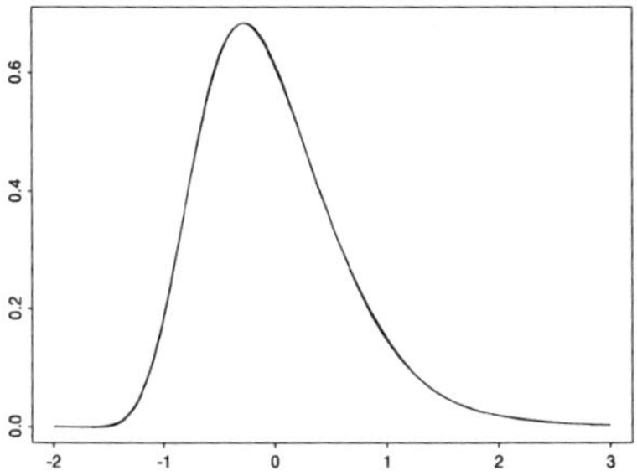

FIGURE 9.12. Smoothed empirical density of quicksort. Simultaneously, a lognormal approximation is given. (M. Cramer; *RAIRO – Informatique Théorique et Applications*, 3, 1996; © from Gauthier-Villars Editeur)

9.2.5 *Limiting Behavior of Random Maxima*

A sample of size n is divided into two parts of random size I_n and $n - I_n$, where I_n is a random variable. We consider the recursion of "maxima type"

$$L_n \stackrel{d}{=} c_n + L_{I_n} \vee \overline{L}_{n-I_n}, \tag{9.2.51}$$

where $L_n, \overline{L}_n$ are independent and identically distributed r.v.s, (I_n) are independent, and (c_n) is a sequence of real numbers.

Given $\alpha > 0$, let us introduce the normalizations

$$Y_n = n^{-1/\alpha} L_n, \quad \overline{Y}_n = n^{-1/\alpha}\overline{L}_n. \tag{9.2.52}$$

By (9.2.51),

$$Y_n \stackrel{d}{=} c_n n^{-1/\alpha} + \left(\frac{I_n}{n}\right)^{1/\alpha} Y_{I_n} \vee \left(\frac{n-I_n}{n}\right)^{1/\alpha} \overline{Y}_{n-I_n}. \qquad (9.2.53)$$

Suppose that $n^{-1/\alpha}$ is the right normalization to obtain the weak convergence results, $Y_n \xrightarrow{\mathcal{D}} Z$, $\overline{Y}_n \xrightarrow{\mathcal{D}} \overline{Z}$, and moreover, let $\frac{I_n}{n} \to \tau$, where τ a random variable independent of Z, $\overline{Z}$. Then, in the limit, we obtain the fixed-point equation

$$Z \stackrel{d}{=} \tau^{1/\alpha} Z \vee (1-\tau)^{1/\alpha} \overline{Z}. \qquad (9.2.54)$$

It is easy to check that, for example, the extreme value distribution

$$F_Z(x) = \begin{cases} e^{-x^{-\alpha}}, & x > 0, \\ 0, & x \le 0 \end{cases} \qquad (9.2.55)$$

satisfies (9.2.53).

As a motivation for the study of equation (9.3.46), consider $L_n = \max\{X_1, \dots, X_n\}$, $c_n = 0$, and assume that (X_i) are i.i.d. r.v.s of Paretian type ($F(x) \sim 1 - x^{-\alpha}$ for $x \to \infty$). Then by Gnedenko's extreme-value theorem, $Y_n = n^{-1/\alpha} L_n \xrightarrow{\mathcal{D}} Z$, with F_Z as in (9.3.50). Note also that formula (9.3.46) concerns some modifications of this recursion, where the maxima are produced by a (random) scheme determined by I_n (for example $I_n \stackrel{d}{=} \mathcal{B}(n,p)$) and c_n corresponding to some weighting of the number of steps in this reduction (for example $c_n = 1$). Furthermore, note that (9.2.51) with $c_n = 1$ also describes the maximum search length of a search algorithm dividing a slot of size n succesively into two parts of size I_n and $n - I_n$, respectively.

Define next $a_k := \ell_r(Y_k, Z)$ for $0 < \alpha < r \le 1$ or $a_k := \varrho_r(Y_k, Z)$, ϱ_r the weighted Kolmogorov metric (cf. (9.1.16)) for $1 \le \alpha < r < \alpha + 1$, and consider the following assumptions:

$$\begin{aligned} \overline{\lim}\, a_k &< \infty, \\ c_n n^{-1/\alpha} &\to 0, \\ \frac{I_n}{n} &\to \tau \quad \text{a.s. with} \quad E\tau^{r/\alpha} + (1-\tau)^{r/\alpha} < 1. \end{aligned} \qquad (9.2.56)$$

The first assumption corresponds to the condition that $n^{-1/\alpha}$ is the right normalization for L_n (as for example in the case $L_n = \max(X_1, \dots, X_n)$). If $I_n \stackrel{d}{=} \mathcal{B}(n,p)$, then $\frac{I_n}{n} \to p$, and for $\alpha < r$ we have $p^{r/\alpha} + (1-p)^{r/\alpha} < 1$.

Theorem 9.2.20 *Let* (L_n) *satisfy the recursion* (9.2.51). *Define* $a_k := \ell_r(Y_k, Z)$ *if* $0 < \alpha < r \leq 1$, *or* $a_k := \varrho_r(Y_k, Z)$ *if* $0 \leq \alpha < r < \alpha + 1$, *and let* F_Z *be as in* (9.2.55). *Then assumption* (9.2.56) *implies*

$$\lim_k a_k = 0.$$

Proof: We consider first the case $0 < \alpha < r \leq 1$ and $a_k = \ell_r(Y_k, Z)$.

Let (Z_i) be an i.i.d. sequence with common extreme-value distribution (9.2.55), and define

$$Z_n^* = n^{-1/\alpha} c_n + \left(\frac{I_n}{n}\right)^{1/\alpha} Z_{I_n} \vee \left(\frac{n - I_n}{n}\right)^{1/\alpha} \overline{Z}_{n-I_n}. \tag{9.2.57}$$

Then

$$\begin{aligned} \ell_r(Y_n, Z_n^*) &= \ell_r\left(\left(\frac{I_n}{n}\right)^{1/\alpha} Y_{I_n} \vee \left(\frac{n-I_n}{n}\right)^{1/\alpha} \overline{Y}_{n-I_n}, \right. \qquad (9.2.58) \\ &\qquad \left. \left(\frac{I_n}{n}\right)^{1/\alpha} Z_{I_n} \vee \left(\frac{n-I_n}{n}\right)^{1/\alpha} \overline{Z}_{n-I_n}\right) \\ &\leq \sum_{k=1}^n P(I_n = k) \left[\ell_r\left(\left(\frac{k}{n}\right)^{1/\alpha} Y_k \vee \left(\frac{n-k}{n}\right)^{1/\alpha} \overline{Y}_{n-k}, \right.\right. \\ &\qquad \left.\left. \left(\frac{k}{n}\right)^{1/\alpha} Z_k \vee \left(\frac{n-k}{n}\right)^{1/\alpha} \overline{Z}_{n-k}\right)\right] \\ &\leq \sum_{k=1}^n P(I_n = k) \left[\left(\frac{k}{n}\right)^{r/\alpha} a_k + \left(\frac{n-k}{n}\right)^{r/\alpha} a_{n-k}\right] \\ &= E\left(\frac{I_n}{n}\right)^{r/\alpha} a_{I_n} + E\left(\frac{n-I_n}{n}\right)^{r/\alpha} a_{n-I_n}. \end{aligned}$$

The arguments in deriving the above bounds rely on the "ideality" of ℓ_r with respect to the maxima scheme.(1)

Define $b_n := \ell_r(Z_n^*, Z_n)$ and let us use the bound

$$\ell_r(X, Y) \leq (\ell_1(X, Y))^r \tag{9.2.59}$$

---

(1) Recall that a metric $\mu(X, Y) = \mu(F_X, F_Y)$ in the space of distribution functions is called *ideal* with respect to the maxima scheme (or max-*ideal*) of order $r > 0$ if for any $c > 0$ and independent X, Y, and Z,

$$\mu(cX \vee Z, cY \vee Z) \leq c^r \mu(X, Z);$$

see Rachev (1991) and Rachev and Rüschendorf (1991).

for any $r \le 1$ that is a single consequence of the Monge–Kantorovich theorem; recall that $\ell_1(X,Y) = \int |F_X(x) - F_Y(x)|\,dx$.

Claim 1. $b_n \to 0$.

To show the claim, we apply (9.2.57) to obtain

$$\begin{aligned} b_n^{1/r} &\le \ell_1\left(n^{-1/\alpha}c_n + \left(\frac{I_n}{n}\right)^{1/\alpha} Z_{I_n} \vee \left(\frac{n-I_n}{n}\right)^{1/\alpha} \overline{Z}_{n-I_n}, Z_n\right) \\ &\le \sum_{k=1}^{n} P(I_n = k)\ell_1\left(n^{-1/\alpha}c_n + \left(\frac{k}{n}\right)^{1/\alpha} Z_k \vee \left(\frac{n-k}{n}\right)^{1/\alpha} \overline{Z}_{n-k}, Z_n\right) \\ &= \sum_{k=1}^{n} P(I_n = k)\ell_1\left(c_n n^{-1/\alpha} + \left(\frac{k}{n} + \frac{n-k}{n}\right)^{1/\alpha} Z, Z\right). \end{aligned}$$

In the above bound we have used that the extreme value distributions Z_n satisfy the max-stability property:

$$\left(\frac{k}{n}\right)^{1/\alpha} Z_k \vee \left(\frac{n-k}{n}\right)^{1/\alpha} \overline{Z}_{n-k} \stackrel{d}{=} \left(\frac{k}{n} + \frac{n-k}{n}\right)^{1/\alpha} Z = Z.$$

Therefore, $b_n^{1/r} \le c_n n^{-1/\alpha} \to 0$, proving the claim.

Applying the triangle inequality and (9.2.58), (9.2.59), we have

$$a_n \le \ell_r(Y_n, Z_n^*) + \ell_r(Z_n^*, Z_n), \tag{9.2.60}$$

and therefore,

$$\overline{a} \le \overline{a}E(\tau^{r/\alpha} + (1-\tau)^{r/\alpha}) + 0, \quad \text{with } \overline{a} := \overline{\lim}\, a_n, \tag{9.2.61}$$

implying that $\overline{a} = 0$.

Next, we shall make use of the weighted Kolmogorov metric ϱ_r (cf. (9.1.16)). It is easy to check that for $\varepsilon \ge 1$ and $X, Y \ge 0$,

$$\varrho_r(X + a, Y + \varepsilon) \le (\varepsilon a)^r + \varepsilon^r \varrho_r(X,Y). \tag{9.2.62}$$

Define then $a_k = \varrho_r(Y_k, Z), b_k = \varrho_r(Z_k^*, Z_k)$. By (9.2.62) (with $\varepsilon = 1$),

$$\begin{aligned} \varrho_r(Y_n, Z_n^*) &\le c_n^r n^{-r/\alpha} + \varrho_r\left(\left(\frac{I_n}{n}\right)^{1/\alpha} Z_{I_n} \vee \left(\frac{n-I_n}{n}\right)^{1/\alpha} \overline{Z}_{n-I_n},\right. \\ &\qquad \left.\left(\frac{I_n}{n}\right)^{1/\alpha} Y_{I_n} \vee \left(\frac{n-I_n}{n}\right)^{1/\alpha} \overline{Y}_{n-I_n}\right) \\ &\le c_n^r n^{-r/\alpha} + E\left(\frac{I_n}{n}\right)^{r/\alpha} a_{I_n} + E\left(\frac{n-I_n}{n}\right)^{r/\alpha} a_{n-I_n}. \end{aligned}$$

This implies that

$$\overline{\lim}\, \varrho_r(Y_n, Z_n^*) \;\leq\; E(\tau^{r/\alpha} + (1-\tau)^{r/\alpha})\overline{a} \quad \text{for } r > \alpha. \tag{9.2.63}$$

If $\alpha < r \leq \alpha + 1$, then

$$\begin{aligned}\varrho_r(Z_n, Z_n^*) &\leq \sum_{k=1}^{n} P(I_n = k)\varrho_r(c_n n^{-1/\alpha} + Z_n, Z_n)\\ &= \varrho_r(c_n n^{-1/\alpha} + Z, Z).\end{aligned}$$

We next prove that $\varrho_r(a + Z, Z) \to 0$ for $a \to 0$. Let

$$\chi(x,a) \;:=\; x^r|F_Z(x) - F_{Z+a}(x)| \;=\; x^r|e^{-x^{-\alpha}} - e^{-(x-a)^{-\alpha}}|.$$

Then $\sup_{0\leq x\leq a}\chi(x,a) \leq a^r$, $\sup_{a\leq x\leq 2a}\chi(x,a) \leq (2a)^r e^{-a^{-\alpha}}$. Furthermore,

$$\begin{aligned}\sup_{2a\leq x\leq 1} x^r|e^{-x^{-\alpha}} - e^{-(x-a)^{\alpha}}| &= \sup_{2a\leq x\leq 1} x^r\alpha\int_{x-a}^{x} y^{-\alpha-1}e^{-y^{-\alpha}}\,dy\\ &\leq \sup_{2a\leq x\leq 1} x^r\alpha a\frac{1}{(x-a)^{\alpha+1}}e^{-x^{-\alpha}}\\ &\leq \sup_{2a\leq x\leq 1}\left(\frac{x}{x-a}\right)^r\frac{\alpha a}{(x-a)^{\alpha+1-r}}e^{-1}\\ &\leq 2^r\alpha a^{r-\alpha}e^{-1}\end{aligned}$$

and

$$\begin{aligned}\sup_{1\leq x<\infty} x^r\alpha\int_{x-a}^{x} y^{-\alpha-1}e^{-y^{-\alpha}}\,dy &\leq \sup_{1\leq x<\infty}\frac{x^{\alpha+1}}{(x-y)^{\alpha+1}}\alpha a\, e^{-x^{-\alpha}}\\ &\leq 2^{\alpha+1}\alpha a.\end{aligned}$$

Combining all this we obtain $\varrho_r(a + Z, Z) \leq C\,a^{r-\alpha}$ for $\alpha < r \leq \alpha + 1$. (Note that $\sup_x \chi(x,u) = \infty$ for $r > \alpha + 1$.) Applying again the triangle inequality, we finally obtain

$$\begin{aligned}\overline{a} \;=\; \overline{\lim}\,\varrho_r(Y_n, Z) &\leq \overline{\lim}\,\varrho_r(Y_n, Z_n^*) + \overline{\lim}\, b_n\\ &\leq E(\tau^{r/\alpha} + (1-\tau)^{r/\alpha})\overline{a},\end{aligned}$$

which indeed implies $\overline{a} = 0$. □

A similar study of logarithmic normalizations for max-search algorithms is provided by Cramer (1995a).

9.2.6 Random Recursion Arising in Probabilistic Modeling: Limit Laws

In this and the next section we study various random recursions arising in probabilistic modeling. In 9.2.6 we shall discuss the limiting behavior of these recursions and describe the limit distributions, and in the next section we estimate the rate of convergence to the corresponding limit.(2)

Let $\{(Y_n, Z_n)\}_{n\geq 1}$ be a sequence of i.i.d. random vectors in $\mathbb{R}^2$. Define the random recursion (S_n^*) by

$$S_n^* = S_{n-1}^* Z_n + Y_n Z_n, \quad n = 1, 2, \ldots; \; S_0^* = 0. \tag{9.2.64}$$

The processes $\{S_n\}_{n\geq 1}$ and $\{S_n^*\}_{n\geq 1}$ have appeared in a variety of situations. The random recursion (9.2.64) is often written in an equivalent form,

$$S_n^* = S_{n-1}^* A_n + B_n, \quad n = 1, 2, \ldots, \tag{9.2.65}$$

for a sequence of i.i.d. random vectors $\{(A_n, B_n)\}_{n\geq 1}$.(3) Alternatively to (9.2.64) we can introduce the process(4)

$$S_n = \sum_{i=1}^{n} Y_i \prod_{j=1}^{i} Z_j, \quad n = 1, 2, \ldots. \tag{9.2.66}$$

---

(2) The results of Sections 9.2.6 and 9.2.7 are due to Rachev and Samorodnitsky (1995); see also the references therein.

(3) Typically, the Markov chain (9.2.65) is supplied with an initial state $S_0^* = B_0$ for a random variable B_0 independent of the sequence $\{(A_n, B_n)\}_{n\geq 1}$. In the ergodic case the limit distribution of S_n^* is, of course, independent of the distribution of B_0. The recursion (9.2.65) arises in many applications, and as pointed out by Vervaat (1979), S_n^* can be regarded as the "wealth" at time n; A_n is the relative change of the "wealth" between times n and $n-1$ due to a "quality" change in the "environment": inflation, change of an exchange rate, erosion, spoilage, decay, etc. B_n represents the added (or removed) "wealth" just prior to time n. Applications of this model are abundant in the literature. Uppuluri et al. (1967) and Paulson and Uppuluri (1972) used this model to represent the evolution of a stock of a radioactive material. Chandrasekhar and Munch (1950) studied the fluctuations in brightness of the Milky Way. Cavalli-Sforza and Feldman (1973) and Cavalli-Sforza (1975) modeled evolution and cultural inheritance. Applications to investment models can be found in Lassner (1974a) and Perrakis and Henin (1974). A particular subclass of random recursions (9.2.65), the so-called ARCH (*autoregressive conditional heteroskedastic*) processes, has been used in mathematical finance to model data exhibiting clusters; see Domowitz and Hakkio (1985) and Hsieh (1988) for modeling exchange rate yields, and Engle et al. (1987) and Bollerslev (1987) for modeling stock returns.

(4) The stochastic process (9.2.66) has been used by Todorovic and Gani (1987) and Todorovic (1987) to model the effect of environmental changes on crop production; see also Puri (1987).

It is obvious that the two models (9.2.66) and (9.2.64) are closely related. The stochastic processes $\{S_n\}_{n\geq 1}$ and $\{S_n^*\}_{n\geq 1}$, although not equal, in general, in finite-dimensional distributions, have equal two-dimensional distributions: More precisely, $(S_n, S_{n+1}) \stackrel{d}{=} (S_n^*, S_{n+1}^*)$ for each $n = 1, 2, \ldots$. However, we may have $(S_n, S_{n+1}, S_{n+2}) \stackrel{d}{\neq} (S_n^*, S_{n+1}^*, S_{n+2}^*)$.[5]

A related pair of processes can be defined by replacing the sum with the maximum:

$$
\begin{aligned}
M_n^* &= \max(M_{n-1}^* Z_n, Y_n Z_n); \quad M_0^* = 0; &(9.2.67)\\
M_n &= \bigvee_{i=1}^{n} Y_i \prod_{j=1}^{i} Z_j, \quad n = 1, 2, \ldots .
\end{aligned}
$$

These models can be regarded as describing the evolution of the highest up-to-date adjusted change in the "wealth" associated with the summation models (9.2.64)–(9.2.66).[6]

Further, we prove limit theorems for the processes $\{S_n\}_{n\geq 1}$ and $\{M_n\}_{n\geq 1}$ stopped at random times. Thinking in terms of "wealth" and "environment" changes described above, suppose that in each time period a disastrous event may occur with probability $p \in (0, 1)$. As a result of the disastrous event (bankruptcy, drought, etc., depending on the application) the whole wealth could be lost. The time of the disastrous event $\tau = \tau(p)$ is assumed to be geometrically distributed,

$$
P(\tau(p) = k) = (1-p)p^{k-1}, \quad k = 1, 2, \ldots, \qquad (9.2.68)
$$

and independent of the sequence $\{(Y_n, Z_n)\}_{n\geq 1}$. We will discuss the limiting behavior of the total "wealth" (until time τ), S_τ, as $p \to 0$.

---

[5] Typically, of interest have been conditions for ergodicity of the Markov chain (9.2.65) and characterization of the limiting distribution. The key reference here is Vervaat (1979). An earlier work of Kesten (1973) studies the multidimensional version of (9.2.65), i.e., S_n^*'s and B_n's are d-dimensional (random) vectors and A_n's are $d \times d$ (random) matrices. This level of generality allows one, for example, to treat one-dimensional recursions of a higher order. It turns out, in particular, that under certain moment conditions on (A_n, B_n) in (9.2.65), the stationary distribution of S_n^* has Pareto-like tails. This phenomenon has been studied further by Goldie (1991) in the case of more general recursions.

[6] For applications of these and related models see Helland and Nilsen (1976), Hooghiemstra and Keane (1985), and Hooghiemstra and Scheffer (1986). An interesting time-reversibility relation between special cases of the models (9.2.64) and (9.3.68) has been noted by Chernick et al. (1988). Extrema of the processes arising in the random recursion (9.2.65) and, in particular, of ARCH processes are studied in de Haan et al. (1989). As general references on random recursions see Letac (1986) and the introduction of Kifer (1986). Brandt et al. (1990, Chapter 9) establishes a continuous dependence of the stationary distribution of the Markov chain (9.2.65) on certain parameters of the recursion.

The disastrous event could be caused by the cumulative effect of a large number of "bad" events with high success probability, $p \approx 1$. This leads to a negative binomial model for the time of the disastrous event $\tau = \tau_1 + \cdots + \tau_r$, where the τ_i's are i.i.d. geometric with mean $1/p$. For $p \to 1$, $r \to \infty$, $r(1-p) \to \lambda > 0$ we have the Poisson approximation $P(\tau = n + r) \approx e^{-\lambda}\lambda^n/n!$. This, in turn, leads to a Poisson model for the time of the disastrous event. Assuming that the N_i are i.i.d. Poiss(λ) r.v.s, $T_i = N_1 + \cdots + N_i$ is viewed as the time of the ith disastrous event. We shall study in the framework of the model (9.2.66) the distributions of the sums

$$S_{T_1} = \sum_{i=0}^{T_1} Y_i \prod_{j=0}^{i} Z_j, \tag{9.2.69}$$

and

$$S_{T_{k+1}} - S_{T_k} = \sum_{i=T_k+1}^{T_{k+1}} Y_i \prod_{j=0}^{i} Z_j, \qquad k = 1, 2, \ldots .$$

Here for the sake of convenience we start the sequence $\{(Y_n, Z_n)\}$ at $n = 0$.

Similarly to (9.2.69) we shall be interested in the laws of

$$M_\tau = \bigvee_{i=1}^{\tau} Y_i \prod_{j=1}^{i} Z_j \tag{9.2.70}$$

and

$$M_{T_1} = \bigvee_{i=0}^{T_1} Y_i \prod_{j=0}^{i} Z_j$$

and

$$\bigvee_{i=T_k+1}^{T_{k+1}} Y_i \prod_{j=0}^{i} Z_j, \qquad k = 1, 2, \ldots .$$

Note that if S_n (or S_n^*) converges in distribution, then the limiting (in distribution) random variable S satisfies the equation

$$S \stackrel{d}{=} (S + Y)Z. \tag{9.2.71}$$

Here $(Y, Z) \stackrel{d}{=} (Y_n, Z_n)$, and the random quantities S and (Y, Z) in the right-hand side of (9.2.71) are independent. In many cases the solution of the equation (9.2.71) turns out to be an infinitely divisible random variable.

Similarily, the distributional limit M of M_n satisfies

$$M \stackrel{d}{=} (M \vee Y)Z. \tag{9.2.72}$$

It is interesting to note that the total "wealth" until the disastrous geometrical event S_τ also satisfies a distributional equation:

$$S_\tau \stackrel{d}{=} (Y + \delta S_\tau)Z. \tag{9.2.73}$$

Here, as before, $(Y, Z) \stackrel{d}{=} (Y_n, Z_n)$, δ is Bernoulli with $P(\delta = 1) = 1-p$); S, δ, and (Y, Z) in the right-hand side of (9.2.73) are independent.[7] Similarily to (9.2.73),

$$M_\tau \stackrel{d}{=} (Y \vee \delta M_\tau)Z; \tag{9.2.74}$$

if $Z \equiv 1$, M_τ is said to be *max-geometric infinitely divisible.*[8]

We start with results on the limiting behavior of the recursions defined above. In the next five theorems $\{(Y_n, Z_n)\}_{n\geq 1}$ is, unless explicitely stated otherwise, a sequence of nonnegative i.i.d. random vectors such that $P(Y_n > 0) > 0$ and $P(Z_n > 0) = 1$. Set S_n and M_n as in (9.2.66) and (9.2.67), and let

$$X_n = Y_n \prod_{j=1}^{n} Z_j.$$

Set $\xi_n = \log Z_n$, $\nu = E\xi_n$ (when they exist).

Lemma 9.2.21 *Let $\{(Y_n, Z_n)\}_{n\geq 1}$ be a sequence of random vectors living on a common probability space such that $\{Y_n\}_{n\geq 1}$ is a sequence of nonnegative i.i.d. random variables with $P(Y_n > 0) > 0$, and $\{Z_n\}_{n\geq 1}$ is a sequence of positive i.i.d. random variables. Suppose that $E\log(1 + Z_n) < \infty$.*

(a) *If $\nu > 0$, then with probability 1, X_n does not converge to 0, and thus $S_n \to \infty$. The same is true in the case $\nu = 0$, provided that the sequence $\{Y_n, Z_n\}_{n\geq 1}$ is a sequence of i.i.d. random vectors. Moreover, in both cases $M_n \to \infty$ (unless $P(Z_n = 1) = 1$).*

(b) *If $-\infty < \nu < 0$, the following are equivalent as $n \to \infty$:*

 (b-i) $X_n \to 0$ *a.s..*

 (b-ii) S_n *converges to a finite limit S a.s..*

---

[7] See Rachev and Todorovich (1990) for some examples of distributions of S_T; if $Z \equiv 1$, S_T is said to be geometrically infinitely divisible; see Klebanov, Maniya, and Melamed (1984).

[8] See Rachev and Resnick (1991).

(b-iii) *M_n converges to a finite limit a.s.*

(b-iv) $0 < \mathrm{E}\log(1+Y_n) < \infty$.

Moreover, (b-iv) *implies* (b-i)–(b-iii) *even if* $\nu = -\infty$.

The proofs of this and the further assertions in this section can be found in Rachev and Samorodnitsky (1995).

Remark 9.2.22 *Given a sequence of nonnegative i.i.d. random vectors*

$$(Y_n, Z_n) = \left(Y_n^{(1)}, \dots, Y_n^{(d)}, Z_n^{(1)}, \dots, Z_n^{(d)}\right) \in \mathbb{R}^{2d},$$

we consider the vector of "wealths" $S_n = (S_n^{(1)}, \dots, S_n^{(d)})$ *given by*

$$S_n^{(k)} = \sum_{i=1}^{n} Y_i^{(k)} \prod_{j=1}^{i} Z_j^{(k)}, \quad n = 1, 2, \dots. \tag{9.2.75}$$

Then Lemma 9.2.21 *applied componentwise yields convergence.*

Our next theorem is the CLT for the "total wealth" S_n in (9.2.66). We assume that $\xi_n = \log Z_n$ belongs to the domain of attraction of an α-stable r.v. η_α $(1 < \alpha \le 2)$; i.e., there exist $a_n > 0$ and $b_n \in \mathbb{R}$ such that

$$a_n \sum_{i=1}^{n} \xi_i + b_n \stackrel{D}{\Longrightarrow} \eta_\alpha; \qquad a_n = n^{-1/\alpha} L(n), \tag{9.2.76}$$

where $L(n)$ is a slowly varying function.

Theorem 9.2.23 *Suppose that* $E\log(1+Z_n) < \infty$.

(a) *If* $\nu > 0$ *and* $E\log(1+Y_n) < \infty$, *then* $a_n \log S_n + b_n \stackrel{D}{\Longrightarrow} \eta_\alpha$.

(b) *If* $\nu < 0$ *and* $E\log(1+Y_n) < \infty$, *then* $a_n \log(S - S_n) + b_n \stackrel{D}{\Longrightarrow} \eta_\alpha$.

(c) *Let* $\nu = 0$, *and assume (without loss of generality) that* $b_n \equiv 0$. *Suppose also that the sequences* $\{Y_n\}_{n\ge1}$ *and* $\{Z_n\}_{n\ge1}$ *are independent and that*

$$P(\log Y_1 > 1/a_n) = o(n^{-1}), \quad n \to \infty. \tag{9.2.77}$$

Then, as $n \to \infty$,

$$a_n \log S_n \stackrel{D}{\Longrightarrow} \sup_{0\le t\le 1} \mathcal{L}(t), \tag{9.2.78}$$

where $\mathcal{L}$ *is a Levy stable motion on* $[0,1]$ *with* $\mathcal{L}(1) \stackrel{d}{=} \eta_\alpha$.

Remark 9.2.24 *The results of Theorem 9.2.23 can be extended both to the multivariate setting and to the form of a functional CLT. We give just one example of such an extension, which is obtainable using Theorem 1 of Resnick and Greenwood (1979).*

In the notation of Remark 9.2.22 let $d = 2$ and set $\xi_n = \left(\xi_n^{(1)}, \xi_n^{(2)}\right) = \left(\log Z_n^{(1)}, \log Z_n^{(2)}\right)$. Assume that there exists $a_n \in \mathbb{R}_+^2$ and $b_n \in \mathbb{R}^2$ such that

$$\left(a_n^{(1)}\xi_n^{(1)}, a_n^{(2)}\xi_n^{(2)}\right) + b_n \stackrel{D}{\Longrightarrow} \eta_\alpha.$$

Here $\alpha = (\alpha_1, \alpha_2)$, $1 < \alpha_i \le 2$, $i = 1, 2$ and $a_n^{(i)} = n^{-1/\alpha_i} L_i(n)$, $i = 1, 2$, where the L_i's are slowly varying functions. If $E\log(1 + Z_n^{(1)}) < \infty$ and $\nu^{(i)} := E \log Z_n^{(i)} > 0$ for $i = 1, 2$, then

$$\left\{\left(a_n^{(1)} \log S_{[nt]}^{(1)}, a_n^{(2)} \log S_{[nt]}^{(2)}\right) + b_n\right\}_{t\ge 0} \stackrel{D}{\Longrightarrow} \{\mathcal{L}(t)\}_{t\ge 0};$$

the weak convergence is in the space $D\left([0,\infty), \mathbb{R}^2\right)$. $\{\mathcal{L}(t) = \left(\mathcal{L}^{(1)}(t), \mathcal{L}^{(2)}(t)\right), t \ge 0\}$ is a Lévy process with $\mathcal{L}(1) \stackrel{d}{=} \eta_\alpha$ and such that

$$\{\mathcal{L}(t), t \ge 0\} \stackrel{d}{=} \left\{\left(t^{1/\alpha_1}\mathcal{L}^{(1)}(1), t^{1/\alpha_2}\mathcal{L}^{(2)}(1)\right) + \beta(t),\ t \ge 0\right\}$$

for some $\beta(t) \in \mathbb{R}^2$ prescribed by the marginal convergence. Moreover,

(i) if $(\alpha_1, \alpha_2) = (2, 2)$, then $\mathcal{L}$ is an $\mathbb{R}^2$-valued Wiener process;

(ii) if $(\alpha_1, \alpha_2) = (\alpha, 2), 1 < \alpha < 2$, then $\{\mathcal{L}(t) = (\mathcal{L}^{(1)}(t), \mathcal{L}^{(2)}(t)), t \ge 0\}$, where $\mathcal{L}^{(1)}$ is an α-stable process and $\mathcal{L}^{(2)}$ is a Wiener process independent of $\mathcal{L}^{(1)}$;

(iii) if $1 < \alpha_i < 2, i = 1, 2$, then $\mathcal{L}$ has Lévy measure ℓ defined by $\ell \circ T = \widetilde{\ell}$, where

$$TX = \left((\operatorname{sign} x_1)|x_1|^{1/\alpha_1}, (\operatorname{sign} x_2)|x_2|^{1/\alpha_2}\right).$$

The measure $\widetilde{\ell}$ is determined by

$$\widetilde{\ell}\{x \in \mathbb{R}^2; |x| > r, \theta(x) \in H\} = r^{-1}S(H)$$

for $r > 0$; H is a Borel subset of $[0, 2\pi]$, where $|x|$ and $\theta(x)$ are the polar coordinates of $x \in \mathbb{R}^2$, and S is a finite Borel measure on $[0, 2\pi]$.[(9)]

[(9)] A more detailed analysis of $\mathcal{L}$ can be obtained using further the results of Resnick and Greenwood (1979) and de Haan et al. (1984). An even more general case where a_n

Propositions (b) and (c) of Theorem 9.2.23 can be extended in a similar fashion.

As far as the maximal "wealth change" (9.2.65) for n years is concerned, we have the following analogue of Theorem 9.2.23.(10)

Theorem 9.2.25 *Under the assumptions of Theorem* 9.2.23 *the following hold:*

(a) *If* $\nu > 0$, *then* $a_n \log M_n + b_n \overset{D}{\Longrightarrow} \eta_\alpha$.

(b) *If* $\nu < 0$, *then* $a_n \log(\vee_{j>n} X_j) + b_n \overset{D}{\Longrightarrow} \eta_\alpha$.

(c) *If* $\nu = 0$ *and* (9.2.78) *hold, then* $a_n \log M_n \overset{D}{\Longrightarrow} \sup_{0\le t\le 1} \mathcal{L}(t)$.

Next, we examine the geometric random sum S_τ as defined above. We say that $\xi_n = \log Z_n$ belongs to the domain of attraction of a *geometric* α*-stable r.v.* G_α if there exist functions $a = a(p) > 0$ and $b = b(p)$ on $[0, 1]$ such that

$$a \sum_{i=1}^{\tau} (\xi_i + b) \overset{D}{\Longrightarrow} G_\alpha \quad \text{as } p \to 0. \tag{9.2.79}$$

Here $a(p) = p^{1/\alpha} L(1/p)$, where L is slowly varying function.(11)

Theorem 9.2.26 *Suppose that* $E \log(1 + Z_n) < \infty$ *and* (9.2.79) *holds.*

(a) *If* $\nu > 0$ *and* $E \log(1+Y_n) < \infty$, *then* $a(\log S_\tau + \tau b) \overset{D}{\Longrightarrow} G_\alpha$ *as* $p \to 0$.

(b) *If* $\nu < 0$ *and* $E \log(1+Y_n) < \infty$, *then* $a(\log \sum_{j\ge\tau+1} X_j + \tau b) \overset{D}{\Longrightarrow} G_\alpha$ *as* $p \to 0$.

(c) *Let* $\nu = 0$ *and* $b \equiv 0$. *Assume also that the sequences* $\{Y_n\}_{n\ge1}$ *and* $\{Z_n\}_{n\ge1}$ *are independent and*

$$P(\log Y_1 > n^{1/\alpha} L(n)^{-1}) = o(n^{-1}), \quad n \to \infty.$$

---

is a (2×2) matrix can be treated using the theory of operator stable random vectors; see Meerschaert (1991).

(10) Extensions similar to the ones discussed in Remark 9.2.24 are possible here as well; we may use the multivariate extreme value theory as in de Haan and Resnick (1977).

(11) The ch.f. of G_α admits the representation $f_{G_\alpha}(t) = 1/(1 - \log \phi_\alpha(t))$, where ϕ_α is the ch.f. of an α-stable r.v. (Klebanov, Maniya, and Melamed (1984)). Similarly, $f_{\xi_n} = 1/(1 - \log \psi)$, where ψ is the ch.f. of a distribution in the domain of attraction of an α-stable r.v. with ch.f. ϕ_α (Mittnik and Rachev (1991)). Examples of geometric α-stable distributions are the exponential law ($\alpha = 1$) and the Laplace law ($\alpha = 2$).

Then

$$a \log S_\tau \overset{D}{\Longrightarrow} \sup_{0 \le t \le 1} \mathcal{G}(t) \quad \text{as } p \to 0. \tag{9.2.80}$$

Here $\mathcal{G}$ is a "geometric Lévy stable motion"; i.e., the weak limit in $\mathcal{D}[0,1]$ of $\mathcal{G}_p(t) = a \sum_{j=1}^{[\tau t]} \xi_j$, $0 \le t \le 1$.

Remark 9.2.27 *Regarding the existence of the process $\mathcal{G}$ as the weak limit of $\mathcal{G}_p$, one can check the following:*

(a) *The finite-dimensional distributions $\mathcal{G}_p(t_1), \ldots, \mathcal{G}_p(t_d)$ $(0 \le t_1 < \cdots < t_d \le 1)$ converge to "geometric strictly stable distributions" $\mathcal{G}(t_1), \ldots, \mathcal{G}(t_d)$ with ch.f. $g(\theta)$ of the form $1/(1 - \log \psi(\theta))$, where $\psi(\theta)$ is the ch.f. of a strictly α-stable random vector on $\mathbb{R}^d$.*

(b) *The set of laws of $\mathcal{G}_p(\cdot)$ $(0 < p < 1)$ is tight.*

Remark 9.2.28 *Under the assumptions listed in Remark* 9.2.24 *we also have*

$$\left\{ \left(a_n^{(1)} S_{[\tau t]}^{(1)}, a_n^{(2)} S_{[\tau t]}^{(2)} \right) + \tau_n t b_n \right\}_{t \ge 0} \overset{w}{\Longrightarrow} \{\mathcal{L}(\nu t)\}_{t \ge 0},$$

where $\tau_n = \tau(1/n)$ and ν is an exponential random variable with mean 1 independent of the bivariate Lévy process $\mathcal{L}$. An important observation is that the above limit relation remains true if we choose any sequence of positive integer-valued random variables τ_n such that $\tau_n/n \overset{D}{\Rightarrow} \tau$, where τ is a positive random variable. Choosing therefore different laws for τ, we arrive at different models for the total "asset value process." We list below some of these models, assuming that $\mathcal{L}$ is a zero mean bivariate Wiener process; that is, we are in case (i) *discussed in Remark* 9.2.24.

(a) *If τ_n is a mixture (by the values of a fixed random variable U) of Poisson random variables with mean nU, then we may take $\tau = U$, and $\mathcal{L}(\tau\cdot)$ is a mixture of Wiener processes; see* Boness et al. *(1979).*

(b) *If $\tau = 1/\sqrt{\mathcal{X}_m}$, where $\mathcal{X}_m$ is a chi-square random variable with m degrees of freedom, then the one-dimensional marginals of $\mathcal{L}(\tau\cdot)$ are Student's t distributed; this model was used in Blattberg and Genodes (1974) to model stock prices.*

(c) *If τ is positive strictly stable with index $\alpha/2, 0 < \alpha < 2$, then $\mathcal{L}(\tau\cdot)$ is an α-stable motion. This subordinated process was used in Mandelbrot and Taylor (1967) to explain the nonnormality of stock price changes.*

(d) *If τ is a lognormal random variable, then $\mathcal{L}(\tau\cdot)$ is the Clark (1973) alternative to the Mandelbrot and Taylor (1967) subordinated process. Note that in contrast to* (c), *here $\mathcal{L}(\tau\cdot)$ has finite variances.*

Similarly to Theorems 9.2.25 and 9.2.26, we obtain the following limit theorem for the distribution of the maximal "wealth change."

Theorem 9.2.29 *Under the assumptions of Theorem* 9.2.26 *the following holds:*

(a) *If $\nu > 0$, then $a(\log M_\tau + \tau b) \stackrel{D}{\Longrightarrow} G_\alpha$ as $p \to 0$.*

(b) *If $\nu < 0$, then $a(\log \bigvee_{j \geq \tau+1} X_j + \tau b) \stackrel{D}{\Longrightarrow} G_\alpha$ as $p \to 0$.*

(c) *Suppose that the conditions of Theorem* 9.2.26(c) *hold. Then as $p \to 0$,*

$$a \log M_\tau \stackrel{D}{\Longrightarrow} \sup_{0 \leq t \leq 1} \mathcal{G}(t).$$

Finally, let us consider the total "wealth" until a Poisson (λ) random moment $T = T(\lambda)$. Let the sequence $\{Y_n, Z_n\}_{n \geq 0}$ be as before and independent of T. Suppose also that the ch.f. f_{ξ_n} of $\xi_n = \log Z_n$ satisfies

$$\lim_{u \to 0} |u|^{-\alpha}(1 - f_{\xi_n - a}(u)) = \mu \tag{9.2.81}$$

for some $\mu > 0$, real a, and $1 < \alpha \leq 2$. Note that $a = E\xi_n$, and at least when $\alpha \neq 2$, (9.2.81) is equivalent to assuming that the ξ_n's are in the domain of normal attraction of an α-stable distribution (Feller (1971, p. 596)).

Theorem 9.2.30 *Suppose that $E \log(1 + Z_n) < \infty$ and* (9.2.81) *holds. Let*

$$S_T = \sum_{i=0}^{T} X_i = \sum_{i=0}^{T} Y_i \prod_{j=0}^{i} Z_j.$$

(a) *If $\nu = E\xi_n > 0$ and $E \log(1 + Y_n) < \infty$, then as $\lambda \to \infty$,*

$$\lambda^{-1/\alpha}(\log S_T - aT) \stackrel{D}{\Longrightarrow} Y_{(\alpha)},$$

where $Y_{(\alpha)}$ is a symmetric stable r.v. with ch.f. $\exp(-\mu|\theta|^\alpha)$.

(b) *If* $\nu < 0$ *and* $E\log(1+Y_n) < \infty$, *then as* $\lambda \to \infty$,

$$\lambda^{-1/\alpha}\left(\log \sum_{j=T+1}^{\infty} X_j - aT\right) \stackrel{D}{\Longrightarrow} Y_{(\alpha)},$$

$$\lambda^{-1/\alpha}\left(\log \sum_{j=T_1+1}^{T_k} X_j - aT_1\right) \stackrel{D}{\Longrightarrow} Y_{(\alpha)},$$

where T_1, T_k *are as in* (9.2.69).

(c) *Let* $\nu = 0$ *and suppose that the sequences* $\{Y_n\}_{n\geq 0}$ *and* $\{Z_n\}_{n\geq 0}$ *are independent and* $P(\log Y_n > u) = o(u^{-\alpha})$ *as* $u \to \infty$. *Then as* $\lambda \to \infty$,

$$\lambda^{-1/\alpha}\log S_T \stackrel{D}{\Longrightarrow} \sup_{0\leq t\leq 1} \mathcal{L}(t),$$

where $\mathcal{L}(\cdot)$ *is a Lévy stable motion on* $[0,1]$ *with* $\mathcal{L}(1) \stackrel{d}{=} Y_{(\alpha)}$.

Analogous theorems can be established for the limit distributions of M_{T_1} and $\bigvee_{i=T_k+1}^{T_{k+1}} X_i$.

The remaining results in this section deal with characterizations of the limit laws of S_n (cf. (9.2.66) and (9.2.64)) and M_n (cf. (9.2.67)), which can arise for any given distribution of Z_n's in a given parametric family of distributions. We will assume that the sequences $\{Y_n\}_{n\geq 1}$ and $\{Z_n\}_{n\geq 1}$ are independent. Also, we will concentrate our attention on the distributions of Z_n supported by (0.1).[(12)] Invoking Lemma 9.2.21(b), we conclude that S_n (resp. M_n) converges to a finite limit S (resp. M) if (and only if, in the case $E\log Z_n > -\infty$)

$$0 \leq E\log(1+Y_n) < \infty. \tag{9.2.82}$$

Given (9.2.82), the limits S and M satisfy the equations (9.2.71) and (9.2.72).[(13)]

We start with a characterization of the class $\mathcal{S}_1$ (resp. $\mathcal{M}_1$) of laws $\mathcal{L}(S)$ of S (resp. $\mathcal{L}(M)$) such that for any $\mathcal{L}(Z) \in \mathcal{Z}_1$ there exist $Y = Y(Z)$

[(12)] The case $Z_n \in (0,1)$ a.s. corresponds to "deteriorating environment," the case being close to the soil erosion model of Todorovic and Gani (1987).

[(13)] Moreover, the converse is also true. Namely, if $S,(Y,Z)$ $(M,(Y,Z)$ respectively) is a solution of (9.2.71) (or (9.2.72) respectively), then the distribution of $S(M)$ is equal to the limiting distribution in the model (9.2.66) ((9.2.67) respectively) with $(Y_n, Z_n) \stackrel{d}{=} (Y,Z)$. This is a simple consequence of the uniquenes principle (the so-called Letac principle); see Letac (1986) or Goldie (1991).

such that (9.2.71) (resp. (9.2.72)) holds. The class[14] $\mathcal{Z}_1$ of Z-laws $\mathcal{L}(Z)$ consists of distributions on $(0,1)$ with densities

$$f_\alpha(u) = (1+\alpha)z^\alpha, \quad 0 < z < 1, \alpha \geq 0. \tag{9.2.83}$$

In the sequel, for any $0 < \beta < 1$ and an r.v. Y, define

$$Y_\beta := \begin{cases} 0 & \text{with probability } 1-\beta, \\ Y & \text{with probability } \beta. \end{cases} \tag{9.2.84}$$

A complete description of the class $\mathcal{Z}_1$ is given in the following theorem.

Theorem 9.2.31 *The class $\mathcal{S}_1$ of the laws $L(S)$ solving $S \stackrel{d}{=} (S+Y)Z$ consists of all nonnegative infinitely divisible r.v.s S with Laplace transform*

$$\phi_S(\theta) = \exp\left\{-\int_0^\infty \frac{1}{x}(1-e^{-\theta x})M_S(\mathrm{d}x)\right\}.$$

Moreover, the Lévy measure M_S is of the following form:

$$M_S \ll Leb \quad \text{and} \quad M_S(\mathrm{d}x) = H(x)\,\mathrm{d}x,$$

where $H(0) \in [0,1]$, H is nonincreasing on $[0,\infty)$ and vanishing at ∞. The corresponding Y has $1-H$ as its distribution function.

Remark 9.2.32 *Suppose that S is a solution of (9.2.71) for a given Y with $0 \leq E\log(1+Y) < \infty$ and Z uniform. Then then S is also a solution of (9.2.71) with Z having density (9.2.83) and Y replaced by $Y_{1/(1+\alpha)}$.*

Note that $\mathcal{Z}_1$ is a subclass of the class of *self-decomposable* random variables; see Vervaat (1979). Also, allowing α in (9.2.83) to take values in the whole range $(-1,\infty)$ would have made the class $\mathcal{Z}_1$ degenerate (consisting of $Z = 0$ a.s.).

Remark 9.2.32, in particular, has no counterpart for α's in the range $(-1,0)$.

Our next task is the characterization of the class $\mathcal{M}_1$ of laws $\mathcal{L}(M)$ such that for every $\mathcal{L}(Z) \in \mathcal{Z}_1$ there exists $Y = Y(Z)$ such that (9.2.72) holds.

Theorem 9.2.33 *The class $\mathcal{M}_1$ consists of all absolutely continuous laws $\mathcal{L}(M)$ with density f_M and d.f. F_M satisfying the following conditions:*

---

[14] The class $\mathcal{Z}_1$ was considered by Vervaat (1979) (who discussed a wider family, allowing $\alpha > -1$ in (9.2.83)) and Todorovich and Gani (1987). Some particular examples of laws $\mathcal{L}(S) \in \mathcal{S}_1, \mathcal{L}(M) \in \mathcal{M}_1$, were studied by Todorovich and Gani (1987), Todorovich (1987), and Rachev and Todorovich (1990).

(i) $f_M(x)$ *is nonincreasing on* $(0,\infty)$.

(ii) $x\, f_M(x)/F_M(x)$ *is nonincreasing on* $(0,\infty)$.

Suppose that $\mathcal{L}(M) \in \mathcal{M}_1$ and let $Z^{(\alpha)}$ have density $f_{Z^\alpha}(z) = (1+\alpha)z^\alpha, 0 < z < 1$. Then $M \stackrel{d}{=} (M \vee Y)Z^{(\alpha)}$ is equivalent to

$$\overline{F}_Y(x) = \frac{1}{1+\alpha}\frac{x\, f_M(x)}{F_M(x)}, \quad x > 0. \tag{9.2.85}$$

By (9.2.85), for any $\mathcal{L}(M) \in \mathcal{M}_1$ and $0 < \alpha < 1$,

$$M \stackrel{d}{=} (M \vee Y_\alpha)Z^{(\alpha)} \iff M \stackrel{d}{=} (M \vee Y_0)Z^{(0)},$$

where Y_α is determined by (9.2.84). The last relation is parallel to the corresponding relation in the scheme of summation (cf. Remark 9.2.32). Note also that gamma $\Gamma(p,\lambda)$-distributions with $0 < p \le 1$ belong to $\mathcal{M}_1$, while those with $p > 1$ do not.

Next, we consider the class $\mathcal{S}_2$ (resp. $\mathcal{M}_2$)[(15)] of laws $\mathcal{L}(S)$ (resp. $\mathcal{L}(M)$) such that for every $\mathcal{L}(Z) \in \mathcal{Z}_2 \equiv \{\delta_z, 0 < z < 1\}$ there is a $Y = Y(Z)$ such that (9.2.71) (resp. (9.2.72)) holds.

Theorem 9.2.34 *The class $\mathcal{S}_2$ coincides with the family of all nonnegative infinitely divisible r.v.s with Laplace tranform of the form*

$$\phi_S(t) = \exp\left\{-at - \int_0^\infty \frac{1}{x}(1-e^{-tx})M_S(\,dx)\right\},$$

where $a \ge 0$ and the Lévy measure $M_S \ll Leb$ is absolutely continuous, whose Radon-Nikodym derivative is nonincreasing a.s.

For any $S \in \mathcal{S}_2$ and $z \in (0,1)$, the corresponding Y in the equation

$$S \stackrel{d}{=} (S+Y)z$$

is a nonnegative infinitely divisible r.v. with Laplace transform

$$\phi_Y(t) = \exp\left\{-\frac{at(1-z)}{z} - \int_0^\infty \frac{1}{x}(1-e^{-tx})M_Y(\,dx)\right\}.$$

---

(15) It turns out that the class $\mathcal{S}_2$ coincides with the class L of Khinchine (cf. Feller (1971, Sect. 8, Chapter XVII) of nonnegative r.v.s. We shall state here a more explicit description of $\mathcal{S}_2$ than that in Feller (1971, Theorem XVII.8). Moreover, $\mathcal{S}_1 \subset \mathcal{S}_2$; see also Vervaat (1979, Remark 4.9). The class of $\mathcal{M}_2$ coincides with the class of the laws of max self-decomposable r.v.s (see Balkema et al. (1990) and the references there). The next theorem, similar to the Mejzler (1956) result, is based on a characterization of the weak limits of the normalized maxima $a_n\{\max(X_1, X_2, \ldots, X_n) - b_n\}$ when the X_i's are independent and nonidentically distributed.

Moreover, $M_Y \ll Leb$, *and*

$$\frac{dM_Y}{d\lambda}(x) = \frac{dM_S}{d\lambda}(zx) - \frac{dM_S}{d\lambda}(x),$$

where $\lambda = Leb$.

Theorem 9.2.35 *The class* $\mathcal{M}_2$ *consists of the laws of positive absolutely continuous r.v.s* M *such that* $xf_M(x)/F_M(x)$ *is a nonincreasing function on* $(0, \infty)$. *Also,* $\mathcal{M}_1 \subset \mathcal{M}_2$.

9.2.7 Random Recursion Arising in Probabilistic Modeling: Rate of Convergence

Throughout this section, $(B, ||\cdot||)$ is the separable Banach space $C(T)$ of continuous mappings $x : T \to \mathbb{R}$, where T is compact and $||\cdot||$ is the usual supremum norm in $C(T)$. For any $x, y \in B$ we set $(x \cdot y)(t) = x(t) \cdot y(t)$, $(x \vee y)(t) = x(t) \vee y(t)$, $t \in T$. Given a nonatomic probability space, let $\mathcal{X}(B)$ be the space of all random fields (r.f.s) X of B-valued random variables, and let $\mathcal{L}(B)$ be the space of all laws P_X. Suppose $\{(Y_n, Z_n)\}_{n\geq 1}$ is a sequence of i.i.d. pairs of r.f.s, and define

$$S_n = \sum_{i=1}^{n} X_i, \quad X_i = Y_i \prod_{j=1}^{i} Z_j. \tag{9.2.86}$$

The r.f. S_n can be interpreted as the "wealth" accumulated in different commodities $\{A_t, t \in T\}$ for a period of n years. We take $T = T(U)$, where U is a compact metric space, $U = (U, \varrho)$, and $T(U)$ is the set of all closed subsets (think, for example, of crop-producing areas) t of U endowed with the Hausdorff metric

$$h(t_1, t_2) = \inf\{\varepsilon > 0;\ t_1 \subset t_2^{\varepsilon},\ t_2 \subset t_1^{\varepsilon}\}.$$

Here t^{ε} stands for the ε-neighborhood of t (cf. Hausdorff (1957, Sect. 29)).(16) Similarly, we define the maximal "wealth changes"

$$M_n = \bigvee_{i=1}^{n} X_i. \tag{9.2.87}$$

Next, we are interested in conditions providing an exponential rate of convergence of S_n and M_n to finite limits S and M, respectively. The rate

(16)Then (T, h) is a compact metric space (cf. Hausdorff (1957, Sect. 29); see also Kuratowski (1966, §21), Kuratowski (1969, §31), and Matheron (1975)).

of convergence of the laws P_{X_n} to P_X will be expressed, as usual in the Banach space setting, in terms of the Prohorov metric

$$\begin{aligned}\pi(X, Y) &:= \pi(P_X, P_Y) \\ &:= \inf\{\varepsilon > 0;\ P(X \in A) \leq P(Y \in A^{\varepsilon}) + \varepsilon \\ &\qquad \text{for all Borel subsets } A \text{ in } B\},\end{aligned} \tag{9.2.88}$$

where A^{ε} is the open ε-neighborhood of A. Further, we shall use also the following metrics and functions in $\mathcal{X}(B)$ and $\mathcal{L}(B)$:[17]

(i) χ_p-metric in $\mathcal{X}(B)$:

$$\chi_p(X, Y) := \left[\sup_{t>0} t^p P(\|X - Y\| > t)\right]^{\frac{1}{1+p}}, \quad p > 0;$$

(ii) χ_p-minimal metric in $\mathcal{L}(B)$:

$$\begin{aligned}\widehat{\chi}_p(X, Y) &= \widehat{\chi}_p(P_X, P_Y) \\ &:= \inf\{\chi_p(\widetilde{X}, \widetilde{Y});\ \widetilde{X}, \widetilde{Y} \in \mathcal{X}(B), \widetilde{X} \stackrel{d}{=} X,\ \widetilde{Y} \stackrel{d}{=} Y\}, p > 0;\end{aligned}$$

(iii)
$$\begin{aligned}\omega_{p,N}(X)^{p+1} &:= \sup_{t>N} t^p P(\|X\| > t), \\ \omega_p(X) &:= \omega_{p,0}(X), \\ N_p(X) &:= \{E\|X\|^p\}^{1/(1+p)}, \quad p > 0.\end{aligned}$$

Note that $\widehat{\chi}_p$ is a metric in $\mathcal{L}(\mathcal{B})$, $\widehat{\chi}_p \geq \pi$, and the following convergence criterion holds: if

$$\omega_{p,N}(X_n) + \omega_{p,N}(X) \to 0 \quad \text{as } N \to \infty \text{ for any } n = 1, 2, \ldots, \tag{9.2.89}$$

then

$$\widehat{\chi}_p(X_n, X) \to 0 \text{ as } n \to \infty \text{ if and only if } X_n \xrightarrow{D} X \text{ as } n \to \infty$$

and

$$\lim_{N\to\infty} \limsup_{n\to\infty} \omega_{p,N}(X_n) = 0.$$

---

[17] (cf. Pisier and Zinn (1977), de Haan and Rachev (1989), and Rachev (1991c).

Similarly, if (9.2.21) holds, then

$$\chi_p(X_n,\ X) \to 0 \text{ as } n \to \infty \text{ if and only if } X_n \xrightarrow{P} X \text{ as } n \to \infty$$

and

$$\lim_{N\to\infty} \limsup_{n\to\infty} \omega_{p,N}(X_n) = 0.$$

Theorem 9.2.36 (a) *If for some $p > 0$, $N_p(Z_1) < 1$ and $\omega_p(Y_1 Z_1) < \infty$, then S_n converges in probability to a P-a.e. finite limit $S = \sum_{i=1}^{\infty} X_i$. Moreover,*

$$\chi_p(S_n,\ S) \ \le\ \phi_n(Z_1)\omega_p(Y_1 Z_1), \tag{9.2.90}$$

where $\phi_n(Z_1) := N_p(Z_1)^n/(1 - N_p(Z_1))$.
(b) *Under the above conditions, suppose additionally that $\{(Y_i^*, Z_i)\}_{i\ge1}$ is a sequence of i.i.d. pairs of r.f.s satisfying the "tail" condition $\omega_p(Y_1^* Z_1) < \infty$. Let S^* be the limit of S_n^*, i.e.,*

$$S_n^* \ :=\ \sum_{i=1}^{n} Y_i^* \prod_{j=1}^{i} Z_j \ \xrightarrow{P}\ S^*.$$

Suppose now that the $\widetilde{Z}_i$'s are i.i.d. copies of the Z_1 independent of the (Y_i, Z_i)'s and let $\widehat{\chi}_p(Y_1 Z_1,\ Y_1^* Z_1) < \infty$. Then as $n \to \infty$,(18)

$$\pi(S^* - S_n^*, \widetilde{Z}_1 \cdots \widetilde{Z}_n \cdot S) \ \le\ \phi_n(Z_1)\widehat{\chi}_p(Y_1 Z_1,\ Y_1^* Z_1) \to 0. \tag{9.2.91}$$

Proof: a) First note that $\chi_p \ge \kappa$, where κ is the distance in probability (the Ky Fan metric):

$$\kappa(X,\ Y) \ =\ \inf\{u > 0;\ P(||X - Y|| > u) < u\}.$$

Indeed, to prove $S_n \xrightarrow{P} S$ it is enough to show that S_n is χ_p-fundamental. Actually, for $k = 1, 2, \ldots,$

$$\begin{aligned} &\chi_p(S_{n+k},\ S_n) \\ &= \ \omega_p\left(\sum_{i=n+1}^{n+k} X_i\right) \ \le\ \sum_{i=n+1}^{n+k} \omega_p(X_i) \\ &\le\ \sum_{i=n+1}^{n+k} \left\{ E_{Z_1=z_1,\ldots,Z_{i-1}=z_{i-1}} \omega_p \left(Y_i \prod_{j=1}^{i-1} z_j Z_i \right)^{p+1} \right\}^{\frac{1}{1+p}} \end{aligned} \tag{9.2.92}$$

---

(18) The factor $\widetilde{Z}_1 \cdots \widetilde{Z}_n$ plays the same role as the normalizing scaling in the usual CLT.

$$= \sum_{i=n+1}^{n+k} \left[E_{Z_1=z_1,\ldots,Z_{i-1}=z_{i-1}} \left(\prod_{j=1}^{i-1} \|z_j\| \right)^p \right]^{\frac{1}{1+p}} \omega_p(Y_i Z_i)$$

$$\le \sum_{i=n+1}^{n+k} N_p(Z_1)^{i-1} \omega_p(Y_1 Z_1)$$

$$= \phi_n(Z_1)\omega_p(Y_1 Z_1) \to 0 \text{ as } n \to \infty,$$

which indeed implies that S_n is χ_p-fundamental. The bound (9.2.90) follows by the same arguments we used to show (9.2.92).

(b) By definition, $\widehat{\chi}_p$ is the minimal metric with respect to χ_p, and thus for any joint distribution of Y_1 and Y_1^*,

$$\widehat{\chi}_p(S^* - S_n^*,\ \widetilde{Z}_1 \cdots \widetilde{Z}_n \cdot S) \le \chi_p \left(\sum_{i>n} Y_i^* \prod_{j=1}^{i} Z_j,\ \sum_{i>n} Y_i \prod_{j=1}^{i} Z_j \right).$$

Now proceed as in (9.2.92) to obtain that the right-hand side is not greater than

$$\phi_n(Z_1)\widehat{\chi}_p(Y_1 Z_1,\ Y_1^* Z_1).$$

We next take the infimum in the last inequality over all joint distributions of $(Y_1,\ Y_1^*)$ with fixed marginals, and use the inequality $\widehat{\chi}_p \ge \pi$ to complete the proof of (9.2.92). □

Theorem 9.2.37 *Suppose the Y_i's and Z_i's are nonnegative r.v.s. Then under the assumptions of Theorem* 9.2.36a), $M_n \xrightarrow{P} M$, *and moreover,*

$$\chi_p(M_n, M) \le \phi_n(Z_1)\omega_p(Y_1 Z_1) \to 0 \text{ as } n \to \infty.$$

If also $\widehat{\chi}_p(Y_1, Y_1^*) < \infty$, then under the assumptions of Theorem 9.2.36(b),

$$\pi \left(\bigvee_{i=n+1}^{\infty} Y_i^* \prod_{j=1}^{i} Z_j, \widetilde{Z}_1 \cdots \widetilde{Z}_n \cdot M \right) \le \phi_n(Z_1)\widehat{\chi}_p(Y_1 Z_1, Y_1^* Z_1) \to 0.$$

Proof: (a) For any $k = 1, 2, \ldots$,

$$\chi_p(M_{n+k},\ M_n) \le \omega_p \left(\sum_{i=n+1}^{n+k} X_i \right) \le \sum_{i=n+1}^{n+k} \omega_p(X_i),$$

and therefore, $M_n \xrightarrow{P} M$ follows by the same arguments as in the proof of Theorem 9.2.36, and the required bound for $\chi_p(M_n, M)$ is obtained in the same way as we did in (9.2.92).

b) With $X_i^* = Y_i^* \prod_{j=1}^i Z_j$ we have

$$\begin{aligned}\widehat{\chi}_p\left(\bigvee_{i>n} X_i^*,\ \widetilde{Z}_1\cdots\widetilde{Z}_n\cdot M\right) &\le \chi_p\left(\bigvee_{i>n} X_i^*,\ \bigvee_{i>n} X_i\right)\\ &\le \left(\sup_{t>0} t^p P\left(\sum_{i>n}\|X_i^* - X_i\| > t\right)\right)^{1/1+p}\\ &\le \sum_{i>n}\omega_p(X_i^* - X_i).\end{aligned}$$

The last inequality follows from the triangle inequality for χ_p in the space of real-valued random variables. Conditioning as in the proof of Theorem 9.2.36(a), we obtain

$$\sum_{i>n}\omega_p(X_i^* - X_i) \ \le\ \phi_n(Z_1)\chi_p(X_1^*,\ X_1).$$

Passing to the minimal metrics and using again $\pi \le \widehat{\chi}_p$, we obtain the necessary bound. □

Suppose N is an integer valued r.v. independent of the Y_i's and Z_i's. Then under the conditions of Theorem 9.2.36,

$$\pi(S_N,\ S) \ \le\ \psi_N(Z_1)\omega_p(Z_1),$$

where $\psi_N(Z_1) = (EN_p(Z_1)^N)/(1 - N_p(Z_1))$, and moreover,

$$\pi(S^* - S_N^*,\ \widetilde{Z}_1\cdots\widetilde{Z}_N\cdot S) \ \le\ \psi_N(Z_1)\widehat{\chi}_p(Y_1,\ Y_1^*).$$

Similar results on the limiting behavior of of the maximum $\bigvee_{i=1}^N X_i$ can be obtained as a consequence of Theorem 9.2.36.

Remark 9.2.38 *Vervaat (1979) showed the following limiting result for S_n^* (see (9.2.65)): let $u_n \uparrow \infty$ as $n \to \infty$ be a sequence of reals, and assume, in addition, that*

(i) $E\log^+|B_1| < \infty$, $E|\log|A_1||^{2+\eta} < \infty$ *for some* $\eta > 0$, $\mu := E\log|A_1| < 0$;

(ii) *the solution S^* of the equation $S^* \stackrel{d}{=} A_1S^* + B_1$, S^* and (A_1, B_1) independent, has a density f that is ultimately nonincreasing and such that $f(t) = O(t^{-1})$ as $t \to \infty$;*

(iii) *there are positive reals b and $\varepsilon < |\mu|$ and a positive nonincreasing integrable function ϕ on $[1, \infty)$ such that the function*

$$T^{\leftarrow}(T_+(x)\phi(y))x^{-1}e^{(\mu+\varepsilon)y}$$

(where $T_+(x) = P(|S^*| > x)$, $T(x) = P(S^* > x)$, and $T^{\leftarrow}$ is its generalized inverse) is bounded on the set $\{(x,y);\ x \geq b,\ y \geq 1\}$.

Then

$$\sum_{n=1}^{\infty} P(S^* > u_n) \begin{cases} < & \infty, \\ = & \infty, \end{cases}$$

implies

$$P(S_n^* > u_n \text{ i.o.}) = \begin{cases} 0, \\ 1. \end{cases}$$

Following our Theorem 9.2.36, let us compare the tail behavior of the distribution of the S_n^*'s and their limit S^*. We consider again the Banach space setting for A_n, B_n, S_n^*, and S^*. Let $\ell_p = \widehat{L}_p$ be the minimal metric with respect to $L_p(X,Y)$, $0 \leq p \leq \infty$,

$$\begin{aligned} L_0(X,\ Y) &= P(X \neq Y), \\ L_p(X,\ Y) &= \{E\|X - Y\|^p\}^{\min(1,1/p)}, \quad 0 < p < \infty, \end{aligned}$$

and

$$L_\infty(X,\ Y) = \operatorname{ess\,sup}\|X - Y\|.$$

Then, as in Theorem 9.2.36(a), if for some $p \in [0, \infty]$, $N_p^*(A_1) := L_p(Z, 0) < 1$ and $N_p^*(B_1) < \infty$, then as $n \to \infty$,

$$\ell_p(S_n^*,\ S^*) \leq \frac{N_p(A_1)^n}{1 - N_p(A_1)} N_p(B_1) \to 0.$$

In the case of real-valued S_n^* and S^*, the last bound gives us conditions for exponential rate of convergence in the total variation metric and in the ℓ_p-Kantorovich metrics:

$$\ell_0(S_n^*,\ S^*) = \sup_{A \text{ Borel}} |P(S_n^* \in A) - P(S^* \in A)| \to 0;$$

$$\ell_1(S_n^*,\ S^*) = \int_{-\infty}^{\infty} |F_{S_n^*}(x) - F_{S^*}(x)|\, dx \to 0;$$

$$\ell_p(S_n^*,\ S^*) \;=\; \left(\int_{-\infty}^{\infty} |F_{S_n^*}^{\leftarrow}(x) - F_{S^*}^{\leftarrow}(x)|^p\, dx\right)^{1/p} \;\to\; 0, \quad 1 \le p < \infty;$$

and

$$\ell_\infty(S_n^*,\ S^*) \;=\; \sup_{0\le x\le 1} |F_{S_n^*}^{\leftarrow}(x) - F_{S^*}^{\leftarrow}(x)| \;\to\; 0.$$

Here as usual, $F_{S_n^*}$ and F_{S^*} are the corresponding distribution functions, and $F^{\leftarrow}$ stands for the generalized inverse of F.

9.3 Extensions of the Contraction Method

A well-known problem in the theory of probability metrics is the extension of the method of ideal metrics to limit theorems for sums or maxima with "nonregular" normalizations of logarithmic type. Moreover, this problem is quite typical in a wide range of stochastic algorithms, since the logarithmic-type normalization is not reflected in the regularity structure of probability metrics, while power normalizations n^a can be captured easily by ideal metrics of order a. The second difficulty arises when the contraction factors converge to one.

In this section we study several examples that show solutions to this problem by the use of a modified version of the contraction method. In Sections 9.3.1 and 9.3.2 we consider the number of inversions for random permutations and the "MAX"-algorithm. In Sections 9.3.3 and 9.3.4 we study successful and unsuccessful searching in binary random trees. Each of these examples needs some special arguments in order to achieve approximation by a limit distribution; so in general, the contraction method cannot be considered an "automatic" method. The advantage of the contraction method is its generality, which allows us, for example, to consider recursions in very general spaces, as well as the fact that it often allows us to obtain quantitative approximations. The examples in this section are due to Cramer and Rüschendorf (1996a).

9.3.1 The Number of Inversions of a Random Permutation

Given a permutation $\sigma = (a_1, \dots, a_n)$, the pair (a_i, a_j), $i < j$, is called an *inversion* if $a_i > a_j$. Denote by I_n the number of inversions in a random permutation of size n. Then the following recursion holds:

$$I_n \overset{d}{=} I_{n-1} + X_n, \quad I_1 \;=\; 0, \tag{9.3.1}$$

where $X_n \sim \mathrm{U}(\{0,\dots,n-1\})$ is uniformly distributed on $0,\dots,n-1$ and the r.v.s I_{n-1}, X_n are independent. This leads to explicit expressions for the moment generating function, the mean, and the variance:

$$G_n(z) = E z^{I_n} = \frac{1}{n!}\cdot\frac{(1-z^2)\cdots(1-z^n)}{(1-z)^{n-1}}, \tag{9.3.2}$$

$$E\,I_n = \frac{n\,(n-1)}{4}, \quad \operatorname{Var} I_n = \frac{(n-1)\,n\,(2n+5)}{72} \tag{9.3.3}$$

(cf. Hofri (1987, pp. 122–124)). For the normalized version

$$\widehat{I}_n := \frac{I_n - E\,I_n}{\sqrt{\operatorname{Var} I_n}} \tag{9.3.4}$$

we obtain the following Berry–Esséen-type result. (Note that we assume that all the occurring random variables are defined on one and the same probability space.)

Theorem 9.3.1 *For $n \geq 7$,*

$$\varrho\left(\widehat{I}_n, \mathcal{N}(0,1)\right) \leq C \cdot n^{-\frac{1}{2}}, \tag{9.3.5}$$

with $C = 2.75 \cdot \frac{8^4}{6\cdot 128}\sqrt{\frac{7}{6}}$.

Proof: Without loss of generality, we assume that $I_n = \sum_{i=1}^n X_i$, where the X_i are independent, $X_i \sim \mathrm{U}(\{0,\dots,i-1\})$. By the Berry–Esséen theorem (cf. Bhattacharya and Ranga Rao (1976, Th. 12.4)),

$$\varrho\left(\widehat{I}_n, \mathcal{N}(0,1)\right) \leq 2.75\,\frac{S_{n,3}}{(S_{n,2})^{3/2}}, \tag{9.3.6}$$

where

$$S_{n,m} := \sum_{k=1}^n E|X_k - E\,X_k|^m. \tag{9.3.7}$$

We have, for $k \geq 2$,

$$E|X_k - E\,X_k|^3 \leq \frac{k^3}{32}, \quad \operatorname{Var} X_k = \frac{k^2-1}{12}.$$

by some tedious calculations. This implies that $\sum_{k=1}^n \operatorname{Var} X_k \geq \frac{(n-1)^3}{36}$ and $\sum_{k=1}^n E|X_k - E\,X_k|^3 \leq \frac{(n+1)^4}{128}$. Thus, from (9.3.6), we obtain, for $n \geq 7$,

$$\varrho\left(\widehat{I}_n, \mathcal{N}(0,1)\right) \;\le\; 2.75\,\frac{6^3}{128}\left(\frac{n+1}{n-1}\right)^4 \sqrt{\frac{n}{n-1}}\; n^{-\frac{1}{2}} \;\le\; C\, n^{-\frac{1}{2}}.$$

□

Recursion (9.3.1) leads to a sum of independent variables and therefore allows the application of the classical tools for the central limit theorem. On the other hand, it is an interesting "test" rate of convergence example for the contraction method, since the contraction factors of the normalized recursion converge to one. Furthermore, the approximation result (in terms of the ζ_3-metric) is of independent interest. It gives the same convergence rate as in Theorem 2.1 uniformly on the set of functions $f(I_n)$ with $\|f^{(3)}\|_\infty \le 1$, when we study the limiting behavior of

$$\widetilde{I}_n \;:=\; \frac{I_n - E\,I_n}{n^{3/2}}. \tag{9.3.8}$$

Theorem 9.3.2 *Let* $\sigma_n^2 := \operatorname{Var}(\widetilde{I}_n)$ *and* $Z_n \sim \mathcal{N}(0,\sigma_n^2)$. *Then for some* $C > 0$, *and for all* $n \in \mathbb{N}$,

$$\zeta_3(\widetilde{I}_n, Z_n) \;\le\; C\, n^{-\frac{1}{2}}. \tag{9.3.9}$$

Proof: First note that $\widetilde{I}_n$ satisfies the modified recursion

$$\widetilde{I}_n \;\stackrel{d}{=}\; \left(\frac{n-1}{n}\right)^{3/2} \widetilde{I}_{n-1} \;+\; \widetilde{X}_n, \tag{9.3.10}$$

where $\widetilde{X}_n := \frac{X_n - E\,X_n}{n^{3/2}}$. Let the sequence (Z_n) be independent, $Z_n \sim \mathcal{N}(0,\sigma_n^2)$, and define the accompanying sequence

$$Z_n^* \;:=\; \left(\frac{n-1}{n}\right)^{3/2} Z_{n-1} \;+\; \widetilde{X}_n. \tag{9.3.11}$$

Let $Y_i \sim \mathcal{N}(0,\tau_i^2)$ be independent of $(\widetilde{X}_i, Z_{i-1})$, where $\tau_i^2 := \sigma_i^2 - \left(\frac{i-1}{i}\right)^3 \sigma_{i-1}^2 = \frac{\operatorname{Var} X_i}{i^3} \ge 0$. Then

$$Z_i \;\stackrel{d}{=}\; \left(\frac{i-1}{i}\right)^{3/2} Z_{i-1} \;+\; Y_i. \tag{9.3.12}$$

Using the homogeneity of order three of the ideal metric ζ_3, we obtain

$$\begin{aligned}
\zeta_3(\widetilde{I}_n, Z_n) \;&\le\; \zeta_3\left(\left(\frac{n-1}{n}\right)^{3/2} \widetilde{I}_{n-1} + \widetilde{X}_n, \left(\frac{n-1}{n}\right)^{3/2} Z_{n-1} + \widetilde{X}_n\right) \\
&\quad + \zeta_3(Z_n^*, Z_n) \\
&\le\; \left(\frac{n-1}{n}\right)^{9/2} \zeta_3(\widetilde{I}_{n-1}, Z_{n-1}) \;+\; \zeta_3(Z_n^*, Z_n).
\end{aligned}$$

By iteration, using $Z_1 = \widetilde{I}_1 = 0$, we obtain the "ground estimate"

$$\zeta_3(\widetilde{I}_n, Z_n) \le \sum_{i=2}^{n} \left(\frac{i}{n}\right)^{9/2} \zeta_3(Z_i^*, Z_i). \tag{9.3.13}$$

Note that $E Z_i = E Z_i^* = 0$ and $E Z_i^2 = E(Z_i^*)^2$. Therefore, by making use of the estimate $\zeta_r \le \frac{T(1+\frac{1}{\alpha})}{\Gamma(1+r)} \kappa_r$ for $r = m+2$, by (9.3.12) and some calculations (cf. Cramer (1995)) we have

$$\begin{aligned}
\zeta_3(Z_i^*, Z_i) &= \zeta_3\left(\widetilde{X}_i + \left(\frac{i-1}{i}\right)^{3/2} Z_{i-1}, Y_i + \left(\frac{i-1}{i}\right)^{3/2} Z_{i-1}\right) \\
&\le \zeta_3\left(\widetilde{X}_i, Y_i\right) \le \frac{\Gamma(2)}{\Gamma(4)} \kappa_3\left(\widetilde{X}_i, Y_i\right) \\
&= \int_{-\infty}^{0} x^2 \left|F_{\widetilde{X}_i}(x) - F_{Y_i}(x)\right| \mathrm{d}x \\
&\le \frac{7}{2^6 \cdot 3^2} i^{-3/2} + \frac{1}{2^5} i^{-5/2}.
\end{aligned}$$

Therefore, by some additional calculations,

$$\begin{aligned}
\zeta_3(\widetilde{I}_n, Z_n) &\le \sum_{i=2}^{n} \left(\frac{i}{n}\right)^{9/2} \zeta_3(Z_i^*, Z_i) \\
&\le \sum_{i=2}^{n} \left(\frac{i}{n}\right)^{9/2} \left(\frac{1}{2^5} i^{-5/2} + \frac{7}{2^6 \cdot 3^2} i^{-3/2}\right) \\
&\le \frac{1}{n^{9/2}} \left[\frac{1}{2^5}\left(n^2 + \frac{1}{3} n^3\right) + \frac{7}{2^6 \cdot 3^2}\left(n^3 + \frac{1}{4} n^4\right)\right] \\
&= \frac{7}{2^8 \cdot 3^2} \cdot \frac{1}{\sqrt{n}} + O\left(n^{-\frac{3}{2}}\right).
\end{aligned}$$

□

Note that the contraction factor in this example is of order $\left(\frac{n-1}{n}\right)^{\frac{3}{2}}$ only, and consequently we cannot obtain a uniform bound, implying that we need to estimate more precisely the individual terms. The exponential convergence rate is reduced to the rate $\sqrt{n}$.

9.3.2 *The Number of Records*

The "MAX"-algorithm determines the maximum element of a random sequence (cf. Hofri (1987, pp. 112–113)). Its complexity is essentially given by the number of records in a random permutation. Let M_n denote the number of maxima of a random permutation read from left to right. Then M_n satisfies the recursion

$$M_n \stackrel{d}{=} M_{n-1} + X_n, \tag{9.3.14}$$

where X_n has a Bernoulli distribution with success probability $\frac{1}{n}$, $X_n \sim B\left(1, \frac{1}{n}\right)$, and X_n, M_{n-1} are assumed independent. Define $M_1 = 0$. Then

$$M_n \stackrel{d}{=} \sum_{i=2}^{n} X_i, \tag{9.3.15}$$

when the (X_i) are independent. Furthermore,

$$E\,M_n = H_n - 1, \quad \operatorname{Var} M_n = H_n - H_n^{(2)}, \tag{9.3.16}$$

where $H_n^{(k)} = \sum_{j=1}^{n} \frac{1}{j^k}$, $\quad H_n = H_n^{(1)} = \ln n + \gamma + O\left(n^{-1}\right)$, and $H_n^{(2)} \underset{n\to\infty}{\longrightarrow} \zeta(2) = \frac{\pi^2}{6}$ (cf. Hofri (1987)).

Define next the normalized sum

$$\widehat{M}_n := \frac{M_n - E\,M_n}{\sqrt{\operatorname{Var} M_n}}. \tag{9.3.17}$$

Then as in Section 9.3.1, we obtain the normal approximation, but with a "very slow" logarithmic rate of convergence.

Theorem 9.3.3 *For all $n \in \mathbb{N}$ and some absolute constant $C > 0$, the following uniform rate of convergence holds:*

$$\varrho\left(\widehat{M}_n, \mathcal{N}(0,1)\right) \leq \frac{C}{\sqrt{\ln n}}. \tag{9.3.18}$$

Proof: We invoke the Berry–Esséen bound (9.3.6), where $E\,X_k = \frac{1}{k}$, $\operatorname{Var} X_k = \frac{k-1}{k^2}$, and $E|X_k - E\,X_k|^3 = \frac{k^3 - 3\,k^2 + 4\,k - 2}{k^4}$.

Therefore, $\sum_{k=2}^{n} E|X_k - EX_k|^3 \sim \ln n$, and $\sum_{k=2}^{n} \operatorname{Var} X_k \sim \ln n$, leading to (9.3.18). The constant C can be easily explicitly calculated. □

The normalization of M_n is logarithmic in n. To get a rate of convergence result similar to that in (9.3.18), we shall make use of the ζ_3-metric. It turns out that in this example we obtain contraction factors of order $\sqrt{\frac{\ln(n-1)}{\ln n}}$ that converge to one. Nevertheless, the method described in the proof of Theorem 9.3.2 can also be applied in this case. To this end, define

$$\widetilde{M}_n := \frac{M_n - E\,M_n}{\sqrt{\ln n}}. \tag{9.3.19}$$

Theorem 9.3.4 *For $\sigma_n^2 := \operatorname{Var} \widetilde{M}_n$ and $Z_n \sim \mathcal{N}(0, \sigma_n^2)$, we have*

$$\zeta_3(\widetilde{M}_n, Z_n) = O\left(\frac{1}{\sqrt{\ln n}}\right). \tag{9.3.20}$$

Proof: Indeed, $\widetilde{M}_n$ satisfies the recursion

$$\widetilde{M}_n \stackrel{d}{=} \sqrt{\frac{\ln(n-1)}{\ln n}}\, \widetilde{M}_{n-1} + \widetilde{X}_n, \tag{9.3.21}$$

where $\widetilde{X}_n := \frac{X_n - E\,X_n}{\sqrt{\ln n}}$. Let (Z_n) be independent normally distributed r.v.s, $Z_n \sim \mathcal{N}(0, \sigma_n^2)$, and let

$$Z_n^* := \sqrt{\frac{\ln(n-1)}{\ln n}}\, Z_{n-1} + \widetilde{X}_n \tag{9.3.22}$$

be the accompanying sequence. Further, let

$$Y_n \sim \mathcal{N}(0, \tau_n^2), \text{ and } \tau_n^2 := \sigma_n^2 - \frac{\ln(n-1)}{\ln n}\sigma_{n-1}^2 = \operatorname{Var} \widetilde{X}_n. \tag{9.3.23}$$

Then

$$Z_n \stackrel{d}{=} \sqrt{\frac{\ln(n-1)}{\ln n}}\, Z_{n-1} + Y_n, \tag{9.3.24}$$

and using the same arguments as in Section 9.2, we get

$$\begin{aligned}
\zeta_3(\widetilde{M}_n, Z_n) &\le \zeta_3(\widetilde{M}_n, Z_n^*) + \zeta_3(Z_n^*, Z_n) \\
&\le \left(\frac{\ln(n-1)}{\ln n}\right)^{3/2} \zeta_3(\widetilde{M}_{n-1}, Z_{n-1}) + \zeta_3(Y_n, \widetilde{X}_n).
\end{aligned}$$

By iteration, this yields the bound

$$\zeta_3(\widetilde{M}_n, Z_n) \le \left(\frac{\ln 2}{\ln n}\right)^{3/2} \zeta_3(\widetilde{M}_2, Z_2) + \sum_{i=3}^{n} \left(\frac{\ln i}{\ln n}\right)^{3/2} \zeta_3(Y_i, \widetilde{X}_i). \tag{9.3.25}$$

By the moment estimate $\zeta_3(\widetilde{M}_2, Z_2) < \infty$, and since $\operatorname{Var} \widetilde{X}_i = \operatorname{Var} Y_i = \tau_i^2 = \frac{1}{\ln i} \cdot \frac{i-1}{i^2}$, we have

$$\begin{aligned}
\zeta_3(Y_i, \widetilde{X}_i) &\le \frac{1}{6}\left(E|Y_i|^3 + E|\widetilde{X}_i|^3\right) \\
&\le \frac{1}{(\ln i)^{3/2}} \cdot \frac{1}{6}\left(\frac{1}{i} + \frac{\sqrt{8}}{\sqrt{\pi}} \cdot \frac{1}{i\sqrt{i}}\right);
\end{aligned} \tag{9.3.26}$$

here we also used the estimate $E\left|X_i - \frac{1}{i}\right|^3 \le \frac{1}{i}$.

From (9.3.25) we finally obtain

$$\begin{aligned}
\zeta_3(\widetilde{M}_n, Z_n) &\le \frac{1}{6}\cdot\left(\frac{\ln 2}{\ln n}\right)^{3/2} \\
&\quad + \sum_{i=3}^{n}\left(\frac{\ln i}{\ln n}\right)^{3/2}\cdot\frac{1}{(\ln i)^{3/2}}\cdot\frac{1}{6}\left(\frac{1}{i}+\frac{\sqrt{8}}{\sqrt{\pi}}\cdot\frac{1}{i\sqrt{i}}\right) \\
&= \frac{1}{6\,(\ln n)^{3/2}}\left[(\ln 2)^{3/2}+\sum_{i=3}^{n}\frac{1}{i}+\sum_{i=3}^{n}\frac{1}{i\sqrt{i}}\cdot\frac{\sqrt{8}}{\sqrt{\pi}}\right] \\
&\le \frac{1}{6\,(\ln n)^{3/2}}\left[(\ln 2)^{3/2}+2\ln n\right] \\
&= \frac{1}{3}\cdot\frac{1}{\sqrt{\ln n}}+O\left((\ln n)^{-3/2}\right).
\end{aligned}$$

□

9.3.3 Unsuccessful Searching in Binary Search Trees

In this and the following section we deal with the analysis of inserting and retrieving randomly ordered data in binary search trees by the contraction method; we refer to Mahmoud (1992) for an introduction to random search tree algorithms.

Let U_n denote the number of comparisons that are necessary in order to insert a new random element in a random search tree. A *search tree* is called *random* if it arises from a random permutation. An element (to be inserted in a tree) is called random if each of the $n+1$ free leaves of the tree has probability $\frac{1}{n+1}$ of being chosen.

U_n satisfies the recursion

$$U_n \stackrel{d}{=} U_{n-1} + Y_n, \quad U_0 = 0, \tag{9.3.27}$$

where U_{n-1}, Y_n are independent, $Y_n \sim B(1, \frac{2}{n+1})$. For $n = 1$, one comparison with the root is necessary. For $n \ge 2$, insertion of the $(n+1)$th element needs as many comparisons in the n-tree as in the $(n-1)$-tree except in the case that one comparison with the nth element is necessary. The probability that no comparison with this element is necessary equals $\frac{n-1}{n+1}$.

From (9.3.27) we have

$$E\,U_n = 2\,(H_{n+1}-1), \quad \operatorname{Var} U_n = 2\,H_{n+1} - 4\,H_{n+1}^{(2)} + 2. \tag{9.3.28}$$

Brown and Shubert (1984) (cf. Mahmoud (1992, p. 76)) proved a central limit theorem for U_n making use of the Lyapunov theorem and the method generating functions. Since by (9.3.27),

$$U_n \stackrel{d}{=} \sum_{i=1}^{n} Y_i, \quad Y_i \sim B\left(1, \frac{2}{i+1}\right), \quad (Y_i) \text{ independent}, \tag{9.3.29}$$

this argument can be simplified to yield the following theorem.

Theorem 9.3.5 *Define*

$$\widehat{U}_n := \frac{U_n - E\,U_n}{\sqrt{\operatorname{Var} U_n}}.$$

Then for some constant $C > 0$ and all n,

$$\varrho\left(\widehat{U}_n, \mathcal{N}(0,1)\right) \leq \frac{C}{\sqrt{\ln n}}. \tag{9.3.30}$$

Proof: Observe that $\dfrac{S_{n,3}}{S_{n,2}^{3/2}} \sim \dfrac{1}{\sqrt{2\,\ln n}}$ (cf. Mahmoud (1992, p. 77)). Therefore, (9.3.30) is a consequence of (9.3.6). □

Applying the results of Deheuvels and Pfeifer (1988) we obtain that $\frac{1}{\ln n}$ is the exact order of approximation of $\widehat{U}_n$ by a Poisson distribution. This indicates that the logarithmic rate in the Berry–Esseen bound (9.3.30) should give essentially the right order of approximation. The following rate of convergence result, obtained by the contraction method, supports the fact that the logarithmic order is sharp.

The contraction method can be applied in the theorem below in much the same way as in Section 9.3.2. We therefore only give a sketch of the proof. For more details we refer to Cramer (1995a).

Theorem 9.3.6 *Define* $\widetilde{U}_n := \frac{U_n - E\,U_n}{\sqrt{\ln n}}$, $\sigma_n^2 := \operatorname{Var} \widetilde{U}_n$, *and* $Z_n \sim \mathcal{N}(0, \sigma_n^2)$. *Then, for some $C > 0$ and all $n \in \mathbb{N}$, we have*

$$\zeta_3\left(\widetilde{U}_n, Z_n\right) \leq \frac{C}{\sqrt{\ln n}}. \tag{9.3.31}$$

Proof: $\widetilde{U}_n$ satifies the recursion

$$\widetilde{U}_n \stackrel{d}{=} \sqrt{\frac{\ln(n-1)}{\ln n}}\, \widetilde{U}_{n-1} + \widetilde{Y}_n, \quad \widetilde{Y}_n := \frac{Y_n - E\,Y_n}{\sqrt{\ln n}}. \tag{9.3.32}$$

Define then

$$Z_n^* \;:=\; \sqrt{\frac{\ln(n-1)}{\ln n}}\, Z_{n-1} + \widetilde{Y}_n \tag{9.3.33}$$

and

$$\tau_n^2 \;:=\; \sigma_n^2 - \frac{\ln(n-1)}{\ln n}\,\sigma_{n-1}^2 \;=\; \operatorname{Var}\widetilde{Y}_n. \tag{9.3.34}$$

Let the normal random variables $W_n \sim \mathcal{N}(0,\tau_n^2)$ be independent of the sequences (Z_n), (Y_n). Then the sequences

$$Z_n \;\stackrel{d}{=}\; \sqrt{\frac{\ln(n-1)}{\ln n}}\, Z_{n-1} + W_n. \tag{9.3.35}$$

Consequently, as in Section 9.3.2, we have the bound

$$\zeta_3\left(\widetilde{U}_n, Z_n\right) \le \left(\frac{\ln 2}{\ln n}\right)^{3/2} \zeta_3\left(\widetilde{U}_2, Z_2\right) + \sum_{i=3}^{n}\left(\frac{\ln i}{\ln n}\right)^{3/2} \zeta_3\left(W_i, \widetilde{Y}_i\right). \tag{9.3.36}$$

Next, since $E\,\widetilde{U}_2 = 0 = E\,Z_2$, $\operatorname{Var}\widetilde{U}_2 = \sigma_2^2 = \operatorname{Var} Z_2$, it follows that

$$\zeta_3\left(\widetilde{U}_2, Z_2\right) \;\le\; \frac{1}{6}\left(E\left|\widetilde{U}_2\right|^3 + E\left|Z_2\right|^3\right) \;<\; \infty.$$

Furthermore,

$$\begin{aligned}
\zeta_3\left(W_i, \widetilde{Y}_i\right) &\le \frac{1}{6}\left(E|W_i|^3 + E\left|\widetilde{Y}_i\right|^3\right)\\
&= \frac{1}{6}\left[\frac{2\sqrt{2}}{\sqrt{\pi}}\tau_i^3 + \frac{1}{(\ln i)^{3/2}}\left(\frac{i-1}{i+1}\left(\frac{2}{i+1}\right)^3 + \frac{2}{i+1}\left(\frac{i-1}{i+1}\right)^3\right)\right]\\
&\le \frac{2}{6\,(\ln i)^{3/2}\,(i+1)}\left[1 + \frac{4}{\sqrt{\pi}\,(i+1)}\right].
\end{aligned}$$

Therefore,

$$\begin{aligned}
\zeta_3\left(\widetilde{U}_n, Z_n\right) \;&\le\; \frac{1}{(\ln n)^{3/2}}\frac{1}{6}\left(\frac{10}{81} + \frac{8}{27\sqrt{\pi}}\right)\\
&\quad + \frac{1}{(\ln n)^{3/2}}\sum_{i=3}^{n}\frac{1}{i+1}\cdot\frac{1}{3}\left(1 + \frac{4}{\sqrt{\pi(i+1)}}\right)\\
&\le\; \frac{1}{\sqrt{\ln n}} \qquad \text{for } n \ge n_0,
\end{aligned} \tag{9.3.37}$$

as required. □

Remark 9.3.7 *Studying the recursion (9.3.32) we can also obtain rate of convergence under alternative distributional assumptions on $\widetilde{Y}_n$ (resp. Y_n). For example, if μ_r is any $(r,+)$-ideal, simple metric, then (as in (9.3.36))*

$$\mu_r\left(\widetilde{U}_n, Z_n\right) \le \left(\frac{\ln 2}{\ln n}\right)^{r/2} \mu_r\left(\widetilde{U}_2, Z_2\right) + \sum_{i=3}^{n} \left(\frac{\ln i}{\ln n}\right)^{r/2} \mu_r\left(W_i, \widetilde{Y}_i\right). \tag{9.3.38}$$

This indeed implies that

$$\mu_r\left(\widetilde{U}_n, Z_n\right) \underset{n\to\infty}{\longrightarrow} 0, \tag{9.3.39}$$

provided that the following conditions hold:

$$\text{(a)} \quad \mu_r\left(\widetilde{U}_2, Z_2\right) < \infty, \quad \mu_r\left(W_i, \widetilde{Y}_i\right) < \infty, \quad i \ge 3.$$

$$\text{(b)} \quad \mu_r\left(W_i, \widetilde{Y}_i\right) = o\left(\frac{1}{i \ln i}\right). \tag{9.3.40}$$

To show (9.3.39) for $\varepsilon > 0$, choose $k_0 \in \mathbb{N}$ such that $\mu_r(W_k, \widetilde{Y}_k) \le \dfrac{\varepsilon}{k \ln k}$, for $k \ge k_0$. Then

$$\begin{aligned}
\limsup_{n\to\infty} \mu_r\left(\widetilde{U}_n, Z_n\right) &\le \limsup_{n\to\infty} \left(\frac{\ln 2}{\ln n}\right)^{r/2} \mu_r\left(\widetilde{U}_2, Z_2\right) \\
&\quad + \limsup_{n\to\infty} \frac{1}{(\ln n)^{r/2}} \sum_{i=3}^{k_0-1} (\ln i)^{r/2} \mu_r\left(W_i, \widetilde{Y}_i\right) \\
&\quad + \limsup_{n\to\infty} \frac{1}{(\ln n)^{r/2}} \sum_{i=k_0}^{n} (\ln i)^{r/2-1} \frac{1}{i}\, \varepsilon \\
&\le 0 + 0 + \limsup_{n\to\infty} \varepsilon \frac{1}{\ln n} \sum_{i=k_0}^{n} \frac{1}{i} \le \varepsilon.
\end{aligned}$$

In the preceding example of unsuccessful searching, the estimate of the rate of "merging" of the sequences (W_i) and $(\widetilde{Y}_i)$ in terms of $\mu_r(W_i, \widetilde{Y}_i)$ is of order $1/i(\ln i)^{3/2}$, allowing us to reach the convergence rate $1/\sqrt{\ln n}$.

9.3.4 Successful Searching in Binary Search Trees

Given a random binary search tree as in Section 9.3.3, let S_n denote the number of comparisons to retrieve a randomly chosen element in the tree. Brown and Shubert (1984) derived a formula for $P(S_n = k)$, and Louchard (1987) proved a central limit theorem for S_n using the generating function method in Mahmoud (1992, pp. 78–82). We shall next derive a quantitative version of the central limit theorem. Our main tool will be the contraction method and moment formulas based on the following recursion for S_n:

$$S_n \stackrel{d}{=} 1 + S_{I_n}, \quad S_0 = 0, \quad S_1 = 1. \tag{9.3.41}$$

Here I_n is independent of (S_i), and $P(I_n = 0) = \frac{1}{n}$, $P(I_n = j) = \frac{2j}{n^2}$, $1 \le j \le n-1$.

It can be shown that this recursion does not transform itself to a sum of independent random variables as was done in the random search algorithm in Rachev and Rüschendorf (1991) (cf. (9.3.59)). Therefore, (9.3.41) does not allow the application of the Berry–Esseen-type or Poisson-type approximation result. In fact, it arises from the recursion

$$\begin{aligned} P(S_n = k) &= \sum_{j=1}^{n} P(S_n = k,\ j \text{ chosen}) \qquad (9.3.42) \\ &= \sum_{j=1}^{n} \sum_{i=1}^{n} \frac{1}{n^2} \big(1_{\{i=j\}}\, \delta_{1k} + 1_{\{i<j\}}\, P(S_{n-i} = k-1) \\ &\quad + 1_{\{i>j\}}\, P(S_{i-1} = k-1)\big) \\ &= \frac{\delta_{1k}}{n} + \sum_{i=1}^{n} \frac{n-i}{n^2} P(S_{n-i} = k-1) \\ &\quad + \sum_{i=1}^{n} \frac{i-1}{n^2} P(S_{i-1} = k-1) \\ &= \frac{\delta_{1k}}{n} + \sum_{j=1}^{n-1} P(S_j = k-1) \cdot \frac{2j}{n^2}. \end{aligned}$$

An explicit formula for $P(S_n = k)$ is due to Brown and Shubert (1984) (cf. Mahmoud (1992, p. 79)). Making use of the Brown–Shubert result, Mahmoud (1992, p. 80) desired formulas for the first two moments of S_n. The recursion (9.3.41) leads to a direct calculation of those moments, as we shall see in the next proposition.

Proposition 9.3.8

(a) $$E\, S_n = 2\left(1 + \frac{1}{n}\right) H_n - 3. \tag{9.3.43}$$

(b) $$\operatorname{Var} S_n = \left(2 + \frac{10}{n}\right) H_n - 4\left(1 + \frac{1}{n}\right)\left[\frac{H_n^2}{n} + H_n^{(2)}\right] + 4. \tag{9.3.44}$$

Proof:

(a)
$$E\,S_n = 1 + E\,(E(S_{I_n}|I_n)) = 1 + \sum_{k=0}^{n-1} P(I_n = k)\; E\,S_k \qquad (9.3.45)$$
$$= 1 + \sum_{k=0}^{n-1} \frac{2\,k}{n^2}\; E\,S_k.$$

With $Q_n := n \cdot E\,S_n$, the recursion (9.3.45) leads to $Q_n = n + \frac{2}{n}\sum_{k=1}^{n-1} Q_k$, $Q_1 = 1$, which implies $Q_{n+1} = \frac{2\,n+1}{n+1} + \frac{n+2}{n+1}\,Q_n$. Iteratively,

$$\begin{aligned} Q_n &= \frac{2\,n-1}{n} + \sum_{k=1}^{n-1} \frac{2\,k-1}{k} \cdot \frac{n+1}{k+1} \\ &= (n+1)\left[\sum_{k=1}^{n} \frac{2}{k+1} - \sum_{k=1}^{n}\left(\frac{1}{k} - \frac{1}{k+1}\right)\right] \\ &= (n+1)\left[2\,(H_{n+1}-1) \;-\; 1 \;+\; \frac{1}{n+1}\right] \\ &= 2(n+1)\,H_n - 3\,n. \end{aligned}$$

(b)
$$\begin{aligned} E\,S_n^2 &= 1 + 2\,E\,S_{I_n} + E\,S_{I_n}^2 \\ &= 1 + 2\,(E\,S_n - 1) + \sum_{j=1}^{n-1} \frac{2\,j}{n^2}\; E\,S_j^2. \end{aligned}$$

With $P_n := n \cdot E\,S_n^2$, we obtain $P_n = -n + 2\,Q_n + \frac{2}{n}\sum_{j=1}^{n-1} P_j$.

This yields $\frac{n+1}{2}\,P_{n+1} - \frac{n}{2}\,P_n = \frac{2\,n+1}{2} + 2\,Q_n + P_n$. By (a), we now have

$$P_{n+1} = 8\,H_n - \frac{10\,n-1}{n+1} + \frac{n+2}{n+1}\,P_n,$$

and iterating the above expression, we get

$$P_n = \sum_{j=1}^{n}\left(8\,H_j - \frac{10\,j-3}{j}\right)\frac{n+1}{j+1}.$$

The relation

$$\sum_{j=1}^{n} \frac{H_j}{j} = \frac{H_n^{(2)} + H_n^2}{2}$$

leads to an explicit calculation of P_n, which yields (9.3.44). □

Our next step is to show that (S_n) after a logarithmic normalization merges to a sequence of normal r.v.s.

Define the following normalized version of (S_n):

$$\widetilde{S}_n := \frac{S_n - E\,S_n}{\sqrt{2\,\ln n}}, \quad \widetilde{S}_0 = \widetilde{S}_1 = 0. \tag{9.3.46}$$

Let

$$a(k,n) := 1 - E\,S_n + E\,S_k, \; b(k) := \operatorname{Var} S_k, \; \sigma_n^2 := \operatorname{Var} \widetilde{S}_n. \tag{9.3.47}$$

For our derivation we need the following (so far unchecked):

$$\text{(C)} \quad \limsup_{n\to\infty} \int y^2 \left| \sum_{k=2}^{n-1} \frac{2\,k}{n^2} \left[\Phi\left(\frac{y}{\sqrt{b(n)}}\right) - \Phi\left(\frac{y-a(k,n)}{\sqrt{b(k)}}\right)\right]\right| dy < \infty. \tag{9.3.48}$$

Here, Φ is the standard normal d.f.

Let (Z_n) be independent of (S_n), and $Z_n \sim \mathcal{N}(0,\sigma_n^2)$.

Theorem 9.3.9 *Suppose that* (C) *holds. Then there exists a constant* $K < \infty$ *such that*

$$\zeta_3\left(\widetilde{S}_n, Z_n\right) \le \frac{K}{\sqrt{\ln n}}. \tag{9.3.49}$$

Proof: Note first that $(\widetilde{S}_n)$ satisfies the recursion

$$\widetilde{S}_n \stackrel{d}{=} \sqrt{\frac{\ln I_n}{\ln n}}\, \widetilde{S}_{I_n} + c_n(I_n), \tag{9.3.50}$$

where $c_n(k) := 1 - E\,S_n + E\,S_k/\sqrt{2}\,\ln n$. Define then the accompanying sequence

$$Z_n^* \stackrel{d}{=} \sqrt{\frac{\ln I_n}{\ln n}}\, Z_{I_n} + c_n(I_n). \tag{9.3.51}$$

Applying the "ideality" properties of the metric ζ_3, we obtain the following recursive bound for $\zeta_3(\widetilde{S}_n, Z_n)$;

$$\begin{aligned}
\zeta_3\left(\widetilde{S}_n, Z_n\right) &\le \zeta_3\left(\widetilde{S}_n, Z_n^*\right) + \zeta_3\left(Z_n^*, Z_n\right) \qquad (9.3.52)\\
&\le \sum_{k=0}^{n-1} P(I_n = k)\, \zeta_3\left(\sqrt{\frac{\ln k}{\ln n}}\, \widetilde{S}_k + c_n(k), \sqrt{\frac{\ln k}{\ln n}}\, Z_k + c_n(k)\right)\\
&\quad + \zeta_3(Z_n^*, Z_n)\\
&\le \sum_{k=2}^{n-1} \frac{2\,k}{n^2}\left(\frac{\ln k}{\ln n}\right)^{3/2} \zeta_3\left(\widetilde{S}_k, Z_k\right) + \zeta_3\left(Z_n^*, Z_n\right).
\end{aligned}$$

To estimate the (ζ_3)-distance between Z_n^*, and Z_n we compute the first two moments of Z_n^*:

$$\begin{aligned} E\, Z_n^* &= \sum_{k=0}^{n-1} P(I_n = k)\, E\left(\sqrt{\frac{\ln k}{\ln n}}\, Z_k + c_n(k)\right) \\ &= \sum_{k=0}^{n-1} P(I_n = k)\, \frac{1 - E\, S_n + E\, S_k}{\sqrt{2\,\ln n}} \\ &= (2\,\ln n)^{-1/2}\, [1 - E\, S_n + E\, S_{I_n}] \;=\; 0 \;=\; E\, Z_n, \end{aligned}$$

and similarly, $E\,(Z_n^*)^2 = \frac{1}{2\,\ln n}\ \mathrm{Var}\, S_n \;=\; \mathrm{Var}\, \widetilde{S}_n$. Now we obtain

$$\zeta_3\,(Z_n^*, Z_n) \;\leq\; \frac{1}{6}\,\kappa_3\,(Z_n^*, Z_n) \;=\; \frac{1}{2}\int x^2\, \left|F_{Z_n^*}(x) - F_{Z_n}(x)\right|\, dx.$$

Furthermore, $F_{Z_n}(x) = \Phi(x/\sigma_n) = \Phi\left(x\,\sqrt{2\,\ln n}/\sqrt{b(n)}\right)$, and

$$\begin{aligned} F_{Z_n^*}(x) &= \sum_{k=0}^{n-1} P(Z_n^* \leq x \,|\, I_n = k)\cdot P(I_n = k) \\ &= \sum_{k=2}^{n-1} \frac{2\,k}{n^2}\, \Phi\left(\frac{x\,\sqrt{2\,\ln n} - a(k,n)}{\sqrt{b(k)}}\right) + \frac{1}{n}\, 1_{[1 - E\, S_n, \infty)}(x\,\sqrt{2\,\ln n}) \\ &\quad + \frac{2}{n^2}\, 1_{[2 - E\, S_n, \infty)}(x\,\sqrt{2\,\ln n}). \end{aligned}$$

Applying the substitution $y = x\cdot\sqrt{2\,\ln n}$, the above implies

$$\zeta_3\,(Z_n^*, Z_n) \;\leq\; \frac{1}{2\cdot(2\,\ln n)^{3/2}}\, [A_n + B_n + C_n]\,, \tag{9.3.53}$$

where

$$\begin{aligned} A_n &:= \frac{1}{n}\int y^2 \left| 1_{[1 - E\, S_n, \infty)}(y) - \Phi\left(\frac{y}{\sqrt{b(n)}}\right)\right|\, dy, \\ B_n &:= \frac{2}{n^2}\int y^2 \left| 1_{[2 - E\, S_n, \infty)}(y) - \Phi\left(\frac{y}{\sqrt{b(n)}}\right)\right|\, dy, \end{aligned}$$

and

$$C_n \;:=\; \int y^2 \left| \sum_{k=2}^{n-1} \frac{2\,k}{n^2} \left[\Phi\left(\frac{y - a(k,n)}{\sqrt{b(k)}}\right) - \Phi\left(\frac{y}{\sqrt{b(n)}}\right)\right]\right|\, dy.$$

Invoking the assumption (C), we obtain $C_n \leq M_C$ for all $n \in \mathbb{N}$ and a fixed constant M_C. For $n \geq n_0$ we have $E\,S_n \geq 1$ and

$$\begin{aligned} A_n &\leq \frac{1}{n}\int y^2 \left|\Phi\left(\frac{y}{\sqrt{b(n)}} - 1_{[0,\infty)}(y)\right)\right| \, dy + \frac{1}{n}\int y^2\, 1_{[1-E\,S_n,0)}(y)\, dy \\ &\leq \frac{1}{n}\cdot\frac{1}{3}2\frac{\sqrt{2}}{\sqrt{\pi}}\sqrt{b(n)}^{\,3} + \frac{1}{n}\cdot\frac{1}{3}(E\,S_n - 1)^3 \underset{n\to\infty}{\longrightarrow} 0. \end{aligned}$$

The last bound follows from the follwoing asymptotics: $b(n) = \operatorname{Var} S_n \sim 2\,\ln n$, and $E\,S_n \sim 2\,\ln n$. Therefore, $A_n \leq M_A$, and similarly $B_n \leq M_B$, for all n, and we obtain

$$\zeta_3(Z_n, Z_n^*) \leq \frac{M}{(\ln n)^{3/2}}, \tag{9.3.54}$$

where M is a fixed constant. Next, we need to apply the Euler summation formula (cf. Hofri (1987, p. 19)) to the function $f(x) = x\,\ln x$, $x \geq 1$:

$$\sum_{j=1}^{n-1} f(j) = \int_1^n f(x)\,dx + \sum_{k=1}^{m}\frac{B_k}{k!}\left[f^{(k-1)}(n) - f^{(k-1)}(1)\right] + R_m, \tag{9.3.55}$$

where (B_k) are the Bernoulli numbers. In (9.3.55) the term R_m has the form

$$R_m = \frac{(-1)^{m+1}}{m!}\int_1^n B_m(\{x\})\,f^{(m)}(x)\,dx, \quad \{x\} = x - \lfloor x\rfloor,$$

where $B_m(x) = \sum_{k\geq 0}\binom{m}{k} B_k\,x^{m-k}$ is the mth Bernoulli polynomial. After some calculations, (9.3.55) with $m = 2$ yields

$$\sum_{j=2}^{n-1} j\,\ln j = \frac{1}{2}n^2\ln n - \frac{1}{4}n^2 - \frac{1}{2}n\,\ln n + O(\ln n). \tag{9.3.56}$$

Consider a sufficiently large n_0 such that for $n \geq n_0$, $\sum_{j=2}^{n-1} j\,\ln j \leq \frac{1}{2}n^2\ln n - \frac{1}{4}n^2$. Choose $\widetilde{M}$ large enough that (9.3.49) (with $\widetilde{M}$ instead of K) holds for $n < n_0$ and define $K := \max\left(\widetilde{M}, 2\,M\right)$. So from (9.3.52), (9.3.54), using inductive arguments and assuming (9.3.49) for all $k < n$, we obtain the final bound:

$$\zeta_3\left(\widetilde{S}_n, Z_n\right) \leq \sum_{k=2}^{n-1}\frac{2\,k}{n^2}\left(\frac{\ln k}{\ln n}\right)^{3/2}\frac{K}{\sqrt{\ln k}} + \frac{M}{(\ln n)^{3/2}}$$

$$
\begin{aligned}
&= \frac{1}{(\ln n)^{3/2}} \cdot \frac{2}{n^2} \cdot K \sum_{k=2}^{n-1} k \ln k + \frac{M}{(\ln n)^{3/2}} \\
&\leq \frac{1}{(\ln n)^{3/2}} \left[\frac{2K}{n^2} \left(\frac{1}{2} n^2 \ln n - \frac{1}{4} n^2 \right) + M \right] \\
&\leq \frac{1}{(\ln n)^{3/2}} \left[K \ln n - \frac{K}{2} + \frac{K}{2} \right] = \frac{K}{\sqrt{\ln n}}.
\end{aligned}
$$

□

Remark 9.3.10 *In the preceding example, a direct proof of the convergence of* $\widetilde{S}_n$ *based on direct application of the method of probability metric seems impossible. We were able to obtain the rate of convergence by induction arguments that use the Euler summation formula in a crucial way. This extension of the contraction technique seems to be potentially useful also for other examples in the theorey of probability metrics.*

Remark 9.3.11 *Numerical simulations (for* $n \leq 10,000$*) indicate that* (C) *is correct. Let us denote the integral in* (9.3.48) *for* $n \in \mathbb{N}$ *by* $f(n)$*. Numerical calculation in the range* -25 *to* 25 *(with a Newton–Cote algorithm with precision* 10^{-5}*) leads to the graphs in Figures* 9.13 *and* 9.14 *of* $f(n)$ *against* n*, respectively against* $\ln(\ln(\ln n))$*. These graphs indicate the boundedness of* f*.*

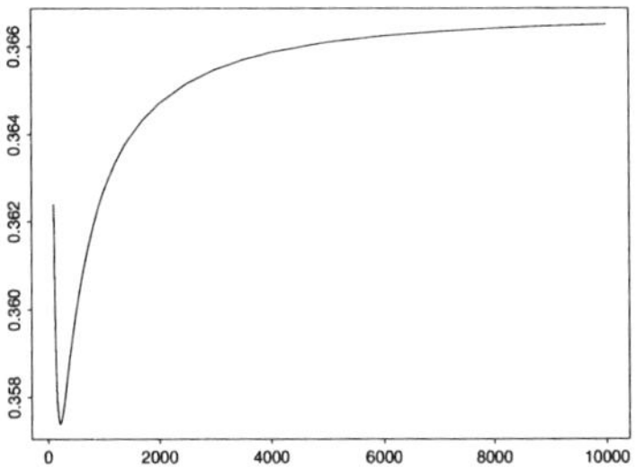

FIGURE 9.13. $f(n)$ against n

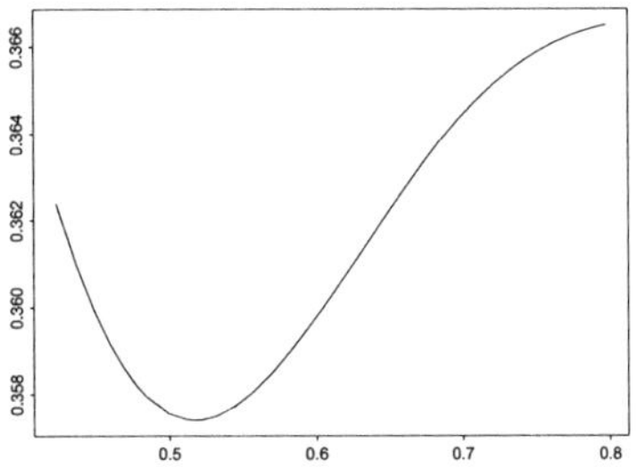

FIGURE 9.14. $f(n)$ against $\ln(\ln(\ln(n)))$

9.3.5 A Random Search Algorithm

In this section we consider a random search in a set of n ordered states $\{1, 2, \ldots, n\}$, starting in the largest state n. Let (T_n) be an independent sequence of random natural numbers, $T_n \leq n - 1$. After one step of the search we reach state $T_n \le n-1$. The search is continued in the smaller set $\{1,\dots,T_n\}$ in the same way, reaching in the next step the state $T_{T_n} \le T_n - 1$. The search ends if state 1 is reached. Let S_n denote the number of steps needed for this random search to reach this final state 1. Then S_n satisfies the recursion

$$S_n \stackrel{d}{=} 1 + S_{T_n}, \quad S_1 = 1. \tag{9.3.57}$$

With r.v.s T_n being uniformly distributed on $\{1,\dots,n-1\}$, this model has been used by Ross (1982, p. 118) and Bickel and Freeman (1981) in a search for an estimate of mean number of steps in the simplex method (with n extreme points). For applications to max-search problems we refer to Nevzorov (1988), and Pfeifer (1991). In their setting there are given independent r.v.s $X_1,\dots,X_n$, and T_n is the largest index $k \le n-1$ such that $X_k > X_n$. We add the index 0 to the state space, $T_n = 0$ meaning that no value larger than X_n occurs.

Consider now the r.v.s $I_1,\dots,I_n$, where I_k is defined as 1 or 0 as state k is visited by the search process or not. Then

$$S_n \stackrel{d}{=} \sum_{k=1}^{n} I_k. \tag{9.3.58}$$

Let $a_i \in [0,1], i \ge 1, a_1 = 1$, and consider the special search strategy

$$P(T_n = k) = \left(\prod_{m=k+1}^{n-1} b_m\right) a_k, \quad 1 \le k \le n-1, \tag{9.3.59}$$

where $b_m = 1 - a_m$ and $\prod_{m=n}^{n-1} b_m = 1$.

Special cases:

(a) If $a_k = 1/k, b_k = (k-1)/k$, then

$$\alpha_{n,k} = \left(\prod_{m=k+1}^{n-1} b_m\right) a_k = \frac{1}{k}\frac{k}{k+1}\cdots\frac{n-2}{n-1} = \frac{1}{n-1},$$

that is, this special case corresponds to the uniform search on $\{1,\dots,n-1\}$;

(b) If $a_k = 1 - e^{-\alpha_k}$, $b_k = e^{-\alpha_k}$, $(\alpha_1 = -\infty)$, then

$$\alpha_{n,k} = e^{-\sum_{m=k+1}^{n-1} \alpha_k}(1 - e^{-\alpha_n}).$$

With our choice of the search probabilities in (9.3.59) we can easily see that the random variables

$$I_1, \ldots, I_n \quad \text{are independent, and} \quad I_i \stackrel{d}{=} B(1, a_i). \tag{9.3.60}$$

The above implies that

$$S_n \stackrel{d}{=} \sum_{i=1}^{n} I_i \tag{9.3.61}$$

is a sum of independent binomial random variables; in particular,

$$ES_n = \sum_{i=1}^{n} a_i, \quad \mathrm{Var}(S_n) = \sum_{i=1}^{n} a_i b_i. \tag{9.3.62}$$

In the uniform search case this leads to

$$ES_n = \log n + \gamma + O\left(\frac{1}{n}\right) \tag{9.3.63}$$

and

$$\mathrm{Var}(S_n) = \log n + \gamma - \frac{\pi^2}{6} + O\left(\frac{1}{n}\right),$$

where $\gamma = 0.5772$ is the Euler constant.

Suppose that $\lambda_n = \sum_{i=2}^{n} a_i$, and $\sum_{i=2}^{n} a_i^2 / \sum_{i=2}^{n} a_i = r_n$ is small for $n \to \infty$. Consider then for the Kolmogorov distance

$$\varrho(X, Y) = \sup_x |F_X(x) - F_Y(x)| \tag{9.3.64}$$

between $S_n - 1$ and a Poisson distributed random variable Z_n with mean λ_n. From the results of Deheuvels and Pfeifer (1988) we have the following asymptotic approximation:

$$\varrho(S_n - 1, Z_n) = \frac{1}{2\sqrt{2\pi e}} r_n + O\left(\max\left\{\sum_2^n p_i^2 / \left(\sum p_i\right)^{3/2}, r_n^2\right\}\right); \tag{9.3.65}$$

that is,

$$P(S_n = k + 1) = e^{-\lambda_n} \lambda_n^k / k! + O(r_n). \tag{9.3.66}$$

Some alternative approximations of S_n in terms of various probability metrics were studied in Rachev and Rüschendorf (1990).

9.3.6 Bucket Algorithms

Consider now n i.i.d. r.v.s $X_1, \ldots, X_n$ with density f on $[0,1]$, and let us divide $[0,1]$ into m intervals $A_i = [\frac{i-1}{m}, \frac{i}{m}]$, $1 \le i \le m$. Let $N = (N_1, \ldots, N_m)$ be the vector of the number of r.v.s in the "m-buckets" $A_1, \ldots, A_m$; in other words, $N_i = \sum_{j=1}^n 1_{A_i}(X_j)$. The total number of comparisons needed to sort n random numbers by the *bucket algorithm* is given by

$$C_n = \sum_{i=1}^m \binom{N_i}{2} = \frac{1}{2}(T_n - n), \qquad T_n = \sum_{i=1}^m N_i^2 \tag{9.3.67}$$

(cf. Devroye (1986)). Since N is multinomial $M(m; p_1, \ldots, p_m)$-distributed with $p_i = \int\limits_{A_i} f(x)\,dx$, we obtain

$$EC_n = \frac{n(n-1)}{2} \sum_{i=1}^m p_j^2. \tag{9.3.68}$$

Therefore, in the case of a uniform distribution $p_j = 1/m$ and for $m/n \to \alpha \in (0,1)$, we have $EC_n \approx n/2\alpha$. In the general case we have the following asymptotics for the first two moments of C_n:

$$EC_n \approx \frac{n}{2\alpha} \int_0^1 f^2(x)\,dx \tag{9.3.69}$$

and

$$\begin{aligned} \frac{1}{n} \operatorname{Var} C_n &\to \frac{4}{\alpha^2}\left(\int f^3(x)\,dx - \left(\int f^2(x)\,dx\right)^2\right) + \frac{2}{\alpha}\int f^2(x)\,dx. \\ &\ge \frac{2}{\alpha}\int f^2(x)\,dx \end{aligned} \tag{9.3.70}$$

We shall demonstrate the method of probability metrics in order to obtain the asymptotic distribution of C_n in the special case $m = 2$ and $n \to \infty$. We have

$$C_n = \frac{1}{n}(T_n - n), \qquad T_n = \left(\sum_{i=1}^{n} \zeta_i\right)^2 + \left(n - \sum_{i=1}^{n} \zeta_i\right)^2, \tag{9.3.71}$$

where the ζ_i's are i.i.d. Bernoulli random variables with success probability p. Define the approximating U-statistic based on a normal sample:

$$D_n = \frac{1}{2}(S_n - n), \quad S_n = \left(\sum_{i=1}^{n} \eta_i\right)^2 + \left(n - \sum_{i=1}^{n} \eta_i\right)^2, \tag{9.3.72}$$

where $\eta_i \sim N(p, pq)$, $q := 1 - p$. (A detailed analysis of the distribution of S_n can be found in Seidel (1988).) Next, by use of $\circ$ we shall denote the normalized quantities $\mathring{\zeta}_i = (\zeta_i - p)/\sqrt{pq}$, $\mathring{\eta}_i = (\eta_i - p)/\sqrt{pq}$, $\mathring{T}_n = (\sum \mathring{\zeta}_i)^2 + (n - \sum \mathring{\zeta}_i)^2$, etc.

The next theorem provides estimates of closeness of $\mathring{C}_n$ and $\mathring{D}_n$ in terms of the Kantorovich ℓ_p-metrics for $p = 1$ and $p = 2$.

Theorem 9.3.12 *For $m = 2$ and $n \to \infty$,*

$$\ell_2\left(\frac{\mathring{C}_n}{n^{3/2}}, \frac{\mathring{D}_n}{n^{3/2}}\right) = O(n^{-1/4}), \tag{9.3.73}$$

and

$$\ell_1\left(\frac{\mathring{C}_n}{n^{3/2}}, \frac{\mathring{D}_n}{n^{3/2}}\right) = O(n^{-1/2}). \tag{9.3.74}$$

Proof: To show (9.3.73) note that the normalization $n^{3/2}$ is of the right order. In fact,

$$\begin{aligned} \operatorname{Var}\left(\frac{\mathring{D}_n}{n^{3/2}}\right) &= \frac{1}{4}\operatorname{Var}(n^{-3/2}\mathring{S}_n) \\ &= \frac{1}{4}\operatorname{Var}\left(\frac{2\sum_{i=1}^{n}\sum_{j=1}^{n} \mathring{\eta}_i\mathring{\eta}_j - 2n\sum \mathring{\eta}_i}{n^{3/2}}\right) \end{aligned}$$

$$\approx \text{ constant Var}\left(\frac{2n\sum \mathring{\eta}_i}{n^{3/2}}\right) \approx \text{ constant } > 0.$$

Since ℓ_2 is the minimal L_2-metric, it follows that

$$\ell_2(\mathring{C}_n, \mathring{D}_n) \leq L_2\left(\frac{1}{2}(\mathring{T}_n - n), \frac{1}{2}(\mathring{S}_n - n)\right) = \frac{1}{2}L_2(\mathring{T}_n, \mathring{S}_n).$$

Thus

$$\begin{aligned}
\ell_2\left(\frac{\mathring{C}_n}{n^{3/2}}, \frac{\mathring{D}_n}{n^{3/2}}\right) &\leq \frac{1}{2}L_2\left(\frac{\mathring{T}_n}{n^{3/2}}, \frac{\mathring{S}_n}{n^{3/2}}\right) \\
&= \frac{1}{2}L_2\left(\left(2\sum_i\sum_j \mathring{\zeta}_i\mathring{\zeta}_j + n^2 - 2n\sum \mathring{\zeta}_i\right)n^{-3/2},\right. \\
&\qquad \left.\left(2\sum_i\sum_j \mathring{\eta}_i\mathring{\eta}_j + n^2 - 2n\sum \mathring{\eta}_i\right)n^{-3/2}\right) \\
&\leq L_2\left(n^{-3/2}\sum_i\sum_j \mathring{\zeta}_i\mathring{\zeta}_j, n^{-3/2}\sum_i\sum_j \mathring{\eta}_i\mathring{\eta}_j\right) \\
&\quad + L_2\left(n^{-1/2}\sum \mathring{\zeta}_i, n^{-1/2}\sum \mathring{\eta}_i\right) =: I_1 + I_2.
\end{aligned}$$

Assuming that the pairs $(\mathring{\zeta}_i, \mathring{\eta}_i)$ are independent, we obtain

$$\begin{aligned}
I_1 &= n^{-3/2}L_2\left(\sum_i \mathring{\zeta}_i^2, \sum_i \mathring{\eta}_i^2\right) + n^{-3/2}L_2\left(\sum_{i\neq j} \mathring{\zeta}_i\mathring{\zeta}_j, \sum_{i\neq j}\mathring{\eta}_i\mathring{\eta}_j\right) \qquad (9.3.75) \\
&\leq n^{-1/2}L_2\left(\mathring{\zeta}_1^2, \mathring{\eta}_1^2\right) + n^{-3/2}\left(E\left(\sum_{i\neq j}\left(\mathring{\zeta}_i\mathring{\zeta}_j - \mathring{\eta}_i\mathring{\eta}_j\right)\right)^2\right)^{1/2} \\
&= n^{-1/2}L_2\left(\mathring{\zeta}_1^2, \mathring{\eta}_1^2\right) + n^{-3/2}\left(\sum_{i\neq j}\left(E\left(\mathring{\zeta}_i\mathring{\zeta}_j - \mathring{\eta}_i\mathring{\eta}_j\right)\right)^2\right)^{1/2} \\
&= n^{-1/2}L_2\left(\mathring{\zeta}_1^2, \mathring{\eta}_1^2\right) + n^{-3/2}\left(n(n-1)L_1\left(\mathring{\zeta}_1\mathring{\zeta}_2, \mathring{\eta}_1\mathring{\eta}_2\right)\right)^{1/2} \\
&\leq n^{-1/2}\left(\text{Var}\left(\mathring{\zeta}_1\right)^2 + \text{Var}\left(\mathring{\eta}_1\right)^2\right) + n^{-1/2}\left(\text{Var}\left(\mathring{\zeta}_1\mathring{\zeta}_2\right) + \text{Var}\left(\mathring{\eta}_1\mathring{\eta}_2\right)\right) \\
&\leq cn^{-1/2}.
\end{aligned}$$

Here and in the sequel c stands for an absolute constant, which may be different at different places. Similarly,

$$\begin{aligned} I_2^2 &= L_2^2\left(n^{-1/2}\sum \mathring{\zeta}_i, n^{-1/2}\sum \mathring{\eta}_i\right) \qquad (9.3.76)\\ &\le L_1\left(n^{-1/2}\sum \mathring{\zeta}_i, n^{-1/2}\sum \mathring{\eta}_i\right)\left(E\left|n^{-1/2}\sum \mathring{\zeta}_i\right| + E\left|n^{-1/2}\sum \mathring{\eta}_i\right|\right)\\ &\le L_1\left(n^{-1/2}\sum \mathring{\zeta}_i, n^{-1/2}\sum \mathring{\eta}_i\right)\left(\operatorname{Var}\mathring{\zeta}_1 + \operatorname{Var}\mathring{\eta}_1\right)\\ &= 2\,L_1\left(n^{-1/2}\sum \mathring{\zeta}_i, n^{-1/2}\sum \mathring{\eta}_i\right). \end{aligned}$$

Passing to the minimal metric, this yields

$$\ell_2(\mathring{C}_n, \mathring{D}_n) \;\le\; cn^{-1/2} + \sqrt{2}\sqrt{\ell_1(n^{-1/2}\sum \mathring{\zeta}_i, \mathring{\eta}_2)}\,. \qquad (9.3.77)$$

The rate of convergence in the CLT for the $\ell_1 = \zeta_1$-metric has been discussed by Zolotarev (1986, Theorem 5.4.7) (see also Rachev and Rüschendorf (1990, Lemma 3.3)). It is given by

$$\ell_1(n^{-1/2}\sum \mathring{\zeta}_i, \mathring{\eta}_1) \;\le\; 11.5 \max\left(\ell_1(\mathring{\zeta}_1, \mathring{\eta}_1), \zeta_2(\mathring{\zeta}_1, \mathring{\eta}_1)\right) n^{-1/2}, \qquad (9.3.78)$$

where ζ_r is the Zolotarev ideal metric of order $r > 0$. This implies $\ell_2(\mathring{C}_n, \mathring{D}_n) \le C n^{-1/4}$.

We can argue similarly to show (9.3.76). The bound (9.3.75) can be replaced by

$$\begin{aligned} I_1 &\le n^{-1/2}\left(E\left|\mathring{\zeta}_1\right|^2 + E\left|\mathring{\eta}_1\right|^2\right) + n^{-1/2}\left(E\left|\mathring{\zeta}_1\mathring{\zeta}_2\right| + E\left|\mathring{\eta}_1\mathring{\eta}_2\right|\right) \qquad (9.3.79)\\ &\le cn^{-1/2}. \end{aligned}$$

Invoking (9.3.78), we see that the term I_2 is of order $n^{-1/2}$. □

We can extend our results to the cases $m = 3, 4$. However, the proofs are computationally quite involved. For some general results on the asymptotic distributions of quadratic forms we refer to de Jong (1989).

10

Stochastic Differential Equations and Empirical Measures

10.1 Propagation of Chaos and Contraction of Stochastic Mappings

In this section we use contraction properties of stochastic mappings with respect to suitably chosen metrics in order to study some new examples of propagation of chaos. In particular, systems of stochastic differential equations (SDEs) with mean field type interactions and the corresponding nonlinear SDEs of McKean–Vlasov type for the limiting cases will be considered. We shall also study the rate of convergence to the corresponding limit. Assumptions on the smoothness and growth properties of the coefficients of the SDEs are to be reflected in the choice of the probability metric in order to obtain the required contraction properties. This allows us to investigate new types of interactions as well as to consider systems with relaxed Lipschitz assumptions.

10.1.1 Introduction

The notion "propagation of chaos" was introduced by Kac in his investigation of the relationship between simple Markov models of interacting particles and nonlinear Boltzmann-type equations; for an introduction to the propagation theory of chaos we refer to Sznitman (1989). A formal definition follows. Let (u_N) be a sequence of symmetric probability measures on E^N, E a separable metric space, and let u be a probability on E. Then (u_N) is called *u-chaotic* if $\pi_k\, u_n \xrightarrow{w} u^{(k)}$. Here π_k stands for the k-marginal distribution, $u^{(k)}$ is the k-fold product, and $\xrightarrow{w}$ denotes weak convergence.

A basic example for chaotic sequences is *McKean's interacting diffusion* (cf. the "laboratory" example in Sznitman (1989, p. 172)). Consider a system of interacting diffusions

$$\begin{aligned} dX_t^{i,N} &= dW_t^i + \frac{1}{N}\sum_{j=1}^N b(X_t^{i,N}, X_t^{j,N})dt, \quad i = 1,\dots,N, \qquad (10.1.1) \\ X_0^{i,N} &= x_0^i, \end{aligned}$$

where the W^i are independent Brownian motions and b satisfies a certain Lipschitz condition. Let u_N denote the distribution of $(X^{1,N},\dots,X^{N,N})$. The nonlinear limiting equation is given by the McKean–Vlasov equation

$$dX_t = dB_t + \int b(X_t, y)u_t(dy)\,dt, \qquad (10.1.2)$$

when B_t is a Brownian motion and u_t is the distribution of X_t. Then u_N is u-chaotic, where u is the distribution of X on $C(\mathbb{R}_+, \mathbb{R}^d)$.

An alternative example of chaotic behavior of particles, not described by SDEs, are uniform distributions on "p-spheres." Let u_N denote the uniform distribution on the p-sphere of radius N in $\mathbb{R}_+^N$, that is, on $S_{p,N} := \{x \in \mathbb{R}_+^N;\ \Sigma x_i^p = N\}$. Let u denote the probability measure on $\mathbb{R}_+$ with density

$$f_p(x) = \frac{p^{1-1/p}}{\Gamma(1/p)} e^{-x^p/p}, \quad x \geq 0.$$

Then for $N > k + p$, and k and N big enough,

$$\|\pi_k u_N - u^{(k)}\| \leq \frac{2(k+p)+1}{N-k-p}, \qquad (10.1.3)$$

where $\|\cdot\|$ denotes the total variation distance (cf. Kuelbs (1977), Rachev and Rüschendorf (1991)). In particular, we obtain that u_N is u-chaotic. This example has its origin in Poincaré's theorem on the asymptotic behavior of particle systems. More general examples of this kind have been developed in the statistical physics literature in connection with the "equivalence of ensembles" but typically without a quantitative estimate as in (10.1.3).

The main goal of this section is the study of propagation of chaos in several modifications of McKean's example. We shall be concerned with the form of the interaction and the regularity assumptions on the coefficients. To this end we introduce suitable probability metrics, allowing us to derive contraction properties of the stochastic equations defined by the corresponding linear equations. Dobrushin (1979) introduced the use of the Kantorovich metric for the interacting diffusions in the model (10.1.1), (10.1.2). The success of this metric is based on a coupling argument inherent in its definition. This metric has been applied since then in several other papers. For some modifications of the model (10.1.1), (10.1.2) we shall need alternatives to the Kantorovich metric that provide suitable regularity and ideality properties for the equations considered. In particular, we need metrics that are "ideal" of higher order when we relax the Lipschitz conditions in equations (10.1.1), (10.1.2).

Our modifications allow us to treat much more complicated forms of interactions than those in the McKean example. In particular, we consider nonlinear interactions via some general energy function, as for example the p-norm of the vector of all pair interactions. We also consider interactions with "outside" particles over the whole past (history) of the process, describing some non-Markovian systems. We demonstrate the flexibility of the approach based on suitable probability metrics to analyze with nonstandard forms of interactions and develop the tools to study complex physical systems.

10.1.2 Equations with p-Norm Interacting Drifts

Consider a system of N interacting diffusions with p-norm interacting drifts; that is, the drift is given by the pth norm of the vector of all pair interactions (which can be viewed as the driving force in the system):

$$\begin{aligned} dX_t^{i,N} &= dW_t^i + \left\{ \frac{1}{N} \sum_{j=1}^{N} b^p \left(X_t^{i,N}, X_t^{j,N} \right) \right\}^{1/p} dt \qquad (10.1.4) \\ X_0^{i,N} &= X_0^i, \quad 1 \le i \le N, \end{aligned}$$

$b \geq 0$, $p \geq 1$. $((W_t^i), X_0^i)$ are independent, identically distributed for all i.) We shall establish that each $X^{i,N}$ has a natural limit $\overline{X}^i$, where the $(\overline{X}^i)$ are independent copies of the solutions of a nonlinear equation

$$\begin{cases} dX_t &= dB_t + \left(\int b(X_t, y)^p \, u_t(dy) \right)^{1/p} dt, \\ X_{t=0} &= X, \end{cases} \tag{10.1.5}$$

with $B \stackrel{d}{=} W^1$ a process on C_T, and $u_t = P^{X_t}$. In order to obtain the necessary contraction properties of these equations, we consider the L_p^*-metric and the corresponding minimal L_p^*-metric (ℓ_p^*), defined for processes X, Y (or the corresponding probability measures $m_1, m_2 \in M^1(C_T)$). Here and in what follows $M^1(C_T)$ denotes the class of all probability distributions on C_T,

$$L_{p,t}^*(X,Y) := (E \sup_{s \leq t} |X_s - Y_s|^p)^{1/p}, \tag{10.1.6}$$

and

$$\ell_{p,t}^*(m_1, m_2) := \inf\{L_{p,t}^*(X,Y);\ X \stackrel{d}{=} m_1,\ Y \stackrel{d}{=} m_2\}. \tag{10.1.7}$$

In (10.1.7) we tacitly assumed that the underlying probability space is rich enough to support all possible couplings of m_1, m_2, which is true, for example, in the case of atomless probability spaces. Define, for $m_0 \in M^1(C_T)$,

$$M_p(C_T, m_0) := \{m_1 \in M^1(C_T);\ \ell_{p,T}^*(m_0, m_1) < \infty\}, \tag{10.1.8}$$

and let $\mathcal{X}_p(C_T, m_0)$ be the class of processes on C_T with distribution $m \in M_p(C_T, m_0)$.

For $m_0 = \delta_a$ (the one-point measure in $a \in C_T$), this is the class of all distributions on C_T with finite pth moment of the process norm. For $m \in M_p(C_T, m_0)$ consider the linear equation corresponding to (10.1.5)

$$X_t = B_t + \int_0^t \left(\int_{C_T} b(X_s, y_s)^p \, dm(y) \right)^{1/p} ds, \tag{10.1.9}$$

where y_s is the value of y at time s. Let (B_t) be a real-valued process on $C_T = C[0,T]$ with finite pth absolute moment ($E\sup_{s\le T}|B_s|^p < \infty$), and let $b \ge 0$ be a Lipschitz function in x; that is,

$$|b(x_1,y) - b(x_2,y)| \;\le\; c|x_1 - x_2|, \quad \text{for all } x_1, x_2 \text{ and } Y \text{ in } C_T. \quad (10.1.10)$$

As usual, a strong solution of the SDE (10.1.9) means a solution measurable with respect to the augmented filtration of the process (B_t). In constrast, a weak solution of (10.1.9) is defined on a suitable filtered space of distributions.

Lemma 10.1.1 *Assume that* (10.1.10) *holds, and let* $\int \left(\int b(0,y_s)^p \,\mathrm{d}m(y)\right)^{1/p} \mathrm{d}s < \infty$.

(a) *Then equation* (10.1.9) *has a unique strong solution* X.

(b) *If* $\Phi(m)$ *is the law of* X*, then* $\Phi(m) \in M_p(C_T, m_0)$*; that is,* $\Phi : M_p(C_T, m_0) \to M_p(C_T, m_0)$.

Proof: Let $X \in \mathcal{X}_p(C_T, m_0)$, and define

$$(SX)_t \;:=\; B_t + \int_0^t \left(\int b(X_s, y_s)^p \,\mathrm{d}m(y)\right)^{1/p} \mathrm{d}s.$$

Then, for $Y \in \mathcal{X}_p(C_T, m_0)$,

$$\begin{aligned}
|(SX)_t - (SY)_t| &\le \int_0^t \mathrm{d}s \left| \left(\int b(X_s, y_s)^p \,\mathrm{d}m(y)\right)^{1/p} - \left(\int b(Y_s, y)^p \,\mathrm{d}m(y)\right)^{1/p} \right| \\
&\le \int_0^t \left(\int |b(X_s, y_s) - b(Y_s, y_s)|^p \,\mathrm{d}m(y)\right)^{1/p} \mathrm{d}s \\
&\le c\int_0^t |X_s - Y_s| \,\mathrm{d}s.
\end{aligned}$$

This implies $\sup_{s\le t}|(SX)_s-(SY)_s| \le c\int_0^t \sup_{u\le s}|X_u-Y_u|\,\mathrm{d}s$, and furthermore,

$$\begin{aligned}
L^*_{p,t}(SX,SY) &= \left(E\sup_{s\le t}|(SX)_s-(SY)_s|^p\right)^{1/p}\\
&\le c\left(E\left(\int_0^t \sup_{u\le s}|X_u-Y_u|\,\mathrm{d}s\right)^p\right)^{1/p}\\
&\le c\int_0^t L^*_{p,s}(X,Y)\,\mathrm{d}s.
\end{aligned}$$

Define inductively $X^0 := B$ and $X^n := SX^{n-1}$. Then the above bound yields

$$L^*_{p,T}(X^n,X^{n-1}) \le c^n\frac{T^n}{n!}L^*_{p,T}(X^1,X^0).$$

For the $L^*_{p,T}$-distance in the right-hand side we have the following estimate:

$$\begin{aligned}
L^*_{p,T}(X^1,X^0) &\le c'\int_0^T\left[E|B_s|^p+\int b(0,y_s)^p\,\mathrm{d}m(y)\right]^{1/p}\mathrm{d}s\\
&\le c'\int_0^T\left(E\sup_{u\le s}|B_u|^p\right)^{1/p}\mathrm{d}s + c'\int_0^T\left(\int b(0,y_s)^p\,\mathrm{d}m(y)\right)^{1/p}\mathrm{d}s\\
&\le c'T(E\sup_{s\le T}|B_s|^p)^{1/p} + c'\int_0^T\left(\int b(0,y_s)^p\,\mathrm{d}m(y)\right)^{1/p}\mathrm{d}s.
\end{aligned}$$

From the assumptions on B and b, the above bound implies that $L^*_{p,T}(X^1,X^0)<\infty$. Consequently,

$$\sum_{n=1}^{\infty} L^*_{p,T}(X^n,X^{n-1}) \le e^{cT}\,L^*_{p,T}(X^1,X^0) < \infty,$$

by the Gronwall lemma. This results in $\sum_{n=1}^{\infty}\sup_{s\le T}|X^n_s-X^{n-1}_s|<\infty$ a.s., and therefore, X^n converges to some process X a.s., uniformly on bounded intervals. The limiting process X is a.s. continuous, has finite pth moments (i.e., $\|X\|_{p,t}^* := E\sup_{s\le t}|X_s|^p < \infty$), and is a fixed point of the mapping S. So, $\Phi(m) = P^X \in M_p(C_T, m_0)$; this holds because $\|B\|_{p,T}^* < \infty$ and $L_{p,T}^*(X, B) < \infty$. □

In addition, suppose that b is Lipschitz in both arguments; that is, for all x_1, x_2, y_1, and y_2 in C_T,

$$|b(x_1,y_1) - b(x_2,y_2)| \ \le \ c[|x_1 - y_1| + |x_2 - y_2|], \tag{10.1.11}$$

and consider the mapping $\Phi : M_p(C_T, m_0) \to M_p(C_T, m_0)$.

Lemma 10.1.2 (Contraction of Φ with respect to the $\ell_{p,t}^*$-minimal metric) *Under the Lipschitz condition* (10.1.11) *and the assumptions of Lemma* 10.1.1, *for $t \le T$ and $m_1, m_2 \in M_p(C_T, m_0)$, the following holds:*

$$\ell_{p,t}^*(\Phi(m_1), \Phi(m_2)) \ \le \ ce^{ct} \int_0^t \ell_{p,u}^*(m_1, m_2)\, du. \tag{10.1.12}$$

Proof: Let for $i = 1, 2$ and $t \le T$,

$$X_t^{(i)} \ := \ B_t + \int_0^t \left(\int_{C_T} b(X_s^{(i)}, y_s)^p \, dm_i(y) \right)^{1/p} ds,$$

and let $m \in M^1(m_1, m_2)$, the class of probability measures on $C_T \times C_T$ with marginals m_1, m_2. Then

$$\begin{aligned}
&\sup_{s\le t} |X_s^{(1)} - X_s^{(2)}| \\
&\le \int_0^t ds \left| \left[\int_{C_T} b(X_s^{(1)}, y_s^{(1)})^p \, dm_1(y^{(1)}) \right]^{1/p} - \left[\int_{C_T} b(X_s^{(2)}, y_s^{(2)})^p \, dm_2(y^{(2)}) \right]^{1/p} \right| \\
&\le \int_0^t ds \left[\int_{C_T \times C_T} \left| b(X_s^{(1)}, y_s^{(1)}) - b(X_s^{(2)}, y_s^{(2)}) \right|^p dm(y^{(1)}, y^{(2)}) \right]^{1/p} \\
&\le c \int_0^t ds \left\{ |X_s^{(1)} - X_s^{(2)}| + \left[\int |y_s^{(1)} - y_s^{(2)}|^p \, dm(y^{(1)}, y^{(2)}) \right]^{1/p} \right\}.
\end{aligned}$$

Minimizing the right-hand side with respect to all couplings m, we obtain

$$\sup_{s\le t}|X_s^{(1)}-X_s^{(2)}|\le c\int_0^t ds \sup_{u\le s}|X_u^{(1)}-X_u^{(2)}|+c\int_0^t ds\,\ell_{p,s}^*(m_1,m_2). \quad (10.1.13)$$

Consequently, by Gronwall's lemma,

$$\sup_{s\le t}|X_s^{(1)}-X_s^{(2)}| \;\le\; c\,e^{ct}\int_0^t \ell_{p,s}^*(m_1,m_2)\,ds. \quad (10.1.14)$$

Finally, passing to the pth norm in the left-hand side of (10.1.14) and then to the corresponding minimal metric $\ell_{p,t}^*$ proves the lemma. □

Theorem 10.1.3 *Under the Lipschitz condition* (10.1.11) *and assuming that* $\int_0^T (\int b(0,y_s)^p\, dm_0(y))^{1/p}\, ds < \infty$, *equation* (10.1.1) *has a unique weak and strong solution in* $\mathcal{X}_p(C_T, m_0)$.

Proof: From Lemma 10.1.2 we obtain that for $m\in M_p(C_T,m_0)$,

$$\begin{aligned}\ell_{p,T}^*(\Phi^{k+1}(m),\Phi^k(m)) &\le c_T^k\frac{T^k}{k!}\ell_{p,T}^*(\Phi(m),m) \qquad (c_T=c\,e^{cT})\\ &\le c_T^k\frac{T^k}{k!}(\ell_{p,T}^*(\Phi(m),m_0)+\ell_{p,T}^*(m,m_0)) \;<\;\infty.\end{aligned}$$

Consequently, $(\Phi^k(m))$ is a Cauchy sequence in $(C_T,\ell_{p,T}^*)$ and thus converges to a fixed point of Φ. Let X^{k+1},X^k denote the couplings of $\Phi^{k+1}(m)$, $\Phi^k(m)$. Then, by (10.1.12), we have that

$$L_{p,T}^*(X^{k+1},X^k)\;\le\; c_T^k\frac{T^k}{k!}\ell_{p,T}^*(\Phi(m),m),$$

and therefore, we determine a unique strong solution with finite pth moment. □

Remark 10.1.4 *While the linear equation in Lemma* 10.1.1 *can be handled with the L_1-metric, in Lemma* 10.1.2 *we obtain only a contraction with respect to the minimal ℓ_p-metric $\ell_{p,T}^*$ (cf. equation* (10.1.12)*).*

Remark 10.1.5 *The result of Theorem* 10.1.3 *can be extended to the case* $p = \infty$ *by applying the metric*

$$L^*_{\infty,T}(X,Y) = \operatorname{ess\,sup} \sup_{s \le T} |X_s - Y_s| \tag{10.1.15}$$

and the corresponding minimal metric

$$\ell^*_{\infty,T}(m_1, m_2) = \inf\{L^*_{\infty,T}(X,Y);\ X \stackrel{d}{=} m_1, Y \stackrel{d}{=} m_2\}. \tag{10.1.16}$$

Then the equation

$$X_t = B_t + \int_0^t \operatorname*{ess\,sup}_{u_s(dy)} b(X_s, y)\,\mathrm{d}s \tag{10.1.17}$$

has a unique solution in $M_\infty(C_T, m_0)$ *if* B *is a.s. bounded, that is, if* $\operatorname{ess\,sup}_{s \le T} |B_s| < \infty$.

Remark 10.1.6 *Several extensions of equation* (10.1.5) *can be handled in a similar way, as for example*

$$X_t = B_t + \int_0^t \left(\int b(X_s, y)^p\, u_s^{(k)}(\mathrm{d}y) \right)^{1/p} \mathrm{d}s, \tag{10.1.18}$$

where $u_s^{(k)} = \bigotimes_{i=1}^k P^{X_s}$ *stands for the k-fold product of* u_s *and* $y = (y_1, \dots, y_k) \in \mathbb{R}^k$. *More generally,* $b = b(s, x, y)$ *can be dependent upon* s *and the past of the process* $y = (y_u)_{u \le s}$. *In this case,* u_s *has to be replaced by* $u_{(s)} := P^{(X_u)_{u \le s}}$ *(the distribution of the past), and we need to assume a functional Lipschitz condition on* b. *In a similar way one can also investigate the d-dimensional case.*

Taking as a starting point Theorem 10.1.3, we next investigate the system of interacting equations in (10.1.4). The following theorem asserts that as N goes to infinity, each $X^{i,N}$ has a natural limit $\overline{X}^i$. Here, the $(\overline{X}^i)$ are independent copies of the solutions of the nonlinear equation (10.1.5).

Theorem 10.1.7 *Let* b *satisfy the Lipschitz condition* (10.1.11) *and suppose that* $\int |b(\overline{X}_s^1, y_s)|^{2p}\, u_s(\mathrm{d}y_s) < \infty$, *a.s. Then*

$$\sup_N \sqrt{N}\, E^{1/p} \sup_{t \le T} |X_t^{i,N} - \overline{X}_t^i|^p < \infty \qquad \textit{for} \quad p \ge 2, \tag{10.1.19}$$

and

$$N^{p-1}\, E \sup_{t\le T} |X_t^{i,N} - \overline{X}_t^i|^p \;=\; o(1) \quad \textit{for} \quad 1 \le p \le 2.$$

Proof: For notational convenience we drop the superscript N; then

$$\begin{aligned}
X_t^i - \overline{X}_t^i &= \int_0^t \left(\left\{ \frac{1}{N} \sum_{j=1}^N b(X_s^i, X_s^j)^p \right\}^{1/p} - \left\{ \int b(\overline{X}_s^i, y)^p \, u_s(dy) \right\}^{1/p} \right) ds \\
&= \int_0^t ds \left\{ \left[\left(\frac{1}{N} \sum_j b(X_s^i, X_s^j)^p \right)^{1/p} - \left(\frac{1}{N} \sum_j b(\overline{X}_s^i, X_s^j \right)^{1/p} \right] \right. \\
&\quad + \left[\left(\frac{1}{N} \sum_j b(\overline{X}_s^i, X_s^j)^p \right)^{1/p} - \left(\frac{1}{N} \sum_j b^p(\overline{X}_s^i, \overline{X}_s^j) \right)^{1/p} \right] \\
&\quad \left. + \left[\left(\frac{1}{N} \sum_j b(\overline{X}_s^i, \overline{X}_s^j)^p \right)^{1/p} - \left(\int b(\overline{X}_s^i, y)^p \, u_s(dy) \right)^{1/p} \right] \right\}.
\end{aligned}$$

Set $|X|_T := \sup_{s\le T} |X_s|$. Then by the Minkowski inequality and the Lipschitz condition on b, the above equality implies

$$\begin{aligned}
\|X^i - \overline{X}^i\|_{T,p} &:= \left(E\left|X^i - \overline{X}^i\right|_T^p \right)^{1/p} \\
&\le \int_0^T ds \left\{ c\|X_s^i - \overline{X}_s^i\|_p + c\frac{1}{N} \sum_{j=1}^N \|X_s^j - \overline{X}_s^j\|_p \right. \\
&\quad \left. + \left(E \left| \left(\frac{1}{N} \sum_j b(\overline{X}_s^i, \overline{X}_s^j)^p \right)^{1/p} - \left(\int b(\overline{X}_s^i, y)^p \, u_s(dy) \right)^{1/p} \right|^p \right)^{1/p} \right\}.
\end{aligned}$$

Summing up over i and using the symmetry, we find

$$N\|X^1 - \overline{X}^1\|_{T,p} \;=\; \sum_{i=1}^N \|X^i - \overline{X}^i\|_{T,p}$$

$$\leq 2c \int_0^T \mathrm{d}s \left\{ \sum_{i=1}^N \|X_s^i - \overline{X}_s^i\|_p + \sum_{i=1}^N E \left| \left(\frac{1}{N} \sum_{j=1}^N b(\overline{X}_s^i, \overline{X}_s^j)^p \right)^{1/p} - \left(\int b(\overline{X}_s^i, y)^p \, u_s(\mathrm{d}y) \right)^{1/p} \right|^p \right\}.$$

This amounts to

$$\|X^i - \overline{X}^i\|_{T,p} \leq 2c \int_0^T \mathrm{d}s \left\{ \|X^i - \overline{X}^i\|_{s,p} + \frac{1}{N} \sum_{i=1}^N \left[E \left| \left(\frac{1}{N} \sum_j b(\overline{X}_s^i, \overline{X}_s^j)^p \right)^{1/p} - \left(\int b(\overline{X}_s^i, y)^p \, u_s(\mathrm{d}y) \right)^{1/p} \right|^p \right] \right\},$$

and consequently, by the Gronwall lemma,

$$\begin{aligned} \|X^i - \overline{X}^i\|_{T,p} &\leq 2c\, e^{2cT} \int_0^T \mathrm{d}s \frac{1}{N} \sum_{i=1}^N \left[E \left| \left(\frac{1}{N} \sum_{j=1}^N b(\overline{X}_s^i, \overline{X}_s^j)^p \right)^{1/p} \right.\right. \\ &\qquad \left.\left. - \left(\int b(\overline{X}_s^i, y)^p \, u_s(\mathrm{d}y) \right)^{1/p} \right|^p \right] \\ &= 2c\, e^{2cT} \int_0^T \mathrm{d}s \, E \left| \left(\frac{1}{N} \sum_j b(\overline{X}_s^1, \overline{X}_s^j)^p \right)^{1/p} - \left(\int b(\overline{X}_s^1, y)^p \, u_s(\mathrm{d}y) \right)^{1/p} \right|^p . \end{aligned}$$

By the Taylor expansion and with $Y_j := b(\overline{X}_s^i, \overline{X}_s^j)^p$ (conditionally on $\overline{X}_s^i$) we obtain

$$E \left| \left(\frac{S_N}{N} + a \right)^{1/p} - a^{1/p} \right|^p \leq \frac{1}{p^p} a^{1-p} E \left| \frac{S_N}{N} \right|^p , \qquad (10.1.20)$$

where $S_N = \sum(Y_j - a)$, $a = EY_j > 0$. Therefore, from the Marcinkiewicz–Zygmund inequality (cf. Chow and Teicher, (1978, p. 357)), we conclude that

$$\sqrt{N} E^{1/p} \left| \left(\frac{S_N}{N} + a \right)^{1/p} - a^{1/p} \right|^p \leq \text{const. } E^{1/p} \left| \frac{S_N}{\sqrt{N}} \right|^p = O(1).$$

This yields (10.1.19) for $p \geq 2$. For $1 \leq p < 2$, the claim follows from the moment bounds of Pyke and Root (1968), giving $E|\frac{S_N}{N^{1/p}}|^p = o(1)$. Therefore,

$$N^{p-1} E \left| \left(\frac{S_N}{N} + a \right)^{1/p} - a^{1/p} \right|^p = o(1). \tag{10.1.21}$$

□

We next interpret Theorem 10.1.7 as a chaotic property of the diffusions governed by (10.1.4). Recall that by Proposition 2.2 in Sznitman (1989), a sequence (u_N) of symmetric probability measures on $E^{(N)}$ is *u-chaotic*, $u \in M^1(E)$, if for $(X_1, \ldots, X_N)$ distributed as u_N,

$$\frac{1}{N} \sum_{i=1}^{N} \delta_{X_i} \xrightarrow{w} u. \tag{10.1.22}$$

For $\overline{X}_N := \frac{1}{N} \sum_{i=1}^{N} \delta_{X^{i,N}}$ we obtain from Theorem 10.1.7 that

$$\overline{X}_N \xrightarrow{w} \overline{X}, \tag{10.1.23}$$

where $\overline{X}$ is the solution of equation (10.1.5). Therefore, with m denoting the law of $\overline{X}$ and m_N denoting the law of $(X^{1,N}, \ldots, X^{N,N})$ we obtain from (10.1.22) the following corollary.

Corollary 10.1.8 *Under the assumptions of Theorems* 10.1.3 *and* 10.1.7, *the sequence (m_N) is m-chaotic.*

Remark 10.1.9 *For $p = \infty$ (see (10.1.17)) the propagation of chaos property does not hold. Also, the case $0 < p < 1$ does not lead to propagation of chaos, and there does not exist a unique strong solution of*

$$X_t = B_t + \int_0^t \int b(X_s, y_s)^p \, \mathrm{d}m(y) \, \mathrm{d}s. \tag{10.1.24}$$

Remark 10.1.10 (An example leading to a Burger type equation.) *Consider the stochastic system*

$$dX_t^i = dW_t^i + \left(\frac{1}{N} \sum_{j=1}^{N} b(X_t^i, X_t^j)^p \right)^{1/p} \mathrm{d}t, \quad i = 1, \dots, N, \tag{10.1.25}$$

with Lipschitz (in both arguments) interactive term $b(\cdot, \cdot)$. Then the instantaneous drift term seen by particle i is

$$\Delta_i = \left(\frac{1}{N} \sum_{j=1}^{N} b(X_t^i, X_t^j)^p \right)^{1/p}.$$

Under the assumptions of Theorem 10.1.7, *we have*

$$\lim_{N \to \infty} E \left[\frac{1}{N} \sum_{i=1}^{N} \left(\Delta_i^p - \left(\int b^p(X_t^i, y) u_t(\mathrm{d}y) \right)^{1/p} \right)^2 \right] = 0,$$

as well as

$$\lim_{N \to \infty} E \left[\frac{1}{N} \sum_{i=1}^{N} \left(\Delta_i^p - \int b^p(X_t^i, y) u_t(\mathrm{d}y) \right) \right]^2 = 0.$$

Similarly to the above limit relations we shall examine the average behavior of the "pseudo drift" $\frac{1}{N} \sum_{i=1}^{n} Z_i^p$. Here, $Z_i^p := \frac{1}{N-1} \sum_{j \neq i} \phi_{N,a}^p (X_t^i - X_t^j)$, and $\phi_{N,a}(x - y) = N^{ad/p} \phi(N^a(x - y))$, where $\phi(\cdot) \geq 0$ is smooth, compactly supported on $\mathbb{R}^d$, and $\int \phi(x) \, \mathrm{d}x = 1$. We consider the vector-valued case here. Note that

$$Z_1^p := \frac{1}{N-1} \sum_{j=2}^{N} \phi_{N,a}^p (X_t^1 - X_t^j)$$

$$= \frac{1}{N-1}\sum_{j=1}^{N} N^{ad}\phi^p(N^a(X_t^i - X_t^j)),$$

and consequently,

$$\begin{aligned} EZ_1^p &= N^{ad}(E\phi^p(N^a(X_t^1 - X_t^2))) \\ &= \left(\frac{N^{ad}E\phi^p(N^a(X_t^1 - X_t^2))}{\int \phi^p}\right)\int \phi^p \underset{N\to\infty}{\longrightarrow} \|u_t\|_{L^2}^2 \int \phi^p =: u_{t,p}(X_t). \end{aligned}$$

Consider next

$$\begin{aligned} a_n &:= E\left[\frac{1}{N}\sum_{i=1}^{N}(Z_i^p - u_{t,p}(X_t^1))\right]^2 \\ &= E\left[\frac{1}{N}\sum_{i=1}^{N}\left(\frac{1}{N-1}\sum_{j\neq i}\phi_{N,a}^p(X_t^i - X_t^j) - u_{t,p}(X_t^1)\right)\right]^2 \\ &= E\left[\frac{1}{N-1}\sum_{j=1}^{N}\phi_{N,a}^p(X_t^1 - X_t^j) - u_{t,p}(X_t^1)\right]^2. \end{aligned}$$

Arguing as in Sznitman (1989, p. 196), we find that

$$a_N \to \begin{cases} 0 & \text{if } 0 < a < \frac{1}{d}, \\ \left(\int \phi^{2p}\,\mathrm{d}x\right)\|u_t\|_{L^2}^2 \frac{\left(\int \phi^p\right)^2}{\int \phi^{2p}} & \text{if } a = \frac{1}{d}, \\ \infty & \text{if } a > \frac{1}{d}. \end{cases}$$

Therefore, only in the case of moderate interaction do we obtain Burger's equation in the limit.

10.1.3 A Random Number of Particles

Let $(W^i)_{i\in\mathbb{N}}$ be a sysetm of i.i.d. real-valued processes (as in (10.1.4)) with finite pth moments and let $(N_n)_{n\geq 1}$ be an i.i.d. integer-valued sequence of r.v.s independent of (W^i). Consider the following system of SDE with a random number of particles and interactions:

$$dX_t^{i,n} = dW_t^i + \left(\frac{1}{n}\sum_{j=1}^{N_n} b(X_t^{i,n}, X_t^{j,n})^p\right)^{1/p} dt, \quad i = 1, \dots, N_n. \quad (10.1.26)$$

We assume that the following asymptotic stability condition holds:

$$\frac{N_n}{n} \to Y \quad \text{a.s. as } n \to \infty. \quad (10.1.27)$$

As in Section 10.1.2, it turns out that $X^{i,N}$ has a natural limit $\overline{X}^i$ that is a solution of the nonlinear SDE

$$dX_t = dB_t + Y^{1/p}\left(\int b(X_t, y)^p \, u_t(\mathrm{d}y)\right)^{1/p}. \quad (10.1.28)$$

Here $B \stackrel{d}{=} W^1$, and Y is assumed to be independent of B. For $m_0 \in M^1(C_T)$ let $M_p(C_T, m_0)$, $L^*_{p,T}$, $\ell^*_{p,T}$ be defined as in Section 10.1.2.

Lemma 10.1.11 *Suppose that* $\int_0^t |B_s| \, \mathrm{d}s < \infty$ *a.s. Then for any* $m \in M_p(C_T, m_0)$, *there exists a unique strong solution of the equation*

$$X_t = B_t + Y^{1/p} \int_0^t \left(\int_{C_T} b(X_s, y_s)^p \, \mathrm{d}m(y) \right)^{1/p} \mathrm{d}s. \quad (10.1.29)$$

Proof: Set $(SX)_t := Y^{1/p} \int_0^t \left(\int_{C_T} b(X_s, y_s)^p \, \mathrm{d}m(y) \right)^{1/p} \mathrm{d}s$. Then arguing in a similar fashion as in the proof of Lemma 10.1.1, we obtain the bound

$$\sup_{s \le t} |(SX)_s - (SY)_s| \le cY^{1/p} \int_0^t \sup_{0 \le u \le s} |X_u - Y_u| \, du.$$

Defining inductively $X^0 := B,\ X^n := SX^{n-1}$, we have

$$\begin{aligned}\sup_{s\le t}|X^n_s - X^{n-1}_s| &\le c^n Y^{n/p}\frac{t^n}{n!}\sup_{s\le t}|X^1_s - X^0_s| \\ &\le c^n Y^{(n+1)/p}\frac{t^n}{n!}\left[\int_0^t \left|B_s\,\mathrm{d}s + \int_0^t \left(\int |y_s|^p\,\mathrm{d}m(y)\right)^{1/p}\right| \mathrm{d}s\right] \\ &< \infty.\end{aligned}$$

This indeed implies the existence of a unique strong solution. □

Given $m \in M_p(C_T, m_0)$, let $\Phi(m)$ denote the distribution of the solution of (10.1.29). Then we have the following contraction-type property for the mapping Φ.

Lemma 10.1.12 *Suppose that* $A_p := \|cY^{1/p}\,e^{cY^{1/p}}\|_p < \infty$. *Then for* $t \le T$, $m_1, m_2 \in M_p(C_T, m_0)$,

$$\ell^*_{p,t}(\Phi(m_1), \Phi(m_2)) \;\le\; A_p \int_0^t \ell^*_{p,s}(m_1, m_2)\,\mathrm{d}s. \tag{10.1.30}$$

Proof: Let $X^{(i)}$ be the solution of the SDE

$$X^{(i)}_t \;=\; B_t + Y^{1/p}\int_0^t \left(\int_{C_T} b(X^{(i)}_s, y_s)^p\,\mathrm{d}m_i(y)\right)^{1/p} \mathrm{d}s.$$

Then, as in the proof of Lemma 10.1.2,

$$\sup_{s\le t}|X^{(1)}_s - X^{(2)}_s| \;\le\; cY^{1/p}\int_0^t \sup_{u\le s}|X^{(1)}_u - X^{(2)}_u| + c\int_0^t \mathrm{d}s\,\ell_{p,s}(m_1, m_2).$$

By the Gronwall lemma, $\sup_{s\le t}|X^1_s - X^2_s| \le cY^{1/p}\,e^{cY^{1/p}}\int_0^t \ell_{p,s}(m_1, m_2)\,\mathrm{d}s$, implying that

$$\ell^*_{p,t}(\Phi(m_1), \Phi(m_2)) \;\leq\; \left\| cY^{1/p}\, e^{cY^{1/p}} \right\|_p \int_0^t \ell^*_{p,s}(m_1, m_2)\, \mathrm{d}s.$$

□

From Lemmas 10.1.11 and 10.1.12 we conclude that (10.1.28) has a unique solution. The proof is similar to that of Theorem 10.1.3.

Theorem 10.1.13 *Under the assumptions of Lemmas* 10.1.11 *and* 10.1.12, *equation* (10.1.28) *has a unique solution, provided that*

$$\|B\|^*_{p,T} \;<\; \infty \qquad \textit{and} \qquad \int \left(\int b(0, y_s)^p \, \mathrm{d}m_0(y) \right)^{1/p} \mathrm{d}s \;<\; \infty.$$

10.1.4 pth Mean Interactions in Time: A Non-Markovian Case

Suppose $(X_t^{i,N})_{i=1,\ldots,N}$ determines a system of N particles and let $b(X_s^{i,N}, \cdot) := (b(X_s^{i,N}, X_s^{j,N}))_{1\leq i\leq N}$ describe the interaction vector. Recall that in Section 10.1.2 we considered equation (10.1.4) with a drift of the form $\|b(X_s^{i,N}, \cdot)\|_p$ corresponding to the pth norm of the interaction vector. In this section we shall study SDEs with mean interactions in time. In fact, let

$$F_i(s) \;:=\; \left| \frac{1}{N} \sum_{j=1}^{N} b\left(X_s^{i,N}, X_s^{j,N}\right) \right| \tag{10.1.31}$$

be the average of the interaction vector and consider the equations

$$X_t^{i,N} \;=\; W_t^i + \left(\int_0^t |F_i(s)|^p \, \mathrm{d}s \right)^{1/p}; \tag{10.1.32}$$

$$X_0^{i,N} \;=\; X_0^i,\; 1 \leq i \leq N, \quad \text{for } 1 \leq p < \infty;$$

$$X_t^{i,N} \;=\; W_t^i + \operatorname*{ess\,sup}_{s\leq t, \lambda\!\!\setminus} |F_i(s)|; \tag{10.1.33}$$

$$X_0^{i,N} \;=\; X_0^i,\; 1 \leq i \leq N, \quad \text{for } p = \infty;$$

$$
\begin{aligned}
X_t^{i,N} &= W_t^i + \int_0^t |F_i(s)|^p \, \mathrm{d}s; && (10.1.34)\\
X_0^{i,N} &= X_0^i \,; 1 \leq i \leq N, \quad \text{for } 0 < p < 1.
\end{aligned}
$$

In other words, we consider SDEs with a drift resulting from the pth mean in time of the average of the interaction vector. From the definition it is clear that this model describes a system that no longer behaves as a Markovian one, since the instantaneous drift $|F_i(t)|^p$ is weighted by the mean interaction $\frac{1}{p}\left(\int_0^t |F_i(s)|^p \, \mathrm{d}s\right)^{1/p-1}$ over the whole past of the process. From this point of view the propagation of chaos property seems to be not so obvious in this model.

First we consider the case $1 \leq p < \infty$. The nonlinear limiting equation is given by

$$
X_t = B_t + \left(\int_0^t \left|\int b(X_s, y) u_s(\mathrm{d}y)\right|^p \mathrm{d}s\right)^{1/p}, \quad u_s = P^{X_s}. \qquad (10.1.35)
$$

Here X_t, B_t, b are real-valued, B_t is a process in $C_T = C[0,T]$, and

$$
|b(x_1, y) - b(x_2, y)| \leq c|x_1 - x_2| \quad \text{for some } c > 0. \qquad (10.1.36)
$$

Define, for $m_0 \in M^1(\ell_T)$,

$$
M_p(C_T, m_0) := \{m_1 \in M^1(C_T);\ \ell^*_{p,t}(m_1, m_0) < \infty\}. \qquad (10.1.37)
$$

Then, for $m \in M_p(C_T, m_0)$, consider the linear equation

$$
X_t = B_t + \left(\int_0^t \left|\int_{C_T} b(X_s, y_s) \, \mathrm{d}m(y)\right|^p \mathrm{d}s\right)^{1/p}, \qquad (10.1.38)
$$

where y_s is the value of y at time s.

Lemma 10.1.14 *Assume that the Lipschitz condition* (10.1.36) *holds, and furthermore,*

$$\int_0^T \left| \int_{C_T} b(0, y_s) m_s(\mathrm{d}y) \right|^p \mathrm{d}s < \infty,$$

where m_s is the distribution of X_s at time s under m. Then

(a) *Equation* (10.1.38) *has a unique strong solution X.*

(b) *If $\Phi(m)$ is the law of X, then $\Phi(m) \in M_p(C_T, m_0)$; that is, $\Phi : M_p(C_T, m_0) \to M_p(C_T, m_0)$.*

Proof: Let $X \in \mathcal{X}_p(C_T, m_0)$ and define

$$(SX)_t \;:=\; B_t + \left(\int_0^t \left| \int b(X_s, y) m_s(\mathrm{d}y) \right|^p \mathrm{d}s \right)^{1/p}. \qquad (10.1.39)$$

Then

$$\begin{aligned}
&|(SX)_t - (SY)_t|^p \\
&= \left(\left(\int_0^t \left| \int b(X_s, y) m_s(\mathrm{d}y) \right|^p \mathrm{d}s \right)^{1/p} \right. \\
&\qquad \left. - \left(\int_0^t \left| \int b(Y_s, y) m_s(\mathrm{d}y) \right|^p \mathrm{d}s \right)^{1/p} \right)^p \\
&\leq \left(\int_0^t \left| \int (b(X_s, y) - b(Y_s, y)) m_s(\mathrm{d}y) \right|^p \mathrm{d}s \right)^{1/p} \\
&\qquad\qquad \text{(Minkowski inequality)} \\
&\leq \int_0^t c^p |X_s - Y_s|^p \,\mathrm{d}s \quad \text{(by the Lipschitz condition (10.1.36))}.
\end{aligned}$$

This implies

$$\sup_{s\le t}|(SX)_s-(SY)_s|^p \;\le\; c^p\int_0^t \sup_{u\le s}|X_u-Y_u|^p\,\mathrm{d}s, \tag{10.1.40}$$

and furthermore, $L^{*p}_{p,t}(SX,SY) \le c^p\int_0^t L^{*p}_{p,s}(X,Y)\,\mathrm{d}s$. Define, inductively, $X^0 := B$, $X^n := SX^{n-1}$. Then $L^{*p}_{p,t}(X^n,X^{n-1}) \le c^{pn}\frac{T^n}{n!}L^{*p}_{p,T}(X^1,X^0)$. By (10.1.36), the integral $\int_{C_T} b(X_s,y_s)m(\mathrm{d}y)$ is a Lipschitz function of X_s. Thus

$$\begin{aligned}
L^{*p}_{p,T}(X^1,X^0) &= E\sup_{t\le T}\int_0^t\left|\int b(B_s,y)m_s(\mathrm{d}y)\right|^p\mathrm{d}s\\
&\le E\int_0^T\left(\int(|b(0,y)|+c|B_s|)m_s(\mathrm{d}y)\right)^p\mathrm{d}s\\
&\le c'\int_0^T\left(\int|b(0,y)|m_s(\mathrm{d}y)\right)^p\mathrm{d}s+c'E\int_0^T|B_s|^p\,\mathrm{d}s \;<\; \infty,
\end{aligned}$$

as by the assumptions the integrals in the right-hand side are finite. Therefore,

$$\sum_{n\ge1}L^*_{p,T}(X^n,X^{n-1}) \;\le\; \sum_{n\ge1}c^n\left(\frac{T^n}{n!}\right)^{1/p} L^*_{p,T}(X^1,X^0) \;<\; \infty.$$

This implies $\sum_{n\ge1}L^*_{p,T}(X^n,X^{n-1})<\infty$. Then

$$\sum_{n\ge1}L^*_{1,T}(X^n,X^{n-1}) \;<\; \infty.$$

In consequence, X^n converges to some process X a.s. uniformly on bounded intervals. X is a.s. continuous, and $E\sup_{s\le t}|X_s|^p<\infty$, since $E\sup_{s\le t}|B_s|^p$ $<\infty$. This yields $\Phi(m)\in M_p(C_T,m_0)$. □

In addition, suppose that b is Lipschitz in both arguments; that is,

$$|b(x_1,y_1)-b(x_2,y_2)| \;\le\; c[|x_1-x_2|+|y_1-y_2|] \tag{10.1.41}$$

for all x_1, x_2, y_1, and y_2 in C_T, and consider the map $\Phi : M_p(X_T, m_0) \to M_p(C_T, m_0)$.

Lemma 10.1.15 (Contraction of Φ with respect to $\ell^*_{p,t}$) *Suppose that* (10.1.41) *and the assumption of Lemma* 10.1.14 *hold. Then for* $t < T$ *and* $m_1, m_2 \in M_p(C_T, m_0)$,

$$\ell^*_{p,t}(\Phi(m_1), \Phi(m_2)) \;\le\; c_p \, e^{c_p t} \int_0^t \ell^{*p}_{p,s}(m_1, m_2)\, ds, \tag{10.1.42}$$

where $c_p := c\, 2^{p-1}$.

Proof: For $i = 1, 2$ and $t \le T$, set

$$X_t^{(i)} \;=\; B_t \left(\int_0^t \left| \int_{C_T} b\left(X_s^{(i)}, y_s\right) dm_i(y) \right|^p ds \right)^{1/p},$$

and let $m \in M^1(m_1, m_2)$, the class of probabilities on $C_T \times C_T$ with marginals m_1 and m_2. Then

$$\begin{aligned}
\sup_{s \le t} |X_s^{(1)} - X_s^{(2)}|^p \;&=\; \left| \left(\int_0^t \left| \int_{C_T} b\left(X_s^{(1)}, y_s^{(1)}\right) dm_1(y^{(1)}) \right|^p ds \right)^{1/p} \right. \\
&\qquad \left. - \left(\int_0^t \left| \int_{C_T} b\left(X_s^{(2)}, y_s^{(2)}\right) dm_2(y^{(2)}) \right|^p ds \right)^{1/p} \right|^p \\
&\le \int_0^t ds \left[\int_{C_T \times C_T} \left| b\left(X_s^{(1)}, y_s^{(1)}\right) - b\left(X_s^{(2)}, y_s^{(2)}\right) \right| dm\left(y^{(1)}, y^{(2)}\right) \right]^p \\
&\le \int_0^t ds \left[c \left| X_s^{(1)} - X_s^{(2)} \right| + \int \left| y_s^{(1)} - y_s^{(2)} \right| dm\left(y^{(1)}, y^{(2)}\right) \right]^p .
\end{aligned}$$

Minimizing the right-hand side over all couplings, we get

$$\sup_{s \le t} \left| X_s^{(1)} - X_s^{(2)} \right|^p$$

$$\leq \underbrace{c \cdot 2^{p-1}}_{=:c_p} \int_0^t \mathrm{d}s \sup_{u \leq s} \left| X_u^{(1)} - X_u^{(2)} \right|^p + \underbrace{c \cdot 2^{p-1}}_{=:c_p} \int_0^t \mathrm{d}s \, \ell_{1,s}^{*p}(m_1, m_2).$$

Consequently, for $p \geq 1$, by the Gronwall lemma and $\ell_{1,s}^* \leq \ell_{p,s}^*$,

$$\sup_{s<t} \left| X_s^{(1)} - X_s^{(2)} \right|^p \leq c_p \, e^{c_p t} \int_0^t \mathrm{d}s \, \ell_{p,s}^{*p}(m_1, m_2).$$

This yields the desired contractive inequality

$$\ell_{p,t}^{*p}(\Phi(m_1), \Phi(m_2)) \leq c_p \, e^{c_p t} \int_0^t \ell_{p,s}^{*p}(m_1, m_2) \, \mathrm{d}s.$$

□

Theorem 10.1.16 *Under* (10.1.41) *and* $\int_0^T (\int_{C_T} b(0, y_s) \, \mathrm{d}m_0(y))^p \, \mathrm{d}s < \infty$, *equation* (10.1.38) *has a unique weak and strong solution in* $\mathcal{X}_p(C_T, m_0)$.

Proof: From Lemma 10.1.15 we obtain that for $m \in M_p(C_T, m_0)$,

$$\begin{aligned}
\ell_{p,T}^{*p}(\Phi^{k+1}(m), \Phi^k(m)) &\leq C_T \frac{T^k}{k!} \ell_{p,T}^{*p}(\Phi(m), m) \\
&\leq 2^{p-1} C_T \frac{T^k}{k!} \left[\ell_{p,T}^{*p}(\Phi(m), m_0) + \ell_{p,T}^{*p}(m, m_0) \right] \\
&< \infty.
\end{aligned}$$

The remaining part of the proof is similar to that of Theorem 10.1.3. □

In the next step we turn our attention to the system of interacting particles defined in (10.1.32), where $((W_t^i), X_0^i)$ are independent processes and identically distributed for all i. The following theorem asserts that as $N \to \infty$, every $X^{i,N}$ has a natural limit $\overline{X}^i$. In fact, the $(\overline{X}^i)$ are independent copies of the solution of the nonlinear equation of McKean–Vlasov type,

$$\begin{aligned} X_t &= B_t + \left(\int_0^t \left| \int_{C_T} b(X_s, y) u_s(\mathrm{d}y) \right|^p \mathrm{d}s \right)^{1/p}, \\ X_{t=0} &= X_0, \end{aligned} \tag{10.1.43}$$

considered in Theorem 10.1.16 with $B \stackrel{d}{=} W^{(1)}$. Let b satisfy the Lipschitz condition (10.1.36).

Theorem 10.1.17 *Suppose that*

$$\int b(\overline{X}_s^1, y)^p \, u_s(\mathrm{d}y) \ < \ \infty \ a.s. \tag{10.1.44}$$

Then for any $i \geq 1$, $T > 0$,

$$\sup_N \sqrt{N} \left(E \sup_{t \leq T} |X_t^{i,N} - \overline{X}_t^i|^p \right)^{1/p} \ < \ \infty \qquad \text{for } p \geq 2 \tag{10.1.45}$$

and

$$N^{(1/p)-1} \left(E \sup_{t \leq T} |X_t^{i,N} - \overline{X}_t^i|^p \right)^{1/p} \ = \ o(1) \qquad \text{for } 1 \leq p < 2.$$

Proof: We drop further the superscript N. Then

$$\begin{aligned} & X_t^i - \overline{X}_t^i \\ = & \left(\int_0^t \left| \frac{1}{N} \sum_{j=1}^N b\left(X_s^i, X_s^j\right) \right|^p \mathrm{d}s \right)^{1/p} - \left(\int_0^t \left| \int b\left(\overline{X}_s^i, y\right) u_s(\mathrm{d}y) \right|^p \mathrm{d}s \right)^{1/p} \\ = & \left[\left(\int_0^t \left| \frac{1}{N} \sum_{j=1}^N b\left(X_s^i, X_s^j\right) \right|^p \mathrm{d}s \right)^{1/p} - \left(\int_0^t \left| \frac{1}{N} \sum_{j=1}^N b\left(\overline{X}_s^i, X_s^j\right) \right|^p \mathrm{d}s \right)^{1/p} \right] \\ & + \left[\left(\int_0^t \left| \frac{1}{N} \sum_{j=1}^N b\left(\overline{X}_s^i, X_s^j\right) \right|^p \mathrm{d}s \right)^{1/p} - \left(\int_0^t \left| \frac{1}{N} \sum_{j=1}^N b\left(\overline{X}_s^i, \overline{X}_s^j\right) \right|^p \mathrm{d}s \right)^{1/p} \right] \end{aligned}$$

$$+\left[\left(\int_0^t \left|\frac{1}{N}\sum_{j=1}^N b\left(\overline{X}_s^i,\overline{X}_s^j\right)\right|^p ds\right)^{1/p} - \left(\int_0^t \left|\int b\left(\overline{X}_s^i,y\right)u_s(dy)\right|^p ds\right)^{1/p}\right].$$

Applying the Minkowski inequality and setting $\|X\|_T := \sup_{s\le T}|X_s|$, we obtain

$$\begin{aligned}
\|X^i-\overline{X}^i\|_{T,p}^p &= E\|X^i-\overline{X}^i\|_T^p \\
&\le 4^{p-1}\left[E\int_0^t ds\left|\frac{1}{N}\sum_{j=1}^N \left[b\left(X_s^i,X_s^j\right)-b\left(\overline{X}_s^i,X_s^j\right)\right]\right|^p\right. \\
&\qquad + E\int_0^T ds\left|\frac{1}{N}\sum_{j=1}^N \left[b\left(\overline{X}_s^i,X_s^j\right)-b\left(\overline{X}_s^i,\overline{X}_s^j\right)\right]\right|^p \\
&\qquad \left.+ E\int_0^T ds\left|\frac{1}{N}\sum_{j=1}^N b\left(\overline{X}_s^i,\overline{X}_s^j\right)-\int b\left(\overline{X}_s^i,y\right)u_s(dy)\right|^p\right] \\
&\le 4^{p-1}\int_0^T ds\left\{c^p\,E|X_s^i-\overline{X}_s^i|^p + c^p\,E\left[\frac{1}{N}\sum_{j=1}^N |X_s^j-\overline{X}_s^j|\right]^p\right. \\
&\qquad \left.+ E\int_0^T ds\left|\frac{1}{N}\sum_{j=1}^N b(\overline{X}_s^i,\overline{X}_s^j)-\int b(\overline{X}_s^i,y)u_s(dy)\right|^p\right\}.
\end{aligned}$$

Summing up over i and using the symmetry, we find that

$$\begin{aligned}
N\|X^i-\overline{X}^i\|_{T,p}^p &= \sum_{i=1}^N \|X^i-\overline{X}^i\|_{T,p}^p \\
&\le 4^{p-1}\int_0^T ds\left\{c^p\sum_{i=1}^N E\|X_s^i-\overline{X}_s^i\|_p^p + c^p\,N\,E\left[\frac{1}{N}\left(\sum_{j=1}^N |X_s^j-\overline{X}_s^j|^p\right)^{1/p}\right]^p\right. \\
&\qquad \left.+ c^p\sum_{i=1}^N E\left|\frac{1}{N}\sum_{j=1}^N b(\overline{X}_s^i,\overline{X}_s^j)-\int b(\overline{X}_s^i,y)u_s(dy)\right|^p\right\}.
\end{aligned}$$

Therefore,

$$\begin{aligned}
\|X^i - \overline{X}^i\|_{T,p}^p \;\le\; & 4^{p-1}c^p \int_0^T \mathrm{d}s \Bigg\{ \|X^i - \overline{X}^i\|_{s,p}^p + c^p \|X^i - \overline{X}^i\|_{s,p}^p \\
& + \frac{1}{N}\sum_{i=1}^N E \left| \frac{1}{N}\sum_{j=1}^N b(\overline{X}_s^i, \overline{X}_s^j) - \int b(\overline{X}_s^i, y) u_s(\mathrm{d}y) \right|^p \Bigg\}.
\end{aligned}$$

Consequently, by the Gronwall lemma,

$$\begin{aligned}
& \|X^i - \overline{X}^i\|_{T,p}^p \\
\le\; & C_p e^{C_p T} \int_0^T \mathrm{d}s \left[\frac{1}{N}\sum_{i=1}^N E \left| \frac{1}{N}\sum_{j=1}^N b(\overline{X}_s^i, \overline{X}_s^j) - \int b(\overline{X}_s^i, y) u_s(\mathrm{d}y) \right|^p \right] \\
& \quad (\text{with } C_p = 2\cdot 4^{p-1}c^p) \\
\le\; & C_p e^{C_p T} \int_0^T \mathrm{d}s\, E \left| \frac{1}{N}\sum_{j=1}^N b(\overline{X}_s^i, \overline{X}_s^j) - \int b(\overline{X}_s^i, y) u_s(\mathrm{d}y) \right|^p \\
=\; & C_p e^{C_p T} T \cdot E\left[0\left(\frac{1}{\sqrt{N}}\right)\right]^p ;
\end{aligned}$$

here we also used the Marcinkiewicz–Zygmund inequality (cf. Chow and Teicher (1978, p. 357)) for $p \ge 2$, and the Pyke and Root (1968) inequality for $1 \le p < 2$. □

Corollary 10.1.18 *Let m denote the law of $\overline{X}$ satisfying* (10.1.43), *and let m_N be the law of $(X^{1,N}, \ldots, X^{N,N})$. Then, under the assumptions of Theorems* 10.1.16 *and* 10.1.17, *m_N is m-chaotic.*

We next study the limiting case $p = \infty$ (cf.(10.1.33)). In contrast to the limiting case in Section 10.1.4 of pth norm interaction, we obtain the propagation of chaos property for pth mean interaction in time under a stronger Lipschitz condition. Consider for $m \in M^1(C_T)$,

$$X_t = B_t + \operatorname*{ess\,sup}_{s \le t} \left| \int_{C_T} b(X_s, y) m(\mathrm{d}y) \right| . \tag{10.1.46}$$

Here X_t, B_t, and b are real-valued, B_t is a process on C_T having finite pth moment, $E\operatorname{ess\,sup}_{s\le T}|B_s|^p < \infty$, and b satisfies the Lipschitz condition for all x_1, x_2, and $y \in C_T$:

$$|b(x_1,y) - b(x_2,y)| \;\le\; c|x_1 - x_2|, \quad \text{with } 0 < c < 1. \tag{10.1.47}$$

We shall use the following L_p-type metric for $p \ge 1$:

$$\widetilde{L}^*_{p,t}(X,Y) \;:=\; (E\operatorname*{ess\,sup}_{s\le t}|X_s - Y_s|^p)^{1/p} \quad \text{in} \quad \mathcal{X}(C_T). \tag{10.1.48}$$

Let

$$\widetilde{\ell}^*_{p,t}(m_1,m_2) \;=\; \widehat{\widetilde{L}}^*_{p,t}(m_1,m_2) \tag{10.1.49}$$

be the corresponding minimal metric. Consider the set of measures on $M^2(C_T)$:

$$\widetilde{M}_p(C_T,m_0) \;=\; \{m_1 \in M^1(C_T);\; \widetilde{\ell}^*_{p,T}(m_1,m_0) < \infty\}, \tag{10.1.50}$$

and let $\widetilde{\mathcal{X}}_p(C_T,m_0)$ denote the corresponding class of processes. For $m_0 \in M^1(C_T)$ and $m \in \widetilde{M}_p(C_T,m_0)$, consider the linear equation

$$X_t \;=\; B_t + \operatorname*{ess\,sup}_{s\le t}\left|\int_{C_T} b(X_s,y_s)\,dm(y)\right|. \tag{10.1.51}$$

Lemma 10.1.19 *Assume that the Lipschitz condition* (10.1.47) *holds, and let*

$$\operatorname*{ess\,sup}_{s\le T}\left|\int_{C_T} b(0,y_s)m(\,dy)\right| \;<\; \infty.$$

Then

(a) *Equation* (10.1.51) *has a unique solution X.*

(b) *If $\Phi(m)$ is the law of X, then $\Phi(m) \in M_p(C_T,m_0)$; that is,* $\Phi : \widetilde{M}_p(C_T,m_0) \to \widetilde{M}_p(C_T,m_0)$.

Proof: Let $X \in \widetilde{\mathcal{X}}_p(C_T, m_0)$ and define

$$(SX)_t \ := \ B_t + \operatorname*{ess\,sup}_{s \le t} \left| \int_{C_T} b(X_s, y_s) m(\,dy) \right| .$$

Then

$$\begin{aligned}
&|(SX)_t - (SY)_t|^p \\
&= \ \left| \operatorname*{ess\,sup}_{0 \le s < t} \left| \int_{C_T} b(X_s, y_s) m(\,dy) \right| - \operatorname*{ess\,sup}_{0 \le s \le t} \left| \int_{C_T} b(Y_s, y_s) m(\,dy) \right| \right|^p \\
&\le \ \operatorname*{ess\,sup}_{0 \le s \le t} c^p |X_s - Y_s|^p
\end{aligned}$$

by the Lipschitz condition (10.1.47). This amounts to

$$\operatorname*{ess\,sup}_{s \le t} |(SX)_s - (SY)_s|^p \ \le \ c^p \operatorname*{ess\,sup}_{0 \le s \le t} |X_s - Y_s|^p,$$

and $\widetilde{L}^*_{p,t}(SX, SY) \le c\, \widetilde{L}^*_{p,t}(X, Y)$.

Define, inductively, $X^0 = B, \ X^n = SX^{n-1}$. Then

$$\widetilde{L}^*_{p,t}(X^n, X^{n-1}) \ \le \ c^n\, \widetilde{L}^*_{p,t}(X^1, X^0).$$

Furthermore,

$$\begin{aligned}
\widetilde{L}^*_{p,T}(X^1, X^0) \ &= \ \left(E \operatorname*{ess\,sup}_{s \le T} \left| \int b(B_s, y_s) m(\,dy) \right|^p \right)^{1/p} \\
&\le \ \left(E \operatorname*{ess\,sup}_{s \le T} \left[\left| \int b(0, y_s) m(\,dy) \right| + c \int |B_s| m(\,dy) \right]^p \right)^{1/p} \\
&\le \ c' \left(\operatorname*{ess\,sup}_{s \le T} \left| \int b(0, y_s) m(\,dy) \right| + E \operatorname*{ess\,sup}_{s \le t} |B_s|^p \right)^{1/p} \ < \ \infty.
\end{aligned}$$

This yields

$$\sum_{n \ge 1} \widetilde{L}^*_{p,T}(X^n, X^{n-1}) \ \le \ \sum_{n \ge 1} c^n\, \widetilde{L}^{*p}_{p,T}(X^1, X^0) < \infty.$$

Therefore, $X^n \xrightarrow{a.s.} X$, uniformly on bounded intervals, and $E\,\text{ess}\sup_{s\le t}|X_s|^p < \infty$. □

In addition, suppose that b is a Lipschitz function in both arguments; for all x_1, x_2, y_1, and y_2 in C_T,

$$|b(x_1, y_1) - b(x_2, y_2)| \;\le\; c[|x_1 - x_2| + |y_1 - y_2|], \tag{10.1.52}$$

where we assume that $0 < c < \frac{1}{2}$. Consider the mapping $\Phi : \widetilde{M}_p(C_T, m_0) \to \widetilde{M}_p(C_T, m_0)$.

Lemma 10.1.20 (Contraction of Φ with respect to the minimal metric $\widetilde{\ell}^*_{p,t}$) *Under* (10.1.52) *and the assumptions of Lemma* 10.1.14, *for* $t < T$ *and* $m_1, m_2 \in \widetilde{M}_p(C_T, m_0)$, *the following contraction property holds:*

$$\widetilde{\ell}^*_{p,t}(\Phi(m_1), \Phi(m_2)) \;\le\; \frac{c}{1-c}\widetilde{\ell}^*_{p,t}(m_1, m_2). \tag{10.1.53}$$

Proof: For $i = 1, 2$, and $t \le T$, define

$$X_t^{(i)} \;=\; B_t + \operatorname*{ess\,sup}_{0<s<t}\left|\int_{C_T} b(X_s^{(i)}, y_s)\, dm_i(y)\right|,$$

and let $m \in M^1(m_1, m_2)$. Then

$$\begin{aligned}
&E\operatorname*{ess\,sup}_{s\le t}|X_s^{(1)} - X_s^{(2)}|^p \\
&= E\left|\operatorname*{ess\,sup}_{s\le t}\left|\int_{C_T} b\left(X_s^{(1)}, y_s^{(1)}\right)\, dm_1\left(y^{(1)}\right)\right| \right.\\
&\qquad\qquad \left. - \operatorname*{ess\,sup}_{s\le t}\left|\int_{C_T} b\left(X_s^{(1)}, y_s^{(2)}\right)\, dm_2\left(y^{(2)}\right)\right|\right|^p \\
&\le E\left|\operatorname*{ess\,sup}_{s\le t} c\left[\left|X_s^{(1)} - X_s^{(2)}\right| + \int_{C_T\times C_T}\left|y_s^{(1)} - y_s^{(2)}\right|\, dm\left(y^{(1)}, y^{(2)}\right)\right]\right|^p.
\end{aligned}$$

Therefore, passing to minimal metrics on the right-hand side,

$$\begin{aligned}&\left(E \operatorname*{ess\,sup}_{s\le t} \left|X_s^{(1)} - X_s^{(2)}\right|^p\right)^{1/p} \\ &\le \; c\left(E \operatorname*{ess\,sup}_{s\le t} \left|X_s^{(1)} - X_s^{(2)}\right|^p\right)^{1/p} \\ &\quad + c\left[\left(\inf_{m\in M^1(m_1,m_2)} \int_{C_T\times C_T} \operatorname*{ess\,sup}_{s\le t} \left|y_s^{(1)} - y_s^{(2)}\right| \, dm(y_1,y_2)\right)^p\right]^{1/p};\end{aligned}$$

that is,

$$(1-c)\left(E \operatorname*{ess\,sup}_{s\le t} \left|X_s^{(1)} - X_s^{(2)}\right|^p\right)^{1/p} \;\le\; c\widetilde{\ell}^*_{1,s}(m_1,m_2) \;\le\; c\widetilde{\ell}^*_{p,s}(m_1,m_2)$$

Passing to the minimal metrics in the left-hand side, we obtain

$$\widetilde{\ell}^{*p}_{p,T}(\Phi(m_1),\Phi(m_2)) \;\le\; \frac{c}{1-c}\widetilde{\ell}^*_{p,T}(m_1,m_2),$$

as desired. □

Next, we conclude the existence of a unique solution of the McKean–Vlasov-type equation

$$X_t \;=\; B_t + \operatorname*{ess\,sup}_{s\le t}\left|\int b(X_s,y_s)u_s(\,dy_s)\right|, \quad X_{t=0} \;=\; X_0. \quad (10.1.54)$$

Theorem 10.1.21 *Under* (10.1.52), *and assuming*

$$\operatorname*{ess\,sup}_{s\le T}\left|\int_{C_T} b(0,y_s)\, dm_0(y)\right| \;<\; \infty,$$

equation (10.1.54) *has (for $m \in \widetilde{M}_p(C_T,m_0)$) a unique weak and strong solution in $\widetilde{\mathcal{X}}_p(C_T,m_0)$.*

Proof: From Lemma 10.1.20 with $C := \dfrac{c}{1-c}$, and $m \in \widetilde{M}_p(C_T, m_0)$, we conclude that

$$\widetilde{\ell}^{*p}_{p,T}(\Phi^{k+1}(m), \Phi^k(m)) \;\le\; C^k\, \widetilde{\ell}^{*p}_{p,T}(\Phi(m), m) \;<\; \infty,$$

which implies the theorem. □

Consider next a system of N interacting particles driven by the equation (10.1.33), namely

$$X^{i,N}_t \;=\; W^i_t + \operatorname*{ess\,sup}_{s\le t} \left| \frac{1}{N} \sum_{j=1}^{N} b\left(X^{i,N}_s, X^{j,N}_s\right) \right| \tag{10.1.55}$$

and

$$X^{i,N}_0 \;=\; X^i_0, \qquad 1 \le i \le N.$$

We shall show that $X^{i,N}$ has a natural limit $\overline{X}^i$, where the $\overline{X}^i$ are i.i.d. copies of the solution of (10.1.43).

Theorem 10.1.22 *Suppose that* (10.1.52) *holds and that the r.v.* $Y_{s,j} := b(\overline{X}^1_s, \overline{X}^j_s)$ *on* $C[0,T]$ *are either* (i) *in the domain of normal attraction (dna) a Gaussian law, or* (ii) *satisfy the bounded law of the iterated logarithm (BLIL). Suppose also that* $E\|b(\overline{X}^1_s, \overline{X}^j_s)\|^2_\infty < \infty$. *Then for any* $i \ge 1$,

$$\sup_N a_N\, E \left\| X^{i,N}_t - \overline{X}^i_t \right\|_\infty \;<\; \infty, \tag{10.1.56}$$

where in case (i) $a_N = \sqrt{N}$, *while* $a_N = \sqrt{N \log\log N}$ *in case* (ii).

Proof: Similarly to the proof of Theorem 10.1.17, we obtain from the condition $2c < 1$ that for $\alpha \ge 1$,

$$\begin{aligned}
&\widetilde{L}^*_{\alpha,T}(X^i, \overline{X}^i)\\
&\quad\le\; \frac{1}{1-2c}\frac{1}{N}\sum_{i=1}^{N} E \operatorname*{ess\,sup}_{s\le T} \left| \frac{1}{N}\sum_{j=1}^{N} b(\overline{X}^i_s, \overline{X}^j_s) - \int b(\overline{X}^i_s, y) u_s(dy) \right|^\alpha.
\end{aligned}$$

If $(Y_{s,j})$ are in the dna (cf. Hoffmann–Jörgensen (1977)), then (10.1.56) follows with $a_N = \sqrt{N}$. If $(Y_{s,j})$ satisfy the BLIL, then for the corresponding centered sum we have $S_N \ \overline{\lim}\, E\|\frac{S_N}{a_N}\|_\infty \leq \overline{\lim}\, \frac{\|S_N\|_\infty}{a_N} < \infty$ a.s. (cf. Kuelbs (1977)), and thus (10.1.56) follows. □

We remark that invoking Corollary 5.7 of Hoffmann–Jörgensen (1977), a sufficient condition for the dna of $S_N = \sum_{j=1}^N X_j$ is given by

$$E\|X_1\|_{bL}^2 \ < \ \infty, \tag{10.1.57}$$

where $\|\cdot\|_{bL}$ is the bounded Lipschitz norm with respect to a uniform distance to a Gaussian law.

Corollary 10.1.23 *Suppose the assumptions of Theorems* 10.1.21 *and* 10.1.22 *hold. Let m denote the law of $\overline{X}$, and m_N stands for the law of $(X^{1,N},\ldots,X^{N,N})$. Then m_N is m-chaotic.*

Remark 10.1.24 *Applying a similar technique in the case $0 < p < 1$, we see that there exists no unique solution of the linear equation, and furthermore, there is no propagation of chaos.*

10.1.5 Minimal Mean Interactions in Time

Next, we study the analogue of equation (10.1.33) with minimal mean interaction in time:

$$\begin{aligned} X_t^{i,N} &= W_t^i + \operatorname*{ess\,inf}_{s\leq t}\left|\frac{1}{N}\sum_{j=1}^N b(X_s^{i,N},X_s^{j,N})\right|, \\ X_0^{i,N} &= X_0^i, \qquad 1 \leq i \leq N. \end{aligned} \tag{10.1.58}$$

The corresponding Boltzmann type equation is

$$\begin{aligned} X_t &= B_t + \operatorname*{ess\,inf}_{s\leq t}\left|\int b(X_s,y)u_s(\,\mathrm{d}y)\right|, \\ X_{t=0} &= X_0. \end{aligned} \tag{10.1.59}$$

We obtain the following results. (The proofs are similar to those in section 10.1.4 and are therefore omitted.)

Theorem 10.1.25 *Suppose that* $m_0 \in M^1(C_T)$ *and for all* x_1, x_2, y_1, *and* y_2 *in* C_T,

$$|b(x_1, y_1) - b(x_2, y_2)| \leq c[|x_1 - x_2| + |y_1 - y_2|], \tag{10.1.60}$$

where $0 < c < \frac{1}{2}$. *Suppose also that*

$$\operatorname*{ess\,sup}_{s \leq T} \left| \int_{C_T} b(0, y_s)\, dm_0(y) \right| < \infty. \tag{10.1.61}$$

Then (10.1.59) *has a unique strong solution in* $\mathcal{X}_p(C_T, m_0)$.

The system $(X^{i,N})$ in (10.1.58) has a natural limiting process $(\overline{X}^i)$, where $\overline{X}^i$, $i \geq 1$, are i.i.d. copies of the solution X of (10.1.59).

Theorem 10.1.26 *Suppose the assumptions of Theorem* 10.1.25 *and Theorem* 10.1.22 *hold. Then for any* $i \geq 1$,

$$\sup_N a_N\, E \sup_{t \leq T} \left| X_t^{i,N} - \overline{X}_t^{i,N} \right| < \infty. \tag{10.1.62}$$

Corollary 10.1.27 *Under the conditions of Theorems* 10.1.25 *and* 10.1.26, *the system* (10.1.58) *admits the propagation of chaos.*

10.1.6 Interactions with a Normalized Variation of the Neighbors: Relaxed Lipschitz Conditions

Consider the following stochastic system:

$$\begin{aligned} X_t^{i,N} &= W_t^i + \int_0^t \left(\frac{1}{N} \sum_{j=1}^N b\left(X_s^{i,N}, \mathring{X}_s^{j,N} \right) \right) ds, \qquad (10.1.63) \\ X_0^{i,N} &= X_0^i, \qquad 1 \leq i \leq N. \end{aligned}$$

Here

$$\mathring{X}_s^i := \frac{X_s^i - EX_s^i}{E|X_s^i - EX_s^i|} \tag{10.1.64}$$

is the normalized variation of particle i, and $((W_t^i), X_0^i)$ are independent identically distributed processes on $C_T \times \mathbb{R}$. The drift is given by the mean of the interactions with the normalized variation of all particles. We assume that

$$b(x,0) = 0, \quad \text{for all } x; \tag{10.1.65}$$

that is, the interaction is zero if the relative variation is zero.

The McKean–Vlasov-type equation corresponding to (10.1.62) is given by

$$\begin{aligned} X_t &= B_t + \int_0^t \left(\int b(X_s, y) \, \mathrm{d}P^{\mathring{X}_s}(y) \right) \mathrm{d}s, \\ X_{t=0} &= X_0, \end{aligned} \tag{10.1.66}$$

where $B \stackrel{d}{=} W^i$. Note that B in this section is not necessarily a Brownian motion. We study these equations under the following relaxed Lipschitz condition on b. Assume that b has a partial derivative

$$b_2' := \frac{\partial b}{\partial y} \tag{10.1.67}$$

with respect to the second argument, and consider the following Lipschitz-type assumptions: For all $x_1, x_2, y \in C_T$,

(L1) $|b_2'(x_1, y) - b_2'(x_2, y)| \le c|x_1 - x_2|$;

or, for all x_1, x_2, y_1, and y_2,

(L2) $|b_2'(x_1, y_1) - b_2'(x_2, y_2)| \le c\left[|x_1 - x_2| + |y_1 - y_2|\right]$.

(L2) allows a quadratic growth of b with respect to the second component. To obtain contraction properties in this case, we have to switch to a suitable probability metric with regularity conditions of higher order. This makes necessary an essential change in the method of the proofs given so far.

Let $m \in M_1(C_T)$ be the distribution of a process (ξ_s), and denote by $\mathring{m}$ the distribution of the normalized process $(\mathring{\xi}_s)$ assuming an absolute first moment of the marginal measure m_s. Define

$$N_s := \mathring{m}_s - \delta_0 = N_s^m, \quad \text{and} \tag{10.1.68}$$

$$F_{N_s}^{(-1)}(y) \quad := \quad \int_{-\infty}^{y} F_{N_s}(u)\,\mathrm{d}u. \tag{10.1.69}$$

Following the common derivates notation of a function f, $f^{(s)}$, $s \geq 1$, we define the s-fold integrated function by $f^{(-s)}$, and thus $(f^{(-s)})^{(s)} = f$.

Note that due to (10.1.65), we can replace the integration of $\mathring{m}_s$ in (10.1.66) by integration of N_s. Consider then the linear equation

$$X_t \;=\; B_t + \int_0^t \left(\int b(X_s, y)\,\mathrm{d}N_s(y_s) \right) \mathrm{d}s. \tag{10.1.70}$$

Integration by parts in (10.1.70) leads to the equivalent equation

$$X_t \;=\; B_t + \int_0^t \left(\int b_2'(X_s, y)\,\mathrm{d}F_{N_s}^{(-1)}(y) \right) \mathrm{d}s. \tag{10.1.71}$$

Theorem 10.1.28 *Suppose that $m \in M^1(C_T)$ has a finite first moment and $E\sup_{s\leq T}|B_s| < \infty$. Furthermore, let (L1) be satisfied and suppose that*

$$\int_0^T \mathrm{d}s \int |b_2'(0, y)|\,|F_{N_s}(y)|\,\mathrm{d}y < \infty.$$

Then

$$\int_0^T \mathrm{d}s \int |b_2'(0, y)|\,|F_{N_s}(y)|\,\mathrm{d}y \;=\; \int_0^T E \int_0^{\mathring{\xi}_s} |b_2'(0, t)|\,\mathrm{d}t \;<\; \infty, \tag{10.1.72}$$

(10.1.70) has a unique strong solution X, and moreover, $E\sup_{s\leq T}|X_s| < \infty$.

Proof: Let

$$(SX)_t \;:=\; B_t + \int_0^t \mathrm{d}s \left(\int_{\mathbb{R}} b(X_s, y_s)\,\mathrm{d}N_s(y_s) \right)$$

$$= \quad B_t + \int_0^t \mathrm{d}s \left(\int_{\mathbb{R}} b_2'(X_s, y_s) \, \mathrm{d}F_{N_s}^{(-1)}(y_s) \right) .$$

Then by the Lipschitz condition (L1),

$$\begin{aligned} |(SX)_t - (SY)_t| \quad &\leq \quad \int_0^t \mathrm{d}s \left| \int_{\mathbb{R}} (b_2'(X_s, y_s) - b_2'(Y_s, y_s)) \, \mathrm{d}F_{N_s}^{(-1)}(y_s) \right| \\ &\leq \quad \int_0^t \mathrm{d}s \int_{\mathbb{R}} c|X_s - Y_s| \, |F_{N_s}(y_s)| \, \mathrm{d}y_s . \end{aligned}$$

Observe that the total variation norm of the measure $F_{N_s}^{(-1)}(dy)$ is 1:

$$\mathrm{Var}(F_{N_s}^{(-1)}) = \int_{\mathbb{R}} |F_{N_s}(y)| \, \mathrm{d}y = \int_{-\infty}^{0} F_{\mathring{\xi}_s}(y) \, \mathrm{d}y + \int_0^{\infty} (1 - F_{\mathring{\xi}_s}(y)) \, \mathrm{d}y = E|\mathring{\xi}_s| = 1.$$

Therefore,

$$|(SX)_t - (SY)_t| \quad \leq \quad c \int_0^t |X_s - Y_s| \, \mathrm{d}s, \tag{10.1.73}$$

implying $L_{1,t}^*(SX, SY) \leq c \int_0^t L_{1,s}^*(X, Y) \, \mathrm{d}s$. Define inductively $X^0 = B$, $X^n = SX^{n-1}$. Then

$$L_{1,T}^*(X^n, X^{n-1}) \quad \leq \quad c^n \frac{T^n}{n!} L_{1,T}^*(X^1, X^0). \tag{10.1.74}$$

Let us estimate the term on the right-hand side of (10.1.74):

$$L_{1,T}^*(X^1, X^0) \quad = \quad E \sup_{s \leq T} \left| \int_0^s \mathrm{d}s \left(\int_{\mathbb{R}} b_2'(B_s, y_s) \, \mathrm{d}F_{N_s}^{(-1)}(y_s) \right) \right| \tag{10.1.75}$$

$$\leq \quad E \int_0^T \mathrm{d}s \int_{\mathbb{R}} |b_2'(B_s, y_s)| \, |F_{N_s}(y_s)| \, \mathrm{d}y_s$$

$$\leq \; E \int_0^T \mathrm{d}s \int_{\mathbb{R}} (c|B_s| + |b_2'(0, y_s)|)|F_{N_s}(y_s)| \, \mathrm{d}y_s$$

$$\leq \; E \int_0^T \mathrm{d}s \, c|B_s| + \int_0^T \mathrm{d}s \int_{\mathbb{R}} |b_2'(0, y_s)| \, |F_{N_s}(y_s)| \, \mathrm{d}y_s \; < \infty.$$

Now the equality in (10.1.72) results from the following integration by parts arguments:

$$\int_{\mathbb{R}} |b_2'(0, y)||F_{N_s}(y)| \, \mathrm{d}y \;=\; \int_{\mathbb{R}} |b_2'(0, y)||F_{\overset{\circ}{\xi}_s}(y) - F_0(y)| \, \mathrm{d}y$$

$$= \int_{-\infty}^{0} |b_2'(0, y)| F_{\overset{\circ}{\xi}_s}(y) \, \mathrm{d}y + \int_0^{\infty} |b_2'(0, y)|(1 - F_{\overset{\circ}{\xi}_s}(y)) \, \mathrm{d}y$$

$$= \int_{-\infty}^{+\infty} \left(\int_0^y |b_2'(0, t)| \, \mathrm{d}t \right) \mathrm{d}F_{\overset{\circ}{\xi}_s}(y) \;=\; E \int_0^{\overset{\circ}{\xi}_s} |b_2'(0, t)| \, \mathrm{d}t \;<\; \infty.$$

Consequently, $L^*_{1,T}(X^1, X^0) < \infty$. Combining (10.1.74), (10.1.75) implies the existence and the uniqueness of a strong solution X. Moreover,

$$L^*_{1,T}(X, B) \;\leq\; \sum_{n \geq 1}^{\infty} L^*_{1,T}(X^n, X^{n-1}) \;\leq\; e^{cT} L^*_{1,T}(X_1, B) \;<\; \infty;$$

that is, $E \sup_{s \leq T} |B_s| < \infty$ provides that $E \sup_{s \leq T} |X_s| < \infty$. □

We next extend the result of Theorem 10.1.28 to the case where pth moments exist, $p \geq 1$. Define $\|X\|^*_{T,p} = (E \sup_{t \leq T} |X(t)|^p)^{1/p}$, $1 \leq p < \infty$, and $\|X\|^*_{T,\infty} = E \operatorname{ess\,sup}_{0 < t \leq T} |X(t)|$.

Theorem 10.1.29 *Suppose that* $\|B\|^*_{T,p} < \infty$ *for some* $1 \leq p \leq \infty$. *Suppose also that* (L1) *holds. Finally, assume that*

(i) *if* $1 \leq p < \infty$, *then*

$$\int_0^T \mathrm{d}s \left(\int_{\mathbb{R}} |b_2'(0, y_s) F_{N_s}(y_s)|^p \, \mathrm{d}y_s \right)^{1/p} \;<\; \infty, \tag{10.1.76}$$

and

(ii) *if* $p = \infty$, *then*

$$\int_0^T \mathrm{d}s(\operatorname*{ess\,sup}_{y_s} |b_2'(0, y_s)||F_{N_s}(y_s)|) \quad < \quad \infty \quad (p = \infty). \qquad (10.1.77)$$

Under these assumptions, the SDE (10.1.70) *has a unique solution* X*, and furthermore,* $\|X\|_{T,p}^* < \infty$*. In particular, if* $\Phi(m)$ *is the distribution of the solution of* (10.1.70)*, then* $\Phi(m)$ *maps* $M_p(C_T, \delta_0)$ *into* $M_p(C_T, \delta_0)$*.*

Proof: As in Theorem 10.1.28, we have $|(SX)_t - (SX)_t| \leq c\int_0^t |X_s - Y_s|\,\mathrm{d}s$.

Thus for any $1 < p \leq \infty$, $L_{p,T}^*(SX, SY) \leq \int_0^t L_{p,T}^*(X, Y)\,\mathrm{d}s$.

Further, for $1 \leq p < \infty$ (the case $p = \infty$ is similar),

$$\begin{aligned}
L_{p,T}^*(X, B) &= \left(E \sup_{s \leq T} \left| \int_0^s \mathrm{d}s \left(\int_{\mathbb{R}} b_2'(B_s, y_s)\,\mathrm{d}F_{N_s}^{(-1)}(y_s) \right|^p \right)\right)^{1/p} \\
&\leq \left(E \left(\int_0^T \mathrm{d}s \int_{\mathbb{R}} |b_2'(B_s, y_s)|\,|F_{N_s}(y_s)|\,\mathrm{d}y_s \right)^p \right)^{1/p} \\
&\leq \int_0^T \mathrm{d}s \left[E \left(\int_{\mathbb{R}} |b_2'(B_s, y_s)|\,|F_{N_s}(y_s)|\,\mathrm{d}y_s \right)^p \right]^{1/p} \\
&\leq \int_0^T \mathrm{d}s \left[E \left(\int_{\mathbb{R}} (c|B_s| + |b_2'(0, y_s)|)\,|F_{N_s}(y_s)|\,\mathrm{d}y_s \right)^p \right]^{1/p} \\
&\leq c \int_0^T \mathrm{d}s(E|B_s|^p)^{1/p} + \int_0^T \mathrm{d}s \left(\int_{\mathbb{R}} |b_2'(0, y_s) F_{N_s}(y_s)|^p\,\mathrm{d}y_s \right)^{1/p} \quad < \quad \infty.
\end{aligned}$$

Now we continue as in Theorem 10.1.28 to complete the proof. □

Denote by $M_2^*(C_T, \delta_0)$ the space of all $m \in M_2(C_T, \delta_0)$ such that

$$\inf_{0 < s \leq T} E|\xi_s - E\xi_s| \quad =: \quad A_T^* > 0, \quad \xi \ \stackrel{d}{=} \ m. \qquad (10.1.78)$$

Condition (10.1.78) postulates that the L^1-variation does not converge to 0 for $0 < s < T$. In the case of B being a Brownian motion, this means that we do not start (at time $s = 0$) deterministically at a fixed point. Let $\Phi(m)$ be the solution of (10.1.70),

$$X_t = B_t + \int_0^t ds \left(\int_{\mathbb{R}} b(X_s, y_s) \, d\mathring{m}_s(y_s) \right)$$

under the assumptions of Theorem 10.1.29 with $p = 2$. Then by Theorem 10.1.29, Φ maps $M_2(C_T, \delta_0)$ into $M_0(C_T, \delta_0)$.

Theorem 10.1.30 (Contraction of Φ) *Suppose that the Lipschitz condition (L2) holds, and $m_1, m_2 \in M_2^*(C_T, \delta_0)$. Then the following contraction inequality for Φ in terms of $\ell_{2,t}^*$ is valid:*

$$\ell_{2,t}^*(\Phi(m_1), \Phi(m_2)) \le c_t \int_0^t \ell_{2,u}^*(m_1, m_2) \, du. \qquad (10.1.79)$$

Proof: For $m_1, m_2 \in M_2^*(C_T, \delta_0)$ let

$$\begin{aligned} X_t^{(i)} &= B_t + \int_0^t \left(\int_{\mathbb{R}} b\left(X_s^{(i)}, y_s^{(i)}\right) dF_{N_s^{(i)}}\left(y_s^{(i)}\right) \right) ds \\ &= B_t + \int_0^t \left(\int_{\mathbb{R}} b_2'\left(X_s^{(i)}, y_s^{(i)}\right) dF_{N_s^{(i)}}^{(-1)}\left(y_s^{(i)}\right) \right) ds. \end{aligned}$$

Then

$$\begin{aligned} X_t^{(1)} - X_t^{(2)} = \int_0^t \Bigg[&\int_{\mathbb{R}} b_2'\left(X_s^{(1)}, y_s^{(1)}\right) dF_{N_s^{(1)}}^{(-1)}\left(y_s^{(1)}\right) \\ &- \int_{\mathbb{R}} b_2'\left(X_s^{(2)}, y_s^{(2)}\right) dF_{N_s^{(2)}}^{(-1)}\left(y_s^{(2)}\right) \Bigg] ds. \end{aligned}$$

The total variation norm of $F_{N_s}^{(-1)}$ is 1 and the total mass is 0. Then the Jordan decomposition has the form $F_{N_s}^{(-1)}(dx) = \mu_s^+(dx) - \mu_s^-(dx)$, where

$\mu_s^+(\mathbb{R}) + \mu_s^-(\mathbb{R}) = 1$, $\mu_s^+(\mathbb{R}) - \mu_s^-(\mathbb{R}) = 0$. In other words, $\mu_s^+(\mathbb{R}) = \mu_s^-(\mathbb{R}) = \frac{1}{2}$.

We write $F^{(-1)}_{N_s^{(i)}}(ds) = \mu_s^{(i)+}(dx) - \mu_s^{(i)-}(dx)$, and consequently,

$$\begin{aligned} X_t^{(1)} - X_t^{(2)} &= \int_0^t \mathrm{ds} \left[\int_{\mathbb{R}} b_2'\left(X_s^{(1)}, y_s^{(1)}\right) \left(\mu_s^{(1)+} - \mu_s^{(1)-}\right) \left(\mathrm{d}y_s^{(1)}\right) \right. \\ &\qquad \left. - \int_{\mathbb{R}} b_2'\left(X_s^{(2)}, y_s^{(2)}\right) \left(\mu_s^{(2)+} - \mu_s^{(2)-}\right) \left(\mathrm{d}y_s^{(2)}\right) \right]. \end{aligned}$$

Let $dm_s^+\left(y_s^{(1)}, y_s^{(2)}\right)$ be a coupling for $\mu_s^{(1)+}$ and $\mu_s^{(2)+}$; that is, m_s^+ is a positive measure with total mass $\frac{1}{2}$ and such that $\pi_i m_s^+ = \mu_s^{(i)+}$, $i = 1, 2$, π_i the ith component. Similarly, let $\mathrm{d}m_s^-\left(y_s^{(1)}, y_s^{(2)}\right)$ be a coupling for $\mu_s^{(1)-}$ and $\mu_s^{(2)-}$. Then

$$\begin{aligned} X_t^{(1)} - X_t^{(2)} = \int_0^t \mathrm{ds} &\left[\int_{\mathbb{R}} \left(b_2'\left(X_s^{(1)}, y_s^{(1)}\right) - b_2'\left(X_s^{(2)}, y_s^{(2)}\right)\right) \mathrm{d}m_s^+\left(y_s^{(1)}, y_s^{(2)}\right) \right. \\ &\left. - \int_{\mathbb{R}} \left(b_2'\left(X_s^{(1)}, y_s^{(1)}\right) - b_2'\left(X_s^{(2)}, y_s^{(2)}\right)\right) \mathrm{d}m_s^-\left(y_s^{(1)}, y_s^{(2)}\right) \right] \end{aligned}$$

Consequently, by the Lipschitz condition,

$$\begin{aligned} &\left|X_t^{(1)} - X_t^{(2)}\right| \qquad (10.1.80) \\ &\leq \int_0^t \mathrm{ds} \left(\int_{\mathbb{R}^2} \left| b_2'\left(X_s^{(1)}, y_s^{(1)}\right) - b_2'\left(X_s^{(2)}, y_s^{(2)}\right)\right| \mathrm{d}\left(m_s^+ + m_s^-\right)\left(y_s^{(1)}, y_s^{(2)}\right) \right) \\ &\leq \int_0^t \mathrm{ds} \left(\int_{\mathbb{R}^2} \left(c\left|X_s^{(1)} - X_s^{(2)}\right| + c\left|y_s^{(1)} - y_s^{(2)}\right| \right) \mathrm{d}\left(m_s^+ + m_s^-\right)\left(y_s^{(1)}, y_s^{(2)}\right) \right). \end{aligned}$$

Observe that the total mass of $m_s^+ + m_s^-$ is 1, and for $i = 1, 2$, $\pi_i m_s^+ + \pi_i m_s^- = \mu_s^{(i)+} + \mu_s^{(i)-}$ is the variation of $F^{(-1)}_{N_s^{(i)}}$. Minimizing with respect to all couplings $m_s^+ + m_s^-$ with marginals $\mu_s^{(i)+} + \mu_s^{(i)-}$, $i = 1, 2$, we obtain

$$\begin{aligned} &\left|X_t^{(1)} - X_t^{(2)}\right| \\ &\quad \le \int_0^t c\,\mathrm{d}s \left|X_s^{(1)} - X_s^{(2)}\right| + \int_{\mathbb{R}} \left|F_{\mu_s^{(1)+}+\mu_s^{(1)-}}(x) - F_{\mu_s^{(2)+}+\mu_s^{(2)-}}(x)\right| \mathrm{d}x. \end{aligned}$$

As $F_{\mu_s^{(1)+}+\mu_s^{(1)-}}(x) = F_{\mathrm{Var}(F^{(-1)}_{N_s^{(1)}})}(x)$, we have that the integral on the right-hand side can be bound from above by κ_2:

$$\begin{aligned} &\int_{\mathbb{R}} \left| F_{\mathrm{Var}\left(F^{(-1)}_{N_s^{(1)}}\right)}(x) - F_{\mathrm{Var}\left(F^{(-1)}_{N_s^{(2)}}\right)}(x)\right| \mathrm{d}x && (10.1.81) \\ &\quad \le \int_{\mathbb{R}} |x| \,\mathrm{Var}\left(\mathrm{Var}\left(F^{(-1)}_{N_s^{(1)}}\right) - \mathrm{Var}\left(F^{(-1)}_{N_s^{(2)}}\right)\right)(\mathrm{d}x) \\ &\qquad\qquad \left(\text{using } \int |F_{\mu_1}(x) - F_{\mu_2}(x)|\,\mathrm{d}x \le \int |x|\,\mathrm{Var}(\mu_1 - \mu_2)(\mathrm{d}x)\right) \\ &\quad \le \int_{\mathbb{R}} |x|\,\mathrm{Var}\left(F^{(-1)}_{N_s^{(1)}} - F^{(-1)}_{N_s^{(2)}}\right)(\mathrm{d}x) = \int_{\mathbb{R}} |x| \left|\frac{d}{dx}\left(F^{(-1)}_{N_s^{(1)}}(x) - F^{(-1)}_{N_s^{(2)}}(x)\right)\right| \mathrm{d}x \\ &\quad = \int_{\mathbb{R}} |x| \left|F_{N_s^{(1)}}(x) - F_{N_s^{(2)}}(x)\right| \mathrm{d}x \quad \left(\text{as} \quad N_s := P^{\overset{\circ}{\xi}_s} - \delta_0\right) \\ &\quad = \int_{\mathbb{R}} |x| \left|F_{\overset{\circ}{\xi}{}_s^{(1)}}(x) - F_{\overset{\circ}{\xi}{}_s^{(2)}}(x)\right| \mathrm{d}x \quad =: \; \kappa_2\left(\overset{\circ}{\xi}{}_s^{(1)}, \overset{\circ}{\xi}{}_s^{(2)}\right), \end{aligned}$$

where $\overset{\circ}{\xi}{}_s^{(i)}$ are r.v.s with laws $P^{\overset{\circ}{\xi}{}_s^{(i)}} = P^{\left(\xi_s^{(i)} - E\xi_s^{(i)}\right)/E\left|\xi_s^{(i)} - E\xi_s^{(i)}\right|}$, and $P^{\xi_s^{(i)}} = m_s^{(i)}$.

The distance κ_2 has the following representation as a minimal metric:

$$\kappa_2\left(\xi_s^{(1)}, \xi_s^{(2)}\right) = \inf\left\{E\left||\eta_s^{(1)}|\eta_s^{(1)} - |\eta_s^{(2)}|\eta_s^{2}\right|;\ \eta_s^{(i)} \overset{d}{=} \xi_s^{(i)}\right\}. \qquad (10.1.82)$$

This representation allows us to estimate $\kappa_2\left(\overset{\circ}{\xi}{}_s^{(1)}, \overset{\circ}{\xi}{}_s^{(2)}\right)$ by $\kappa_2\left(\xi_s^{(1)}, \xi_s^{(2)}\right)$ making use of the assumption that

$$\sup_{s \le T} E\left|\xi_s^{(i)}\right|^2 \le A_T \quad \text{and} \quad \inf_{s \le T} E\left|\xi_s^{(i)} - E\xi_s^{(i)}\right| =: A_T^* > 0.$$

Then

$$
\begin{aligned}
&\kappa_2\left(\xi_s^{(1)},\xi_s^{(2)}\right) && (10.1.83)\\
&\le \ell_2\left(\xi_s^{(1)},\xi_s^{(2)}\right)\left(E|\xi_s^{(1)}|^2+E|\xi_s^{(2)}|^2\right)\\
&\le 2A_T\ell_2\left(\frac{\xi_s^{(1)}-E\xi_s^{(1)}}{E|\xi_s^{(1)}-E\xi_s^{(1)}|},\frac{\xi_s^{(2)}-E\xi_s^{(2)}}{E|\xi_s^{(2)}-E\xi_s^{(2)}|}\right)\\
&\le 2A_T\ell_2\left(\frac{\xi_s^{(1)}-E\xi_s^{(1)}}{E|\xi_s^{(1)}-E\xi_s^{(1)}|},\frac{\xi_s^{(2)}-E\xi_s^{(2)}}{E|\xi_s^{(1)}-E\xi_s^{(1)}|}\right)\\
&\quad+2A_T\ell_2\left(\frac{\xi_s^{(2)}-E\xi_s^{(2)}}{E|\xi_s^{(1)}-E\xi_s^{(1)}|},\frac{\xi_s^{(2)}-E\xi_s^{(2)}}{E|\xi_s^{(2)}-E\xi_s^{(2)}|}\right)\\
&\le \frac{2A_T}{E|\xi_s^{(1)}-E\xi_s^{(1)}|}\cdot\ell_2\left(\xi_s^{(1)}-E\xi_s^{(1)},\xi_s^{(2)}-E\xi_s^{(2)}\right)\\
&\quad+2A_T\left(E|\xi_s^{(2)}-E\xi_s^{(2)}|^2\right)^{1/2}\frac{|E|\xi_s^{(1)}-E\xi_s^{(1)}|-E|\xi_s^{(2)}-E\xi_s^{(2)}||}{(E|\xi_s^{(1)}-E\xi_s^{(1)}|)(E|\xi_s^{(2)}-E\xi_s^{(2)}|)}\\
&\le c_T\ell_2\left(\xi_2^{(1)},\xi_s^{(2)}\right).
\end{aligned}
$$

In the above derivation we used the fact that $\left|E\xi_s^{(1)}-E\xi_s^{(2)}\right|\le\ell_1\left(\xi_s^{(1)},\xi_s^{(2)}\right)$ $\le\ell_2\left(\xi_s^{(1)},\xi_s^{(2)}\right)$, and $\frac{1}{|E\xi_s^{(i)}-E\xi_s^{(i)}|}\le\frac{1}{A_T^*}$. Combining these estimates, we write

$$
\left|X_t^{(1)}-X_t^{(2)}\right| \le \int_0^t c\,\mathrm{ds}\left|X_s^{(1)}-X_s^{(2)}\right|+c_T\ell_2\left(m_s^{(1)},m_s^{(2)}\right), \qquad (10.1.84)
$$

invoking the assumptions $E(\xi_s^{(i)})^2<\infty$, $i=1,2$, and noticing that $E\left|\xi_s^{(i)}-E\xi_s^{(i)}\right|\ge A_T^*>0$ uniformly on $s\in(0,T]$. Then, by the Gronwall inequality, with $c_T^*=c\vee c_t$, we have the uniform bound

$$
\sup_{s\le t}\left|X_s^{(1)}-X_s^{(2)}\right| \le c_T^*e^{C_T^*T}\int_0^t\ell_{2,s}^*\left(m_s^{(1)},m_s^{(2)}\right)\mathrm{ds}.
$$

By passing to minimal metrics, the above inequality implies that

$$\ell_{2,t}^*(\Phi(m_1), \Phi(m_2)) \;\le\; c_T^*\, e^{c_T^* T} \int_0^t \ell_{2,s}^*(m_1, m_2)\, \mathrm{d}s.$$

□

Theorem 10.1.31 *Suppose that $\|B\|_{T,2}^* < \infty$. Suppose also that the assumption* (L2) *holds. Finally, assume that for some $m_0 \in M_2(C_T, \delta_0)$, the following boundedness assumptions on the interaction term b hold:*

$$\int_0^T \mathrm{d}s \left(\int |b_2'(0,y) F_{N_s}(y)|^2 \, \mathrm{d}y \right)^{1/2} \;<\; \infty, \quad \text{with } N_s = N_s^{m_0} \tag{10.1.85}$$

and

$$\Phi^n(m_0) \in M_2^*(C_T, \delta_0), \quad \forall\, n \in \mathbb{N}. \tag{10.1.86}$$

Then the Boltzmann-type equation (10.1.66) *has a unique weak and strong solution in $M_2(C_T, \delta_0)$.*

Proof: From Theorem 10.1.30,

$$\ell_{2,T}^*(\Phi^{(k+1)}(m), \Phi^{(k)}(m)) \;\le\; C_T^k \frac{T^k}{k!} (\ell_{2,T}^*(\Phi(m), \delta_0) + \ell_{2,T}^*(m, \delta_0)) \;<\; \infty,$$

for $m \in M_2(C_T, \delta_0)$. Therefore, $(\Phi^k(m))$ is a Cauchy sequence in $(C_T, \ell_{2,T}^*)$ and converges to a fixed point. If $X^{(k+1)}, X^{(k)}$ are the optimal couplings of $\Phi^{(k+1)}(m)$ and $\Phi^{(k)}(m)$ respectively, we obtain that $(X^{(k)})$ is an $L_{*,T}^2$-Cauchy sequence, leading to a (unique) $L_{2,T}^*$ fixed point X. □

Remark 10.1.32 *Condition* (10.1.86) *postulates that the solutions of the linear equations corresponding to $\Phi^m(m_0)$ have strictly positive variation. A simple, sufficient condition for that to hold is* $\inf_{s \le T} |B_s - EB_s| \ge TM + \varepsilon$, *provided that b is bounded and $|b| \le M$. This condition is useful only for fixed T but not for $T \to \infty$. However, it might be possible (at least in some examples, as in the construction of solutions of special SDEs) to construct a solution piecewise on small time intervals and to join the pieces to a solution on the whole real line. For special choices of b it is possible to obtain weaker sufficient conditions for* (10.1.86). *Condition* (10.1.78) *is needed in order* *to reconstruct the process. Without this condition we only can reconstruct the normalized process (cf.* (10.1.83)*).*

We now turn our attention to equation (10.1.63). The next theorem asserts that as $N \to \infty$ each $X^{i,N}$ has a limit $\overline{X}^i$. The $(\overline{X}^i)$ are independent copies of the solution of (10.1.67) considered in Theorem 10.1.31.

Theorem 10.1.33 *Suppose that* (L2) *holds, and moreover,* $\|b\|_\infty = \sup_{x,y} |b(x,y)| < \infty$. *Suppose also that uniformly on* i,

$$|W^i|_{T,\infty} \;:=\; \operatorname{ess\,sup} \sup_{0<s<T} |W^i_s| \;\leq\; X \;<\; \infty.$$

Then for any $i \geq 1$, $T > 0$,

$$\sup_N \sqrt{N} E \sup_{0<t\leq T} |X^{i,N}_t - \overline{X}^i_t| \;<\; \infty.$$

Corollary 10.1.34 (Propagation of Chaos) *Let* m *denote the law of* $\overline{X}^i$ *satisfying* (10.1.25) *and let* W_N *denote the law of* $(X^{1,N}, \ldots, X^{N,N})$. *Then, under the assumptions of Theorems* 10.1.31 *and* 10.1.33, W_N *is* m*-chaotic.*

Proof of Theorem 10.1.33: Omitting the index N, we get

$$\begin{aligned} X^i_t - \overline{X}^i_t \;&=\; \int_0^t \mathrm{d}s \frac{1}{N} \sum_{j=1}^N b(X^i_s, \mathring{X}^j_s) - \int_0^t \mathrm{d}s \int_{C_T} b(\overline{X}_s, y_s) P^{\mathring{\overline{X}}_s}(\mathrm{d}y) \\ &=:\; I_1(t) + I_2(t) + I_3(t), \end{aligned}$$

where

$$\begin{aligned} I_1(t) \;&:=\; \left[\int_0^t \mathrm{d}s \frac{1}{N} \sum_{j=1}^N b(X^i_s, \mathring{X}^j_s) - \int_0^t \mathrm{d}s \frac{1}{N} \sum_{j=1}^N b(\overline{X}^i_s, \mathring{X}^j_s)\right], \\ I_2(t) \;&:=\; \left[\int_0^t \mathrm{d}s \frac{1}{N} \sum_{j=1}^N b(\overline{X}^i_s, \mathring{X}^j_s) - \int_0^t \mathrm{d}s \frac{1}{N} \sum_{j=1}^N b(\overline{X}^i_s, \mathring{\overline{X}}^j_s)\right], \\ I_3(t) \;&:=\; \left[\int_0^t \mathrm{d}s \frac{1}{N} \sum_{j=1}^N b(\overline{X}^i_s, \mathring{\overline{X}}^j_s) - \int_0^t \mathrm{d}s \int_{C_T} b(\overline{X}_s, y_s) P^{\mathring{\overline{X}}_s}(\mathrm{d}y)\right], \end{aligned}$$

and

$$E|I_1|_T \;:=\; E \sup_{0<t<T} E|I_1(t)|$$

$$= \quad E\int_0^T ds \left| \frac{1}{N}\sum_{j=1}^N [b(X_s^i, \mathring{X}_s^j) - b(\overline{X}_s^i, \mathring{X}_s^j)] \right|.$$

From (L2),

$$\begin{aligned} |b(x,y) - b(\overline{x},y)| &= |b(x,y) - b(x,0) - (b(\overline{x},y) - b(\overline{x},0))| \\ &= \left| \int_0^y b_2'(x,t)\,dt - \int_0^t b_2'(\overline{x},t)\,dt \right| \\ &\leq \int_0^{|y|} |b_2'(x,t) - b_2'(\overline{x},t)|\,dt \\ &\leq c|x-\overline{x}|\,|y|. \end{aligned}$$

Therefore, $E|I_1|_T \;\leq\; c\,E\int_0^T \frac{1}{N}\sum_{j=1}^N |X_s^i - \overline{X}_s^i|\,|\mathring{X}_s^j|$.

Assuming that $\|b\|_\infty = \sup_{x,y}|b(x,y)| < \infty$ and

$$|W^{i,N}|_{T,\infty} \;:=\; \operatorname{ess\,sup}\sup_{0<s\leq T} |W_s^{i,N}| \;\leq\; K,$$

then $\sup_{i,N}|X_t^{i,N}| \leq K + T\cdot\|b\|_\infty$. Therefore,

$$\begin{aligned} E|I_1|_{T,1} &\leq C\int_0^T ds\, E|X_s^i - \overline{X}_s^i|, \\ E|I_2|_{T,1} &\leq \int_0^T ds \frac{1}{N}\sum_{j=1}^N \left| b\left(\overline{X}_s^i, \mathring{X}_s^j\right) - b\left(\overline{X}_s^i, \mathring{\overline{X}}_s^i\right)\right|. \end{aligned}$$

For $0 < y < \overline{y}$,

$$\begin{aligned} |b(x,y) - b(x,\overline{y})| &= \int_y^{\overline{y}} |b_2'(x,t)|dt \\ &\leq c\int_y^{\overline{y}} |b_2'(x,t) - b_2'(0,t)|\,dt + c\int_y^{\overline{y}} |b_2'(0,t) - b_2'(0,0)|\,dt \end{aligned}$$

$$\leq \; c|x|\,|\overline{y}-y| + \frac{1}{2}|\overline{y}^2 - y^2| \; \leq \; c|\overline{y}-y|\left(|x| + \frac{\overline{y}+y}{2}\right).$$

In general, $|b(x,y) - b(x,\overline{y})| \leq c|y-\overline{y}|\left(|x| + \frac{|y|+|\overline{y}|}{2}\right)$.

Assuming that $X_s^{j,N}$ are bounded a.s., $|X^{j,N}|_{T,\infty} := \operatorname{ess\,sup} \sup\limits_{0<s<T} |X_s^{j,N}| < \infty$, we obtain

$$\begin{aligned}
\|I_2\|_{T,1} \; &:= \; E|I_2|_T \\
&\leq \; \int_0^T \mathrm{d}s \frac{1}{N}\sum_{j=1}^N \left| b\left(\overline{X}_s^i, \mathring{X}_s^j\right) - b\left(\overline{X}_s^i, \mathring{\overline{X}}_s^j\right)\right| \\
&\leq \; c\int_0^T \mathrm{d}s \frac{1}{N}\sum_{j=1}^N E|\mathring{X}_s^j - \mathring{\overline{X}}_s^j| \times \left(|\overline{X}_s^i| + \frac{1}{2}\left(|\mathring{X}_s^j| + |\mathring{\overline{X}}_s^j|\right)\right) \\
&\leq \; c\int_0^T \mathrm{d}s \frac{1}{N}\sum_{j=1}^N E|\mathring{X}_s^j - \mathring{\overline{X}}_s^j|,
\end{aligned}$$

changing the values of the absolute constants c wherever it is necessary. Using the estimates for $|I_1|_T$ and $|I_2|_T$, we have

$$\begin{aligned}
N\|X^i - \overline{X}^i\|_{T,1} \; &= \; \sum_{i=1}^N \|X^i - \overline{X}^i\|_{T,1} \\
\leq \; c\int_0^T \mathrm{d}s &\left\{\sum_{i=1}^N E|X^i - \overline{X}^i|_{s,1} + \sum_{j=1}^N E|X^j - \overline{X}^j|_{s,1}\right\} \\
+ \sum_{i=1}^N \int_0^T \mathrm{d}s &\left|\frac{1}{N}\sum_{j=1}^N b(\overline{X}_s^i, \mathring{\overline{X}}_s^j) - \int_{C_T} b(\overline{X}_s, y_s) P^{\mathring{\overline{X}}_s}(\mathrm{d}y)\right|.
\end{aligned}$$

By the Gronwall lemma and the Pyke and Root (1968) inequality,

$$\|X^i - \overline{X}^i\|_{T,1} \; \leq \; c\int_0^T \mathrm{d}s \left[\frac{1}{N}\sum_{i=1}^N \int_0^T \mathrm{d}s \left|\frac{1}{N}\sum_{j=1}^N b(\overline{X}_s^i, \mathring{\overline{X}}_s^j)\right.\right.$$

$$- \int_{C_T} b(\overline{X}_s, y_s) P^{\overset{\circ}{\overline{X}}_s}(\mathrm{d}y_s) \Bigg\| \Bigg] \leq O(\frac{1}{\sqrt{N}}).$$

□

10.2 Rates of Convergence of Empirical Measures in the Kantorovich Metric

Let μ be a probability measure on $\mathbb{R}^d$ (typically unknown) and let $X_1, X_2, \ldots, X_n$ be i.i.d. r.v.s with common probability law μ. Let

$$\mu_n = \frac{1}{n} \sum_{i=1}^{n} \delta_{X_i} \quad \text{with} \quad \delta_x(A) := \begin{cases} 1 & \text{if} \quad x \in A, \\ 0 & \text{if} \quad x \notin A \end{cases}$$

be the *empirical measure* of $X_1, X_2, \ldots, X_n$. Then it is well known that

$$\mu_n \to \mu \text{ a.s.} \tag{10.2.1}$$

in the topology of weak convergence.[1] If

$$\sigma^2 = \int |x|^2 \mu(\mathrm{d}x) < \infty, \tag{10.2.2}$$

then by the SLLN,

$$\frac{1}{n} \sum_{i=1}^{n} X_i^2 = \int |x|^2 \mu_n(\mathrm{d}x) \to \sigma^2 \tag{10.2.3}$$

a.s. and in $L^1(P)$. We denote by $\mathcal{P}_2 = \mathcal{P}_2(\mathbb{R}^d)$ the space of probability measures on (the Borel sets of) $\mathbb{R}^d$ having finite second moments, i.e., such that $\int |u|^2 \mu(\mathrm{d}u) < \infty$. Recall that the L_2-Kantorovich metric (the Wasserstein metric of order 2) on $\mathcal{P}_2$ is

$$\ell_2^2(\mu, \nu) = \inf \left\{ \int |u - v|^2 P(\mathrm{d}u, \mathrm{d}v); \; P \in M(\mu, \nu) \right\},$$

---

[1] See Dudley (1989) and Rachev (1991) and the references therein.

where $M(\mu, \nu)$ denotes the set of probability measures on $\mathbb{R}^d \times \mathbb{R}^d$ with marginals μ and ν. (Here and below $|\cdot|$ denotes the usual Euclidean norm on the appropriate space.) Equivalently,

$$\ell_2^2(\mu, \nu) \;=\; \inf E|X - Y|^2,$$

where the "inf" is taken over all pairs of r.v.s X, Y having laws μ, ν, respectively, in other words, over all *couplings* of μ and ν. From (10.2.1)–(10.2.3) it follows that $\ell_2(\mu_n, \mu) \to 0$ a.s.

In this section we investigate the rate of convergence to zero of $E\ell_2^2(\mu_n, \mu)$. [2] A similar result is obtained for infinite exchangeable sequences except that the common probability law must be replaced by the directing measure. Finally a mean square uniform rate of convergence is obtained for an i.i.d. sequence of stochastic processes on a finite time interval.

Theorem 10.2.1 *Suppose that the unknown μ has high enough finite absolute moments $c := \int |u|^{d+5}\mu(\mathrm{d}u) < \infty$. Then there is a constant C, depending only on c and the dimension d, such that*

$$E\ell_2^2(\mu_n, \mu) \;\le\; Cn^{-2/(d+4)}.$$

The proof is built up on lemmas that are of some independent interest.

Lemma 10.2.2 (Carlson's lemma) *Let g be a nonnegative, measurable function on $\mathbb{R}^d$. Then for $p > d$,*

$$\int g(x)\,\mathrm{d}x \;\le\; C_{p,d}\sqrt{\left(\int g^2(x)\,\mathrm{d}x\right)^{1-d/p}\left(\int |x|^p g^2(x)\,\mathrm{d}x\right)^{d/p}}. \qquad (10.2.4)$$

where

$$C_{p,d} \;=\; \sqrt{\frac{\omega_d \pi}{\sin(\pi d/p)d^{d/p}(p-d)^{1-d/p}}}$$

---

[2] The results of this section are due to Horowitz and Karandikar (1994); see also Horowitz and Karandikar (1990). Their study presented in this section was motivated by the observation (see also Tanaka (1978)) that the Wasserstein metric is convenient for formulating weak convergence results for the empirical measures of finite interacting particle systems related to the Boltzmann equation.

and ω_d is the surface area of the unit sphere in $\mathbb{R}^d$.

In particular, for $p = d+1$ we have

$$\int g(x)\,\mathrm{d}x \;\leq\; C_d \sqrt{\left(\int g^2(x)\,\mathrm{d}x\right)^{1/(d+1)} \left(\int |x|^{d+1} g^2(x)\,\mathrm{d}x\right)^{d/(d+1)}},$$

from which follows

$$\int g(x)\,\mathrm{d}x \;\leq\; C_d \sqrt{\int (|x|^{d+1}+1) g^2(x)\,\mathrm{d}x}, \tag{10.2.5}$$

where C_d is a constant depending only on d.

Lemma 10.2.3 (Density coupling lemma) *Let f, g be probability densities on $\mathbb{R}^d$ such that*

$$\int |x|^2 (f(x) + g(x))\,\mathrm{d}x \;<\; \infty,$$

and define $\mu(\mathrm{d}x) = f(x)\,\mathrm{d}x$, $\nu(\mathrm{d}x) = g(x)\,\mathrm{d}x$. Then(3)

$$\ell_2^2(\mu,\nu) \;\leq\; 3 \int |x|^2 |f(x) - g(x)|\,\mathrm{d}x.$$

Proof: Let M be a coupling of μ and ν defined by

$$\begin{aligned}
&\int \varphi(x,y) M(\mathrm{d}x,\,\mathrm{d}y)\\
&\quad= \frac{1}{1-A} \iint \varphi(x,y)(f(x) - f\wedge g(x))(g(y) - f \wedge g(y))\,\mathrm{d}x\,\mathrm{d}y\\
&\qquad\quad + \int \varphi(x,x) f \wedge g(x)\,\mathrm{d}x,
\end{aligned}$$

where $A = \int f \wedge g(x)\,\mathrm{d}x$ and $\varphi(x,y)$ is any nonnegative Borel function. Then

$$\int |x-y|^2 M(\mathrm{d}x\,\mathrm{d}y)$$

(3) Zolotarev (1978) proves this lemma with the constant 4 instead of the constant 3 here.

$$
\begin{aligned}
&= \int |x|^2 (f(x) - f\wedge g(x))\,\mathrm{d}x + \int |y|^2 (g(y) - f\wedge g(y))\,\mathrm{d}y \\
&\quad - \frac{2}{1-A}\int x(f(x) - f\wedge g(x))\,\mathrm{d}x \cdot \int y(g(y) - f\wedge g(y))\,\mathrm{d}y \\
&= \int |x|^2 |f(x) - g(x)|\,\mathrm{d}x \\
&\quad - \frac{2}{1-A}\int x(f(x) - f\wedge g(x))\,\mathrm{d}x \cdot \int y(g(y) - f\wedge g(y))\,\mathrm{d}y
\end{aligned}
$$

(the dot $\cdot$ indicates the usual inner product in $\mathbb{R}^d$). Furthermore,

$$
\begin{aligned}
\left|\int x(f(x) - f\wedge g(x))\,\mathrm{d}x\right| &\leq \left(\int |x|^2 |f - f\wedge g|\,\mathrm{d}x\right)^{1/2}\left(\int |f - f\wedge g|\,\mathrm{d}x\right)^{1/2} \\
&= \left(\int |x|^2 |f - g|\,\mathrm{d}x\right)^{1/2} (1-A)^{1/2}.
\end{aligned}
$$

Thus $\int |x-y|^2\,\mathrm{d}M \;\leq\; 3\int |x|^2 |f-g|\,\mathrm{d}x$, and the result follows. □

Lemma 10.2.4 (Pollard (1986)) *For any r.v.s $Z_1, \ldots, Z_N$,*

$$E \max_{1\leq k\leq n} |Z_k| \;\leq\; \sqrt{N}\sqrt{\max EZ_k^2}.$$

The proof is obvious:

$$E \max |Z_k| \;\leq\; \sqrt{E\max Z_k^2} \;\leq\; \sqrt{\sum EZ_k^2} \;\leq\; \sqrt{N \max EZ_k^2}.$$

We next write $\Phi_\sigma \sim N(0, \sigma^2 I)$ to indicate that Φ_σ is the multivariate normal distribution on $\mathbb{R}^d$ with mean vector 0 and dispersion matrix $\sigma^2 I$; here $\sigma^2 > 0$, and I is the $d\times d$ identity matrix. For any probability measure μ on $\mathbb{R}^d$, let $\mu^\sigma := \Phi_\sigma * \mu$ be the convolution of Φ_σ and μ. Thus μ^σ will have density $q_\sigma := \phi_\sigma * \mu$, where ϕ_σ is the density of Φ_σ.

Lemma 10.2.5 *If $\mu \in \mathcal{P}_2$, then $\ell_2^2(\mu^\sigma, \mu) \leq d\sigma^2$.*

Proof: Let X and Y be independent random vectors with laws μ and Φ_σ, respectively. Then $(X, X+Y)$ is a coupling of μ and μ^σ, and $\ell_2^2(\mu^\sigma, \mu) \leq E|Y|^2 = d\sigma^2$. □

Proof of Theorem 10.2.1: Let $X_1, X_2, \ldots$ be i.i.d. r.v.s with law $\mu \in \mathcal{P}_2$, and let μ_n be the corresponding empirical measure. The triangle inequality gives

$$\ell_2^2(\mu_n, \mu) \;\le\; 2\left(\ell_2^2(\mu_n, \mu_n^\sigma) + \ell_2^2(\mu_n^\sigma, \mu^\sigma) + \ell_2^2(\mu^\sigma, \mu)\right).$$

Thus

$$\ell_2^2(\mu_n, \mu) \;\le\; C(\sigma^2 + \ell_2^2(\mu_n^\sigma, \mu^\sigma)). \tag{10.2.6}$$

The constant $\sigma^2 > 0$ will be chosen later.

Let $g^\sigma := \phi_\sigma * \mu$ and $g_n^\sigma := \phi_\sigma * \mu_n$ be the densities of μ^σ and μ_n^σ, respectively; here g_n^σ is given by

$$g_n^\sigma(x) \;=\; \frac{1}{n}\sum_{i=1}^n \phi_\sigma(x - X_i).$$

By Lemma 10.2.3 and inequality (10.2.5), we have

$$\begin{aligned} \ell_2^2(\mu_n^\sigma, \mu^\sigma) \;&\le\; 3\int |x|^2 |g^\sigma(x) - g_n^\sigma(x)|\, \mathrm{d}x \qquad (10.2.7)\\ &\le\; C\sqrt{\int (|x|^{d+5}+1)|g^\sigma(x) - g_n^\sigma(x)|^2\, \mathrm{d}x}. \end{aligned}$$

The above bound yields

$$E\ell_2^2(\mu_n^\sigma, \mu^\sigma) \;\le\; C\sqrt{\int (|x|^{d+5}+1) E|g^\sigma(x) - g_n^\sigma(x)|^2\, \mathrm{d}x}\,. \tag{10.2.8}$$

Since $g_n^\sigma(x)$ is the mean of n i.i.d. r.v.s, the expectation in (10.2.8) is $(1/n)V(\phi_\sigma(x - X))$, since $Eg^{n,\sigma}(x) = g^\sigma(x)$, where V stands for variance and X has law μ, $X \sim \mu$. The indicated variance is dominated by $E\phi_\sigma^2(x - X)$, and we obtain

$$E\ell_2^2(\mu_n^\sigma, \mu^\sigma) \;\le\; \frac{C}{\sqrt{n}}\sqrt{\int (|x|^{d+5}+1)\int \phi_\sigma^2(x-y)\mu(\mathrm{d}y)\, \mathrm{d}x}. \tag{10.2.9}$$

Now observe that

$$\phi_\sigma^2(x) \;=\; 2^{-d/2}(2\pi)^{-d/2}\sigma^{-d}\phi_{\sigma/\sqrt{2}}(x). \tag{10.2.10}$$

Using this, the integral in (10.2.9) is easily seen to be dominated by

$$(4\pi)^{-d/2}\sigma^{-d}\left(1+2^{d+4}\left(\sigma^{d+5}E|Z|^{d+5}+\int|y|^{d+5}\mu(\,dy)\right)\right) \;=\; C\sigma^{-d},$$

where $Z \sim N(0, I)$ and we assume $\sigma \leq 1$. Thus

$$E\ell_2^2(\mu_n^\sigma, \mu^\sigma) \;\leq\; C_n^{-1/2}\sigma^{-d/2}. \tag{10.2.11}$$

Taking expectations in (10.2.6), we get

$$E\ell_2^2(\mu_n, \mu) \;\leq\; C(\sigma^2 + n^{1/2}\sigma^{-d/2}). \tag{10.2.12}$$

Choose $\sigma = n^{-1/(d+4)}$, and Theorem 10.2.1 is proved. □

Theorem 10.2.1[4] is also valid, with a slight modification, for infinite exchangeable sequences. Let $X_1, X_2, \ldots$ be an infinite exchangeable sequence with directing measure μ (Aldous (1985)). Thus μ is now a *random* measure on $\mathbb{R}^d$, and *conditional on* μ, the r.v.s X_n are i.i.d. r.v.s with law μ. Let β be the marginal distribution of X_n, so $\beta(B) = E\mu(B)$. We then have the following rate of convergence result.

Theorem 10.2.6 *Suppose* $c := \int |u|^{d+5}\beta(\,du) < \infty$. *Then there is a constant* C*, depending only on* c *and* d*, such that*

$$E\ell_2^2(\mu_n, \mu) \;\leq\; c_n^{\frac{-2}{d+4}}.$$

Proof: The proof is virtually the same as that of Theorem 10.2.1, except that the notation μ now refers to the directing measure. In (10.2.8) we take conditional expectation given μ instead of the ordinary (unconditional) expectation, and, arguing as in (10.2.9), but conditional on μ, we get

$$E(\ell_2^2(\mu_n^\sigma, \mu^\sigma)|\mu) \;\leq\; C\sigma^{-d/2}n^{-1/2}\sqrt{C_1 + C_2\int|y|^{d+5}\mu(\,dy)},$$

[4] Two results related to Theorem 10.2.1 are (i) If the law μ is the Lebesgue measure on $[0,1]^d$, then for $d \geq 3$, $E\ell_2^2(\mu_n, \mu) = O(n^{-2/d})$; see Yukich (1991). (ii) Rachev (1991c, Theorem 11.1.6) (see also Dudley (1969)) showed that under a metric entropy condition, the rate $E\ell_2^2(\mu_n, \mu) = O(n^{-1/d})$ is optimal.

for some constants C, C_1, C_2. Taking expectation yields (10.2.11), and the proof is completed as before. □

Consider an i.i.d. sequence of *processes* $X_n(t)$ with sample functions in $D := D([0,1], \mathbb{R}^d)$, i.e., the space of cadlag functions (i.e., right continuous and having left limits at each point) on the unit interval, with values in $\mathbb{R}^d$. Let $X(t)$ denote a process having the same law. Set μ_t to be the marginal law of the process X at time t. The empirical measure at time t, based on observations $X_1(t), \ldots, X_n(t)$, is defined by

$$\mu_t^n = \frac{1}{n}\sum_{i=1}^n \delta_{X_i(t)}.$$

In this case we give a bound on the mean square *uniform* rate of convergence, that is of convergence to zero of $E\sup_{0\le t\le 1}\ell_2^2(\mu_t^n,\mu_t)$ under mild assumptions.

Theorem 10.2.7 *Suppose that for some constants $p > 2$ and $c < \infty$,*

(i) $E|X(t)|^{d+5} \le c$ *for* $0 \le t \le 1$,

(ii) $E|X(s)-X(r)|^p\,|X(s)-X(t)|^p \le c|t-r|^2$, *for* $0 \le r < s < t \le 1$,

(iii) $E|X(t)-X(s)|^p \le c|t-s|$, *for* $0 \le s \le t \le 1$,

(iv) $E|X(t)-X(s)|^2 \le c|t-s|$, *for* $0 \le s \le t \le 1$.

Then there is a constant C, depending only on p, c, and the dimension d, such that

$$E\sup_{0\le t\le 1}\ell_2^2(\mu_t^n,\mu_t) \le Cn^{-2/(d+8)}.$$

Proof: Let N be a positive integer, to be chosen later, and let $t_k = k/N, 0 \le k \le N$, so the t_k partition $[0,1]$, and let

$$Z_k := \sup_{t_k\le t\le t_{k+t}} \ell_2^2(\mu_t^n,\mu_{t_k}^n) \wedge \ell_2^2(\mu_t^n,\mu_{t_{k+1}}^n).$$

Then

$$\sup_{0\le t\le 1}\ell_2^2(\mu_t^n,\mu_t) \tag{10.2.13}$$

$$\leq \; 3\left[\max_k Z_k + \max_k \ell_2^2(\mu_{t_k}^n, \mu_{t_k}) + \max_k \sup_{t_k \leq t \leq t_{k+1}} \ell_2^2(\mu_{t_k}, \mu_t)\right].$$

The last term on the right is easy to estimate: μ_{t_k} and μ_t are coupled by $X(t_k)$ and $X(t)$, so by (iv),

$$\ell_2^2(\mu_{t_k}, \mu_t) \; \leq \; E|X(t_k) - X(t)|^2 \; \leq \; C|t_k - t|.$$

Thus the contribution of the last term on the right-hand side of (10.2.13) is at most C/N. Next we consider the middle term. Let $\sigma > 0$ (to be chosen later). With the notation of Theorem 10.2.1, we have

$$E \max_k \ell_2^2(\mu_{t_k}^n, \mu_{t_k}) \tag{10.2.14}$$
$$\leq \; C\left(E\max_k \ell_2^2(\mu_{t_k}^n, \mu_{t_k}^{n,\sigma}) + E\max_k \ell_2^2(\mu_{t_k}^{n,\sigma}, \mu_{t_k}^{\sigma}) + E\max_k \ell_2^2(u_{t_k}^{\sigma}, \mu_{t_k})\right)$$
$$\leq \; C\left(\sigma^2 + E\max_k \ell_2^2(\mu_{t_k}^{n,\sigma}, \mu_{t_k}^{\sigma})\right).$$

The last inequality follows by Lemma 10.2.5.

By Lemma 10.2.4,

$$E\max_k \ell_2^2(\mu_{t_k}^{n,\sigma}, \mu_{t_k}^{\sigma}) \; \leq \; \sqrt{N}\sqrt{\max_k E\ell_2^4(\mu_{t_k}^{n,\sigma}, \mu_{t_k}^{\sigma})}. \tag{10.2.15}$$

Now, as in (10.2.7),

$$\ell_2^4(\mu_{t_k}^{n,\sigma}, \mu_{t_k}^{\sigma}) \; \leq \; C\int (|x|^{d+5} + 1)|g_{t_k}^{n,\sigma}(x) - g_{t_k}^{\sigma}(x)|^2 \, dx,$$

where $g_{t_k}^{n,\sigma}(x), g_{t_k}^{n}(x)$ are the p.d.f.s of $\mu_{t_k}^{n,\sigma}, \mu_{t_k}^{\sigma}$. Arguing as in Theorem 10.2.1 and noting (i), we find $E\ell_2^4(\mu_{t_k}^{n,\sigma}, \mu_{t_k}^{\sigma}) \; \leq \; C\sigma^{-d}n^{-1}$, and (10.2.15) yields

$$E\max_k \ell_2^2(\mu_{t_k}^{n,\sigma}) \; \leq \; C\sigma^{-d/2}\sqrt{N/n}. \tag{10.2.16}$$

Putting everything together in (10.2.13) we have

$$E\sup_k \ell_2^2(\mu_t^n, \mu_t) \; \leq \; C\left(1/N + \sigma^2 + \sigma^{-d/2}\sqrt{N/n} + E\max_k Z_k\right) \tag{10.2.17}$$

$$\leq C\left(1/N+\sigma^2+\sigma^{-d/2}\sqrt{N/n}+\sqrt{N}\sqrt{\max_k EZ_k^2}\right).$$

Finally, we analyze the term involving Z_k. Define a random vector in $(\mathbb{R}^d)^n$ by

$$Y(t) := \frac{1}{\sqrt{n}}(X_1(t),\ldots,X_n(t)).$$

Then for $t_1 \leq t \leq t_2$,

$$\begin{aligned}
&E|Y(t)-Y(t_1)|^p|Y(t_2)-Y(t)|^p \\
&= E\left(\frac{1}{n}\sum_1^n |X_i(t)-X_i(t_1)|^2\right)^{p/2}\left(\frac{1}{n}\sum_1^n |X_j(t_2)-X_j(t)|^2\right)^{p/2} \\
&\leq E\left(\frac{1}{n}\sum_1^n |X_i(t)-X_i(t_1)|^p\right)\left(\frac{1}{n}\sum_1^n |X_j(t_2)-X_j(t)|^p\right) \\
&= \frac{1}{n^2}\sum_1^n E(|X_i(t)-X_i(t_1)|^p|X_i(t_2)-X_i(t)|^p) \\
&\quad + \frac{1}{n^2}\sum_{i\neq j} E(|X_i(t)-X_i(t_1)|^p)E(|X_j(t_2)-X_j(t)|^p) \\
&\leq C|t_2-t_1|^2.
\end{aligned}$$

Here we have used the independence of X_i and X_j for $i \neq j$, and conditions (ii) and (iii) of the theorem.

By Billingsley (1968; (15.26)), there is a constant $K = K_p$, depending only on p, such that

$$P\left(\sup_{t_1\leq t\leq t_2}\min(|Y(t)-Y(t_1)|,|Y(t_2)-Y(t)|)>\lambda\right) \leq CK\lambda^{-2p}|t_2-t_1|^2.$$

Hence

$$E\left[\sup_{t_1\leq t\leq t_2}|Y(t)-Y(t_1)|^4\wedge|Y(t_2)-Y(t)|^4\right] \leq CK|t_2-t_1|^2;$$

that is,

$$E\left[\sup_{t_1 \le t \le t_2}\left(\frac{1}{n}\sum_1^n |X_i(t)-X_i(t_1)|^2 \wedge \frac{1}{n}\sum_1^n |X_i(t_2)-X_i(t)|^2\right)^2\right]$$

$$\le \; CK|t_2-t_1|^2. \qquad (10.2.18)$$

It is easy to check that

$$\ell_2^2(\mu_t^n,\mu_{t_1}^n) \;\le\; \frac{1}{n}\sum_1^n |X_i(t)-X_i(t_1)|^2,$$

so that (10.2.18) implies (replace t_1, t_2 by t_k, t_{k+1})

$$EZ_k^2 \;\le\; CK/N^2. \qquad (10.2.19)$$

Putting this into (10.2.17), we have

$$\begin{aligned} E\sup_t \ell_2^2(\mu_t^n,\mu_t) \;&\le\; C\left(1/N+\sigma^2+\sigma^{-d/2}\sqrt{N/n}+1/\sqrt{N}\right) \qquad (10.2.20)\\ &\le\; C\left(\sigma^2+\sigma^{-d/2}\sqrt{N/n}+1/\sqrt{N}\right). \end{aligned}$$

The choice $\sigma = n^{-1/(d+8)}$, and $N = n^{4/(d+8)}$ now gives the result. □

Remark 10.2.8 *Horowitz and Karandikar also showed that if $X(t)$ is an $\mathbb{R}^d$-valued diffusion with jumps, then under some conditions on the drift and diffusion coefficient, it satisfies moment estimates as in Theorem* 10.2.7. *More precisely, let $X(t)$ be a solution of the SDE*

$$\mathrm{d}X(t) \;=\; \sigma(t,X(t))\,\mathrm{d}W(t)+b(t,X(t))\,\mathrm{d}t+h(t,X(t-),z)\widetilde{N}(\,\mathrm{d}t\;\mathrm{d}z).$$

Here, $W(t)$ is an $\mathbb{R}^d$-valued standard Wiener process; N is a Poisson random measure on $[0,T]\times\mathbb{R}^d$ with intensity measure Λ given by

$$\Lambda(\,\mathrm{d}t,\,\mathrm{d}z) \;=\; \lambda_t(\,\mathrm{d}z)\,\mathrm{d}t,$$

$\widetilde{N} = N-\Lambda$, $\sigma(t,x)$, $b(t,x)$ are measurable functions on $[0,T]\times\mathbb{R}^d$ taking values in $d\times d$ matrices and $\mathbb{R}^d$, respectively; and $h(t,x,z)$ is an $\mathbb{R}^d$-

valued measurable function on $[0, T] \times \mathbb{R}^d \times \mathbb{R}^d$. *It is assumed that* $X(0)$ *is independent of* W, N, *and that* $X(t)$ *is* $\mathcal{F}_t$*-adapted, where*

$$\mathcal{F}_t = \sigma\Big(X(0), W(s), N([0,s] \times A), \ s \le t, A \text{ Borel in } \mathbb{R}^d\Big)$$

(see Jacod (1979)).

Suppose that

$$|\sigma(t,x)|^2 \le C(1+|x|^2), \quad |b(t,x)|^2 \le C(1+|x|^2), \quad E|X(0)|^{d+5} < \infty,$$

and for $2 \le q \le d+5$, $0 \le r \le T$,

$$\int h^q(r,x,z)\lambda_r(\,dz) \le C(1+|x|^q).$$

Under these conditions, the process $X(t)$ *satisfies the moment conditions imposed in Theorem* 10.2.7. *For details we refer to Horowitz and Karandikar (1994, Section 5).*

10.3 Wasserstein Metric and Approximation of Stochastic Differential Equations

In this section we shall study weak approximations of stochastic differential equations (SDEs) of Itô type. Moreover, we shall estimate the convergence rates of the approximate solutions using the L_p-distance $\mathrm{E}\|\cdot\|^p_{C[t_0,T]}$, $p \in [2,\infty)$.[5] These results can also be interpreted as convergence rates for the minimal L^p-(Wasserstein) metric, $p \in [1,\infty)$, between the distributions of exact and approximate solutions. Two approximation schemes will be considered. They represent a combination of the time discretization methods of Euler and Milshtein with a chance discretization based on the invariance principle, and they work on a grid constructed to tune both discretizations.

The methods investigated here are based on the evaluation of the drift and diffusion coefficients in grid points, and they combine the time discretization of the SDE—as done, for instance, by the stochastic analogue of Euler's method—with the discretization of the stochastic input, the Wiener

---

[5] The results in this section are due to Gelbrich (1995).

process. This combination of time and chance discretization is necessary for a computer simulation of the solution of Itô SDEs.

Another idea for discretizing such SDEs without using the Wiener process can be found in Pardoux and Talay (1985) and Talay (1988) and is based on the approach of Doss (1977) and Sussmann (1978). In fact, Doss (1977) and Sussmann (1978) use a partial and an ordinary differential equation for constructing a pathwise solution of the SDE; that is, a pathwise convergence in the supremum norm is considered; see also Milshtein (1978), Newton (1986), Wagner (1988). A broad survey over various approximations of solutions for SDEs is given in the monograph Kloeden and Platen (1992), see also Platen (1981), Maruyama (1955), Milshtein (1978), Janssen (1984), Dudley (1968), Doss (1977), Sussmann (1978), and Römisch and Wakolbinger (1985).

Kanagawa (1986) used a method derived from the stochastic Euler method by replacing the increments of the Wiener process by other "simpler" i.i.d. r.v.s. He uses L^p-Wasserstein metrics ($p \geq 2$) between the distributions of exact and approximate solutions, thus achieving convergence rates, see also Rachev (1991), Givens and Shortt (1984), and Gelbrich (1990). Gelbrich (1995) uses the same metrics but generalizes the method of Kanagawa. For that he uses as a basis the stochastic Euler method (see further the method (E1)) (proposed by Maruyama (1955)) and Milshtein's method (M1*) (proposed by Milshtein (1978)) having orders 1 and 2, respectively, with respect to the mean square of the supremum norm of the difference between exact and approximate solutions. Since these methods use values of the drift and diffusion coefficients and of the Wiener process only in grid points t_k, the order 2 is optimal, as shown by Clark and Cameron (1980). The orders of these methods (see further (E1) and (M1*)) are proved in Platen (1981), as well as higher orders for methods using also iterated integrals of the Wiener process.

Consider a stochastic differential equation of Itô type written in integral form:

$$\text{(SE)} \qquad \begin{aligned} x(t) - x_0 &= \int_{t_0}^{t} b(x(s))\,\mathrm{d}s + \int_{t_0}^{t} \sigma(x(s))\,\mathrm{d}w(s) \\ &= \int_{t_0}^{t} b(x(s))\,\mathrm{d}s + \sum_{j=1}^{q} \int_{t_0}^{t} \sigma_j(x(s))\,\mathrm{d}w_j(s), \\ &\qquad\qquad t \in [t_0, T],\ \ x_0 \in \mathbb{R}^d, \end{aligned}$$

where $w = (w_1, \ldots, w_q)^T$ is a q-dimensional standard Wiener process, $b \in C(\mathbb{R}^d; \mathbb{R}^d)$, and $\sigma \in C(\mathbb{R}^d; \mathcal{L}(\mathbb{R}^q; \mathbb{R}^d))$, and where $\sigma_j \in C(\mathbb{R}^d; \mathbb{R}^d)$, $j = 1, \ldots, q$, denote the columns of the matrix function $\sigma = (\sigma_1, \ldots, \sigma_q)$.

In the sequel we denote by C and C^i spaces of continuous and i times differentiable functions, respectively, and by $\mathcal{L}$ spaces of linear mappings. By $\|\cdot\|$ we shall denote the Euclidean norm on $\mathbb{R}^n$ ($n \in \mathbb{N}$) and the corresponding induced norm on a space $\mathcal{L}$.

For any random variable (r.v.) ζ mapping a probability space $(\Omega, \mathcal{A}, P)$ into a separable metric space (X, d) with the Borel σ-algebra $\mathcal{B}(X)$, $\mathcal{L}(\zeta)$ denotes the distribution $P \circ \zeta^{-1}$ induced on X by ζ. $\mathcal{P}(X)$ is the set of all Borel probability measures on X.

The case that b and σ explicitly depend on the time t can be written in the form (SE) by taking t as another component of x. A direct treatment of this case follows the same lines as in this section and is carried out—for equidistant grids and bounded b and σ—in Gelbrich (1989). It allows us to relax the eventually required second-order t-differentiability to first-order t-differentiability. For $p \in [1, \infty)$, recall the minimal L_p-metric ℓ_p on the set $\mathcal{M}_p(X) := \left\{\mu \in \mathcal{P}(X);\ \int\limits_X (d(x,\theta))^p \,\mathrm{d}\mu(x) < \infty,\ \theta \in X\right\}$:

$$\ell_p(\mu, \nu) \;:=\; \left[\inf \int\limits_{X \times X} (d(x,y))^p \,\mathrm{d}\eta(x,y)\right]^{1/p}, \qquad \mu, \nu \in \mathcal{M}_p(X);$$

the infimum is taken over all measures $\eta \in \mathcal{P}(X \times X)$ with marginal distributions μ and ν.

Theorem 10.3.1 (Kanagawa (1986)) *Let $\{\zeta^k, k = 1, \ldots, N\} \in \mathbb{R}^q$ be a set of bounded i.i.d. q-dimensional r.v.s with mean value 0 and covariance matrix I_q (unit matrix), with finite $(2+\delta)$th absolute moments for some $\delta \in (0,1]$, and with a quadratically integrable density. If b and σ are Lipschitz continuous, then the method*

$$\text{(K)}\quad \begin{cases} \widehat{y}_N(0) &= x_0, \\ \widehat{y}_N(\frac{k}{N}) &= x_0 + \sum\limits_{j=1}^{k} \sigma\left(\frac{j-1}{N}, \widehat{y}_N\left(\frac{j-1}{N}\right)\right) \frac{\zeta^j}{\sqrt{N}} + \sum\limits_{j=1}^{k} b\left(\frac{j-1}{N}, \widehat{y}_N\left(\frac{j-1}{N}\right)\right) \frac{1}{N}, \\ & \qquad k = 1, \ldots, N, \\ \widehat{y}_N(t) &= \widehat{y}_N\left(\frac{k-1}{N}\right) + (N \cdot t - k + 1)\left(\widehat{y}_N\left(\frac{k}{N}\right) - \widehat{y}_N\left(\frac{k-1}{N}\right)\right) \\ & \qquad \textit{for} \quad t \in \left[\frac{k-1}{N}, \frac{k}{N}\right), \ \ k = 1, \ldots, N, \end{cases}$$

converges for any $\varepsilon > \frac{1}{2}$ *and every* $p \in [2, 2+\delta)$ *at the rate*

$$\ell_p(D(\widehat{y}_N), D(x)) = O(N^{-\delta/2(2+\delta)}(\ln N)^{\varepsilon}) \quad \textit{for} \quad N \to \infty.$$

In the sequel we shall use the following assumptions concerning (SE):

(AS1) There exists a constant $M > 0$ such that for all $j = 1, \ldots, q$ and $x \in \mathbb{R}^d$,

$$\|b(x)\| \leq M(1 + \|x\|) \quad \text{and} \quad \|\sigma_j(x)\| \leq M.$$

(AS2) There exists a constant $L > 0$ such that, for all $j = 1, \ldots, q$ and $x, y \in \mathbb{R}^d$,

$$\|b(x) - b(y)\| \leq L\|x - y\| \quad \text{and} \quad \|\sigma_j(x) - \sigma_j(y)\| \leq L\|x - y\|.$$

(AS3) $b, \sigma_j \in C^2(\mathbb{R}^d; \mathbb{R}^d)$, $j = 1, \ldots, q$, and there exists a constant $B > 0$ such that for all $j = 1, \ldots, q$ and $x, y \in \mathbb{R}^d$,

$$\|b'(x) - b'(y)\| \leq B\|x - y\| \text{ and } \|\sigma_j'(x) - \sigma_j'(y)\| \leq B\|x - y\|.$$

(AS2*) $b, \sigma_j \in C^2(\mathbb{R}^d; \mathbb{R}^d)$, $j = 1, \ldots, q$, and there exists a constant $L > 0$ such that for all $j = 1, \ldots, q$ and $x \in \mathbb{R}^d$,

$$\|b'(x)\| \leq L \quad \text{and} \quad \|\sigma_j'(x)\| \leq L.$$

(AS3*) There exists a constant $B > 0$ such that for all $j = 1, \ldots, q$ and $x \in \mathbb{R}^d$,

$$\sup\{ b''(x)[h, k]; \; h, k \in \mathbb{R}^d, \|h\| \leq 1, \|k\| \leq 1\} \leq B$$

and

$$\sup\{\sigma_j''(x)[h, k]; \; h, k \in \mathbb{R}^d, \|h\| \leq 1, \|k\| \leq 1\} \leq B.$$

(AS4) $\sigma_i'\sigma_j = \sigma_j'\sigma_i$ for all $i, j = 1, \ldots, q$.

The construction of the approximate solutions in Theorem 10.3.1 will be generalized by considering—instead of *one* equidistant grid for both time and chance discretization—a not necessarily equidistant *coarse* grid for the time discretization and a *fine* grid. The fine grid will be a refinement of the coarse grid and will be needed for the chance discretization via the invariance principle, which yields a lower convergence speed than the time discretization. To this end, we consider a grid class $\mathcal{G}(m, \Lambda, \alpha, \beta)$. Here let $m : (0, T - t_0] \to [1, \infty)$ be a monotonically decreasing function and let $\Lambda, \alpha, \beta > 0$ be constants. Then each element G of $\mathcal{G}(m, \Lambda, \alpha, \beta)$ is constructed in the following way and has the following properties:

G consists of two kinds of grid points:

- the time discretization points $t_k, k = 0 \ldots, n$, with $t_0 < t_1 < \cdots < t_n = T$ and
- the chance discretization points $u_i^k, i = 0, \ldots, m_k;\ k = 0, \ldots, n-1$, with $t_k = u_0^k < u_1^k < \cdots < u_{m_k}^k = t_{k+1};\ \ k = 0, \ldots, n-1$.

Hence, G is a combination of a coarse subgrid consisting of all points t_k relevant for the pure time discretization and of a fine grid consisting of all points u_i^k needed for the discretization of the Wiener process. Denote by

$$h_k := t_{k+1} - t_k;\ \ k = 0, \ldots, n-1, \quad \text{and} \quad h := \max_{0 \le k \le n-1} h_k$$

the step sizes and the maximal step size of the coarse subgrid. Now, G is required to satisfy the following assumptions:

(G1) $h \cdot n \le \Lambda$ and $n \in \mathbb{N}$, $h \le 1$.

(G2) $1 \le m_k \le m(h)^\alpha$ and $m_k \in \mathbb{N}$ for all $k = 0, \ldots, n-1$.

(G3) $u_i^k - u_{i-1}^k = \frac{h_k}{m_k} \le \beta \frac{h}{m(h)}$ for all $k = 0, \ldots, n-1$; $i = 1, \ldots, m_k$.

Here (G1) restricts the number of intervals of the coarse subgrid with given h, which is bounded by 1 only for convenience (in order to write simpler upper bounds later). (G2) and (G3) say that each interval of the coarse subgrid is subdivided in an equidistant way by the points u_i^k, both the number of the subdivisions and the step size of the full grid being bounded by functions of h. As an example, it is easy to see that all equidistant grids that also have an equidistant coarse subgrid and satisfy $m_k = [m(h)]$, $k = 0, \ldots, n-1$, belong to $\mathcal{G}(m, T - t_0, 1, 2)$.

For a grid G of $\mathcal{G}(m, \Lambda, \alpha, \beta)$ we define

$$\begin{aligned} [t]_G &:= t_k \text{ and } i_G(t) := k, \quad \text{if } t \in [t_k, t_{k+1});\ k = 0, \ldots, n-1; \\ [t]_G^* &:= u_i^k \quad \text{if } t \in [u_i^k, u_{i+1}^k);\ i = 1, \ldots, m_k;\ k = 0, \ldots, n-1. \end{aligned}$$

We construct the approximate solutions in (E3) and (M3) in three steps. The first step is a *pure time discretization* using the stochastic Euler method (E1) and the method (M1) corresponding to Milshtein's method (M1*). Here only the coarse subgrid is involved. We define these two methods as follows:

$$\text{(E1)} \qquad y^E(t) = x_0 + \int_{t_0}^{t} b(y^E([s]_G))\,\mathrm{d}s + \sum_{j=1}^{q}\int_{t_0}^{t}\sigma_j(y^E([s]_G))\,\mathrm{d}w_j(s) \quad \text{for all } t \in [t_0, T],$$

and

$$\text{(M1)} \quad \begin{cases} y^M(t) = x_0 + \displaystyle\int_{t_0}^{t} b(y^M([s]_G))\,\mathrm{d}s + \sum_{k=1}^{q}\left\{ \int_{t_0}^{t}\sigma_k(y^M([s]_G))\,\mathrm{d}w_k(s) \right. \\ \qquad \left. + \displaystyle\sum_{j=1}^{q}\int_{t_0}^{t}\int_{[s]_G}^{s}(\sigma_k'\sigma_j)(y^M([s]_G))\,\mathrm{d}w_j(u)\,\mathrm{d}w_k(s)\right\} \\ \qquad\qquad \text{for all } t \in [t_0, T]. \end{cases}$$

If (AS4) holds and $\widetilde{b} := b - \frac{1}{2}\sum_{j=1}^{q}\sigma_j'\sigma_j$, then (M1) is equivalent to the following method (M1*) proposed by Milshtein (1978). This equivalence is an immediate consequence of Itô's formula.

$$\text{(M1*)} \quad \begin{cases} y^M(t) = \\ x_0 + \displaystyle\sum_{r=0}^{i_G(t)-1} h_r\widetilde{b}(y^M(t_r)) + \widetilde{b}(y^M([t]_G))(t - [t]_G) \\ + \displaystyle\sum_{j=1}^{q}\left[\sum_{r=0}^{i_G(t)-1}\sigma_j(y^M(t_r))(w_j(t_{r+1}) - w_j(t_r)) \right. \\ \qquad \left. + \sigma_j(y^M([t]_G))\,(w_j(t) - w_j([t]_G))\right] \\ + \displaystyle\frac{1}{2}\sum_{j,g=1}^{q}\left[\sum_{r=0}^{i_G(t)-1}(\sigma_j'\sigma_g)(y^M(t_r))\,(w_j(t_{r+1}) - w_j(t_r)) \right. \\ \qquad \cdot (w_g(t_{r+1}) - w_g(t_r)) \\ \qquad + (\sigma_j'\sigma_g)(y^M([t]_G))\,(w_j(t) - w_j([t]_G)) \\ \qquad \left. \cdot (w_g(t) - w_g([t]_G))\right] \\ \text{for all} \quad t \in [t_0, T]. \end{cases}$$

In the second step, a *continuous and piecewise linear interpolation of the trajectories* in (E1) and (M1) between the points of the whole fine grid yields the methods (E2) and (M2), respectively:

$$\text{(E2)}\quad \begin{cases} \widetilde{y}^E \text{ is continuous, and linear in the intervals } [u_{i-1}^k, u_i^k], \\ i = 1, \ldots, m_k;\ k = 0, \ldots, n-1, \\ \text{with } \widetilde{y}^E(u_i^k) = y^E(u_i^k),\ i{=}0, \ldots, m_k;\ k = 0, \ldots, n-1. \end{cases}$$

$$\text{(M2)}\quad \begin{cases} \widetilde{y}^M \text{ is continuous, and linear in the intervals } [u_{i-1}^k, u_i^k], \\ i = 1, \ldots, m_k;\ k = 0, \ldots, n-1, \\ \text{with } \widetilde{y}^M(u_i^k) = y^M(u_i^k),\ i{=}0, \ldots, m_k;\ k = 0, \ldots, n-1. \end{cases}$$

In the third step, *the Wiener process increments over the fine grid are replaced by other i.i.d. r.v.s*: Let $\mu \in \mathcal{P}(\mathbb{R})$ be a measure with mean value 0 and variance 1, and let

$$\{\xi_{js}^k : j = 1, \ldots, q;\ \ s = 1, \ldots, m_k;\ \ k = 0, \ldots, n-1\}$$

be a family of i.i.d. r.v.s with distribution $D(\xi_{11}^0) = \mu$.

Then we shall define the methods (E3) and (M3) yielding continuous trajectories linear between neighboring grid points:

$$\text{(E3)}\quad \begin{cases} z^E(u_0^0) = x_0; \\ z^E(u_i^k) = x_0 + \sum_{r=0}^{k-1} h_r b(z^E(t_r)) + h_k \cdot \frac{i}{m_k} b(z^E(t_k)) \\ \qquad + \sum_{j=1}^{q} \left[\sum_{r=0}^{k-1} \sqrt{\frac{h_r}{m_r}}\, \sigma_j(z^E(t_r)) \sum_{s=1}^{m_r} \xi_{js}^r \right. \\ \qquad \left. + \sqrt{\frac{h_k}{m_k}}\, \sigma_j(z^E(t_k)) \sum_{s=1}^{i} \xi_{js}^k \right] \\ \text{for all } i = 1, \ldots, m_k;\ \ k = 0, \ldots, n-1; \end{cases}$$

and

$$
\text{(M3)}\quad
\begin{cases}
z^M(u_0^0) = & x_0, \text{ and for } \widetilde{b} := b - \dfrac{1}{2}\displaystyle\sum_{j=1}^{q} \sigma_j'\sigma_j; \\
z^M(u_i^k) = & x_0 + \displaystyle\sum_{r=0}^{k-1} h_r \widetilde{b}(z^M(t_r)) + h_k \cdot \tfrac{i}{m_k}\widetilde{b}(z^M(t_k)) \\
& + \displaystyle\sum_{j=1}^{q}\left[\sum_{r=0}^{k-1}\sqrt{\frac{h_r}{m_r}}\sigma_j(z^M(t_r))\sum_{s=1}^{m_r}\xi_{js}^r \right. \\
& \qquad \left. + \sqrt{\frac{h_k}{m_k}}\sigma_j(z^M(t_k))\sum_{s=1}^{i}\xi_{js}^k\right] \\
& + \dfrac{1}{2}\displaystyle\sum_{j,g=1}^{q}\left[\sum_{r=0}^{k-1}\tfrac{h_r}{m_r}(\sigma_j'\sigma_g)(z^M(t_r))\left(\sum_{s=1}^{m_r}\xi_{js}^r\right)\left(\sum_{s=1}^{m_r}\xi_{gs}^r\right)\right. \\
& \qquad \left. + \tfrac{h_k}{m_k}(\sigma_j'\sigma_g)(z^M(t_k))\left(\sum_{s=1}^{i}\xi_{js}^k\right)\left(\sum_{s=1}^{i}\xi_{gs}^k\right)\right] \\
& \text{for all} \quad i = 1,\ldots,m_k; \;\; k = 0,\ldots,n-1.
\end{cases}
$$

For this last step, the Wiener process w and the r.v.s ξ_{ji}^k will have to be defined anew on a *common* probability space. In the following we investigate the convergence rates in terms of the norm $E\sup_{t_0\le t\le T}\|\cdot\|^p$ for $C([t_0,T];\mathbb{R}^d)$-valued r.v.s in each of the three steps.

For convenience we shall denote by K any constant depending only on p, the considered grid class, and on the data of the original SDE (SE). This means that K does not depend on the particular grid. Moreover, K may have different values at different occurrences. The theorems[(6)] in the sequel will be formulated for an arbitrary fixed grid G of the grid class $\mathcal{G}(m,\Lambda,\alpha,\beta)$. Therefore, G fulfils (G1)–(G3) with the construction above. We start with some preliminary results. The first one provides the multidimensional Hölder inequality.

Lemma 10.3.2 (Hölder's inequality)

(a) *Let $p\in[1,\infty)$, $s<t$, and let $g:[s,t]\to\mathbb{R}^d$, $g(u)=(g_1(u),\ldots,g_d(u))^T$ ($u\in[s,t]$), be a Borel measurable function such that $|g_i|^p$ is Lebesgue integrable over $[s,t]$ for $i=1,\ldots,d$. Then*

$$\left\|\int_s^t g(u)\,\mathrm{d}u\right\|^p \le (t-s)^{p-1}\int_s^t \|g(u)\|^p\,\mathrm{d}u.$$

(6) For the proof of the results in this section we refer to Gelbrich (1995).

(b) *Let $p \in [1, \infty)$ and $a_i \in \mathbb{R}^d$ for all $i=1, \ldots, r$. Then*

$$\left\| \sum_{i=1}^{r} a_i \right\|^p \leq r^{p-1} \sum_{i=1}^{r} \|a_i\|^p.$$

Lemma 10.3.3 (The multidimensional martingale inequalities) *Let $p \in [2, \infty)$. Then there exist constants $C_p, A_p > 0$ such that the following assertions hold:*

(a) *Let $(w(t), \mathcal{F}(t))_{t \in [\alpha,\beta]}$ be a one-dimensional standard Wiener process over the probability space $(\Omega, \mathcal{A}, P)$. Then for every function $g = (g_1, \ldots, g_d) : [\alpha, \beta] \times \Omega \to \mathbb{R}^d$ with the properties*

(i) *$g(\cdot, \omega)$ is square-integrable over $[\alpha, \beta]$ for almost all $\omega \in \Omega$,*

(ii) *$g(u) = g(u, \cdot)$ is $\mathcal{F}(u)$-measurable for all $u \in [\alpha, \beta]$,*

we have

$$E \sup_{\alpha \leq s \leq t} \left\| \int_{\alpha}^{s} g(u)\, dw(u) \right\|^p \leq d^{p/2-1} C_p E \left(\int_{\alpha}^{t} \|g(u)\|^2\, du \right)^{p/2}$$

for all $t \in [\alpha, \beta]$.

(b) *Let $(M_s, \mathcal{F}_s)_{s=0,\ldots,r}$ be an $\mathbb{R}^d$-valued martingale (i.e., each component is a martingale), and let $p \in [2, \infty)$. Then with $\Delta M_s := M_s - M_{s-1}$ we have*

$$E \sup_{0 \leq s \leq r} \|M_s\|^p \leq d^{p/2-1} A_p E \left(\sum_{s=1}^{r} \|\Delta M_s\|^2 \right)^{p/2}.$$

Corollary 10.3.4 *Let $p \in [2, \infty)$. Then there exist constants $C_p, A_p > 0$ such that*

(a) *under the assumptions of Lemma* 10.3.3(a), *for all $t \in [\alpha, \beta]$,*

$$E \sup_{\alpha \leq s \leq t} \left\| \int_{\alpha}^{s} g(u)\, dw(u) \right\|^p \leq [d(\beta - \alpha)]^{p/2-1} C_p \int_{\alpha}^{t} E\|g(u)\|^p\, du,$$

(b) *under the assumptions of Lemma* 10.3.3(b),

$$E \max_{0 \leq s \leq r} \|M_s\|^p \leq A_p (dr)^{p/2-1} E \sum_{s=1}^{r} \|\Delta M_s\|^p.$$

Lemma 10.3.5 (Gronwall's lemma)

(a) *Let* $f : [t_0, T] \to [0, \infty)$ *be a continuous function and* c_1, c_2 *positive constants. If for all* $t \in [t_0, T]$, $f(t) \leq c_1 + c_2 \int_{t_0}^{t} f(s)\, \mathrm{d}s$, *then*

$$\sup_{t_0 \leq t \leq T} f(t) \leq c_1 e^{c_2 (T - t_0)}.$$

(b) *Let* $a_0, \dots, a_n$ *and* c_1, c_2 *be nonnegative real numbers. If for all* $k = 0, \dots, n$, $a_k \leq c_1 + c_2 \frac{1}{n} \sum_{i=0}^{k-1} a_i$, *then*

$$\max_{0 \leq i \leq n} a_i \leq c_1 e^{c_2}.$$

Based on Lemmas 10.3.2, 10.3.3, and 10.3.5 one gets the following convergence results for the time discretization step.

Theorem 10.3.6 *Let* $p \in [2, \infty)$. *Then,*

(a) (AS1) *and* (AS2) *imply*

$$E \sup_{t_0 \leq t \leq T} \|x(t) - y^E(t)\|^p \leq K \cdot h^{p/2},$$

(b) (AS1), (AS2), *and* (AS3) *imply*

$$E \sup_{t_0 \leq t \leq T} \|x(t) - y^M(t)\|^p \leq K \cdot h^p.$$

Next, the solutions in (E1) and (M1*)—which behave like the Wiener process between two neighboring points t_{k-1} and t_k of the coarse subgrid of G—will be smoothened by linear interpolation with vertices in *all* grid points of G, that means in all u_i^k. This will be the contents of Theorem 10.3.10. For its proof we need the following three lemmas.

Lemma 10.3.7 *Let* v_i, $i = 1, \dots, r$, *be i.i.d. standard-normally distributed real-valued r.v.s. Then for all* $p \in [0, \infty)$,

$$E \max_{1 \leq i \leq r} |v_i|^p \leq K(1 + \ln r)^{p/2}.$$

Lemma 10.3.8 *Let* $(\widetilde{w}(t))_{t \in [\tau_0, \infty]}$ *be a one-dimensional standard Wiener process and* x *a standard-normally distributed random variable. Then for*

$\tau_0 \le a < \bar{a} < \infty$, the random variables $\sqrt{\frac{1}{\bar{a}-a}}\sup_{a\le t\le \bar{a}}(\widetilde{w}(t) - \widetilde{w}(a))$ and $|x|$ have the same distribution.

Lemma 10.3.9 *Let $a_0 < a_1 < \cdots < a_r$ be a partition of $[a_0, a_r]$ with maximal step size $\Delta := \max_{0\le i\le r-1}(a_{i+1} - a_i)$ and $(\widetilde{w}(t))_{t\in[a_0,a_r]}$ a one-dimensional standard Wiener process. Then*

$$E \max_{0\le i\le r-1} \sup_{a_i\le t\le a_{i+1}} |\widetilde{w}(t) - \widetilde{w}(a_i)|^p \;\le\; K\cdot \Delta^{p/2}(1+\ln r)^{p/2}.$$

Now, upper bounds for the L^p-norm of the differences between the approximate solutions in (E1) and (E2), and in (M1*) and (M2), respectively, can be obtained:

Theorem 10.3.10 *Let $p \in [2,\infty)$. Then*

(a) (AS1) *and* (AS2) *imply*

$$E \sup_{t_0\le t\le T} \|y^E(t) - \widetilde{y}^E(t)\|^p \;\le\; K\Big(\tfrac{h}{m(h)}\Big)^{p/2}\Big(1+\ln\Big(\tfrac{m(h)}{h}\Big)\Big)^{p/2};$$

(b) (AS1)–(AS4) *imply*

$$E \sup_{t_0\le t\le T} \|y^M(t) - \widetilde{y}^M(t)\|^p \;\le\; K\Big(\tfrac{h}{m(h)}\Big)^{p/2}\Big(1+\ln\Big(\tfrac{m(h)}{h}\Big)\Big)^{p/2}.$$

Proof:(7)

(a) First, consider the process $\overline{y}^E$ with $\overline{y}^E(t_0) = x_0$, $\overline{y}^E(u_i^k) = \widetilde{y}^E(u_i^k)$, $\overline{y}^E(t) = \overline{y}^E(u_{i-1}^k)$ for $t \in [u_{i-1}^k, u_i^k)$ $(k = 0,\ldots,n-1;\ i = 1,\ldots,m_k)$. Then, with Lemma 10.3.2(b), (AS1), Lemma 10.3.9, (G2), and (G3), we have

$$E \sup_{t_0\le t\le T} \|y^E(t) - \overline{y}^E(t)\|^p \tag{10.3.1}$$

$$\le\; K\Bigg\{E \sup_{t_0\le t\le T}\Bigg\|\int_{[t]_G^*}^{t} b(y^E([t]_G))\,ds\Bigg\|^p$$

(7)We shall include only this proof, which is typical for the methods used in Gelbrich (1995).

$$+\sum_{j=1}^{q} E \sup_{t_0 \le t \le T} \left\| \sigma_j(y^E([t]_G)) \int_{[t]_G^*}^{t} \mathrm{d}w_j(s) \right\|^p \Bigg\}$$

$$\le K\Big\{ \max_{0\le k\le n-1} \left(\tfrac{h_k}{m_k}\right)^p \left(1 + E \sup_{t_J 0 \le t \le T} \|y^E(t)\|^p\right)$$

$$+\sum_{j=1}^{q} \max_{0\le k\le n-1} \left(\tfrac{h_k}{m_k}\right)^{p/2} (1 + \ln(n \cdot m(h)^\alpha))^{p/2} \Big\}$$

$$\le K \left\{ 1 + E \sup_{t_0 \le t \le T} \|y^E(t)\|^p \right\} \left(\tfrac{h}{m(h)}\right)^{p/2} (1 + \ln n + \ln m(h))^{p/2}.$$

Since we have by Minkowski's inequality that

$$\left(E \sup_{t_0 \le t \le T} \|y^E(t)\|^p\right)^{1/p}$$

$$\le \left(E \sup_{t_0 \le t \le T} \|x(t) - y^E(t)\|^p\right)^{1/p} + \left(E \sup_{t_0 \le t \le T} \|x(t)\|^p\right)^{1/p},$$

where the right-hand side is bounded because of Theorem 10.3.6, it holds that

$$E \sup_{t_o \le t \le T} \|y^E(t)\|^p \quad \le \quad K. \tag{10.3.2}$$

Hence, by (10.3.1) and (G1),

$$E \sup_{t_0 \le t \le T} \|y^E(t) - \overline{y}^E(t)\|^p \tag{10.3.3}$$

$$\le K \left(\tfrac{h}{m(h)}\right)^{p/2} (1 + \ln n + \ln m(h))^{p/2}$$

$$\le K \left(\tfrac{h}{m(h)}\right)^{p/2} \left(1 + \ln\left(\tfrac{m(h)}{h}\right)\right)^{p/2}.$$

On the other hand,

$$E \sup_{t_0 \le t \le T} \|\overline{y}^E(t) - \widetilde{y}^E(t)\|^p \tag{10.3.4}$$

$$= E \max_{\substack{0\le k\le n-1 \\ 0\le i \le m_k - 1}} \sup_{u_i^k \le t \le u_{i+1}^k} \|\overline{y}^E(t) - \widetilde{y}^E(t)\|^p$$

$$= E \max_{\substack{0\le k\le n-1 \\ 0\le i \le m_k - 1}} \|\widetilde{y}^E(u_{i+1}^k) - \widetilde{y}^E(u_i^k)\|^p$$

$$\leq E \max_{\substack{0\leq k\leq n-1\\0\leq i\leq m_k-1}} \sup_{u_i^k\leq t\leq u_{i+1}^k} \|y^E(t)-y^E(u_i^k)\|^p$$
$$= E \sup_{t_0\leq t\leq T} \|y^E(t)-\overline{y}^E(t)\|^p.$$

Now, with (10.3.3) and (10.3.4) we have

$$E \sup_{t_0\leq t\leq T} \|y^E(t)-\widetilde{y}^E(t)\|^p$$
$$\leq K\left\{E \sup_{t_0\leq t\leq T} \|y^E(t)-\overline{y}^E(t)\|^p + E \sup_{t_0\leq t\leq T} \|\overline{y}^E(t)-\widetilde{y}^E(t)\|^p\right\}$$
$$\leq K\left(\tfrac{h}{m(h)}\right)^{p/2}\left(1+\ln\left(\tfrac{m(h)}{h}\right)\right)^{p/2}.$$

(b) As in (a), we first consider the process $\overline{y}^M$ defined by $\overline{y}^M(t_0)=x_0$; $\overline{y}^M(u_i^k) = \widetilde{y}^M(u_i^k)$; $\overline{y}^M(t) = \overline{y}^M(u_{i-1}^k)$ for $t\in[u_{i-1}^k,u_i^k)$ $(k=0,\ldots,n-1;\ i=1,\ldots,m_k)$; and with $\widetilde{b}=b-\frac{1}{2}\sum_{j=1}^q\sigma_j'\sigma_j$ and $\Delta_j w(u,v):=w_j(v)-w_j(u)$ $(j=1,\ldots,q;\ u,v\in[t_0,T])$ we have, using method (M1*),

$$E \sup_{t_0\leq t\leq T} \|y^M(t)-\overline{y}^M(t)\|^p \tag{10.3.5}$$
$$\leq K\left\{E \sup_{t_0\leq t\leq T}\left\|\int_{[t]_G^*}^t \widetilde{b}(y^M([t]_G))\,ds\right\|^p\right.$$
$$+\sum_{j=1}^q E \sup_{t_0\leq t\leq T}\left\|\int_{[t]_G^*}^t \sigma_j(y^M([t]_G))\,ds\right\|^p$$
$$+\sum_{i,j=1}^q E \sup_{t_0\leq t\leq T}\|(\sigma_i'\sigma_j)(y^M([t]_G))[\Delta_i w([t]_G,t)\Delta_j w([t]_G,t)$$
$$\left.-\Delta_i w([t]_G,[t]_G^*)\Delta_j w([t]_G,[t]_G^*)]\|^p\right\}$$
$$\leq K\left(\tfrac{h}{m(h)}\right)^{p/2}\left(1+\ln\left(\tfrac{m(h)}{h}\right)\right)^{p/2}$$
$$+K\sum_{i,j=1}^q E \sup_{t_0\leq t\leq T}|\Delta_i w([t]_G,t)\Delta_j w([t]_G,t)$$
$$-\Delta_i w([t]_G,[t]_G^*)\Delta_j w([t]_G,[t]_G^*)|^p,$$

analogously to (10.3.1)–(10.3.3), but having used the inequalities

$$\|\widetilde{b}(x)\|\leq K(1+\|x\|)\ \ (x\in\mathbb{R}^d)\text{and } E \sup_{t_0\leq t\leq T}\|y^M(t)\|^p\leq K.$$

By the Cauchy–Schwarz inequality and by the relations

$$\sup_{t_0 \le t \le T} |\Delta w_j([t]_G^*, t)|^{2p} = \max_{\substack{0 \le k \le n-1 \\ 0 \le i \le m_k - 1}} \sup_{u_i^k \le t \le u_{i+1}^k} |\Delta w_j(u_i^k, t)|^{2p},$$

$$\begin{aligned} \sup_{t_0 \le t \le T} |\Delta w_j([t]_G, [t]_G^*)|^{2p} &\le \sup_{t_0 \le t \le T} |\Delta w_j([t]_G, t)|^{2p} \\ &= \max_{0 \le k \le n-1} \sup_{t_k \le t \le t_{k+1}} |\Delta w_j(t_k, t)|^{2p}. \end{aligned}$$

By Lemma 10.3.9 and (G3) we obtain

$$\begin{aligned} & E \sup_{t_0 \le t \le T} |\Delta w_i([t]_G, t) \Delta w_j([t]_G, t) - \Delta w_i([t]_G, [t]_G^*) \Delta w_j([t]_G, [t]_G^*)|^p \\ &\le K \Big\{ E \sup_{t_0 \le t \le T} |\Delta w_i([t]_G, t)[\Delta w_j([t]_G, t) - \Delta w_j([t]_G, [t]_G^*)]|^p \\ &\quad + E \sup_{t_0 \le t \le T} |[\Delta w_i([t]_G, t) - \Delta w_i([t]_G, [t]_G^*)] \Delta w_j([t]_G, [t]_G^*)|^p \Big\} \\ &\le K \Big\{ E \Big[\sup_{t_0 \le t \le T} |\Delta w_i([t]_G, t)|^p \sup_{t_0 \le t \le T} |\Delta w_j([t]_G^*, t)|^p \Big] \\ &\quad + E \Big[\sup_{t_0 \le t \le T} |\Delta w_i([t]_G^*, t)|^p \sup_{t_0 \le t \le T} |\Delta w_j([t]_G, [t]_G^*)|^p \Big] \Big\} \\ &\le K \Big\{ \Big(E \sup_{t_0 \le t \le T} |\Delta w_i([t]_G, t)|^{2p} \Big)^{1/2} \Big(E \sup_{t_0 \le t \le T} |\Delta w_j([t]_G^*, t)|^{2p} \Big)^{1/2} \\ &\quad + \Big(E \sup_{t_0 \le t \le T} |\Delta w_i([t]_G^*, t)|^{2p} \Big)^{1/2} \Big(E \sup_{t_0 \le t \le T} |\Delta w_j([t]_G, [t]_G^*)|^{2p} \Big)^{1/2} \Big\} \\ &\le K \Big\{ \max_{0 \le k \le n-1} h_k^{p/2} \max_{0 \le k \le n-1} \left(\tfrac{h_k}{m_k}\right)^{p/2} + \max_{0 \le k \le n-1} \left(\tfrac{h_k}{m_k}\right)^{p/2} \max_{0 \le k \le n-1} h_k^{p/2} \Big\} \\ &\quad \times (1 + \ln n + \ln m(h))^{p/2} (1 + \ln n)^{p/2} \\ &\le K \cdot h^{p/2} \left(\tfrac{h}{m(h)}\right)^{p/2} (1 + \ln n + \ln m(h))^{p/2} (1 + \ln n)^{p/2} \\ &\le K \cdot \left(\tfrac{h}{m(h)}\right)^{p/2} \left(1 + \ln\left(\tfrac{m(h)}{h}\right)\right)^{p/2}. \end{aligned} \tag{10.3.6}$$

Here the last step is based on (G1) and the boundedness of $h(1 + \ln n) \le h(1 + \ln(\Lambda/h))$. Now, (10.3.5) and (10.3.6) yield

$$E \sup_{t_0 \le t \le T} \|y^M(t) - \overline{y}^M(t)\|^p \le K \cdot \left(\tfrac{h}{m(h)}\right)^{p/2} \left(1 + \ln\left(\tfrac{m(h)}{h}\right)\right)^{p/2}. \tag{10.3.7}$$

Analogously to (10.3.4), it follows that

$$E \sup_{t_0 \le t \le T} \|\overline{y}^M(t) - \widetilde{y}^M(t)\|^p \;\le\; E \sup_{t_0 \le t \le T} \|y^M(t) - \overline{y}^M(t)\|^p,$$

which, together with (10.3.7), gives us the estimate (b).

□

In the last discretization step the Wiener process increments shall be replaced by i.i.d. r.v.s with a given distribution μ on $\mathbb{R}$. But the corresponding results hold only in the weak sense; i.e., the Wiener process (and its increments between the points of G) and i.i.d. r.v.s ξ_{ji}^k can be defined on a common probability space such that the estimates hold.

Theorem 10.3.11 (Komlós, Major, Tusnády (1975, 1976)) *Let $\mu \in \mathcal{P}(\mathbb{R})$ have the following properties:*

$$\int_{-\infty}^{\infty} x \, d\mu(x) \;=\; 0, \quad \int_{-\infty}^{\infty} x^2 \, d\mu(x) \;=\; 1, \quad \int_{-\infty}^{\infty} e^{tx} \, d\mu(x) \;<\; \infty,$$

for all t with $\|t\| \le \tau$, $\tau > 0$. (10.3.8)

Then there exist positive constants C, A, λ, depending only on μ, and for each natural number $s > 0$ two s-tuples $(x_1, \ldots, x_s)$ and $(y_1, \ldots, y_s)$, each consisting of i.i.d. real-valued r.v.s, with $\mathcal{L}(x_1) = \mu$ and y_1 being standard-normally distributed, such that for each $a > 0$,

$$P\left(\max_{1 \le k \le s} \left|\textstyle\sum_{i=1}^k (x_i - y_i)\right| \;>\; C \ln s + a\right) < A e^{-\lambda a}.$$

For translating this estimate into an estimate with the distance used in the previous chapters, we need the following lemma.

Lemma 10.3.12 *Assume that there exist constants $C, A, \lambda > 0$ with $\lambda C \ge 1$ and for any two natural numbers $r, s \ge 1$ an r-tuple $(\delta_{1,s}, \ldots, \delta_{r,s})$ of i.i.d. positive real-valued r.v.s satisfying*

$$P(\delta_{1,s} > C \ln s + a) \;<\; A e^{-\lambda a} \quad \text{for all } a > 0. \tag{10.3.9}$$

Then for each $p \in [0, \infty)$, there exists a constant $M_p > 0$ such that for all natural $r, s \ge 1$,

$$E \max_{1 \le i \le r} \delta_{i,s}^p \;\le\; M_p (1 + \ln r + \ln s)^p.$$

The following well-known result (cf., for example, Shortt (1983), Rachev (1991)) is used in the proof of the following theorem.

Lemma 10.3.13 *Let S_1, S_2, and S_3 be Polish spaces (i.e., topological spaces that are metrizable with a complete separable metric), and let $\pi_{12} : S_1 \times S_2 \times S_3 \to S_1 \times S_2$; $\pi_{23} : S_1 \times S_2 \times S_3 \to S_2 \times S_3$; $\pi_2^{12} : S_1 \times S_2 \to S_2$, and $\pi_2^{23} : S_2 \times S_3 \to S_2$ denote the projections defined by dropping one component. Then for any two measures $\nu_{12} \in \mathcal{P}(S_1 \times S_2)$ and $\nu_{23} \in \mathcal{P}(S_2 \times S_3)$ with $\nu_{12} \circ (\pi_2^{12})^{-1} = \nu_{23} \circ (\pi_2^{23})^{-1}$, i.e., with identical marginal distributions on S_2, there exists a measure $\nu_{123} \in \mathcal{P}(S_1 \times S_2 \times S_3)$ with $\nu_{123} \circ (\pi_{12})^{-1} = \nu_{12}$ and $\nu_{123} \circ (\pi_{23})^{-1} = \nu_{23}$.*

Now we can prove the estimates for the chance discretization step:

Theorem 10.3.14 *Let $p \in [2, \infty)$ and $\mu \in \mathcal{P}(\mathbb{R})$ satisfy* (10.3.8). *Then we can define a q-dimensional standard Wiener process $(w(t))_{t\in[t_0,T]}$ and a set of i.i.d. r.v.s $\{\xi_{ji}^k : j = 1, \dots, q; i = 1, \dots, m_k; k = 0, \dots, n-1\}$ with distribution $\mathcal{L}(\xi_{11}^0) = \mu$ on a common probability space such that for the methods* (E2), (E3), (M2), *and* (M3) *constructed with them we have*
(a) *If* (AS1) *and* (AS2) *hold, then*

$$E \sup_{t_0 \le t \le T} \|\widetilde{y}^E(t) - z^E(t)\|^p \le K\left(\frac{1+\ln m(h)}{\sqrt{m(h)}}\right)^p .$$

(b) *If* (AS1), (AS2), (AS3), *and* (AS4) *hold, then*

$$E \sup_{t_0 \le t \le T} \|\widetilde{y}^M(t) - z^M(t)\|^p \le K\left(\frac{1+\ln m(h)}{\sqrt{m(h)}}\right)^p .$$

The preceding results yield the following theorem, which gives bounds for the L^p-norm of the differences between the exact solution x of (SE) and the approximate solutions z^E and z^M defined in (E3) and (M3). Again, as in Theorem 10.3.14, this is a result in the weak sense.

Theorem 10.3.15 *Let $p \in [2, \infty)$ and $\mu \in \mathcal{P}(\mathbb{R})$ satisfy* (10.3.8). *Then we can define a q-dimensional standard Wiener process $(w(t))_{t\in[t_0,T]}$ and a set of i.i.d. r.v.s $\{\xi_{ji}^k, j = 1, \dots, q, i = 1, \dots, m_k, k = 0, \dots, n-1\}$ with distribution $\mathcal{L}(\xi_{11}^0) = \mu$ on a common probability space such that for* (SE) *and the methods* (E3) *and* (M3) *constructed with them we have*

(a) *If* (AS1) *and* (AS2) *hold, then*

$$E \sup_{t_0 \le t \le T} \|x(t) - z^E(t)\|^p \;\le\; K\left\{h^{p/2} + \left(\tfrac{1+\ln m(h)}{\sqrt{m(h)}}\right)^p\right\}.$$

(b) *If* (AS1), (AS2), (AS3), *and* (AS4) *hold, then*

$$E \sup_{t_0 \le t \le T} \|x(t) - z^M(t)\|^p \;\le\; K\left\{h^p + \left(\tfrac{1+\ln m(h)}{\sqrt{m(h)}}\right)^p\right\}.$$

To show that both assertions (a) and (b) follow from the Theorems 10.3.6, 10.3.10, and 10.3.14, it suffices to verify that

$$\tfrac{h}{m(h)}\left(1 + \ln\left(\tfrac{m(h)}{h}\right)\right) \;\le\; K\left(\tfrac{1+\ln m(h)}{\sqrt{m(h)}}\right)^2,$$

which follows easily from (G1).

Since Theorem 10.3.15 provides results in the weak sense, it is appropriate to formulate it as an estimate for the L^p-Wasserstein metric between the *distributions* of the exact solution and the approximate solutions:

Corollary 10.3.16 *Let* $p \in [1,\infty)$ *and* $\mu \in \mathcal{P}(\mathbb{R})$ *have the properties* (10.3.8). *Moreover, let* $(w(t))_{t\in[t_0,T]}$ *be a* q*-dimensional standard Wiener process and* $\{\xi_{ji}^k : j = 1,\dots,q; i = 1,\dots,m_k; k = 0,\dots,n-1\}$ *a set of i.i.d. r.v.s with distribution* $\mathcal{L}(\xi_{11}^0) = \mu$. *Then for* (SE) *and the methods* (E3) *and* (M3) *constructed with them we have*

(a) *If* (AS1) *and* (AS2) *hold, then*

$$\ell_p(x, z^E) \;\le\; K\left\{h^{1/2} + \tfrac{1+\ln m(h)}{\sqrt{m(h)}}\right\}.$$

(b) *If* (AS1), (AS2), (AS3), *and* (AS4) *hold, then*

$$\ell_p(x, z^M) \;\le\; K\left\{h + \tfrac{1+\ln m(h)}{\sqrt{m(h)}}\right\}.$$

For $p \in [2,\infty)$ the assertions follow directly from Theorem 10.3.15 and applying Lemma 10.3.3 to the right-hand sides. Then the assertions are also true for $p \in [1,2)$, since $\ell_{p_1} \le \ell_{p_2}$ for $1 \le p_1 \le p_2 < \infty$. The estimates in Theorem 10.3.15 and Corollary 10.3.16 give convergence rates with respect to h for the methods (E3) and (M3) and for any grid sequence in $\mathcal{G}(m, \Lambda, \alpha, \beta)$. These rates consist of two summands, one depending on h and the other depending on $m(h)$, representing the rates of time and chance discretization, respectively.

It is desirable to tune the rates of both summands, i.e., to equal the powers of h in both summands. This means to choose $m(h)$ to be increasing like $1/h$ for method (E3) and like $1/h^2$ for the method (M3).

Corollary 10.3.17 *Let* $p \in [2, \infty)$ *and* $\mu \in \mathcal{P}(\mathbb{R})$ *satisfy* (10.3.8). *Then we can construct solutions in* (SE), (E3), *and* (M3) *on a common probability space (as in Theorem* 10.3.15*) with the following properties.*

(a) *If* (AS1) *and* (AS2) *hold and* $\max\left\{\sup_{0<s\leq 1} sm(s), \sup_{0<s\leq 1} \frac{1}{sm(s)}\right\} \leq K$, *then*

$$E \sup_{t_0 \leq t \leq T} \|x(t) - z^E(t)\|^p \;\leq\; K \cdot h^{p/2}(1 - \ln h)^p.$$

(b) *If* (AS1), (AS2), (AS3), *and* (AS4) *hold and* $\max\left\{\sup_{0<s\leq 1} s^2 m(s), \sup_{0<s\leq 1} \frac{1}{s^2 m(s)}\right\} \leq K$, *then*

$$E \sup_{t_0 \leq t \leq T} \|x(t) - z^M(t)\|^p \;\leq\; K \cdot h^p(1 - \ln h)^p.$$

Corollary 10.3.18 *Under the general assumptions in Corollary* 10.3.16 *we have*

(a) *If* (AS1) *and* (AS2) *hold and* $\max\left\{\sup_{0<s\leq 1} sm(s), \sup_{0<s\leq 1} \frac{1}{sm(s)}\right\} \leq K$, *then*

$$\ell_p(x, z^E) \leq K \cdot h^{1/2}(1 - \ln h).$$

(b) *If* (AS1), (AS2), (AS3), *and* (AS4) *hold and* $\max\left\{\sup_{0<s\leq 1} s^2 m(s), \sup_{0<s\leq 1} \frac{1}{s^2 m(s)}\right\} \leq K$, *then*

$$\ell_p(x, z^M) \leq K \cdot h(1 - \ln h).$$

In conclusion, suppose that a grid sequence in $\mathcal{G}(m, \Lambda, \alpha, \beta)$ with $h \to 0$ is given. Then using the metric ℓ_p, we have under the assumptions of Corollary 10.3.18(a), for the method (E3), the convergence rate $O(h^{1/2}(1 - \ln h))$. This convergence is in terms of the maximal step sizes h of the coarse subgrids. Similarly, we have the convergence rate $O((\frac{h}{m(h)})^{1/4}(1 - \ln \frac{h}{m(h)}))$ with respect to the maximal step sizes $\frac{h}{m(h)}$ of the whole fine grids. Finally, we have the convergence rate $O(N^{-1/4}(1 + \ln N))$ with respect to the number N of all gridpoints of the whole fine grids. Analogously, under the assumptions of Corollary 10.3.18(b) we have, for the method (M3), the convergence rates $O(h(1 - \ln h))$, $O((\frac{h}{m(h)})^{1/3}(1 - \ln \frac{h}{m(h)}))$, and $O(N^{-1/3}(1 + \ln N))$.

References

[1] T. Abdellaoui. *Distances de deux lois dans les espaces de Banach.* PhD thesis, Université de Rouen, 1993.

[2] T. Abdellaoui. Détermination d'un couple optimal du problème de Monge–Kantorovich. *C.R. Acad. Sci. Paris I*, 319:981–984, 1994.

[3] T. Abdellaoui and H. Heinich. Sur la distance de deux lois dans le cas vectoriel. *C.R. Acad. Sci. Paris I*, 319:397–400, 1994.

[4] M. Abramowitz and I.A. Stegun. *Handbook of Mathematical Functions.* Dover Publications, New York, 9th edition, 1970.

[5] A. Acosta and G. Gine. Convergence of moments and related functionals in the general central limit theorem in Banach spaces. *Z. Wahrscheinlichkeitstheorie Verw. Geb.*, 48(2):213–241, 1979.

[6] N.I. Ahiezer. *Classical Moment Problem and Related Questions of Analysis.* GIFML, Moscow, 1961.

[7] N.I. Ahiezer and M. Krein. *Some Questions in the Theory of Moments.* American Mathematical Society, Providence, 1962.

[8] H. Akaike. Modern development of statistical methods. In P. Eykhoff, editor, *Trends and Progress in System Identification*, pages 169–184. Pergamon Press, 1981.

[9] D.J. Aldous. Exchangeability and related topics. *Lecture Notes in Mathematics*, 1117, 1985.

[10] D.J. Aldous. Ultimate instability of exponential backoff protocol for acknowledgement-based transmission control of random access communication channels. *IEEE Transactions on Information Theory*, IT 33:219–223, 1987.

[11] D.J. Aldous. Asymptotic fringe distribution for general families of random trees. *Annals of Applied Probability*, 1:228–266, 1991.

[12] D.J. Aldous. The continuum random tree II: An overview. In M.T. Barlow and N.H. Bingham, editors, *Stochastic Analysis*, volume 167 of *London Math. Soc. Lecture Notes Series*, pages 23–70. Cambridge University Press, 1991.

[13] D.J. Aldous and J.M. Steele. Introduction to the interface of probability and algorithms. *Statistical Science*, 8:3–9, 1993.

[14] G.A. Anastassiou. *Moments in Probability and Approximation Theory.* Pitman, England, 1993.

[15] G.A. Anastassiou and S.T. Rachev. Approximation of a random queue by means of deterministic queueing models. In C.K. Chui, L.L. Shumaker, and J.D. Ward, editors, *Approximation Theory VI*, volume 1, pages 9–11. Academic Press, 1989.

[16] G.A. Anastassiou and S.T. Rachev. Moment problems and their applications to characterization of stochastic processes, queueing theory, and rounding problems. In *Approximation Theory*, volume 138, pages 1–77, New York, 1992. Proceedings of 6th S.E.A. Meeting, Marcel Dekker Inc.

[17] G.A. Anastassiou and S.T. Rachev. Moment problems and their applications to the stability of queueing models. *Computers and Mathematics with Applications*, 24(8/9):229–246, 1992.

[18] E.J. Anderson and P. Nash. *Linear Programming in Infinite Dimensional Spaces. Theory and Applications.* Wiley, New York, 1987.

[19] E.J. Anderson and A.B. Philpott. An algorithm for a continuous version of the assignment problem. *Lecture Notes in Economics and Mathematical Systems*, 215:108–117, 1983. Semi-Infinite Programming and Applications (Austin, Texas, 1981).

[20] E.J. Anderson and A.B. Philpott. Duality and an algorithm for a class of continuous transportation problems. *Mathematics of Operations Research*, 9:222–231, 1984.

[21] T.W. Anderson. *An Introduction to Multivariate Statistical Analysis.* Wiley, New York, 1984.

[22] W. Apitzsch, B. Fritzsche, and B. Kirstein. A Schur analysis approach to minimum distance problems. *Linear Algebra and its Applications*, 1990.

[23] A. Araujo and E. Giné. *The Central Limit Theorem for Real and Banach Valued Random Variables.* Wiley, New York, 1980.

[24] M.A. Arbeiter. Random recursive constructions of self-similar fractal measures. The non-compact case. *Probability Theory and Related Fields*, 88:497–520, 1991.

[25] R.J. Aumann. Measurable utility and the measurable choice theorem. In Centre Nat. Recherche Sci., Paris, editor, *La Decision*, volume 171, pages 15–26. Actes Coll. Internat., Aix-en-Provence, 1967.

[26] F. Aurenhammer, F. Hoffmann, and B. Arnov. Minkowski-type theorems and least-square partitioning. Reports of the Institute for Computer Science, 1992. Dept. of Mathematics, Freie Universität Berlin.

[27] J. Auslander. Generalized recurrences in dynamical systems. *Contributions to differential equations*, 3(1):65–74, 1964.

[28] M.L. Balinski. Signature des points extrêmes du polyhedre dual du problème de transport. Comptes Rendus de l'Académie des Sciences, Paris, 1983.

[29] M.L. Balinski. The Hirsch conjecture for dual transportation polyhedra. *Mathematics of Operations Research*, 9:629–633, 1984.

[30] M.L. Balinski. Signature methods for the assignment problem. *Operations Research*, 34:125–141, 1985.

[31] M.L. Balinski. A complex (dual) simplex method for the assingment problem. *Mathematical Programming Study*, 34:125–141, 1986.

[32] M.L. Balinski, B. Athanasopoulos, and S.T. Rachev. Some developments on the theory of rounding proportions. In *Bulletin of thi ISI, 49th Session*, volume 1, pages 71–72, Firenze, 1993.

[33] M.L. Balinski and D. Gale. On the core of the assignment game. In *Functional Analysis, Optimization, and Mathematical Economics: A Collection of Papers dedicated to the Memory of L.V. Kantorovich*, pages 274–289, Oxford, 1990. Oxford University Press.

[34] M.L. Balinski and S.T. Rachev. On Monge–Kantorovich problems. Preprint, 1989. SUNY at Stony Brook, Dept. of Applied Mathematics and Statistics.

[35] M.L. Balinski and S.T. Rachev. Rounding proportions: rules of rounding. *Numer. Funct. Anal. Optimization*, 14:475–501, 1993.

[36] M.L. Balinski and S.T. Rachev. Rounding proportions: methods of rounding. *Mathematical Scientist*, 1997.

[37] M.L. Balinski and A. Russakoff. Faces of dual transportation polyhedra. *Mathematical Programming Study*, 22:1–8, 1984.

[38] M.L. Balinski and H.P. Young. Stability, coalitions and schisms in proportional representation systems. *Americal Political Science Review*, 72:848–858, 1978.

[39] M.L. Balinski and H.P. Young. *Fair Representation: Meeting the Ideal of One Man, One Vote.* Yale University Press, New Haven, 1982.

[40] A.A. Balkema, L. de Haan, and R. Karandikar. The maximum of n independent stochastic processes. Preprint, 1990. Erasmus University, Rotterdam.

[41] D.P. Barbu and Th. Precupanu. *Convexity and optimization in Banach spaces.* Sijthoff/Nordhoff, 1978.

[42] R.E. Barlow and F. Proschan. *Statistical Theory of Reliability and Life Testing: Probability Models.* Hold, Rinehart, and Winston, New York, 1975.

[43] E.R. Barnes and A.J. Hoffman. Partitioning spectra and linear programming. In *Proc. Silver Jubilee Conference on Combinations*, Ontario, Canada, June 1982. Univ. Waterloo.

[44] E.R. Barnes and A.J. Hoffman. On transportation problems with upper bounds on leading rectangles. *SIAM Journal of Algebraic and Discrete Methods*, 6:487–496, 1985.

[45] M.F. Barnsley and J.H. Elton. A new class of of Markov processes for image encoding. *Advances in Applied Probability*, 20:14–32, 1988.

[46] D.P. Baron and R.B. Myerson. Regulating a monopolist with unknown cost. *Econometrica*, 50:911–930, 1982.

[47] S.K. Basu. On the rate of convergence to normality of sums of dependent random variables. *Acta Math. Acad. Sci. Hungarica*, 28:261–265, 1976.

[48] S.K. Basu and G. Simons. Moment spaces of IFR distributions, applications and related material. In P.K. Sen, editor, *Contributions to Statistics: Essay in Honor of Norman L. Johnson*, pages 27–46. North-Holland Publishing Company, 1983.

[49] J. Beirlant and S.T. Rachev. The problems in stability in insurance mathematics. *Insurance: Mathematics and Economics*, 6:179–188, 1987.

[50] V. Beneš. The moment problem and its technical application. In *Proc. 30th Int. Wissen. Kolloq.*, pages 11–14. TH Ilmenau, 1985.

[51] V. Beneš. *Moment Problem and Its Application.* PhD thesis, Charles University, 1986.

[52] V. Beneš. Extremal and optimal solutions in the transshipment problem. *Comment. Math. Univ. Carolinae*, 33:97–112, 1992.

[53] V. Beneš. *Extremal and Optimal Solutions of the Marginal and Transshipment Problem.* PhD thesis, Dept. of Mathematics, FSI, Czech Technical University, Praha, Czech Republic, 1995.

[54] V. Beneš and J. Štěpán. The support of extremal probability measure with given marginals. In M.L. Puri, P. Revesz, and W. Werzt, editors, *Mathematical Statistics and Probability Theory*, volume A of *Proc. 6th Pannon Symp.*, pages 33–41. D. Reidel Publ. Comp., 1987.

[55] V. Beneš and J. Štěpán. Extremal solutions in the marginal problem. In G. Dall'Aglio et al., editor, *Advances in Probability Measures with Given Marginals*, pages 189–206. Kluwer, Dordrecht, 1991.

[56] V.Y. Bentkus, F. Götze, V. Paulauskas, and A. Rackauskas. The accuracy of Gaussian approximation in Banach spaces. University of Bielefeld, Preprint 90-100, 1990.

[57] C. Berge. *Théorie générale des jeux à n personnes*, volume 138. Gauthier-Villars, Paris, 1957. Mémorial des science mathématiques.

[58] C. Berge and A. Ghouila-Houri. *Programming, Games and Transportation Networks.* Methnen, John Wiley and Sons, Inc., New York, 1965.

[59] S. Bertino. Su di una sottoclasse della classe di Fréchet. *Statistica*, 28:511–542, 1968.

[60] S. Bertino. Sulla distanza tra distribuzioni. Pubbl. Ist. Calc. Prob. Univ. Roma, 1968.

[61] D. Bertsekas and R. Gallager. *Data Networks.* Prentice-Hall, New Jersey, 1987.

[62] N.P. Bhatia and G.P. Szegö. *Stability theory of dynamical systems.* Number 161 in Dre Grundlehren der mathematischen Wissenschaften. Springer, 1970.

[63] R.M. Bhattacharya and R. Rango Rao. *Normal Approximation and Asymptotoic Expansions.* Wiley, 1976.

[64] P.J. Bickel and D.A. Freedman. Some asymptotic theory for the bootstrap. *Annals of Statistics*, 9:1196–1217, 1981.

[65] P. Billingsley. *Convergence of Probability Measures.* Wiley, New York, 1968.

[66] P. Billingsley. *Probability and Measure.* Wiley, New York, 2nd edition, 1986.

[67] D. Blackwell and L.E. Dubins. An extension of Skorohod's almost sure representation theorem. *Proc. Amer. Math. Soc.*, 89:691–692, 1983.

[68] R.C. Blattberg and N.J. Genodes. A comparison of the stable and student distributions as statistical models for stock prices. *J. Business*, 47:244–280, 1974.

[69] T. Bollerslev. A conditionally heteroscedastic time series model for speculative prices and rates of return. *Review of Economic Studies*, 69:542–547, 1987.

[70] E. Bolthausen. Exact convergence rate in some martingale central limit theorems. *Annals of Probability*, 10:672–688, 1982.

[71] A. Boness, A. Chen, and S. Jatusipitak. Investigations of nonstationary prices. *J. Business*, 48:518–537, 1979.

[72] A.A. Borovkov. *Asymptotic Methods in Queueing Theory.* Wiley, New York, 1984.

[73] A.A. Borovkov. On the ergodicity and stability of the sequence $w_{n+1} = f(w_n, z_n)$: applications to communication networks. *Theory of Probability and its Applications*, 33:595–611, 1988.

[74] A. Brandt, P. Franken, and B. Lisek. *Stationary Stochastic Models.* Wiley, New York, 1990.

[75] Y. Brenier. Polar factorization and monotone rearrangement of vector-valued functions. *Comm. Pure. Appl. Maths.*, XLIV:375–417, 1987.

[76] G. Brown and B. Shubert. On random binary trees. *Mathematics of Operations Research*, 9:43–65, 1984.

[77] R.A. Brualdi and J. Csima. Extremal plane stochastic matrices of dimension three. *Journal of Linear Algebra and its Applications*, 11:105–133, 1975.

[78] R.A. Brualdi and J. Csima. Stochastic patterns. *J. Comb. Theory*, 19:1–12, 1975.

[79] Y.A. Brudnii. A multidimensional analog of a theorem of Whitney. *USSR Math. Sbornik*, 11:157–170, 1970.

[80] R.E. Burkard, B. Klinz, and R. Rudolf. Perspectives of Monge properties in optimization. Bericht 2, 1994. Spezialforschungsbereich F 003, Karl-Franzens-Universität Graz & Technische Universität Graz.

[81] R.M. Burton and U. Rösler. An L_2-convergence theorem for random affine mappings. *Journal of Applied Probability*, 32:183–192, 1995.

[82] P.L. Butzer, L. Hahn, and M.Th. Roeckerath. Central limit theorem and weak law of large numbers with rates for martingales in Banach spaces. *Journal of Multivariate Analysis*, 13:287–301, 1983.

[83] S. Cambanis and G. Simons. Probability and expectation inequalities. *Z. Wahrscheinlichkeitstheorie Verw. Geb.*, 59:285–294, 1982.

[84] S. Cambanis, G. Simons, and W. Stout. Inequalities for $Ek(X,Y)$ when the marginals are fixed. *Z. Wahrscheinlichkeitstheorie Verw. Geb.*, 36:285–294, 1976.

[85] L. Cavalli-Sforza. Cultural and biological evolution: a theoretical inquirey. In S.G. Ghurye, editor, *Proceedings of the Conference on Directions for Mathematical Statistics*, volume 7 of *Suppl. Adv. Appl. Prob.*, pages 90–99, 1975.

[86] L. Cavalli-Sforza and M.W. Feldman. Models for cultural inheritance I. Group mean and within group variation. *Theoret. Popn. Biol.*, 4:42–55, 1973.

[87] S. Chandrasekhar and G. Munch. The theory of the fluctuations in brightness of the milky way. I and II. *Astrophys. J.*, 112:380–398, 1950.

[88] M.R. Chernick, D.J. Daley, and R.P. Littlejohn. A time-revisibility relationship between two markov chains with exponential stationary distributions. *Journal of Applied Probability*, 25:418–422, 1988.

[89] G. Choquet. Forme abstraite du théorème de capacitabilité. *Ann. Inst. Fourier*, 9:83–89, 1959.

[90] Y.S. Chow and H. Teicher. *Probability Theory: Independeance, interchangeability, martingales.* Springer, New York, 1978.

[91] F.H. Clark. Optimization and nonsmooth analysis. *Classics in Appl. Math. SIAM*, 1990.

[92] J.M.C. Clark and R.J. Cameron. The maximum rate of convergence of discrete approximations for stochastic differential equations. *Lecture Notes in Control and Information Science*, 25:162–171, 1980.

[93] P.K. Clark. A subordinated stochastic process model with finite variance for speculative prices. *Econometrica*, 41:135–155, 1973.

[94] M. Cramer. *Stochastische Analyse rekursiver Algorithmen mit idealen Metriken.* PhD thesis, Universität Freiburg, 1995a.

[95] M. Cramer. Convergence of a branching type recursion with non-stationary immigration. *Metrica*, 1995b. To appear.

[96] M. Cramer. A note concerning the limit distribution of the Quicksort algorithm. *Informatique Théoriqué et Appl.*, 30:195–207, 1996.

[97] M. Cramer and L. Rüschendorf. Analysis of recursive algorithms by the contraction method. *Lecture Notes in Statistics*, 114:18–33, 1996a.

[98] M. Cramer and L. Rüschendorf. Convergence of a branching type recursion. *Annales de l'Institut Henri Poincaré*, 32:725–741, 1996b.

[99] J. Csima. Multidimensional stochastic matrices and patterns. *J. Algebra*, 14:194–202, 1970.

[100] J.A. Cuesta-Albertos and C. Matrán. Strong convergence of weighted sums of random elements through the equivalence of sequences of distributions. *Journal of Multivariate Analysis*, 25:311–322, 1988.

[101] J.A. Cuesta-Albertos and C. Matrán. Notes on the Wasserstein metric in Hilbert spaces. *Annals of Probability*, 17:1264–1276, 1989.

[102] J.A. Cuesta-Albertos and C. Matrán. Skorohod representation theorem and Wasserstein metrics. Preprint, 1991.

[103] J.A. Cuesta-Albertos and C. Matrán. A review on strong convergence of weighted sums of random elements based on Wasserstein metrics. *Journal of Stat. Planning Infer.*, 30:359–370, 1992.

[104] J.A. Cuesta-Albertos and C. Matrán. Stochastic convergence through Skorohod representation theorems and Wasserstein metrics. *Suppl. Rendic. Circolo Matem. Palermo II*, 35:89–113, 1994.

[105] J.A. Cuesta-Albertos, C. Matrán, S.T. Rachev, and L. Rüschendorf. Mass transportation problems in probability theory. *Mathematical Scientist*, 21:37–72, 1996.

[106] J.A. Cuesta-Albertos, L. Rüschendorf, and A. Tuero-Diaz. Optimal coupling of multivariate distributions and stochastic processes. *Journal of Multivariate Analysis*, 46:335–361, 1993.

[107] J.A. Cuesta-Albertos and A. Tuero-Diaz. A characterization for the solution of the Monge–Kantorovich mass transference problem. *Statist. Probab. Letters*, 16:147–152, 1993.

[108] G. Dall'Aglio. Sugli estremi dei momenti delle funzioni di ripartizione doppie. *Ann. Scuola Normale Superiore Di Pisa, Cl. Sci.*, 3(1):33–74, 1956.

[109] G. Dall'Aglio. Sulla compatibilita delle funzioni di ripartizione doppia. *Rendiconti di Math.*, 18:385–413, 1959.

[110] G. Dall'Aglio. Les fonctions extrèmes de la classe de Fréchet à 3 dimensions. *Publ. Inst. Stat. Univ. Paris*, IX:175–188, 1960.

[111] G. Dall'Aglio. Sulle distribuzioni doppie con margini assegnati soggette a delle limitazioni. *It. Giorn. 1st. Ital. Attuari*, 94, 1961.

[112] G. Dall'Aglio. Fréchet classes and compatibility of distribution functions. *Symposia Mathematica*, 9:131–150, 1972.

[113] G. Dall'Aglio, S. Kotz, and G. Salinetti. *Advances in Probability Distributions with Given Marginals.* Kluver, Dordrecht, 1991.

[114] G.B. Dantzig and A.R. Ferguson. The allocation of aircraft to routes—an example of linear programming under uncertain demands. *Mang. Science*, 3:45–73, 1956.

[115] A. D'Aristotile, P. Diaconis, and D. Freedman. On a merging of probabilities. No. 301, 1988. Dept. of Statistics, Stanford University.

[116] M.M. Day. *Normed Linear Spaces.* Springer, Berlin–Göttingen–Heidelberg, 1958.

[117] A. de Acosta. Invariance principles in probability for triangle arrays of B-valued random vectors and some applications. *Annals of Probability*, 10:346–373, 1982.

[118] L. de Haan, E. Omey, and S.I. Resnick. Domains of attraction and regular variation in $\mathbb{R}^d$. *Journal of Multivariate Analysis*, 14:17–33, 1984.

[119] L. de Haan and S.T. Rachev. Estimates of the rate of convergence for max-stable processes. *Annals of Probability*, 17:651–677, 1989.

[120] L. de Haan and S.I. Resnick. Limit theory for multivariate sample extremes. *Z. Wahrscheinlichkeitstheorie Verw. Geb.*, 40:317–337, 1977.

[121] L. de Haan, S.I. Resnick, H. Rootzen, and C.G. Vries. Extremal behavior of solutions to a stochastic difference equation with applications to ARCH process. *Stoch. Processes and Applications*, 32:213–224, 1989.

[122] P. de Jong. Central limit theorems for generalized multilinear forms. *CWI Tract, Amsterdam*, 61, 1989.

[123] G. Debreu. Representation of a preference ordering by a numerical function. In *Decision Processes*, pages 159–165. Wiley, New York, 1954.

[124] G. Debreu. Continuity properties of paretian utility. *Intern. Econ. Revue*, 5:285–293, 1964.

[125] P. Deheuvels and D. Pfeifer. On a relationship between Uspensky's theorem and Poisson approximation. *Ann. Inst. Statist. Math.*, 40:671–681, 1988.

[126] C. Dellacherie and P.A. Meyer. *Probabilités et potential*, volume 29 of *North-Holland Mathematics Studies*. Hermann, Paris, 1983. Chapitres IX a XI.

[127] U. Derigs, O. Goecke, and R. Schrader. Monge sequences and a simple assignment algorithm. *Discrete Applied Mathematics*, 15:241–248, 1986.

[128] L. Devroye. *Lecture Notes on Bucket Algorithms*, volume 6. Birkhäuser, Boston, 1986. Progress in computer science.

[129] L. Devroye. *A Course in Density Estimation*, volume 14 of *Progress in probability and statistics*. Birkhäuser, Boston, 1987.

[130] P. Diaconis and D. Freedman. On rounding percentages. *Journal of the American Statistical Association*, 74:359–364, 1979.

[131] P. Diaconis and D. Freedman. A dozen of the Finetti-style results in search of a theory. *Annales de l'Institut Henri Poincaré*, 23:397–423, 1987.

[132] H. Dietrich. Zur c-Konvexität und c-Subdifferenzierbarkeit von Funktionalen. *Optimization*, 19:355–371, 1988.

[133] N. Dinculeanu. *Vector Measures*, volume 95 of *International series of monographs on pure and applied mathematics*. Pergamon Press, Oxford, 1967.

[134] R.L. Dobrushin. Prescribing a system of random variables by conditional distributions. *Theory of Probability and its Applications*, 15:458–486, 1970.

[135] R.L. Dobrushin. Vlasov equations. *Func. Anal. Appl.*, 13:115–123, 1979.

[136] I. Domowitz and C.S. Hakkio. Conditional variance and the risk premium in the foreign exchange market. *Journal of Internat. Economics*, 19:47–66, 1985.

[137] H. Doss. Liens entre équation differentielles stochastiques et ordinaires. *Annales de l'Institut Henri Poincaré*, XIII:99–125, 1977.

[138] R.G. Douglas. On extremal measures and subspace density. *Michigan Math. J.*, 11:243–246, 1964.

[139] D.C. Dowson and B.V. Landau. The Fréchet distance between multivariate normal distributions. *Journal of Multivariate Analysis*, 12:450–455, 1982.

[140] A.Y. Dubovitskii and A.A. Milyutin. *Necessary Conditions for a Weak Extremum in the General Problems of Optimal Management.* Nauka, Moscow, 1971. In Russian.

[141] R.M. Dudley. Convergence of Baire measures. *Studia Mathematica*, 27:251–268, 1966.

[142] R.M. Dudley. Distances of probability measures and random variables. *Annals of Mathematical Statistics*, 39:1563–1572, 1968.

[143] R.M. Dudley. The speed of mean Glivenko–Cantelli convergence. *Annals of Mathematical Statistics*, 40:40–50, 1969.

[144] R.M. Dudley. Speeds of metric probability convergence. *Z. Wahrscheinlichkeitstheorie Verw. Geb.*, 22:323–332, 1972.

[145] R.M. Dudley. Probability and metrics. Convergence of laws on metric spaces, with a view to statistical testing. *Aarhus Univ. Lect. Notes*, 45, 1976.

[146] R.M. Dudley. *Real Analysis and Probability.* Wadsworth & Brooks-Cole, Pacific Grove, California, 1989.

[147] D. Duffie. *Dynamic Asset Pricing Theory.* Princeton University Press, Princeton, 1992.

[148] N. Dunford and J. Schwartz. *Linear Operators. General Theory*, volume Part I. Wiley-Interscience Publication, New York, 1958.

[149] R. Durrett and M. Liggett. Fixed points of the smoothing transformation. *Z. Wahrscheinlichkeitstheorie Verw. Geb.*, 64:275–301, 1983.

[150] A. Dvoretzky. Asymptotic normality for sums of dependent random variables. *Proc. Berkeley Symp. II*, pages 513–535, 1970.

[151] D.A. Edwards. On the existence of probability measures with given marginals. *Ann. Inst. Fourier*, 28:53–78, 1978.

[152] I. Ekeland and R. Teman. *Convex analysis and variational problems.* North Holland, 1976.

[153] K.H. Elster and R. Nehse. Zur Theorie der Polarfunktionale. *Optimization*, 5:3–21, 1974.

[154] R.F. Engle, D.M. Lilien, and R.P. Robins. Estimating time varying risk premia in the term structure: the ARCH model. *Econometrica*, 55:391–407, 1987.

[155] Y. Ermoljev, A. Gaivoronski, and C. Nedeva. Stochastic optimization problem with incomplete information on distribution functions. Report WP-83-113, 1983.

[156] I.V. Evstigneev. Measurable choice theorems and probabilistic control models. *Dokl. Akad. Nauk USSR*, 283(5):1065–1068, 1985.

[157] G. Fayolle, P. Flajolet, and M. Hofri. On a functional equation arising in the analysis of a protocol for a multi-access broadcast channel. *Advances in Applied Probability*, 18:441–472, 1986.

[158] G. Fayolle, P. Flajolet, M. Hofri, and P. Jacquet. Analysis of a stack algorithm for random multiple-access communication. *IEEE Transactions on Information Theory*, 31:244–254, 1985.

[159] M.W. Feldman, S.T. Rachev, and L. Rüschendorf. Limit theorems for recursive algorithms. *Journal of Computational and Applied Mathematics*, 56:69–182, 1994.

[160] W. Feller. *An Introduction to Probability Theory and Its Applications*, volume II. Wiley, New York, 2nd edition, 1971.

[161] R. Ferland and G. Giroux. Cutoff-type Boltzmann equations: Convergence of the solution. *Adv. Appl. Math.*, 8:98–107, 1987.

[162] R. Ferland and G. Giroux. Le modèle Bose–Einstein de l'équation non linéaire de Boltzmann: Convergence vers l'equilibre. *Ann. Sc. Math. Québec*, 15:23–33, 1991.

[163] X. Fernique. Sur le théorème de Kantorovich–Rubinstein dans les espaces polonais. *Lecture Notes in Mathematics*, 850:6–10, 1981.

[164] P.C. Fishburn, J.C. Lagarias, J.A. Reeds, and L.A. Shepp. Sets uniquelly determine by projections on axes. I. Continuous case. *SIAM Journal on Applied Mathematics*, 50:288–306, 1990.

[165] A.T. Fomenko and S.T. Rachev. Volume functions on historical (narrative) texts and the amplitude correlation principle. *Computers and Humanities*, 24(3):187–206, 1990.

[166] P.R. Fortet and B. Mourier. Convergence de la repartition empirique vers la repartition theoretique. *Ann. Sci. Ecole Norm. Sup.*, 70(3):267–285, 1953.

[167] M.J. Frank. Operations arising from copulas. In *Symp. Probab. Measures with Given Marginals*, volume 67 of *Math. Appl.*, pages 75–93, Rome, 1991.

[168] M.J. Frank, R.B. Nelsen, and B. Schweizer. Best possible bounds for the distribution of a sum — a problem of Kolmogorov. *Probability Theory and Related Fields*, 74:199–211, 1987.

[169] M. Fréchet. Sur les tableaux de corrélation dont les marges sont données. *Ann. Univ. de Lyon, Sciences*, 14:53–77, 1951.

[170] M. Fréchet. Les tableaux de correlation dont les marges sont données. *Ann. Univ. de Lyon, Sciences*, 20:13–31, 1957.

[171] M. Fréchet. Sur la distance de deux lois de probabilité. *C.R. Acad. Sci. Paris*, 244:689–692, 1957.

[172] M. Fréchet. Sur les tableaux de corrélation dont les marges et des bornes sont données. *Revue Inst. Int. de Statistique*, 28:10–32, 1960.

[173] N. Gaffke and L. Rüschendorf. On a class of extremal problems in statistics. *Math. Operationsforschung Statist.*, 12:123–135, 1981.

[174] N. Gaffke and L. Rüschendorf. On the existence of probability measures with given marginals. *Statistics & Decisions*, 2:163–174, 1984.

[175] D. Gale. *Theory of Linear Economic Models.* McGraw-Hill, New York, 1960.

[176] D. Gale and A. Mas-Colell. An equilibrium existence theorem for a general model without ordered preferences. *Journal of Mathematical Economics*, 2:9–15, 1975.

[177] W. Gangbo and R.J. McCann. Optimal maps in Monge's mass transport problem. *CRAS*, Ser. I, 321:1653–1658, 1995.

[178] W. Gangbo and R.J. McCann. The geometry of optimal transformations. Preprint, 1996.

[179] M. Gelbrich. On a formula for the L^p Wasserstein metric between measures on Euclidean and Hilbert spaces. Preprint 179, 1988. Sektion Mathematik der Humboldt-Universität zu Berlin.

[180] M. Gelbrich. *L^p-Wasserstein-Metriken und Approximationen stochastischer Differentialgleichungen.* Dissertation A, Humboldt-Universität zu Berlin, Sektion Mathematik, 1989.

[181] M. Gelbrich. On a formula for the L^2-Wasserstein metric between measures on Euclidean and Hilbert spaces. *Math. Nachr.*, 147:185–203, 1990.

[182] M. Gelbrich. Simultaneous time and chance discretization for stochastic differential equations. *Journal of Computational and Applied Mathematics*, 58:255–289, 1995.

[183] M. Gelbrich and S.T. Rachev. Discretization for stochastic differential equations, L^2-Wasserstein metrics, and econometric models. In *Distributions with Given Marginals.* IMS Proc., 1996. To appear.

[184] I. Gelfand, D. Raikov, and G. Shilov. *Kommutative normierte Algebren.* VEB Deutscher Verlag der Wissenschaften, 1964.

[185] C. Genest. A survey of the statistical properties and applications of Archimedean copulas, 1990. Technical Report.

[186] H. Gerber. *An Introducation to Mathematical Risk Theory.* Huebner Foundation Monograph, 1981.

[187] I.I. Gikhman and A.W. Skorokhod. *Introduction to the theory of stochastic processes.* Nauka, Moscow, 1977. In Russian.

[188] C. Gini. Di una misura delle ralazioni tra le graduatorie di due caratteri. Appendix to: A. Hancini. L'Elezioni Generali Politiche del 1913 nel comune di Roma, Ludovic, Cecehini, 1914.

[189] C. Gini. La dissomiglianza. *Matron*, 24:309–331, 1965.

[190] C.R. Givens and R.M. Shortt. A class of Wasserstein metrics for probability distributions. *Michigan Math. J.*, 31:231–240, 1984.

[191] D. Goldfarb. Efficient dual simplex algorithms for the assignment problem. Preprint, 1985.

[192] C.M. Goldie. Implicit renewal theory and tails of solutions of random equations. *Annals of Applied Probability*, 1:126–166, 1991.

[193] C. Graham. McKean–Vlasov Itô–Skorohod equations and nonlinear diffusions with discrete jump sets. *Stoch. Proc. Appl.*, 40:69–82, 1992.

[194] C. Graham. Nonlinear diffusions with jumps. Preprint, 1992.

[195] R.M. Gray, D.L. Neuhoff, and R.L. Dobrushin. Block synchronization, sliding-block coding, invulnerable sources and zero error codes for discrete noisy channels. *Annals of Probability*, 8:315–328, 1980.

[196] R.M. Gray, D.L. Neuhoff, and P.C. Shields. A generalization to Ornstein's $\bar{d}$-distance with applications to information theory. *Annals of Probability*, 3:315–328, 1975.

[197] R.M. Gray and D.S. Ornstein. Block coding for discrete stationary $\bar{d}$-continuous channels. *IEEE Transactions on Information Theory*, 25:292–306, 1979.

[198] N.E. Gretsky, J.M. Ostroy, and W.R. Zame. The nonatomic assignment model. *Journal of Economic Theory*, 2:103–128, 1992.

[199] N.V. Grigorevski and I.S. Shiganov. On some modifications of Duley's metric. *Zap. Nauchnich Sem. LOMI*, 61:17–24, 1976.

[200] F.A. Grünbaum. Propagation of chaos for the Boltzmann equation. *Arch. Rational Mech. Anal.*, 42:323–345, 1971.

[201] P. Gudynas. Approximation by distributions of sums of conditionally independent random variables. *Litovski Mat. Sbornik*, 24:68–80, 1985.

[202] Y. Guivarch. Sur une extension de la notion de loi semi-stable. *Annales de l'Institut Henri Poincaré*, 26:261–286, 1990.

[203] W. Gutjahr and G.Ch. Pflug. The asymptotic contour process of a binary tree is a Brownian excursion. *Stoch. Processes and Applications*, 41:69–89, 1992.

[204] S. Gutmann, J.H.B. Kemperman, and J.A. Reeds. Existence of probability measures with given marginals. *Annals of Probability*, 19:1781–1791, 1991.

[205] S. Gutmann, J.H.B. Kemperman, J.A. Reeds, and L.A. Shepp. Existence of probability measures with given marginals. *Annals of Probability*, 19:1781–1791, 1991.

[206] D.L. Guy. common extension of finitely additive probability measures. *Portugalia Math.*, 20:1–5, 1961.

[207] M.G. Hahn, W.N. Hudson, and J.A. Veeh. Operator stable laws: series representations and domains of normal attraction. *Journal of Multivariate Analysis*, 10:26–37, 1989.

[208] P. Hall. Personal communication, 1985.

[209] J.P. Hammond. Straightforward individual incentive compatiblility in large economies. *Review of Economic Studies*, 46:263–282, 1979.

[210] W.K.K. Haneveld. *Duality in Stochastic Linear and Dynamic Programming.* Centrum voor Wiskunde en Informatica, Amsterdam, 1985.

[211] L.G. Hanin. Kantorovich–Rubinstein duality for Lipschitz spaces defined by differences of arbitrary order. *Soviet Math. Doklady*, 42(1):220–224, 1991.

[212] L.G. Hanin and S.T. Rachev. An extension of the Kantorovich–Rubinstein mass transportation problem., 1991. Dept. of Statistics and Applied Probability, University of California, Santa Barbara.

[213] L.G. Hanin and S.T. Rachev. Mass transshipment problems and ideal metrics. *Journal of Computational and Applied Mathematics*, 56:183–196, 1994.

[214] L.G. Hanin and S.T. Rachev. An extension of the Kantorovich–Rubinstein mass transshipment problem. *Numer. Funct. Anal. Optimization*, 16:701–735, 1995.

[215] G. Hansel and J.P. Troallic. Measures marginales et théorème de Ford–Fulkerson. *Z. Wahrscheinlichkeitstheorie Verw. Geb.*, 43:245–251, 1978.

[216] G. Hansel and J.P. Troallic. Sur le problème des marges. *Probability Theory and Related Fields*, 71:357–366, 1986.

[217] F. Hausdorff. *Set Theory.* Chelsea Publishing Company, New York, 1957.

[218] E. Häussler. On the rate of convergence in the central limit theorem for martingales with discrete and continuous time. *Annals of Probability*, 16:275–299, 1988.

[219] H. Heinich and J.C. Lootgieter. Convergence des fonctions monotones. Preprint, 1993.

[220] I.S. Helland and T.S. Nilsen. On a general random exchange mode. *Journal of Applied Probability*, 13:781–790, 1976.

[221] P.L. Hennequin and A. Tortrat. *Probability Theory and Some of Its Applications.* Nauka, Moscow, 1974. Russian translation.

[222] W. Hildenbrand. On economies with many agents. *Journal of Economic Theory*, 2:161–168, 1970.

[223] C. Hipp and R. Michel. Risikotheorie: Stochastische Modelle und Statistische Methoden. *DGVM*, 24, 1990.

[224] W. Hoeffding. Maßstabinvariante Korrelationstheorie. *Schriften des Mathematischen Instituts und des Instituts für Angewandte Mathematik der Universität Berlin*, 5:181–233, 1940.

[225] W. Hoeffding. The extrema of the expected value of a function of independent random variables. *Annals of Mathematical Statistics*, 26:268–275, 1955.

[226] W. Hoeffding and S.S. Shrikahande. Bounds for the distribution function of a sum of independent, identically distributed random variables. *Annals of Mathematical Statistics*, 27:439–449, 1956.

[227] A.J. Hoffman. On simple linear programming problems. Convexity. In *Proceedings of Symposia in Pure Mathematics*, volume 7, pages 317–327, Providence, R.I, 1961.

[228] A.J. Hoffman. On simple linear programming problems. In V. Klee, editor, *Convexity*, volume 7, pages 317–327, Providence, R.I, 1963. Proc. Symp. Pure Math.

[229] A.J. Hoffman and A.F. Veinott jr. Staircase transportation problems with hyperadditive rewards and cumulative capacities. Preprint, 1990. IBM T.Y. Watson Research Center, Yorktown Heights, New York, 10598.

[230] J. Hoffmann-Jörgensen. Probability in Banach space. *Lecture Notes in Mathematics*, 598:2–186, 1977.

[231] M. Hofri. *Probabilistic Analysis of Algorithms.* Springer, New York, 1987.

[232] R. Holley and M. Liggett. Generalized potlach and smoothing processes. *Z. Wahrscheinlichkeitstheorie Verw. Geb.*, 55:165–195, 1981.

[233] G. Hooghiemstra and M. Keane. Calculation of the equilibrium distribution for a solar energy storage model. *Journal of Applied Probability*, 22:852–864, 1985.

[234] G. Hooghiemstra and C.L. Scheffer. Some limit theorems for an energy storage model. *Stoch. Processes and Applications*, 22:121–127, 1986.

[235] J. Horowitz and R.L. Karandikar. Martingale problems associated with the Boltzmann equation. In E. Çinlar et al., editor, *Seminar on Stochastic Processes 1989*, Boston, 1990. Birkhäuser.

[236] J. Horowitz and R.L. Karandikar. Mean rates of convergence of empirical measures in the Wasserstein metric. *Journal of Computational and Applied Mathematics*, 55:261–273, 1994.

[237] D.A. Hsieh. The statistical properties of daily foreign exchange rates: 1974-1983. *Journal of Internat. Economics*, 24:129–145, 1988.

[238] P.J. Huber. *Robust Statistics.* Wiley, New York, 1981.

[239] W.N. Hudson. Operator-stable distributions and stable marginals. *Journal of Multivariate Analysis*, 10:26–37, 1980.

[240] W.N. Hudson, Z.J. Jurek, and J.A. Veeh. The symmetry group and exponents of operator stable probability measures. *Annals of Probability*, 14:1014–1023, 1986.

[241] W.N. Hudson and J.D. Mason. Operator-stable laws. *Journal of Multivariate Analysis*, 11:434–447, 1981.

[242] W.N. Hudson, J.A. Veeh, and D.C. Weiner. Moments of distributions attracted to operator-stable laws. *Journal of Multivariate Analysis*, 24:1–10, 1988.

[243] J.E. Hutchinson. Fractals and selfsimilarity. *Indiana Univ. Math. Journal*, 30:713–747, 1981.

[244] Z. Ignatov and S.T. Rachev. Minimality of ideal probabilistic metrics. *J. Soviet Math.*, 32(6):595–608, 1986.

[245] N. Ikeda and S. Watanabe. *Stochastic Differential Equations and Diffusion Processes.* North-Holland, Amsterdam, 1981.

[246] A.D. Ioffe and V.M. Tihomirov. *Theory der Extremalaufgaben.* VEB Deutscher Verlag der Wissenschaften, Berlin, 1979.

[247] K. Isii. Inequalities of the type of Chebychev and Cramér-Rao and mathematical programming. *Ann. Inst. Statist. Math.*, 16:247–270, 1964.

[248] E.H. Ivanov and R. Nehse. Relations between generalized concepts of convexity and conjugacy. *Math. Operationsforschung Statist.*, 13:9–18, 1982.

[249] K. Jacobs. *Measure and Integral.* Academic Press, New York, 1987.

[250] J. Jacod. Calcul stochastique et problème de martingales. *Lecture Notes in Mathematics*, 714, 1979.

[251] P. Jacquet and M. Regnier. Normal limiting distribution of the size of tries. In P.J. Courtois and G. Latouche, editors, *Proc. Performance 87*, pages 209–223, Amsterdam, 1988. Elsevier Science Publications B.V. (North Holland).

[252] R. Janssen. Discretization of the Wiener-Process in Difference-Methods for stochastic differential equations. *Stoch. Processes and Applications*, 18:361–369, 1984.

[253] M. Jirina and J. Nedoma. Minimax solution of a sampling inventory process. *Aplikace matematiky*, 1:296–314, 1957. In Czech.

[254] R. Jirousek. A survey of methods used in probabilistic expert systems for knowledge integration. *Knowledge Based Systems*, 3:7–12, 1990.

[255] R. Jirousek. Solution of the marginal problem and decomposable distributions. *Kybernetika*, 27(5):403–412, 1991.

[256] H. Johnen and K. Scherer. On the equivalence of K-functional and moduli of continuity and some applications. *Lecture Notes in Mathematics*, 571:119–130, 1977.

[257] J.P. Kahane and J. Peyrière. Sur certaines martingales de Benoit Mandelbrot. *Adv. Math.*, 22:131–145, 1976.

[258] A.V. Kakosjan, K. Klebanov, and S.T. Rachev. *Quantitative Criteria for Convergence of Probability Measures.* Ayastan Press, Erevan, 1988. (In Russian, Engl. transl.: Springer-Verlag, To appear).

[259] A.V. Kakosjan and L.B. Klebanov. On estimates of the closeness of distributions in terms of characteristic functions. *Theory of Probability and its Applications*, 29:852–853, 1984.

[260] V.V. Kalashnikov and S.T. Rachev. Characterization problems in queueing theory and their stability. *Advances in Applied Probability*, 17:320–348, 1985.

[261] V.V. Kalashnikov and S.T. Rachev. Characterization of inverse problems in queueing and their stability, 1986.

[262] V.V. Kalashnikov and S.T. Rachev. *Mathematical Methods for Construction of Stochastic Queueing Models.* Wadsworth & Brooks/Cole, California, 1990.

[263] T. Kamae, U. Krengel, and G.I. O'Brien. Stochastic inequalities on partially ordered spaces. *Annals of Probability*, 5:899–912, 1977.

[264] S. Kanagawa. The rate of convergence for approximate solutions of stochastic differential equations. *Tokyo J. Math.*, 12:33–48, 1986.

[265] Y. Kannai. Continuity properties of the core of a market. *Econometrica*, 38(6):791–815, 1970.

[266] L.V. Kantorovich. On the transfer of masses. *Dokl. Akad. Nauk USSR*, 37:7–8, 1942.

[267] L.V. Kantorovich. On a problem of Monge. *Uspekhi Mat. Nauk*, 3:225–226, 1948. In Russian.

[268] L.V. Kantorovich and G.P. Akilov. *Functional Analysis*. Nauka, Moscow, 3rd edition, 1984. In Russian.

[269] L.V. Kantorovich and G.Sh. Rubinstein. On a function space in certain extremal problems. *Dokl. Akad. Nauk USSR*, 115(6):1058–1061, 1957.

[270] L.V. Kantorovich and G.Sh. Rubinstein. On the space of completely additive functions. *Vestnic Leningrad Univ., Ser. Mat. Mekh. i Astron.*, 13(7):52–59, 1958. In Russian.

[271] S. Karlin and W.J. Studden. *Tchebycheff Systems*. Interscience, New York, 1966.

[272] T. Kawata. *Fourier Analysis in Probability Theory*. Academic Press, New York, 1972.

[273] H.G. Kellerer. Funktionen auf Produkträumen mit vorgegebenen Marginal-Funktionen. *Math. Ann.*, 144:323–344, 1961.

[274] H.G. Kellerer. Maßtheoretische Marginal Probleme. *Math. Annalen*, 153:168–198, 1964.

[275] H.G. Kellerer. Duality theorems and probability metrics. In *Proc. 7th Brasov Conf.*, pages 211–220, Bucuresti, 1984.

[276] H.G. Kellerer. Duality theorems for marginal problems. In M. Iosifescu, editor, *Proceedings of the 7th Conference on Probability Theory*, Braşov, Romania, 1984.

[277] H.G. Kellerer. Duality theorems for marginal problems. *Z. Wahrscheinlichkeitstheorie Verw. Geb.*, 67:399–432, 1984.

[278] H.G. Kellerer. Ambiguity in bounded moment problems. In *AMS-IMS-SIAM Joint Research Conference: Distributions with fixed marginals, double-stochastic measures and Markov operators*, 1993. To appear.

[279] R. Kemp. *Fundamentals of the Average Case Analysis of Particular Algorithms*. Wiley, New York, 1984.

[280] J.H.B. Kemperman. The general moment problem, a geometric approach. *Annals of Mathematical Statistics*, 19:93–122, 1968.

[281] J.H.B. Kemperman. On a class of moment problems. In *Proceedings 6th Berkeley Symposium on Mathematical Statistics and Probability*, volume 2, pages 101–126, 1972.

[282] J.H.B. Kemperman. On the FKG-inequality for measures on a partially ordered space. *Proc. Nederl. Akad. Wet.*, 80:313–331, 1977.

[283] J.H.B. Kemperman. On the role of duality in the theory of moments. In *Semi-Infinite Programming and Applications 1981*, volume 215, pages 63–92. Springer, 1983.

[284] J.H.B. Kemperman. Geometry of the moment problem. In *Proceedings of Symposia in Applied Mathematics*, volume 27, pages 16–53. American Mathematical Society, 1987.

[285] J.H.B. Kemperman. Moment problems for measures on $\mathbb{R}^n$ with given k-dimensional marginals. In *AMS-IMS-SIAM; Joint Research Conference. Distributions with fixed marginals, double-stochastic measures and Markov operators*, 1993. To appear.

[286] H. Kesten. Random difference equations and renewal theory for products of random matrices. *Acta Math.*, 131:207–248, 1973.

[287] L.A. Khalfin and L.B. Klebanov. A solution of the computer tomography paradox and estimation of the distances between the densities of measures with the same marginals. *Annals of Probability*, 22:2235–2241, 1994.

[288] V. Kifer. *Ergodic Theory of Random Transformations.* Birkhäuser, Boston, 1986.

[289] T. Kim and M.K. Richter. Nontransitive-nontotal consumer theory. *Journal of Economic Theory*, 38, 1986.

[290] A.Y. Kiruta, A.M. Rubinov, and E.B. Yanovskaya. *Optimal choice of distributions in complex socio-economic problems.* Nauka, Leningrad, 1980. In Russian.

[291] L.B. Klebanov, G.M. Maniya, and I.A. Melamed. A problem of Zolotarev and analogs of infinitely divisible and stable distributions in a scheme for summing a random number of random variables. *Theory of Probability and its Applications*, 29:791–794, 1984.

[292] L.B. Klebanov and S.T. Mkrtchian. Estimator of the closeness of distributions in terms of coinciding moments. In *Problems of Stability of Stochastic Models, Proceedings*, pages 64–72, Moscow, 1980.

[293] L.B. Klebanov and S.T. Rachev. The method of moments in computer tomography. *Math. Scientist*, 20:1–14, 1995.

[294] L.B. Klebanov and S.T. Rachev. On a special case of the basic problem in diffraction tomography. In *Stochastic Models*, 1995.

[295] L.B. Klebanov and S.T. Rachev. Closeness of probability measures with common marginals on finite number of direction. In *Proceedings of Distributions with fixed Marginals and Related Topics*, volume 28, pages 162–174. IMS Lecture Notes Monography Series, 1996.

[296] L.B. Klebanov and S.T. Rachev. Proximity of probability with common marginals in a finite number of directions. In *Distributions with Given Marginals*, 1996.

[297] P. Kleinschmidt, C.W. Lee, and H. Schannath. Transportation problems which can be solved by the use of Hirsch paths for the dual problems. *Mathematical Programming Study*, 37:153–168, 1987.

[298] P.E. Kloeden and E. Platen. *Numerical Solution of Stochastic Differential Equations*. Springer-Verlag, Berlin, 1992.

[299] M. Knott and C.S. Smith. On the optimal mapping of distributions. *Journal of Optimization Theory and Applications*, 43:39–49, 1984.

[300] M. Knott and C.S. Smith. Note on the optimal transportation of distributions. *Journal of Optimization Theory and Applications*, 52:323–329, 1987.

[301] M. Knott and C.S. Smith. On Hoeffding–Fréchet bounds and cyclic monotone relations. *Journal of Multivariate Analysis*, 40:328–334, 1992.

[302] M. Knott and C.S. Smith. On a generalization of cyclic monotonicity and distances among random vectors. *Linear Algebra and its Applications*, 199:363–371, 1994.

[303] D.E. Knuth. *The Art of Computer Programming*, volume II. Addison-Wesley, 1969.

[304] J. Komlós, P. Major, and G. Tusnády. An approximation of partial sums of independent r.v.s and the sample d.f., I. *Z. Wahrscheinlichkeitstheorie Verw. Geb.*, 32:111–131, 1975.

[305] J. Komlós, P. Major, and G. Tusnády. An approximation of partial sums of independent r.v.s and the sample d.f., II. *Z. Wahrscheinlichkeitstheorie Verw. Geb.*, 34:33–58, 1976.

[306] M.G. Krein and A.A. Nudelman. The markov moment problem and extremal problems, 1977.

[307] W.M. Kruskal. Ordinal measures of association. *Journal of the American Statistical Association*, 53:814–861, 1958.

[308] J. Kuelbs. Kolmogorov's law of the iterated logarithm for Banach space valued random variables. *Illinois J. Math.*, 21:784–800, 1977.

[309] K. Kuratowski. *Topology*, volume I. Academic Press, New York, 1966.

[310] K. Kuratowski. *Topology*, volume II. Academic Press, New York, 1969.

[311] I. Kuznezova-Sholpo and S.T. Rachev. Explicit solutions of moment problems. *Probability and Mathematical Statistics*, 10:297–312, 1989.

[312] J.J. Laffont and E. Maskin. A differential approach to dominant strategy mechanisms. *Econometrica*, 48:1507–1520, 1980.

[313] T.L. Lai and M. Robbins. Maximally dependent random variables. *Proc. Nat. Acad. Sci. USA*, 73:286–288, 1976.

[314] P. Lancaster. *Theory of Matrices.* Wiley, New York, London, 1969.

[315] D. Landers and L. Rogge. Best approximations in L_ϕ-spaces. *Z. Wahrscheinlichkeitstheorie Verw. Geb.*, 51:215–237, 1980.

[316] F. Lassner. *Sommes de produit de variables aléatoires indépendantes.* Thesis, Université de Paris VI, 1974.

[317] M. Ledoux and M. Talagrand. *Probability in Banach Spaces.* Springer, Berlin, 1991.

[318] S.J. Leese. Multifunctions of Suslin type. *Bull. Austral. Math. Soc.*, 11:395–411, 1975. and 13:159-160.

[319] G. Letac. Représentation des mesures de probabilité sur le produit de deux espaces denombrables, de marges données. *Ann. Inst. Fourier*, 16:497–507, 1966.

[320] G. Letac. A contraction principle for certain Markov chains and its applications. Random matrices and their applications. In H. Kesten J.E. Cohen and C.M. Newman, editors, *Proc. AMS-IMS-SIAM Joint Summer Research Conf. 1984*, volume 50 of *Contemp. Math.*, pages 263–273, Providence, R.I., 1986. Amer. Math. Soc.

[321] V.L. Levin. Application of E. Helly's theorem to convex programming, problems of best approximation and related questions. *USSR Math. Sbornik*, 8:235–248, 1969.

[322] V.L. Levin. Duality and approximation in the problem of mass transfer. In B.S. Mityagin, editor, *Mathematical Economics and Functional Analysis*, pages 94–108. Nauka, Moscow, 1974. In Russian.

[323] V.L. Levin. On the problem of mass transfer. *Soviet Math. Doklady*, 16:1349–1353, 1975.

[324] V.L. Levin. On the theorems in the Monge–Kantorovich problem. *Uspekhi Mat. Nauk*, 32:171–172, 1977. In Russian.

[325] V.L. Levin. The mass transfer problem, strong stochastic domination and probability measures on the product of two compact spaces with given projections. Preprint, TsEMI, Moscow, 1978a. In Russian.

[326] V.L. Levin. The Monge–Kantorovich problem on mass transfer. In *Methods of Functional Analysis in Mathematical Economics*, pages 23–55. Nauka, Moscow, 1978b. In Russian.

[327] V.L. Levin. Measurable selections of multivalued mappings into topological spaces and upper envelopes of Carathéodory integrands. *Soviet Math. Doklady*, 21:771–775, 1980.

[328] V.L. Levin. Some applications of duality for the problem of translocation of masses with a lower semicontinuous cost function. Closed preferences and Choquet theory. *Soviet Math. Doklady*, 2:262–267, 1981.

[329] V.L. Levin. A continuous utility theorem for closed preorders on a compact metrizable space. *Soviet Math. Doklady*, 28:715–718, 1983a.

[330] V.L. Levin. Measurable utility theorems for closed and lexicographic preference relations. *Soviet Math. Doklady*, 27:639–643, 1983b.

[331] V.L. Levin. Lipschitz preorders and Lipschitz utility functions. *Russian Mathematical Surveys*, 39:199–200, 1984a.

[332] V.L. Levin. The mass transfer problem in topological space and probability measures on the product of two spaces with given marginal measures. *Soviet Math. Doklady*, 29:638–643, 1984b.

[333] V.L. Levin. *Convex Analysis in Spaces of Measurable Functions and Its Applications in Mathematics and Economics.* Nauka, Moscow, 1985a. In Russian.

[334] V.L. Levin. Functionally closed preorders and strong stochastic dominance. *Soviet Math. Doklady*, 32:22–26, 1985b.

[335] V.L. Levin. Extremal problems with probability measures, functionally closed preorders and strong stochastic dominance. In *Stochastic Optimization*, volume 81 of *Lecture Notes in Control and Information Science*, pages 435–447, Berlin, New York, 1986. Proc. Int. Conf. Kiev 1984, Springer-Verlag.

[336] V.L. Levin. Measurable selectors of multivalued mappings and the mass transfer problem. *Dokl. Akad. Nauk USSR*, 292:1048–1053, 1987.

[337] V.L. Levin. General Monge–Kantorovich problem and its applications in measure theory and mathematical economics. In L.J. Leifman, editor, *Functional Analysis, Optimization and Mathematical Economics.* Oxford University Press, 1990. A collection of papers dedicated to the Memory of L.V. Kantorovich.

[338] V.L. Levin. Some applications of set-valued mappings in mathematical economics. *Journal of Mathematical Economics*, 20:69–87, 1991.

[339] V.L. Levin. A formula for the optimal value in the Monge–Kantorovich problem with a smooth cost function and a characterization of cyclically monotone mappings. *USSR Math. Sbornik*, 71:533–548, 1992.

[340] V.L. Levin. Private communication, 1994.

[341] V.L. Levin. Quasi-convex functions and quasi-monotone operators. *Journal of Convex Analysis*, 2, 1995a.

[342] V.L. Levin. Reduced cost functions and their applications. *Journal of Mathematical Economics*, 1995b. To appear.

[343] V.L. Levin and A.A. Milyutin. The mass transfer problem with discontinuous cost function and a mass setting for the problem of duality of convex extremum problems. *Trans Russian Math. Surveys*, 34:1–78, 1979.

[344] V.L. Levin and S.T. Rachev. New duality theorems for marginal problems with some applications in stochastics. *Lecture Notes in Mathematics*, 1412:137–170, 1989.

[345] M. Loeve. *Probability Theory.* Van Nostrand, 1977.

[346] G.G. Lorentz. A problem of plane measure. *Amer. J. Math.*, 71:417–426, 1949.

[347] G.G. Lorentz. An inequality for rearrangements. *American Mathematics Monthly*, 60:176–179, 1953.

[348] G. Louchard. Exact and asymptotic distributions in digital and binary search trees. *Theor. Inf. Appl.*, 21:479–495, 1987.

[349] R. Lucchetti and F. Patrone. Closure and upper semicontinuity results in mathematical programming, Nash and economic equilibria, Optimization. *Mathematische Operationsforschung und Statistik-Series Optimization*, 17:619–628, 1986.

[350] N. Lusin. *Leçons sur les Ensembles Analytiques.* Gauthier-Villars, 1930.

[351] M. Maejima. Some limit theorems for summability methods of iid random variables. In V.V. Kalashnikov et al., editor, *Stability problems of stochastic models*, volume 1233 of *Lecture Notes in Mathematics*, pages 57–68, 1985. Varna 1985.

[352] M. Maejima. Some limit theorems for stability methods of i.i.d. random variables. *Lecture Notes in Mathematics*, 1233:57–68, 1988.

[353] M. Maejima and S.T. Rachev. An ideal metric and the rate of convergence to a self-similar process. *Annals of Probability*, 15:708–727, 1987.

[354] M. Maejima and S.T. Rachev. Rates of convergence in the operator-stable limit theorems. *J. Theor. Probability*, 9:37–86, 1996.

[355] H.M. Mahmoud. *Evolution of Random Search Trees.* Wiley, New York, London, 1992.

[356] G.D. Makarov. Estimates for the distributions function of a sum of two random variables when the marginal distributions are fixed. *Theory of Probability and its Applications*, 26:803–806, 1981.

[357] C.L. Mallows. A note on asymptotic joint normality. *Annals of Mathematical Statistics*, 43:508–515, 1972.

[358] B.B. Mandelbrot. Multiplications aléatoires itérées et distributions invariantes par moyenne pondérée aléatorie. *C.R. Acad. Sci. Paris*, 278, 1974.

[359] B.B. Mandelbrot and M. Taylor. On the distribution of stock price differences. *Oper. Res.*, 15:1057–1062, 1967.

[360] M. Marcus. Some properties and applications of doubly stochastic matrices. *American Mathematics Monthly*, 67:215–222, 1960.

[361] A.W. Marshall and I. Olkin. *Theory of majorization and its applications.* Academic Press, New York, 1979.

[362] G. Maruyama. Continuous Markov processes and stochastic equations. *Rend. Circolo Math. Palermo*, 4:48–90, 1955.

[363] A. Mas-Colell. On the continuous representation of preorders. *Intern. Econ. Revue*, 18:509–513, 1977.

[364] E. Maskin and J. Riley. Monopoly with incomplete information. *Rand Journal of Economics*, 15:171–196, 1984.

[365] J.L. Massey. Collision-resolution algorithms and random-access communications, multi-user communication systems. *CISM Courses and Lectures*, 1981.

[366] R. Mathar and D. Pfeifer. *Stochastik für Informatiker.* Teubner, Stuttgart, 1990.

[367] G. Matheron. *Random Sets and Integral Geometry.* Wiley, 1975.

[368] M. Meerschaert. Moments of random vectors which belong to some domain of normal attraction. *Annals of Probability*, 18:870–876, 1989.

[369] M. Meerschaert. Spectral decomposition for generalized domains of attraction. *Annals of Probability*, 19:875–892, 1991.

[370] K. Mehlhorn. *Datenstrukturen und effiziente Algorithmen*, volume I. Teubner, Stuttgart, 1986.

[371] I. Meilijson and A. Nadas. Convex majorization with application to the length of critical paths. *Journal of Applied Probability*, 16:671–677, 1979.

[372] D. Mejzler. On the problem of the limit distributions for the maximal term of a variational series. *Lvov Politechn. Inst. Naucn. Zap. Ser. Fiz.-Mat.*, 38:90–109, 1956. In Russian.

[373] E. Michael. Continuous selections. *Ann. of Math.*, 63:361–382, 1956.

[374] P. Mikusinski, H. Sherwood, and M.D. Taylor. Probabilistic interpretations of copulas and their convex sums. In *Symp. Probab. Measures with Given Marginals*, volume 67 of *Math. Appl.*, pages 95–112, Rome, 1991.

[375] G.N. Milshtein. A method of second-order accuracy integration of stochastic differential equations. *Theory of Probability and its Applications*, 23, 1978.

[376] G.N. Milshtein. Numerical integration of stochastic differential equations. *Izd. Ural. Univ. Sverdlovsk*, 1988. In Russian.

[377] J.A. Mirrlees. Optimal tax theory: a synthesis. *Journal of Public Economics*, 6:327–358, 1976.

[378] S. Mittnik and S.T. Rachev. Alternative multivariate stable distributions and their applications to financial modeling. In S. Cambanis, G. Samordodnitsky, and M.S. Taqqu, editors, *Stable Processes and Related Topics*, pages 107–120, Boston, 1991. Birkhäuser.

[379] S. Mittnik and S.T. Rachev. Modeling assets returns with alternative stable laws. *Econometric reviews*, 12(3):261–330, 1993.

[380] S. Mittnik and S.T. Rachev. Reply on comments on "modeling assets returns with alternative stable laws" and some extensions. *Econometric reviews*, 12(3):347–389, 1993.

[381] S. Mittnik and S.T. Rachev. *Modelling Financial Assets with Alternative Stable Models.* Series in Financial Economics and Quantitative Analysis. Wiley, New York, 1997.

[382] G. Monge. Mémoire sur la théorie des déblais et des remblais, 1781.

[383] F. Mosteller, C. Youtz, and D. Zahn. The distribution of sums of rounded percentages. *Demography*, 4:850–858, 1967.

[384] K.R. Mount and S. Reiter. Construction of a continuous utility function for a class of preferences. *Journal of Mathematical Economics*, 3:227–245, 1976.

[385] L. Nachbin. *Topology and Order.* Van Nostrand, New York, 1965.

[386] R.B. Nelsen. Copulas and association. In *Symp. Probab. Measures with Given Marginals*, pages 51–74, Rome, 1991. Kluwer.

[387] W. Neuefeind. On continuous utility. *Journal of Economic Theory*, 5:174–176, 1972.

[388] J. Neveu. *Mathematical Foundations of the Calculus of Probability.* Holden-Day, San Francisco, 1965.

[389] J. Neveu and R.M. Dudley. On Kantorovich–Rubinstein theorems. (Transcript), 1980.

[390] V.B. Nevzorov. Records. *Theory of Probability and its Applications*, 32:201–228, 1988.

[391] N.J. Newton. An asymptotically efficient difference formula for solving stochastic differential equations. *Stochastics*, 19:175–206, 1986.

[392] I. Olkin and F. Pukelsheim. The distance between two random vectors with given dispersion matrices. *Journal of Linear Algebra and its Applications*, 48:257–263, 1982.

[393] I. Olkin and F. Pukelsheim. Marginal problems with additional constraints. Tech. report, 270, 1990. Department of Statistics, Stanford University, Stanford, CA.

[394] I. Olkin and S.T. Rachev. Distances among random vecotrs with given dispersion matrices. Preprint, 1991. Department of Statistics, Stanford University, Stanford, CA.

[395] I. Olkin and S.T. Rachev. Maximum submatrix traces for positive definite matrices. *SIAM Journal of Matrix Analysis Applications*, 14:390–397, 1993.

[396] J.M. Ortega and W.C. Rheinboldt. *Iterative solution of nonlinear equations in several variables.* Academic Press, New York, 1970.

[397] J. Pachl. Two classes of measures. *Colloq. Math.*, 42:331–340, 1979.

[398] E. Pardoux and D. Talay. Discretization and simulation of stochastic differential equations. *Acta Appl. Math.*, 3:23–47, 1985.

[399] V. Paulauskas and A. Rackauskas. *Approximation Theory in the Central Limit Theorem.* Kluwer Academic Publisher, 1989.

[400] A.S. Paulson and V.R.R. Uppuluri. Limit laws of a sequence determined by a random difference equation governing a one-compartment system. *Math. Biosci.*, 13:325–333, 1972.

[401] A. Perez and R. Jirousek. Constructing an intentional expert system INES. In J.H. van Remmel, F. Gremy, and J. Zvarova, editors, *Medical decision making: Diagnostic Strategies and Expert Systems*, pages 307–315. North-Holland, 1985.

[402] S. Perrakis and C. Henin. Evaluation of risky investments with random timing of cash returns. *Management Sci.*, 21:79–86, 1974.

[403] D. Pfeifer. Some remarks on Nevzorov's record model. *Advances in Applied Probability*, 23:823–834, 1991.

[404] G. Pflug. *Stochastische Modelle in der Informatik.* Teubner, Stuttgart, 1986.

[405] G. Pisier and J. Zinn. On limit theorems for random variables with values in the spaces L^p. *Z. Wahrscheinlichkeitstheorie Verw. Geb.*, 41:286–305, 1977.

[406] B. Pittel. Paths in a random digital tree: Limiting distributions. *Advances in Applied Probability*, 18:139–155, 1986.

[407] E. Platen. An approximation method for a class of Itô processes. *Lietuvos Math. Rink. XXI*, 1:121–133, 1981.

[408] D. Pollard. *Convergence of Stochastic Processes.* Springer, 1984.

[409] C.J. Preston. A generalization of the FKG inequalities. *Comm. Math. Phys.*, 36:233–241, 1974.

[410] P.S. Puri. On almost sure convergence of an erosion process due to Todorovic and Gani. *Journal of Applied Probability*, 24:1001–1005, 1987.

[411] G. Pyatt and J.J. Round, editors. *Social Accounting Matrics: A Basis for Planning.* World Bank, Washington, D.C., 1985.

[412] R. Pyke and D. Root. On convergence in r-mean of normalized partial sums. *Annals of Mathematical Statistics*, 39:379–381, 1968.

[413] S.T. Rachev. On a metric construction of Hausdorff in a space of probability measures. *Zapiski Nauchn. Sem. LOMI*, 87:87–104, 1978.

[414] S.T. Rachev. Minimal metrics in a space of real random variables. *Dokl. Akad. Nauk SSSR*, 257(5):1067–1070, 1981.

[415] S.T. Rachev. On minimal metrics in the space of real-valued random variables. *Soviet Dokl. Math.*, 23(2):425–438, 1981a.

[416] S.T. Rachev. Minimal metrics in the random variables spaces. *Pub. Inst. Stat. Univ. Paris*, 27(1):27–47, 1982a.

[417] S.T. Rachev. Minimal metrics in the random variables spaces. In W. Grossmann et al., editor, *Probability and Statistical Inference Proceedings of the 2nd Pannonian Symp.*, pages 319–327, Dordrecht, 1982b. D. Reidel Company.

[418] S.T. Rachev. Compactness in the probability measures space. In M. Galyare et al., editor, *Proceedings of the 3rd European Young Statisticians Meeting*, pages 136–150, Katholieke Univ., Leuven, 1983a.

[419] S.T. Rachev. Minimal metrics in the real valued random variable spaces. *Lecture Notes in Mathematics*, 982:172–190, 1983b.

[420] S.T. Rachev. Hausdorff metric construction in the probability measures space. *Studia Mathematica*, 7:152–162, 1984a. Pliska.

[421] S.T. Rachev. The Monge–Kantorovich mass transference problem and its stochastic applications. *Theory of Probability and its Applications*, 29:647–676, 1984b.

[422] S.T. Rachev. On a class of minimal functionals on a space of probability measure. *Theory of Probability and its Applications*, 29(1):41–49, 1984c.

[423] S.T. Rachev. On a problem of Dudley. *Soviet Math. Doklady*, 29(2):162–164, 1984d.

[424] S.T. Rachev. Extreme functionals in the space of probability measures. *Lecture Notes in Mathematics*, 1155:320–348, 1985a. Proc. "Stability Problems for Stochastic Models".

[425] S.T. Rachev. Probability metrics and their applications to the stability problems for stochastic models, 1985b. Author's review of doctor of sciences theses, Steklov Mathematical Institute, USSR Academy of Sciences, Moscow. In Russian.

[426] S.T. Rachev. Extreme functional in the space of probability theory and mathematical statistics. *VNU Science Press*, 2:474–476, 1986.

[427] S.T. Rachev. Minimal metrics in a space of random vectors with fixed one-dimensional marginal distributions. *J. Soviet Math.*, 34(2):1542–1555, 1986. Stability Problems for Stochastic Models. Proceedings, Moscow, VNIISI.

[428] S.T. Rachev. The stability of stochastic models. *Applied Probability Newsletter*, 12(2):3–4, 1988.

[429] S.T. Rachev. The problem of stability in queueing theory. *Queueing Systems Theory and Applications*, 4:287–318, 1989.

[430] S.T. Rachev. Mass transshipment problems and ideal metrics. *Numer. Func. Anal. & Optimiz.*, 12(5& 6):563–573, 1991a.

[431] S.T. Rachev. Optimal mass transportation problems. In *Proceedings of XI Congres de Metodologias en Ingenieria de Sistemas*, pages 115–120, Azocar, Santiago de Chile, 1991b.

[432] S.T. Rachev. *Probability Metrics and the Stability of Stochastic Models*. Wiley, Chichester-New York, 1991c.

[433] S.T. Rachev. Theory of probability metrics and recursive algorithms. In S. Joly and G. le Calve, editors, *Distancia 1992, Proceedings of Congres International sur Analyse en Distance*, pages 339–403, Université de haute Bretagne, Rennes, 1992.

[434] S.T. Rachev and G.S. Chobanov. Minimality of ideal probabilistic metrics. *Pliska*, 2:1154–1158, 1986. In Russian.

[435] S.T. Rachev, B. Dimitrov, and Z. Khalil. A probabilistic approach to optimal quality usage. *Computers and Mathematics with Applications*, 24(8/9):219–227, 1992.

[436] S.T. Rachev and Z. Ignatov. Minimality of ideal probabilistic metrics. *J. Soviet Math.*, 32(6):595–608, 1986.

[437] S.T. Rachev and S.I. Resnick. Max-geometric infinite divisibility and stability. *Stoch. Models*, 2:191–218, 1991.

[438] S.T. Rachev and L. Rüschendorf. Approximation of sums by compound Poisson distributions with respect to stop-loss distances. *Advances in Applied Probability*, 22:350–374, 1990.

[439] S.T. Rachev and L. Rüschendorf. A counterexample to a.s. constructions. *Stat. Prob. Letters*, 9:307–309, 1990a.

[440] S.T. Rachev and L. Rüschendorf. A transformation property of minimal metrics. *Theory of Probability and its Applications*, 35:131–137, 1990b.

[441] S.T. Rachev and L. Rüschendorf. Approximate independence of distributions on spheres and their stability properties. *Annals of Probability*, 19:1311–1337, 1991.

[442] S.T. Rachev and L. Rüschendorf. Recent results in the theory of probability metrics. *Statistics & Decisions*, 9:327–373, 1991a.

[443] S.T. Rachev and L. Rüschendorf. A new ideal metric with applications to multivariate stable limit theorems, summability methods and compound Poisson approximation. *Probability Theory and Related Fields*, 94:163–187, 1992.

[444] S.T. Rachev and L. Rüschendorf. Rate of convergence for sums and maxima and doubly ideal metrics. *Theory of Probability and its Applications*, 37:276–289, 1992a.

[445] S.T. Rachev and L. Rüschendorf. On constrained transportation problems. In *Proceedings of the 32nd Conference on Decision and Control*, volume 3, pages 2896–2900. IEEE Control System Society, 1993.

[446] S.T. Rachev and L. Rüschendorf. On the Cox, Ross and Rubinstein model for option pricing. *Theory of Probability and its Applications*, 39:150–190, 1994.

[447] S.T. Rachev and L. Rüschendorf. On the rate of convergence in the CLT with respect to the Kantorovich metric. In J. Kuelbs, M. Marcus, and J. Hoffman-Jorgensen, editors, *9th Conf. on Probability on Banach Spaces*, pages 193–207, Boston–Basel–Berlin, 1994a. Birkhäuser.

[448] S.T. Rachev and L. Rüschendorf. Propagation of chaos and contraction of stochastic mappings. *Siberian Advances in Mathematics*, 4:114–150, 1994b.

[449] S.T. Rachev and L. Rüschendorf. Solution of some transportation problems with relaxed or additional constraints. *SIAM Journal of Control and Optimization*, 32(3):673–689, 1994c.

[450] S.T. Rachev and L. Rüschendorf. Probability metrics and recursive algorithms. *Journal of Applied Probability*, 27:770–799, 1995. Technical Report (1991).

[451] S.T. Rachev and L. Rüschendorf. Propagation of chaos and contraction of stochastic mappings. *Siberian Adv. Math.*, 4:114–150, 1995a.

[452] S.T. Rachev, L. Rüschendorf, and A. Schief. Uniformities for the convergence in law and probability. *Journal of Theoretical Probability*, 5:33–44, 1992.

[453] S.T. Rachev and G. Samorodnitsky. Geometric stable distributions in Banach spaces. *Journal of Theoretical Probability*, 7(29):351–373, 1994.

[454] S.T. Rachev and G. Samorodnitsky. Limit laws for a stochastic process and random recursion arising in probabilistic modelling. *Advances in Applied Probability*, 27:185–203, 1995.

[455] S.T. Rachev and A. Schief. On L_p-minimal metric. *Probability and Mathematical Statistics*, 13(2):311–320, 1992.

[456] S.T. Rachev and A. SenGupta. Geometric stable distributions and Laplace–Weibull mixtures. *Statistics & Decisions*, 10:251–271, 1992.

[457] S.T. Rachev and A. SenGupta. Laplace-Weibull mixtures for modeling price changes. *Management Science*, pages 1029–1038, 1993.

[458] S.T. Rachev and R.M. Shortt. *Classification problem for probability metrics*, volume 94 of *Contemporary Mathematics*, pages 221–262. AMS, 1989.

[459] S.T. Rachev and R.M. Shortt. Duality theorems for Kantorovich–Rubinstein and Wasserstein functionals. *Dissertationes Mathematicae*, 299:647–676, 1990.

[460] S.T. Rachev and M. Taksar. Kantorovich's functionals in space of measures. In I. Karatzas and D. Ocone, editors, *Applied Stochastic Analysis*, volume 77 of *Lecture Notes in Control and Information Science*, pages 248–261, Berlin–New York, 1992. Proceedings of the US–French Workshop, Springer-Verlag.

[461] S.T. Rachev and P. Todorovic. On the rate of convergence of some functionals of a stochastic process. *Journal of Applied Probability*, 28:805–814, 1990.

[462] S.T. Rachev and J.E. Yukich. Rates for the CLT via new ideal metrics. *Annals of Probability*, 17:775–788, 1989.

[463] S.T. Rachev and J.E. Yukich. Smoothing metrics for measures on groups with applications to random motions. *Annales de l'Institut Henri Poincaré*, 25:429–941, 1990.

[464] S.T. Rachev and J.E. Yukich. Rates of convergence of α-stable random motions. *J. Theor. Prob.*, 4:333–352, 1991.

[465] A. Rackauskas. On the convergence rate in martingale CLT in Hilbert spaces. Preprint 90-031, 1990. University of Bielefeld.

[466] D. Ramachandran. *Perfect measures. Part I: Basic theory*, volume 5. Macmillan, New Delhi, 1979.

[467] D. Ramachandran. *Perfect measures. Part II: Special topics*, volume 7. Macmillan, New Delhi, 1979.

[468] D. Ramachandran. Marginal problem in arbitrary product spaces. In *Proceedings of the conference on "Distribution with Fixed Marginals, Double Stochastic Measures and Markov Operators"*, volume 28, pages 260–272, Seattle, August 1993. IMS Lecture Notes Monograph Series 1997.

[469] D. Ramachandran and L. Rüschendorf. A general duality theorem for marginal problems. *Probability Theory and Related Fields*, 101:311–319, 1995.

[470] D. Ramachandran and L. Rüschendorf. Duality and perfect probability spaces. *Proc. Amer. Math. Soc.*, 124:2223–2228, 1996a.

[471] D. Ramachandran and L. Rüschendorf. Duality theorems for assignments with upper bounds. In *'Distributions with Fixed Marginals and Moment Problems'*, pages 283–290. Kluwer, 1997.

[472] D. Ramachandran and L. Rüschendorf. On the validity of the Monge–Kantorovich duality theorem. Preprint, 1997.

[473] F. Ramsey. A mathematical theory of savings. *Economic Journal*, 38:543–559, 1928.

[474] M. Regnier and P. Jacquet. New results on the size of tries. *IEEE Transactions on Information Theory*, 35:203–205, 1989.

[475] S.I. Resnick and P. Greenwood. A bivariate stable characterization and domains of attraction. *Journal of Multivariate Analysis*, 9:206–221, 1979.

[476] M.K. Richter. Duality and rationality. *Journal of Economic Theory*, 20:131–181, 1979.

[477] H. Robbins. The maximum of identically distributed random variables. *I.M.S. Bull.*, March 1975. Abstract.

[478] H. Robbins and D. Siegmund. A convergence theorem for nonnegative almost supermartingales. In Rustagi, editor, *Optimiz. Meth. in Statistics*, pages 233–258. Academic Press, 1971.

[479] J.C. Rochet. The taxation principle and multi-time Hamilton–Jacobi equation. *Journal of Mathematical Economics*, 14:113–128, 1985.

[480] J.C. Rochet. A necessary and sufficient condition for rationalizability in a quasi-linear context. *Journal of Mathematical Economics*, 16:191–200, 1987.

[481] R.T. Rockafellar. Characterization of the subdifferentials of convex functions. *Pacific J. Math.*, 17:497–510, 1966.

[482] R.T. Rockafellar. *Convex Analysis*. Princeton Univ. Press, Princeton, NJ, 1970.

[483] C. Rogers. Coupling of random walks, 1992. Private communication.

[484] W.W. Rogosinski. Moments of non-negative mass. In *Proceedings of Royal Society London, Ser. A*, volume 245, pages 1–27, 1958.

[485] W. Römisch. An approximation method in stochastic optimization and control. In *Optimization techniques*, volume 22, pages 169–178. Proc. 9th IFIP Conf., Warsaw 1979, Part 1, Lecture Notes in Control and Information Science, 1980.

[486] W. Römisch. On discrete approximations in stochastic programming, 1981. Seminarbericht.

[487] W. Römisch and R. Schultz. Stability analysis of stochastic programs. *Ann. Operat. Res.*, 30:241–266, 1991.

[488] W. Römisch and R. Schultz. Stability of solutions for stochastic programs with complete recourse. *Mathematics of Operations Research*, 18:590–609, 1993.

[489] W. Römisch and A. Wakolbinger. On Lipschitz dependence in systems with differentiated inputs. *Math. Ann*, 272:237–248, 1985.

[490] U. Rösler. A limit theorem for quicksort. *Informatique Théorique et Applications*, 25:85–100, 1991.

[491] U. Rösler. A fixed point theorem for distributions. *Stoch. Processes and Applications*, 37:195–214, 1992.

[492] S.M. Ross. A simple heuristic approach to simplex efficiency. *European J. Oper. Res.*, 9:344–346, 1982.

[493] S.M. Ross. *Stochastic Processes.* Wiley, New York, 1983.

[494] B. Rüger. Scharfe untere und obere Schranken für die Wahrscheinlichkeit der Realisation von k unter n Ereignissen. *Metrika*, 26:71–77, 1979.

[495] L. Rüschendorf. Vergleich von Zufallsvariablen bzgl. integralinduzierter Halbordnungen, 1979. Habilitationsschrift.

[496] L. Rüschendorf. Inequalities for the expectiation of $\triangle$-monotone functions. *Z. Wahrscheinlichkeitstheorie Verw. Geb.*, 54:341–349, 1980.

[497] L. Rüschendorf. Ordering of distributions and rearrangement of functions. *Annals of Probability*, 9:276–283, 1980.

[498] L. Rüschendorf. Sharpness of Fréchet-Bounds. *Z. Wahrscheinlichkeitstheorie Verw. Geb.*, 57:293–302, 1981.

[499] L. Rüschendorf. Random variables with maximum sums. *Advances in Applied Probability*, 14:623–632, 1982.

[500] L. Rüschendorf. On the multidimensional assignment problem. *Methods of OR*, 47:107–113, 1983.

[501] L. Rüschendorf. Solution of a statistical optimization problem by rearrangement methods. *Metrika*, 30:55–62, 1983.

[502] L. Rüschendorf. On the minimum discrimination information theorem. *Statistics & Decisions*, 1:263–283, 1984. Suppl. Issue.

[503] L. Rüschendorf. Construction of multivariate distributions with given marginals. *Ann. Inst. Stat. Math.*, 37:225–233, 1985.

[504] L. Rüschendorf. The Wasserstein distance and approximation theorems. *Z. Wahrscheinlichkeitstheorie Verw. Geb.*, 70:117–129, 1985.

[505] L. Rüschendorf. Monotonicity and unbiasedness of tests via a.s. constructions. *Statistics*, 17:221–230, 1986.

[506] L. Rüschendorf. Fréchet-bounds and their applications. In G. Dall'Aglio, S. Kotz, and G. Salinetti, editors, *Advances in Probability Measure with Given Marginals*, pages 151–188. Kluver, Amsterdam, 1991.

[507] L. Rüschendorf. Bounds for distributions with multivariate marginals. In K. Mosler and M. Scarsini, editors, *Stochastic Order and Decision under Risk*, volume 19, pages 285–310. IMS Lecture Notes, 1991a.

[508] L. Rüschendorf. Conditional stochastic ordering of distributions. *Advances in Applied Probability*, 23:46–63, 1991b.

[509] L. Rüschendorf. Stochastic ordering of likelihood ratios and partial sufficiency. *Statistics*, 22:551–558, 1991c.

[510] L. Rüschendorf. Optimal solutions of multivariate coupling problems. *Appl. Mathematicae*, 22:325–338, 1995.

[511] L. Rüschendorf. Developments on Fréchet bounds. In *Proceedings of Distributions with Fixed Marginals and Related Topics*, volume 28, pages 273–296. IMS Lecture Notes Monograph Series, 1996.

[512] L. Rüschendorf. On c-optimal random variables. *Statistics Prob. Letters*, 27:267–270, 1996.

[513] L. Rüschendorf and S.T. Rachev. A characterization of random variables with minimum L^2-distance. *Journal of Multivariate Analysis*, 32:48–54, 1990.

[514] L. Rüschendorf, B. Schweizer, and M.D. Taylor. Distributions with Fixed Marginals and Related Topics. In *Proceedings of Distributions with Fixed Marginals and Related Topics*, volume 28. IMS Lecture Notes Monograph Series, 1996.

[515] L. Rüschendorf and L. Uckelmann. On optimal multivariate couplings. In *Distribution with given marginals and moment problems*, pages 261–274. Kluwer, 1997.

[516] T. Rychlik. Stochastically extremal distributions of order statistics for dependent samples. *Statistics & Probability Letters*, 13:337–341, 1992.

[517] C. Ryll-Nardzewski. On quasi-compact measures. *Fund. Math.*, 40:125–130, 1953.

[518] G. Samorodnitsky and M. Taqqu. *Stable Non-Gaussian Random Processes. Stochastic Models with Infinite Variance.* Chapman & Hall, New York, 1994.

[519] E. Samuel and R. Bachi. Measures of distance of distribution functions and some applications. *Metron*, 23:83–122, 1964.

[520] V.V. Sazonov. Normal approximation - some recent advances. *Lecture Notes in Mathematics*, 879, 1981.

[521] H.H. Schaefer. *Topological Vector Spaces.* Springer, New York, 1966.

[522] M. Schaefer. Note on the k-dimensional Jensen inequality. *Annals of Probability*, 2:502–504, 1976.

[523] G. Schay. Optimal joint distributions of several random variables with given marginals. *Stud. Appl. Math.*, LXI:179–183, 1979.

[524] L. Schwartz. *Radon Measures On Arbitrary Topological Spaces and Cylindrical Measures.* Oxford University Press, London, 1973.

[525] B. Schweizer. Thirty years of copulas. In G. Dall'Aglio, S. Kotz, and G. Salinetti, editors, *Symp. Probab. Measures with Given Marginals*, pages 13–50, Rome, 1990. Kluwer.

[526] B. Schweizer and A. Sklar. *Probabilistic Metric Spaces.* Elsevier, North-Holland, 1983.

[527] L. Seidel. On limit distributions of random symmetric polynomials. *Theory of Probability and its Applications*, 23:266–278, 1988.

[528] V.V. Senatov. Uniform estimates of the rate of convergence in the multi-dimensional central limit theorem. *Theory of Probability and its Applications*, 25:745–759, 1980.

[529] V.V. Senatov. Some lower estimates for the rate of convergence in the multi-dimensional central limit theorem. *Soviet Math. Doklady*, 23:188–192, 1981.

[530] W.J. Shafer and H.F. Sonnenschein. Equilibrium in abstract economics without ordered preferences. *Journal of Mathematical Economics*, 2:345–348, 1975.

[531] L.S. Shapley and M. Shubik. The assignment game, 1: the core. *Int. J. Game Theory*, 1:110–130, 1972.

[532] M. Sharpe. Operator-stable probability distributions on vector groups. *Trans. Amer. Math. Soc.*, 136:51–65, 1969.

[533] A.N. Shiryaev. *Probability Theory.* Springer, 1984.

[534] J.A. Shohat and J.D. Tamarkin. *The Problem of Moments.* American Mathematical Society, Providence, 1943.

[535] I.A. Sholpo. ε-minimal metrics. *Theory of Probability and its Applications*, 28:854–855, 1983.

[536] G.R. Shorack and J.A. Wellner. *Empiricial Processes With Applications to Statistics.* Wiley, New York, 1986.

[537] R.M. Shortt. Private communication.

[538] R.M. Shortt. Combinatorial methods in the study of marginal problems over separable spaces. *Journal of Mathematical Analalysis and its Applications*, 97:462–479, 1983.

[539] R.M. Shortt. Strassen's marginal problems in two or more dimensions. *Z. Wahrscheinlichkeitstheorie Verw. Geb.*, 64:313–325, 1983.

[540] R.M. Shortt. Univerally measurable spaces: An invariance theorem and diverse characterizations. *Fund. Math. Th.*, 121:35–42, 1983.

[541] H.J. Skala. The existence of probability measures with given marginals. *Annals of Probability*, 21:136–142, 1993.

[542] M. Sklar. Fonctions de repartition a dimensions et leurs marges. *Publ. Inst. Stat. Univ. Paris*, 8:229–231, 1959.

[543] C.S. Smith and M. Knott. Note on the optimal transportation of distributions. *Journal of Optimization Theory and Applications*, 52:323–329, 1987.

[544] C.S. Smith and M. Knott. On Hoeffding–Fréchet bounds and cyclic monotone relations. *Journal of Multivariate Analysis*, 40:328–334, 1992.

[545] T.A.B. Snijders. Antithetic variates for Monte Carlo estimation of probabilites. *Statistics Neerlandica*, 38:1–19, 1984.

[546] D. Stoyan. *Comparison Methods for Queues and Other Stochastic Models.* Wiley, 1983.

[547] V. Strassen. The existence of probability measures with given marginals. *Annals of Mathematical Statistics*, 36(2):423–439, 1965.

[548] J. Štěpán. Simplicial measures. In *Memor. Vol. of J. Hájek*, pages 239–251. Academia Prague, 1977.

[549] J. Štěpán. Probability measures with given expectations. In *Proc. of the 2nd Prague Symp. on Asympt. Statistics*, pages 315–320. North Holland, 1979.

[550] V.N. Sudakov. Geometric problems in the theory of infinite dimensional probability distributions. *Proc. Steklov Inst. Math.*, 141(2), 1979.

[551] H. Sussmann. On the gap between deterministic and stochastic differential equations. *Annals of Probability*, 6:19–41, 1978.

[552] A.S. Sznitman. Equations de type de Boltzmann, Spatialement homogènes. *Z. Wahrscheinlichkeitstheorie Verw. Geb.*, 660:559–592, 1984.

[553] A.S. Sznitman. Propagation of chaos. In *Ecole d'Eté Saint-Flour*, volume 1464 of *Lecture Notes in Mathematics*, pages 165–251, 1989.

[554] A. Szulga. On the Wasserstein metric. In *Transactions of the 8th Prague Conference on Information Theory, Statistical Decision Functions and Random Processes*, volume B, pages 267–273, Prague, 1978. Akademia Praha.

[555] A. Szulga. On minimal metrics in the space of random variables. *Theory of Probability and its Applications*, 27:424–430, 1982.

[556] W. Szwarc and M. Posner. The tridiagonal transportation problem. *Operations Research Letters*, 3:25–30, 1984.

[557] M. Talagrand. Matching random samples in many dimensions. *Annals of Applied Probability*, 2:846–856, 1992.

[558] D. Talay. Résolution trajectorielle et analyse numérique des équations differentielles stochastiques. *Stochastics*, 9:275–306, 1988.

[559] H. Tanaka. An inequality for a functional of probabillity distributions and its applications to Kac's one-dimensional modal of a Maxwellian gas. *Z. Wahrscheinlichkeitstheorie Verw. Geb.*, 27:47–52, 1973.

[560] H. Tanaka. Probabilistic treatment of the Boltzmann equation for Maxwellian molecules. *Z. Wahrscheinlichkeitstheorie Verw. Geb.*, 46:67–105, 1978.

[561] A.H. Tchen. Inequalities for distributions with given marginals. *Annals of Probability*, 8:814–827, 1980.

[562] P. Todorovic. *An extremal problem arising in soil erosion modeling*, pages 65–73. Reidel, Dordrecht, 1987. edt.: I.B. MacNeil and G.J. Umphrey.

[563] P. Todorovic and J. Gani. Modeling of the effect of erosion on crop production. *Journal of Applied Probability*, 24:787–797, 1987.

[564] Y.L. Tong. *Probability Inequalities in Multivariate Distributions.* Academic Press, 1980.

[565] D.M. Topkis and A.F. Veinott jr. Monotone solution of extremal problems on lattices (abstract). In *Abstract of 8th International Symposium on Mathematical Programming*, volume 131, Stanford, CA,, 1973. Stanford University.

[566] A. Tuero-Diaz. *Aplicaciones crecientes: Relaciones con las métricas de Wasserstein.* PhD thesis, Universidad de Cantabria, 1991.

[567] A. Tuero-Diaz. On the stochastic convergence of representations based on Wasserstein metrics. *Annals of Probability*, 21:72–85, 1993.

[568] L. Uckelmann. Konstruktion von optimalen Couplings. Universität Münster, 1993. Diplom-Arbeit.

[569] L. Uckelmann. Optimal couplings between one dimensional distributions. In *Distribution with given marginals and moment problems*, pages 275–282. Kluwer, 1997.

[570] V.R.R. Uppuluri, P.I. Feder, and L.R. Shenton. Random difference equations occuring in one-compartment models. *Math. Biosci.*, 2:143–171, 1967.

[571] S.S. Vallander. Calculation of the Wasserstein distance between probability distributions on the line. *Theory of Probability and its Applications*, 18:784–786, 1973.

[572] A.F. Veinott Jr. Representation of general and polyhedral sublattices and sublattices of product spaces. *Journal of Linear Algebra and its Applications*, 114/115:681–704, 1989.

[573] A.M. Vershik. Some remarks on infinite-dimensional linear programming problems. *Russian Math. Surveys*, 25:117–124, 1970.

[574] A.M. Vershik and V. Temelt. Some questions of approximation of the optimal value of infinite-dimensional linear programming problems. *Siberian Math. J*, 9:591–601, 1968.

[575] W. Vervaat. On a stochastic difference equation and a representation of non-negative infinitely divisible random variables. *Advances in Applied Probability*, 11:750–783, 1979.

[576] N.N. Vorobev. Consistent families of measures and their extensions. *Theory of Probability and its Applications*, 7:147–163, 1962.

[577] W. Wagner. Monte Carlo evalutation of functionals of solutions of stochastic differential equations. Variance reduction and numerical examples. *Stoch. Analysis Appl.*, 6:447–468, 1988.

[578] W. Warmuth. Marginal Fréchet-bounds for multidimensional distribution functions. *Statistics*, 19:283–294, 1976.

[579] L.N. Wasserstein. Markov processes over denumerable products of spaces describing large systems of automata. *Problems of Information Transmission*, 1969.

[580] H. von Weizsäcker and G. Winkler. Integral representation in the set of solutions of a generalized moment problem, 1980.

[581] E. Wesley. Borel preference orders in markets with a continuum of traders. *Journal of Mathematical Economics*, 3:155–165, 1976.

[582] A. Wieczorek. On the measurable utility theorem. *Journal of Mathematical Economics*, 7:165–173, 1980.

[583] E. Wild. On Boltzmann's equation in the kinetic theory of gases. *Proc. Camb. Phil. Soc.*, 4:602–609, 1951.

[584] G. Winkler. Choquet order and simplices with applications in probabilistic models. *Lecture Notes in Mathematics*, 1145, 1988.

[585] J. Yukich. Exact order rates of convergence of empirical measures. Preprint, 1991.

[586] J. Yukich. The exponential integrability of transportation cost. Preprint, 1991.

[587] J. Yukich. Some generalizations of the Euclidean two-sample matching problem. *Prob. Banach Spaces*, 8:55–66, 1992.

[588] V.M. Zolotarev. On the continuity of stochastic sequences generated by recursive procedures. *Theory of Probability and its Applications*, 20:819–832, 1975.

[589] V.M. Zolotarev. Approximation of distributions of sums of independent random variables with values in infinite dimensional spaces. *Theory of Probability and its Applications*, 21:721–737, 1976.

[590] V.M. Zolotarev. Metric distances in spaces of random variables and their distributions. *Math. Sb.*, 30(3):393–401, 1976.

[591] V.M. Zolotarev. General problems of the stability of mathematical models. *Bull. Int. Stat. Inst.*, 47(2):382–401, 1977.

[592] V.M. Zolotarev. On pseudomoments. *Theory of Probability and its Applications*, 23:269–278, 1978.

[593] V.M. Zolotarev. On the properties and relationships of certain types of metrics. *Zapiski Nauchn. Sem. LOMI*, 87:18–35, 1978.

[594] V.M. Zolotarev. Ideal metrics in the problems of probability theory and mathematical statistics. *Austral. J. Statist.*, 21(3):193–208, 1979.

[595] V.M. Zolotarev. Probability metrics. *Theory of Probability and its Applications*, 28:278–302, 1983.

[596] V.M. Zolotarev. *Contemporary Theory of Summation of Independent Random Variables.* Nauka, Moscow, 1986. In Russian.

[597] V.M. Zolotarev. *Modern theory of summation of independent random varables.* Nauka, Moscow, 1987. In Russian.

[598] V.M. Zolotarev and S.T. Rachev. Rate of convergence in limit theorems for the max scheme. In *Stability Problems for stochastic models*, volume 1155, pages 415–442. Springer, 1984.

Abbreviations

Bold page numbers refer to this volume, non-bold page numbers to the other volume.

a.e.	almost everywhere	158, 385
ARCH	autoregressive conditional heteroscedasticity	39
a.s.	almost sure	8
BLIL	bounded law of the iterated logarithm	**306**
CLT	central limit theorem	**85**
ch.f.	characteristic function	400
CRI	communication resolution interval	38
CTM	Capetanakis–Tsybakov–Mikhailov	**220**
d.f.(s)	distribution function(s)	8, 107
dna	domain of normal attraction	**306**
DP	dual polyhedron	23
DTP	dual transportation problem	23
GARCH	general ARCH	39
htl	explained on page	433
IFS	iterated function systems	**202**
i.i.d.	independent identically distributed	35
KKR	Kakosjan, Klebanov, and Rachev	**43**
KRP	Kantorovich–Rubinstein transshipment problem	vii, 2
LCFS	last come first served	**220**
LHS	left-hand side	405
LLN	law of large numbers	**81**
lsc	lower semicontinuous	113

MKP	Monge–Kontorovich mass transportation problem	vii, 1, 19, 58
MKTP	classical Monge–Kantorovich transportation problem	374
MTPA	MTP with additional constraints	vii
MTP	mass transportation	vii, 1
MTPP	MTP with partial knowledge of the marginals	4
OTP	optimal transportation plan	3
PDE	partial differential equation	xii, **xvi**
PERT	network model	148
PP	primal polyhedron	23
r.f.(s)	random field(s)	**248**
r.v.(s)	random variable(s)	3
SDE	stochastic differential equation	39
SLLN	strong law of large numbers	30
supp P	support of P	20
TP	transportation problem	21
usc	upper semicontinuous	127

Symbols

Bold page numbers refer to this volume, non-bold page numbers to the other volume.

$B_p(H)$	upper bound for $A_p(H)$ 158
$B(p;x,y)$	quadratic form 280
B_r	ball of radius r 149
$B(S_i)$	72
$B(X,\mathcal{B})$	58
$B_1(x,y)$	**28**
$B_2(x,y)$	**28**
$B(\alpha)$	263
$B_x(\varepsilon)$	**109**
$(B, \|\cdot\|)$	separable Banach space $C(T)$ **248**
b	transshipment 371
b_r	absolute moments **102**
$b_{u_1}(x)$	296
ba $(P_1,\dots,P_n)$	measures with fixed marginals 62
ba $(S,\mathcal{B})$	finitely additive measures 58
C, C^i	spaces of continuous and i times differentiable functions **255**, **333**
$C_{p,d}$	**321**
$C_{s,Z}$	integrable $(s-1)$-fold derivative **115**
$C_{s,\varepsilon Z}$	**117**
$C(g)$	assumption for the solution of a moment problem **62**
$C_b(S), C^b(S)$	Banach space of bounded continuous real-valued functions on S 63, 164
$C(T)$	Banach space **248**
$C^\gamma(c;\sigma_1,\sigma_2)$	307
$C(Q)$	set of continuous functions 384
$\mathring{C}(Q)$	quotient of the space $C(Q)$ 384
$\mathring{C}(Q)^*$	conjugate space 393
$C_b(\mathbb{R}^d)$	set of all bounded continuous functions on $\mathbb{R}^d$ **152**
$C(S)_+$	166
$C(S)^*_+$	166
$C_m(\theta)$	**170**
Cov	covariance 108
Cov $(X_i \mid \mathcal{F}_{i-1})$	conditional covariance **96**
$c\ell$	closure 304
$c(i,j)$	discrete cost **27**, **29**
$c(x,y)$	cost function viii, **10**
$c_1(x,y)$	$= \frac{\partial}{\partial x} c(x,y)$ 128
$c_*(x,y)$	reduced cost function 170
con(supp (P))	129
(D)	duality 76
$D(h,g)$	assumption for a moment problem **62**
DP	dual polyhedron 23
$D(P,Q)$	50
$D_p f(x)$	optimal pairs 103
$D\Phi$	$= (\frac{\partial \Phi_i}{\partial x_j})$ 118
$D_k(\vartheta)$	**112**
$D_m(\theta)$	**170**
d_r	smoothed version of d **137**
$d_{r,k}$	**61**
$d(h)$	divisor criterion **180**
$d_{n,m}(x,t)$	determinant of $A_{n,m}(x,t)$ 397
$d(\overline{x},\overline{y})$	**76**
$d(X,Y)$	uniform metric **137**

Index

Bold page numbers refer to this volume, non-bold page numbers to the other volume.